GeoPlanet: Earth and Planetary Sciences

More information about this series at http://www.springer.com/series/8821

Zbigniew W. Kundzewicz · Markus Stoffel
Tadeusz Niedźwiedź · Bartłomiej Wyżga
Editors

Flood Risk in the Upper Vistula Basin

Springer

Editors
Zbigniew W. Kundzewicz
Department of Climate and Water Resources
Institute for Agricultural
and Forest Environment
Polish Academy of Sciences
Poznań
Poland

Markus Stoffel
Institute of Geological Sciences
University of Bern
Bern
Switzerland

Tadeusz Niedźwiedź
Department of Climatology,
Faculty of Earth Sciences
University of Silesia
Katowice
Poland

Bartłomiej Wyżga
Institute of Nature Conservation
Polish Academy of Sciences
Kraków
Poland

The GeoPlanet: Earth and Planetary Sciences Book Series is in part a continuation of Monographic Volumes of Publications of the Institute of Geophysics, Polish Academy of Sciences, the journal published since 1962 (http://pub.igf.edu.pl/index.php).

ISSN 2190-5193 ISSN 2190-5207 (electronic)
GeoPlanet: Earth and Planetary Sciences
ISBN 978-3-319-41922-0 ISBN 978-3-319-41923-7 (eBook)
DOI 10.1007/978-3-319-41923-7

Library of Congress Control Number: 2016946296

Printed on acid-free paper

This Springer imprint is published by Springer Nature
The registered company is Springer International Publishing AG Switzerland

The rocks ought to stand and threaten
The clouds – to bear rainwaters
The lightnings – to thunder and go
Me – to float and float and float

[Adam Mickiewicz, translated from Polish by Zbigniew W. Kundzewicz and Joanna Pociask-Karteczka]

Since a large part of this book reports results of scientific research achieved within a Polish-Swiss project, it is opportune to propose a relevant motto – a few lines written in Switzerland by a great Polish romantic poet, Adam Mickiewicz (1798–1855). In 1839–1840, Mickiewicz lived in Switzerland and taught Latin literature in Lausanne, a beautiful city situated on the Lake Geneva. The motto, stemming from the poem „Over a great and pure water", being a part of Mickiewicz's „Lausanne lyrics" collection, is directly germane to a hydrological event that may generate a flood in the foothills of the mountains. In this poem, Mickiewicz refers to the large Lake Geneva (Leman) and to the rocky Alps located to the south from the Lake. Due to its large volume, the Lake can buffer an influx of heavy precipitation. Unfortunately, buffering capacities in the northern foothills of the Tatra Mounatins are smaller, by orders of magnitude.

[Zbigniew W. Kundzewicz and Joanna Pociask-Karteczka]

Foreword

River floods are a major natural threat in Central and Eastern Europe, and in particular in Poland. The most destructive floods in the country have not only been responsible for high, and increasing, material damage, up to the level of about 1 % of the national Polish GDP, but have also claimed dozens of fatalities per event. Much of the flood damage in the country occurred in the basin of the Vistula, the longest and largest river in Poland that flows northwards from the Carpathian Mountains through uplands and then lowlands and then empties into the Baltic Sea in northern Poland. The Vistula River Basin is international and most of it is located in Poland but the remainder is in Slovakia, Ukraine, and Belarus.

The book examines the flood risk in the Upper Vistula River Basin, from the source near Barania Góra in the Beskidy Mountains (belonging to the Western Carpathians) to Zawichost. This part of the basin includes the high Tatra Mountains (the highest part of the Carpathians) with the highest, orographically conditioned, precipitation in Poland—up to the Polish records of the annual maximum precipitation of 2770 mm and a daily maximum of 300 mm. High precipitation largely contributes to flood generation. The specific runoff in both the Tatra Mountains and their northern foothills is high as well and this translates into considerable flood hazard. Among the most disastrous floods in the Upper Vistula River Basin were the events in 1813, 1934, 1970, 1997, 2001, and 2010.

In the upper reaches of the mountainous tributaries of the Vistula River (such as Dunajec River), flood events are typically violent and highly erosive. Most of the flood damage there has resulted from erosional action rather than terrain inundations. Farther downstream, floods transform into huge masses of water propagating through lowland river reaches, causing inundation even at distances of hundreds kilometers downstream from the Carpathian Mountains. Processes driving changes in flood hazard at the northern foothills of the Tatra Mountains include climatic and terrestrial mechanisms. The most damaging floods in the region result from intense precipitation lasting incessantly for at least three days. Yet, some floods have been caused by torrential rains of much shorter duration or by intense snowmelt. Trends in hydroclimatic variables of relevance to flood generation (except for strongly increasing temperature) are typically weak, i.e., statistically insignificant, superimposed on strong natural variability. Among environmental changes with impacts on flood hazard are: land-use and land-cover changes (including changes in the extent of arable land and forests, expansion of riparian forests, changes in logging patterns, and urbanization).

Stationarity is dead, i.e., the past is not really a key to the future, as we are entering a situation with no analogy in past records. This general finding is of vast importance also for flood preparedness systems and design rules. What used to be a 100-year river discharge (with exceedance probability of 0.01 in any one year) is projected to be exceeded less frequently in some areas of the Upper Vistula Basin (e.g., where snowmelt is the dominating flood driver) and more frequently in other areas (e.g., where increasing intense precipitation prevails as the flood driver). In the areas of increasing flood hazard, where the level of past 100-year flood is projected to be exceeded more frequently, it will be necessary to strengthen the existing flood preparedness system, in order to maintain the same protection level. Multiple flood protection measures are needed, of both structural and non-structural type, also adjusting the management of the valley floors. In order to adapt to changes, better understanding and more reliable and accurate information is needed.

The first part of the book, setting the stage, consists of three chapters. Pociask-Karteczka (2016) gives a geographical overview of the Upper Vistula Basin. Stoffel et al. (2016) provide introduction to the problem of floods in mountainous basins and then Kundzewicz and Stoffel (2016) discuss the anatomy of flood risk.

The second part of the book, dealing with factors influencing flood hazard, consists of four chapters. Wyżga et al. (2016) provide an introduction to flood generation mechanisms and then review changes in principal flood drivers in the Upper Vistula Basin. Further, Mikuś et al. (2016) examine methods to assess large wood dynamics and associated flood hazard in the region of interest. Ruiz-Villanueva et al. (2016) review the results of numerical modelling of large wood transport, deposition, and remobilization during floods in contrasting morphologies of a mountain river. Finally, Radecki-Pawlik et al. (2016) discuss hydraulic conditions of flood flows in a river subject to variable human impacts.

The third part of the book, attempting to decipher changes in observational records, consists of seven chapters. Łupikasza et al. (2016) present a study of observed changes in temperature and precipitation in the region and the relationship between them. Next, Niedźwiedź and Łupikasza (2016) report on changes detected in atmospheric circulation patterns. Kaczka et al. (2016) inform the readership on the temperature reconstruction from tree-ring data. The following four chapters of part three deal with deciphering changes in floods. Ruiz-Villanueva et al. (2016) report on the results of a study of variability in flood frequency and magnitude in the region. Ballesteros-Cánovas et al. (2016) present information on floods deciphered from tree-ring data. Next, Nachlik and Kundzewicz (2016) elaborate on the history of floods on the Upper Vistula. Finally, Starkel (2016) introduces palaeohydrology of the Upper Vistula Basin.

The last part of the book, dealing with climate change projections and adaptation, consists of six chapters (of which three are devoted to projections and three to adaptation). Pińskwar et al. (2016) commence with a review of projections of precipitation in the northern foothills of the Tatra Mountains. Then, Piniewski et al. (2016) and Romanowicz et al. (2016) present projections of river flow. The former paper embraces projections of river discharge, covering the whole Upper Vistula Basin. In turn, the latter deals with projections of changes in flood hazard in two headwater catchments of the Vistula, placing the problem in the context of results of large-scale studies.

Then, Wyżga et al. (2016) put the flood risk management in the Upper Vistula Basin in perspective: juxtaposing traditional versus alternative measures. Matczak et al. (2016) analyze stability and change of flood risk governance in Poland. In the concluding chapter, Konieczny et al. (2016) discuss flood risk reduction in the context of opportunities for learning.

A considerable part of the material presented in the book stems from the Polish–Swiss research project FLORIST (Flood risk on the northern foothills of the Tatra Mountains) supported by a grant from the Swiss Government through the Swiss Contribution to the enlarged European Union. Nevertheless, the book goes far beyond the project region and the pool of project scientists. Several eminent and highly competent scientists from beyond the FLORIST project have contributed to this book, attracted by the vision of a valuable monograph.

The book largely draws from the results of the FLORIST project, focused around four competence clusters: observation-based climatology, model-based climate change projections and impact assessment, dendrogeomorphology, and impact of large woody debris on fluvial processes.

The knowledge generated by this book is likely to have an impact on understanding and interpretation of flood risk in the Upper Vistula Basin, in the past, present, and future. It can help solving important practical problems related to flood risk reduction strategies and flood preparedness. Hence, it is not only of considerable scientific interest, but also of social and practical relevance, e.g. for implementation of the EU Floods Directive.

The international dimension of the book does not only result from the international nature of the Upper Vistula Basin. Many chapters of the book are jointly authored by scientists from the dendrolab.ch, University of Berne (Switzerland), a partner in the FLORIST project consortium, who have collaborated with Polish scientists (from the Institute for Agricultural and Forest Environment of the Polish Academy of Sciences in Poznań, the University of Silesia in Katowice/Sosnowiec, and the Institute of Nature Conservation of the Polish Academy of Sciences in Cracow) in a project on flood risk in the piedmont areas of the Upper Vistula Basin.

Poznań, Poland Zbigniew W. Kundzewicz
Bern, Switzerland Markus Stoffel
Katowice, Poland Tadeusz Niedźwiedź
Kraków, Poland Bartłomiej Wyżga

References

Ballesteros-Cánovas JA, Spyt B, Janecka K, Kaczka RJ, Stoffel M (2016) Deciphering flood event information from tree-ring data in the Tatra Mountains: Implications for hazard assessment. (This volume, 257–277).

Kaczka RJ, Spyt B, Janecka K, Niedźwiedź T, Bednarz Z (2016) Climate reconstruction from tree-rings in the Tatra Mountains. (This volume, 209–229).

Konieczny R, Kundzewicz ZW, Matczak P (2016) Flood risk reduction – Opportunities for learning. (This volume, 399–418).

Kundzewicz ZW, Stoffel M (2016) Anatomy of flood risk. (This volume, 39–52).

Łupikasza E, Niedźwiedź T, Pińskwar I, Ruiz-Villanueva V, Kundzewicz ZW (2016) Observed changes in temperature and precipitation and relationship between them, in the Upper Vistula Basin. (This volume, 155–187).

Matczak P, Lewandowski J, Choryński A, Szwed M, Kundzewicz ZW (2016) Stability and change of flood risk governance in Poland. (This volume, 381–398).

Mikuś P, Wyżga B, Ruiz-Villanueva V, Zawiejska J, Kaczka RJ, Stoffel M (2016) Methods to assess large wood dynamics and the associated flood hazard in Polish Carpathian watercourses of different size. (This volume, 77–101).

Nachlik E, Kundzewicz ZW (2016) History of floods on the Upper Vistula. (This volume, 279–292).

Niedźwiedź T, Łupikasza E (2016) Change in atmospheric circulation patterns. (This volume, 189–208).

Piniewski M, Szczęśniak M, Mezghani A, Kundzewicz ZW (2016) Hydroclimatic projections for the Upper Vistula Basin. (This volume, 331–339).

Pińskwar I, Choryński A, Kundzewicz ZW, Ruiz-Villanueva V (2016) Projections of precipitation in the northern foothills of the Tatra Mountains. (This volume, 311–329).

Pociask-Karteczka J (2016) The Upper Vistula Basin – a geographical overview. (This volume, 3–21).

Radecki-Pawlik A, Wyżga B, Czech W, Mikuś P, Zawiejska J, Ruiz-Villanueva V (2016) Modelling hydraulic parameters of flood flows for a Polish Carpathian river subjected to variable human impacts. (This volume, 127–151).

Romanowicz R, Kundzewicz ZW, Meresa HK, Stoffel M, Krysanova V, Doroszkiewicz J (2016) Projections of changes in flood hazard in two headwater catchments of the Vistula in the context of European-scale studies. (This volume, 341–359).

Ruiz-Villanueva V, Wyżga B, Kundzewicz ZW, Niedźwiedź T, Łupikasza E, Stoffel M (2016) Variability of flood frequency and magnitude during the late 20th and early 21st centuries in the northern foreland of the Tatra Mountains. (This volume, 231–256).

Ruiz-Villanueva V, Wyżga B, Zawiejska J, Mikuś P, Hajdukiewicz H, Hajdukiewicz M, Stoffel M (2016) Large wood transport, deposition and remobilization during floods in the Czarny Dunajec River: outcomes from numerical modeling. (This volume, 103–125).

Starkel L (2016) Palaeohydrology of the Upper Vistula Basin. (This volume, 293–307).

Stoffel M, Wyżga B, Niedźwiedź T, Ruiz-Villanueva V, Ballesteros-Cánovas JA, Kundzewicz ZW (2016) Floods in mountain basins. (This volume, 23–37).

Wyżga B, Kundzewicz ZW, Ruiz-Villanueva V, Zawiejska J (2016) Flood generation mechanisms and changes in principal drivers. (This volume, 55–75).

Wyżga B, Radecki-Pawlik A, Zawiejska J (2016) Flood risk management in the Upper Vistula Basin in perspective: traditional versus alternative measures. (This volume, 361–380).

Series Editors

Managing Editor

Advisory Board

Kon-Kee Liu
Institute of Hydrological
and Oceanic Sciences
National Central University Jhongli
Jhongli, Taiwan

Teresa Madeyska
Research Centre in Warsaw
Institute of Geological Sciences
Warszawa, Poland

Stanisław Massel
Institute of Oceanology
Polish Academy of Sciences
Sopot, Poland

Antonio Meloni
Instituto Nazionale di Geofisica
Rome, Italy

Evangelos Papathanassiou
Hellenic Centre for Marine Research
Anavissos, Greece

Kaja Pietsch
AGH University of Science and
Technology
Kraków, Poland

Dušan Plašienka
Prírodovedecká fakulta, UK
Univerzita Komenského
Bratislava, Slovakia

Barbara Popielawska
Space Research Centre
Polish Academy of Sciences
Warszawa, Poland

Tilman Spohn
Deutsches Zentrum für Luftund
Raumfahrt in der Helmholtz
Gemeinschaft
Institut für Planetenforschung
Berlin, Germany

Krzysztof Stasiewicz
Swedish Institute of Space Physics
Uppsala, Sweden

Ewa Szuszkiewicz
Department of Astronomy
and Astrophysics
University of Szczecin
Szczecin, Poland

Roman Teisseyre
Department of Theoretical Geophysics
Institute of Geophysics
Polish Academy of Sciences
Warszawa, Poland

Jacek Tronczynski
Laboratory of Biogeochemistry
of Organic Contaminants
IFREMER DCN_BE
Nantes, France

Steve Wallis
School of the Built Environment
Heriot-Watt University
Riccarton, Edinburgh
Scotland, UK

Wacław M. Zuberek
Department of Applied Geology
University of Silesia
Sosnowiec, Poland

Piotr Życki
Nicolaus Copernicus Astronomical
Centre
Polish Academy of Sciences
Warszawa, Poland

Contents

Contributors

Juan Antonio Ballesteros-Cánovas Dendrolab.ch, Institute of Geological Sciences, University of Bern, Bern, Switzerland; Climatic Change and Climate Impacts, Institute for Environmental Sciences, University of Geneva, Geneva, Switzerland

Zdzisław Bednarz Department of Forest Biodiversity, Institute of Forest Ecology and Silviculture, University of Agriculture, Kraków, Poland

Adam Chorynski Institute for Agricultural and Forest Environment, Polish Academy of Sciences, Poznań, Poland

Wiktoria Czech Department of Hydraulics Engineering and Geotechnique, University of Agriculture, Kraków, Poland

Joanna Doroszkiewicz Institute of Geophysics, Polish Academy of Sciences, Warsaw, Poland

Hanna Hajdukiewicz Institute of Nature Conservation, Polish Academy of Sciences, Kraków, Poland

Maciej Hajdukiewicz Department of Environmental Engineering, Geomatics and Energetics, Kielce University of Technology, Kielce, Poland

Karolina Janecka Faculty of Earth Sciences, University of Silesia, Katowice, Poland

Ryszard J. Kaczka Faculty of Earth Sciences, University of Silesia, Katowice, Poland

Roman Konieczny Institute of Meteorology and Water Management—State Research Institute, Cracow, Poland

Valentina Krysanova Potsdam Institute for Climate Impact Research, Potsdam, Germany

Zbigniew W. Kundzewicz Institute for Agricultural and Forest Environment, Polish Academy of Sciences, Poznań, Poland; Potsdam Institute for Climate Impact Research, Potsdam, Germany

Jakub Lewandowski Institute for Agricultural and Forest Environment, Polish Academy of Sciences, Poznań, Poland

Ewa Łupikasza Faculty of Earth Sciences, University of Silesia, Katowice, Poland

Piotr Matczak Institute for Agricultural and Forest Environment, Polish Academy of Sciences, Poznań, Poland; Institute of Sociology, Adam Mickiewicz University, Poznań, Poland

Hadush K. Meresa Institute of Geophysics, Polish Academy of Sciences, Warsaw, Poland

Abdelkader Mezghani Norwegian Meteorological Institute, Oslo, Norway

Paweł Mikuś Institute of Nature Conservation, Polish Academy of Sciences, Kraków, Poland; Faculty of Earth Sciences, University of Silesia, Katowice, Poland

Elżbieta Nachlik Institute of Water Engineering and Water Management, Technical University of Cracow, Cracow, Poland

Tadeusz Niedźwiedź Faculty of Earth Sciences, Department of Climatology, University of Silesia, Katowice, Poland

Mikolaj Piniewski Warsaw University of Life Sciences, Warsaw, Poland; Potsdam Institute for Climate Impact Research, Potsdam, Germany

Iwona Pinskwar Institute for Agricultural and Forest Environment, Polish Academy of Sciences, Poznań, Poland

Joanna Pociask-Karteczka Institute of Geography and Spatial Management, Jagiellonian University, Cracow, Poland

Artur Radecki-Pawlik Department of Hydraulics Engineering and Geotechnique, University of Agriculture, Kraków, Poland

Renata J. Romanowicz Institute of Geophysics, Polish Academy of Sciences, Warsaw, Poland

Virginia Ruiz-Villanueva Dendrolab.ch, Institute of Geological Sciences, University of Bern, Bern, Switzerland; Climatic Change and Climate Impacts, Institute for Environmental Sciences, University of Geneva, Geneva, Switzerland

Barbara Spyt Faculty of Earth Sciences, University of Silesia, Katowice, Poland

Leszek Starkel Department of Geoenvironmental Research, Institute of Geography and Spatial Organization, Polish Academy of Sciences, Kraków, Poland

Markus Stoffel Dendrolab.ch, Institute of Geological Sciences, University of Bern, Bern, Switzerland; Climatic Change and Climate Impacts, Department of Earth Sciences, Institute for Environmental Sciences, University of Geneva, Geneva, Switzerland

Mateusz Szczesniak Warsaw University of Life Sciences, Warsaw, Poland

Malgorzata Szwed Institute for Agricultural and Forest Environment, Polish Academy of Sciences, Poznań, Poland

Bartłomiej Wyżga Institute of Nature Conservation, Polish Academy of Sciences, Kraków, Poland; Faculty of Earth Sciences, University of Silesia, Katowice, Poland

Joanna Zawiejska Institute of Geography, Pedagogical University of Cracow, Kraków, Poland

Part I
Setting the Stage

The Upper Vistula Basin—A Geographical Overview

Joanna Pociask-Karteczka

Abstract The Upper Vistula Basin covers a large part of Southern Poland. This area is very diverse, as it comprises of various geographical regions formed during different geological periods what can be seen in its multifarious composition (Carpathian Mountains, Subcarpathian basins, Lesser Poland Uplands). Culturally, it combines a range of secular traditions formulated by different tribes, in particular by its original inhabitants. The majority of the first cities were founded as trade centers (13th–16th century), later transformed into tourist or industrial centers (19th–20th century). The subregions of the Upper Vistula Basin like Silesia, Cracow, Bielsko-Biała, Tarnów–Rzeszów, Tarnobrzeg–Stalowa Wola, continue to develop coal, chemical, steel, light and food industries. Nowadays, these cities are very well connected by a modern grid of motorways, expressways and railways enabling a faster development of the region. Agriculture has been a pillar of economy for the inhabitants of northern part of the Upper Vistula Basin where soils are rich, whereas in the Carpathians where the soils were poorer, only pastoral farming could be developed. In spite of the industrial activity in the region, the natural environment hasn't been modified much so far what makes it even more attractive for tourists who numerously visit wild and well preserved national parks of the regions, in particular the Tatra National Park. The Upper Vistula Basin is of a great importance in the hydrology of Poland, and its influence goes far beyond its watershed. It embraces the headwater area of the largest river in Poland, i.e. the River Vistula and its tributaries. The water resources (precipitation, river runoff) of the area are the richest in the whole country. The water cycle dynamics in the mountain area is very high due to steep relief, low permeability of the ground, and flood formation develops fast. It has been favoured by intensive surface and subsurface flow in the area with a dense agricultural land fragmentation and field road network.

Keywords Natural environment · Cultural and socio-economic factors · Water circulation · Southern Poland

J. Pociask-Karteczka (✉)
Institute of Geography and Spatial Management, Jagiellonian University, Cracow, Poland
e-mail: joanna.pociask-karteczka@uj.edu.pl

Z.W. Kundzewicz et al. (eds.), *Flood Risk in the Upper Vistula Basin*, GeoPlanet: Earth and Planetary Sciences, DOI 10.1007/978-3-319-41923-7_1

1 Location in Europe

The Upper Vistula Basin is located in southern part of Poland and is a part of the Vistula drainage basin. The River Vistula is the 17th largest river in Europe according to the length, and 11th one according to the catchment area. The Upper Vistula River (at Zawichost gauge) drains a total of 50,731.8 km^2 (Atlas Podziału Hydrograficznego Polski 2005). This area is about 25 % of the total Vistula drainage basin and is shared by Poland (91 %), Ukraine (5 %) and Slovakia (4 %, Table 1). The length of the water divide is 1,450 km and more than half of it overlaps the Main European Water Divide separating the Baltic Sea and the Black Sea basins (Fig. 1). The highest point within the watershed reaches the altitude of 2,438 m a.s.l. in the Tatra Mountains, whereas the lower one is 134 m a.s.l. The Upper Vistula Basin adjoins to the Oder River catchment in the west, the Danube River catchment in the south, and the Dniester River Basin in the east.

The length of the Upper Vistula River channel is 393.8 km and there is 653.7 km distance from downstream limit of the Upper Vistula to the river mouth at the Baltic Sea. The Upper Vistula River flows nearly latitudinally from the west to the east (Atlas Hydrograficzny Polski 1986a). The headwater area is located at the slopes of the Barania Mt. (1,220 m a.s.l.; Silesian Beskid, Carpathians). The Upper Vistula Basin includes several sub-basins. The main left-bank tributaries of the Upper Vistula River are: Przemsza, Dlubnia, Szreniawa, Nidzica, Nida, Czarna, Koprzywianka and the right-bank tributaries: Sola, Skawa, Raba, Dunajec, Wisloka, San (Fig. 2).

The San River and the Dunajec River basins have a substantial share in the whole Upper Vistula Basin, i.e. 33.5 % (16,877.0 km^2) and 13.5 % (6,796.3 km^2) respectively. The watershed increase is significant at the Vistula confluence with Przemsza, Dunajec, and San (Fig. 3). There is a distinctly outlined river network and subbasins asymmetry. A number and length of the left-bank tributaries is lower than the right-ones, and the ratio of the left-bank basin to the right's one is 1:2.8 which is similar to the whole Vistula River basin (1:2.7, Chelmicki 1991).

2 Geographical Diversity of the Upper Vistula Basin

The Upper Vistula Basin environment is geographically varied. The area covers three geographical regions: the Carpathian Mountains (45 % of the basin), the Subcarpathian basins (35 %), the Lesser Poland Uplands (25 %, Chelmicki 1991).

Table 1 The Upper Vistula Basin area (Atlas Podzialu Hydrograficznego Polski 2005)

Basin	Area (in km^2)			
	Total	Poland	Slovakia	Ukraine
Upper Vistula	**50,731.8**	46,308.1	1,952.4	2,451.2

Fig. 1 Location of the Upper Vistula Basin in Poland (Chelmicki 1991, modified)

Carpathians and Uplands include headwater areas of most of the tributaries of the Upper Vistula River, and the Subcarpathian basins constitute a transit area for the River Vistula and lower river courses of its tributaries (Fig. 4). The Carpathians are a very distinct region and a major landmark in the Upper Vistula Basin likewise across the whole country. They are characterized by complex geology and landforms ranging in altitude from 300–500 m in the marginal foothill zone in the north to 2,655 m a.s.l. (the Tatra Mountains) in the south. A considerable portion of the Carpathians shapes Beskid Mountains with an elevation 600–1,700 m a.s.l. stretching latitudinally (the belt of the Carpathian Foothills accompanies them to the north).

The Carpathians were formed during the Alpine orogeny. The mountains take the form of a fold and thrust belt with numerous thin-skinned nappes. The flysch (interbedded sandstones, shale and mudstone) prevails in geology of the region. A small portion of the southern part of the Upper Vistula basin—the Tatra Mountains—is composed of crystalline (granite) and metamorphic rocks.

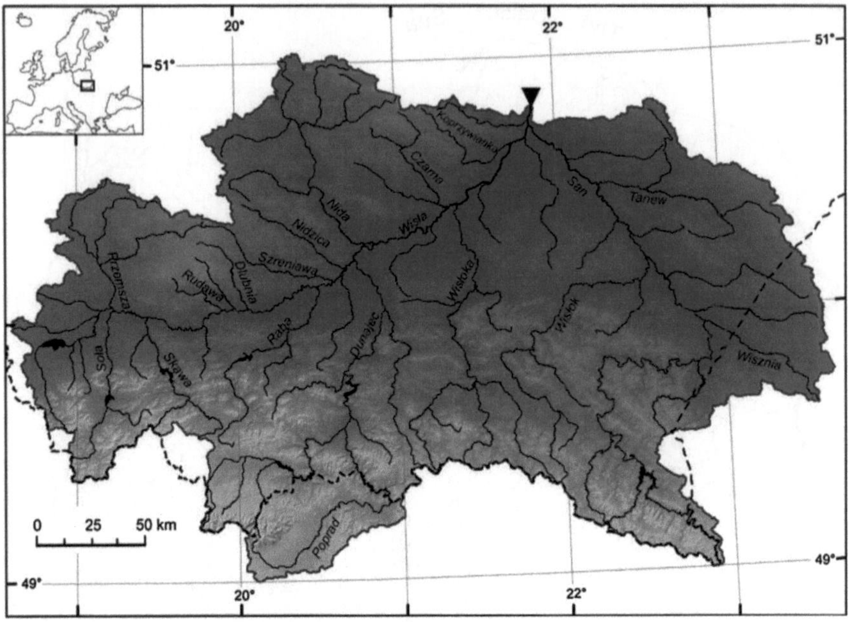

Fig. 2 The river network in the Upper Vistula Basin

There are several climatic-vegetation belts in the Carpathians (Hess 1965):

- temperate warm (foothill belt: 300–600 m a.s.l., multi-species deciduous forest),
- temperate cool (the lower sub-alpine forest belt: 600–1,100 m a.s.l.; Carpathian beech forest, herb layer),
- cool (the upper sub-alpine forest belt: 1,100–1,550 m a.s.l., spruce, shrub poorly developed, mosses),
- rather cold (dwarf mountain pine belt: 1,550–1,850 m a.s.l., *Pinetum mughi carpaticum*, numerous herbs),
- temperate cold (alpine belt: 1,850–2,200 m a.s.l., high mountain grasslands),
- cold (subnival belt: above 2,200 m a.s.l., fells, bare rock and almost no vegetation, lichens).

There are high, medium and low mountains within the drainage basin of the Upper Vistula River (Starkel 1991). High mountains are represented by the Mount Babia, build of flysch and the Tatra Mountains bordered to the north by a series of sedimentary nappes consisting of limestones, sandstones and flysch. The crestline of the Tatra with the altitude exceeding two thousand metres a.s.l. extends for 57 km forming the frontier between Poland and Slovakia, while the part of the Tatra belonging to Polish territory (150 km^2) is much smaller than to Slovakia. The vertical relief between the northern foothills and the main crest is 1800 m, providing for a range of altitudinal geoecological belts from the beech forests on the foothills, across the timberline at about 1,700 m a.s.l., further to alpine meadows

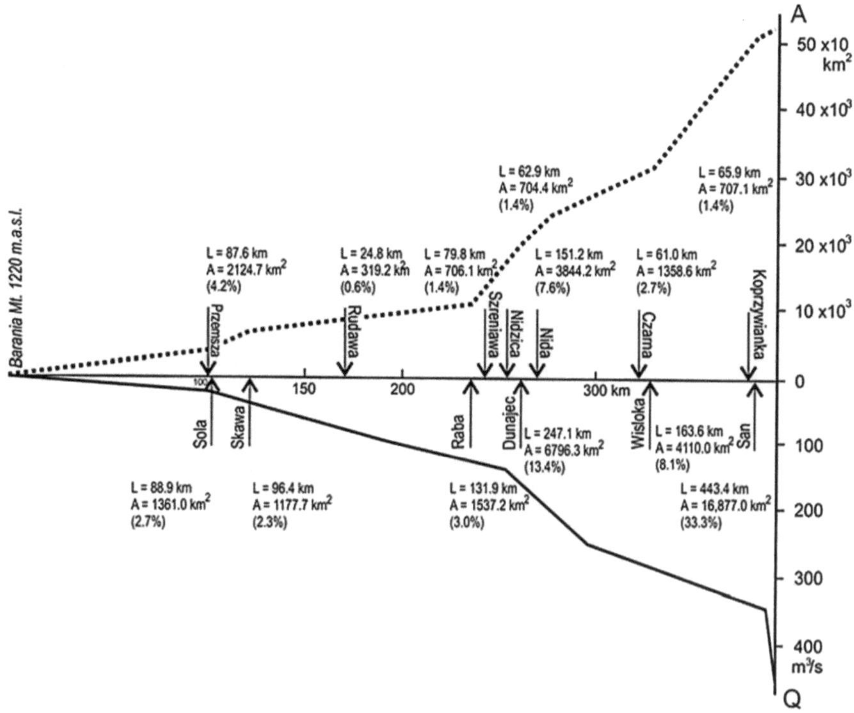

Fig. 3 Increase of watershed (A) and the Vistula discharge (Q) in the Upper Vistula Basin (L—length of rivers, A—area of watershed, (%)—a share of watershed in the Upper Vistula Basin (Atlas Hydrograficzny Polski 1986a, b; Atlas Podzialu Hydrograficznego Polski 2005)

and the uppermost periglacial belt of rock faces, frost-shattered debris and occasional small permanent snow beds. The relief forms of the Tatra Mountains result from the glaciations of the last Ice Age, therefore the Tatra Mountains—apart from the Alps—comprise the only high mountain landscape of Central Europe with the alpine attributes. These affected most of all mountain valleys and gave them their specific relief character. Glacial erosion produced substantial valley widening and over-deepening whereas in the upper parts, series of cirque basins and knife-edged arêtes were formed. Rock walls up to 1,000 m in height give the area high mountain character (Fig. 5). The Tatra Mountains have no glaciers, and the uppermost mountain subnival belt occurs (Kotarba 1998). The relief forms of today have been shaped by the action of running water and slope processes.

Whereas medium mountains in the Carphatians consist of both compact mountain's groups of Beskid, as well as more isolated ridges as Bieszczady where denivelation ranges between 400–800 m and elevation 800–1,300 m a.s.l. Low mountains are usually situated on the border of medium mountains, where

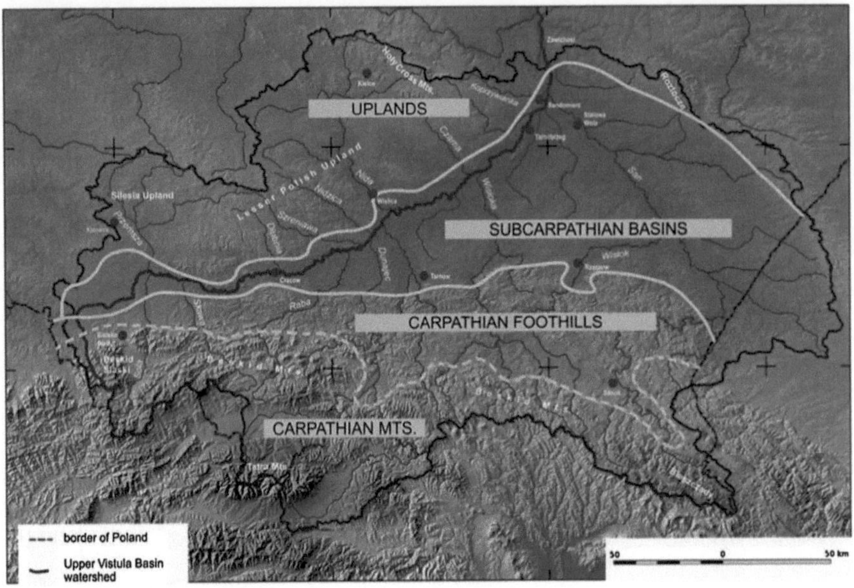

Fig. 4 Geographical regions in the Upper Vistula Basin

Fig. 5 A typical landscape in the Tatra Mountains (*Photo* K. Wojnarowski)

denivelation reaches 200–400 m (some Beskid ranges and the Holy Cross Mountains).

The Beskid Mountains are represented by discontinuous series of forested or partly forested mountain ranges (300–1,200 m a.s.l.) spreading from the east to the west. Along the northern edges of the Beskid Mountains, the Carpathian Foothills extend with an average elevation around 300–500 m a.s.l. They form a hilly area of gentle slopes with a thick cover of loess-like silty deposits particularly prone to the processes of runoff and wash in arable land (Fig. 6).

The Subcarpathian basin with lowland landscape is the lowest region in elevation (200–300 m a.s.l.). It consists of the Oswiecim Plain to the west and the Sandomierz Basin to the east. They both divide the Upper Vistula Basin into two parts: southern—mountainous and northern—upland. The left-bank side of the Upper Vistula basin is comprised of following uplands: Silesian, Lesser Poland and Roztocze. A hilly landscape with Jurassic limestone (karst phenomena), Cretaceous marls covered by loess deposits, is typical for this area (Fig. 7). Sporadic and isolated patches of forests occur locally. The only northern marginal zone of the upland zone differs considerably due to the Holy Cross Mountains—the oldest Caledonian orogeny mountains in Europe elevated up to 612 m a.s.l. The mountains are built of Palaeozoic and Mesozoic metamorphic formations (quartzitic sandstone, mudstone, slate). The oldest evidence of *Dinosauromorpha*, four-legged animals living during the Early Triassic was found there recently. The fossil

Fig. 6 The Beskid Mountains—a direction of plowing accelerate surface flow and soil erosion (*Photo* K. Ostafin)

Fig. 7 A typical landscape of uplands in the Upper Vistula Basin (*Photo* M. Kumon)

footprints are dated to approximately 250 million years ago. Footprints of early bipedal dinosaurs known as *Sphingopus*, from 246 million years ago, have also been found there (Niedźwiedzki and Remin 2008).

3 Tradition and Culture

The Upper Vistula Basin is endowed with a rich tradition and culture which is a combination of secular traditions and customs. They have roots in history, but continue to develop, incorporating old tradition with modern sensibilities. An exceptional role in the cultural landscape in the Upper Vistula Basin was played by the Carpathians. This area acted as a cultural and artistic transmitting zone from the south of Europe—the Danube basin to the north and north-east (Węcławowicz 1995). New patterns were carried by tribes penetrating in particular mountain valleys: the Polish and German speaking settlers, Italian, Greek, Jewish and Armenian merchants and Wallachian shepherds. A coexistence of these various cultures influenced the features of localities. The trace of the past can still be seen in tiny medieval towns with stone churches and brick houses, villages composed of wooden cottages of primitive log construction, Latin wooden Roman Catholic village churches in the west of the Carpathians and Orthodox ones in the east. At the close of the 19th century the Carpathian ethno-cultural mixture was incorporated into the Austrian part of Poland called Galicia. The marvel of the Carpathians cultural landscape lies in the artistic merit of historical monuments as well as in a specific *genius loci*: the surroundings with hills, river valleys, forest and old architecture. So the Carpathian cultural heritage consists of the whole historical structure of monuments, sites, routes, local dialects, even customs.

The folk culture of the Carpathians has been formed by considerable influence of pastoral tribes from Transylvania and the Balkans shepherds (Maj 1995). It is revealed in certain types of folk crafts, local music, wood carving, painting on glass, regional costumes and certain rituals. Local people keep their tradition. Carpathians folk costumes are colorfully decorated and they are often seen during holidays and festivals. Women wear open-work embroidered shirts, flowered skirts, and necklaces of coral beads. White lace aprons complete the outfits. Men wear blue vests tied with a sash; on their heads there are hats trimmed with decoration. The flowered skirts of Polish traditional costumes are topped with vests and blouses.

The Upper Vistula Basin was populated mostly by Roman Catholics. A considerable influence of the Orthodox Church was observed in the Carpathians eastward of the River Dunajec. In the 16th century Lemko tribes accepted the Greek Catholic rite (known as the Ukrainian-Byzantine rite), and the followers of the Orthodox rite formed a considerable religion minority. They were compulsorily resettled to the other regions of Poland or to the former Soviet Union shortly after the World War II. A considerable concentration of the worship centers in the Carpathians is exceptional in Poland and in the world. Tradition of pilgrimage to Carpathian worship centers reaches back up to the old times. By the 16th century about 100 centers has been formed most of them with crowned images of the Virgin Mary (Jackowski 1995). The most important ones are in Kalwaria Zebrzydowska, Kalwaria Pacławska, Tuchow, Ludzmierz. Since Cracow—Lagiewniki has been established as an international center of the Divine Mercy worship, the sanctuary increased the significance. Recently Wadowice has received international recognition as the birthplace of the Pope John Paul II, especially after he became canonised.

Over the centuries (up to 1939) judaism was an important religion in the Upper Vistula Basin. The Jews have been living in Poland since the king Casimir the Great (14th century) allowed them to settle in Poland in great numbers. The king was favorably disposed toward the Jews. This religion developed especially at the end of 18th century when mystic movement called Chassidism expanded. A lot of pilgrimage centers developed in the Carpathians like Lezajsk, Bobowa, Dynow, Stary Sacz, Rymanow. Some of them had become the residence places of charismatic Chassidic leaders.

Cracow is the most important historical and cultural center in the Upper Vistula Basin situated by the River Vistula. This is the second largest and one of the oldest cities in Poland. It is also a leading center of Polish academic, cultural, and artistic life, and one of the most important economic hubs. The former capital of Polish kingdom (until 1596) with its numerous historical, cultural and religious monuments, offers attractions to a few million tourists a year (2012—8 million, 2014—9.9 million).

The cultural qualities are represented by monuments of sacral architecture (churches, monastery complexes, small wooden Roman and Greek Catholic churches, chapels and by-road statues) and the relicts of residential architecture of old urban systems. Fortified castles, cathedrals and museums are still present in the Upper Vistula Basin as a part of its architectural and cultural heritage. The 16th century castles topping white-rocks of Jurassic limestone in the Lesser Poland Uplands are very spectacular. The places of national martyrdom and numerous cemeteries including those from the World Wars I and II are closely associated with the history of the Carpathians—the place of three big military operations in the 20th century: in 1914–1915, in 1939 and in 1944–1945 (Przyboś 1995; Groch and Kurek 1995). The battle at Gorlice in 1914, which ended in a defeat of Russian army as well as the Dukla Pass combat in 1944 (not a very successful operation of the Soviet troop either) which was one of the bloodiest and most relentless battles in the history of the Second World War are especially worth of mentioning.

4 Population and Economy—Land Use and Functional Variety

Until the middle of the 6th century, the Upper Vistula Basin was very sparsely populated—the Slavonic settlers form the north reached the edge of the Carpathians. The mountains with elevation above 350 m a.s.l. were not populated at all (Przyboś 1995). At the end of the 9th century Wislica was founded. It was the most ancient settlement in Polish territory at that time, located at the Krakow Upland very close to the important trade route running from Cracow to Sandomierz. It was probably the capital of the Vistulans, a Slavonic tribe. Inhabitants of ten sub-settlements around Wislica worked for the needs of the city in the 12th century. The city was presumably burnt down by Tatars during the invasion of 1241 and it was never again inhabited by settlers. A gypsum baptismal font found during excavations works is thought to had served as a font for collective baptism in the 9th century. If so, it would be the place of the earliest baptism on the Polish lands. The Wallachian colonization wave characterized by the dominating animal husbandry and shepherding type of economy arrived from the east at the end of the 14th century.

With the laps of time the colonization pushed to the south—mainly along the Carpathian river valleys. Most of cities in the Upper Vistula Basin were founded in the 13th through 16th centuries as trade centers, those cities developed in the 19th and 20th centuries becoming industrial or tourist centers. Agriculture was the main activity of inhabitants north to the Vistula River (uplands). Rich soils (chernozem) and gentle slopes inclination had been favourable land attributes for cultivation of wheat, sugar beet, and tobacco. This portion of the basin is predominantly comprised of cultivated crops with pockets of forest and very little hay lands (flat bottom of river valleys). The predominant role of agriculture in the economy of this part of the Upper Vistula Basin remains till nowadays (Fig. 8).

In the Carpathians there were favourable conditions for pastoral farming and wood industry. Poor soils and steep relief allowed to cultivate primarily rye and

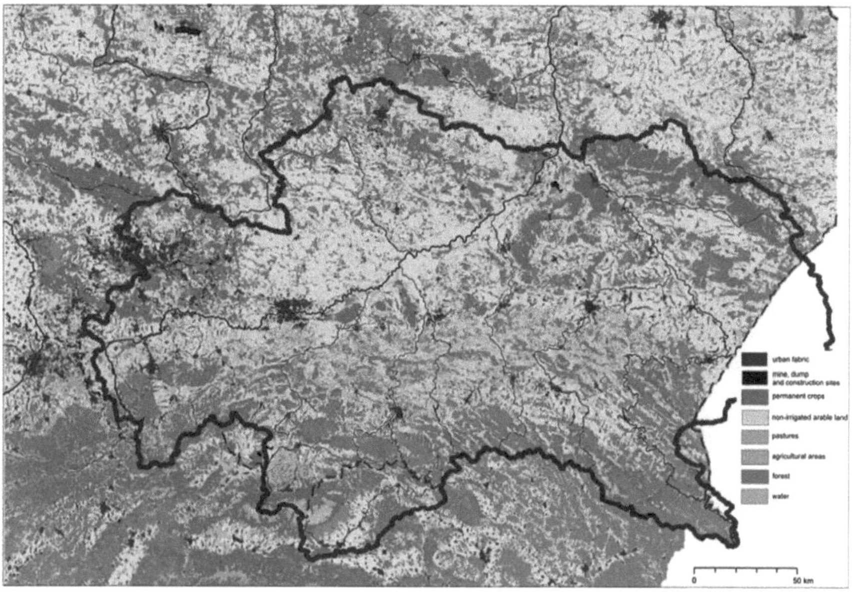

Fig. 8 Land use in the Upper Vistula Basin (CORINE 2006)

potatoes. Small size and plot fragmentation of farms reflected a high-density population in rural areas and a historic process of triple partitions of Poland by Austrian, Russian and Prussian empires (1772, 1793, 1795). Most of the Upper Vistula Basin was under Austrian annexation and the inheritance-related division of fields in high-density population rural areas was possible. The region was separated from the rest of Poland and deprived of the opportunities for progress. Farms were very small, by area, in comparison with farms in the Prussian partition of Polish lands.

Salt had been exploited for more than seven ages in the salt mine established in the 13th century (Wieliczka) and exported to all over Europe via the Vistula River waterway, which had been one of the main trading arteries in Poland for hundreds of years. Except for salt, among the goods shipped *via* this route between the 10th and 13th centuries there were: timber, grain, and building stone.

There were other natural resources in the Upper Vistula Basin in the past. Oil and natural gas industry (vicinity of Sanok) was a traditional branch in the Carpathians. Thanks to native sulfur deposits (vicinity of Tarnobrzeg), Poland was one of the world's largest producers of sulfur. The Silesia is the most industrial region in the Upper Vistula Basin. The coal mine activity and many branches of industry concentrate there. Brine pumped from mines is discharged to the River Vistula and cause a significant pollution of its water. The population density in this region exceeds 500 person per 1 km^2 (Fig. 9). At present, apart from Silesia, there are a few strongly industrialized regions in the Upper Vistula Basin with the following centers: Cracow, Bielsko-Biala, Tarnow–Rzeszow, Tarnobrzeg–Stalowa

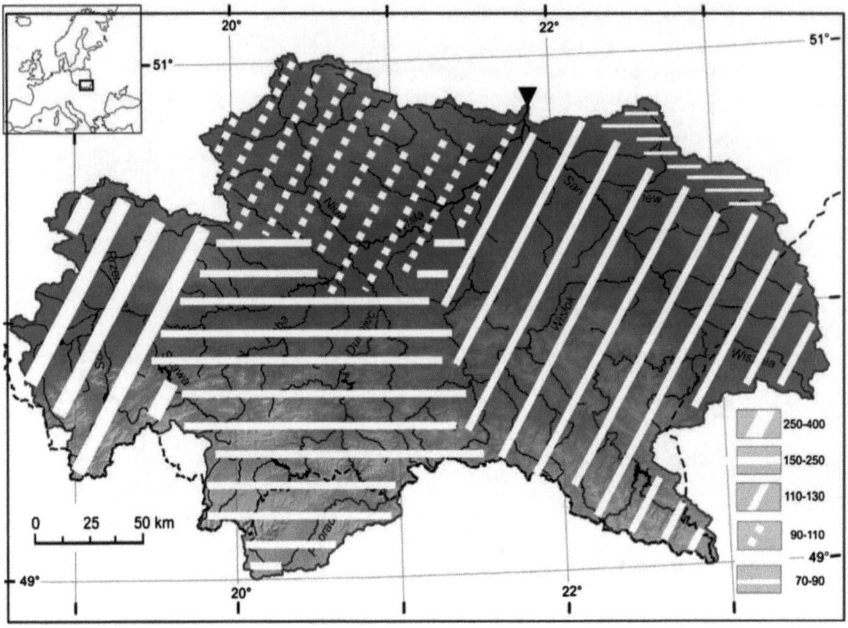

Fig. 9 Population density (person per km^2) in communes in the Upper Vistula Basin (Raport z wyników 2011, 2012)

Wola with the industries like: electromechanical, food, light, chemical and steel. A discovery of abundant mineral water mainly in the Carpathians enabled developing of spas, health and revitalizing resorts for therapeutic healing, as well as production of bottled water.

There is one city, Cracow, with population over half million (761,873 thousand) and a few medium-size cities (100–300 thousand: Katowice, Kielce, Rzeszów, Bielsko-Biała, Tarnów). A basic function of those cities is the industry (especially in the west) as well as services, education, and culture. Together, they constitute a grid of well connected cities linked by numerous latitudinal and meridional lines of communication (Zborowski 1995). Some of the routes are very old, they used to connect Germany with Russia, the Baltic Sea with the Black Sea and Hungary since the Middle Ages (Birek and Janiec 1995). The construction of main Vienna route and several other roads had taken place in the 18th century. The industrial and technological revolution induced new demands for transportation in the 19th century and it resulted in construction of a railway network. Most of railway lines constructed during that time by Austrian government (subcarpathian and transversal lines) are still operating.

The network of public roads has been improving in recent years. Motorways and expressways linking Warsaw, Silesia, Cracow and Ukraine are a part of national roads network. Government's investments on road construction increases, as it is necessary due to rapid development of the country and the inflow of European

Union funds for infrastructure projects. Moreover, Poland as a country located at the "cross-roads" of Europe needs a large and increasingly modern network of transport infrastructure.

In spite of long lasting human activity there are areas in the Upper Vistula Basin where natural environment has not been significantly modified, even in the upland part of the basin. Large parts of the Carpathians are covered with forests, and their share exceeds 70 % in some areas (e.g. south-eastern part of the region). The mountains are known for rich and unique fauna and flora with numerous endemic species and communities. The most intact, undisturbed, wild natural areas have been protected as national parks. There are nine national parks in the Upper Vistula Basin and this constitutes 40 % of total number of national parks in Poland. Three of these nine parks preserve mountainous environment, where wilderness areas prevail. A mountain climate (low air temperature and ample snow cover) encourage winter sports there, so numerous ski resorts are located in the south-western part of the mountains. Tourism is a very important branch of economy in the Carpathians which are the principal touristic region in the country. The touristic traffic is considerably diversified there. The highest concentration of traffic is observed in the Tatra Mountains and their vicinity. The Tatra National Park was established in 1950. Number of visits has reached 3 million per annum, which is considered to be beyond the park's long-term carrying capacity. The following types of recreational use are characteristic of the park: organized group sight-seeing, hiking, ski-tourism, downhill skiing, mountaineering, and caving. Sheep grazing remains an important activity in the Park. First, it had been eliminated, but later in 1981 was reintroduced in order to preserve local culture and maintain biodiversity. The Polish and Slovak Tatra Mountains range was included into the Biosphere Reserves (UNESCO Man and the Biosphere Programme).

All regions in the Upper Vistula Basin have experienced strong political and economic changes after more than 40 years of communism and especially after having joined the European Union on 1 May 2004. Some of these changes are significant for water cycle. Natural reforestation of abandoned agricultural land is a widespread present phenomenon in the Upper Vistula Basin. The increase of the surface of designated protected areas, extended forest cover due to natural succession. Such changes favour slower water circulation and water storage in the basin. Nowadays one may observe a remarkable increase in the awareness of the value of water as a good which should be limited in use.

5 Hydrological Significance of the Upper Vistula Basin in the Hydrology of Poland

The Upper Vistula Basin plays a very important role in the hydrology of Poland. Due to significant water resources the Carpathians secure water for a number of regions far beyond their boundaries being an important water tower for the rest of

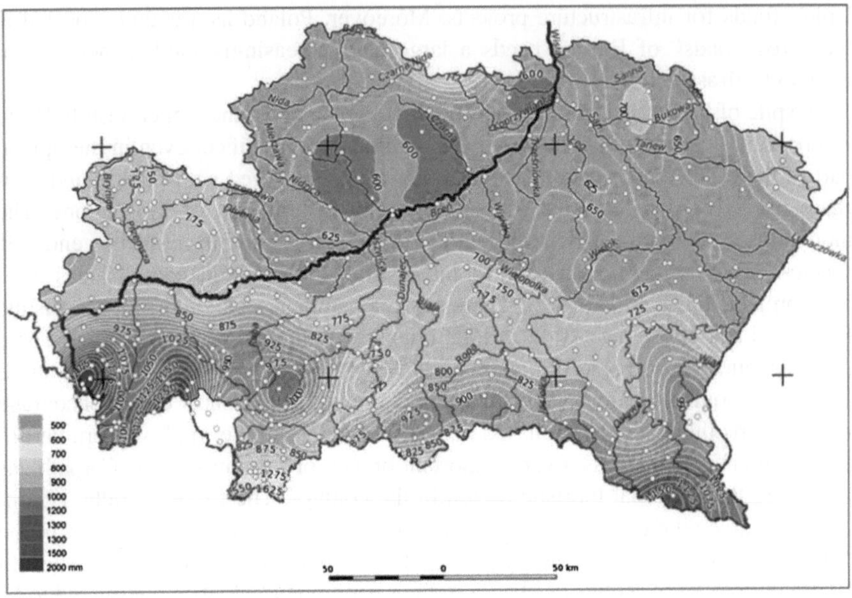

Fig. 10 Mean annual precipitation in the Upper Vistula Basin (1952–1981, after Cebulska et al. 2013, modified)

the country. The spatial distribution of precipitation in the Upper Vistula Basin is uneven. The smallest annual precipitation (550–650 mm) occurs in the Sandomierska Basin, and the highest one—reaching 1,700 mm occurs in the Tatra Mountains. This is the highest precipitation in Poland (Fig. 10). Considering horizontal precipitation (which is not measured) in the Tatra Mountains, there might be even 70 % more of water available in the water cycle. Precipitation on the northern slopes of the Tatra Mountains is approximately 200 mm higher than on southern slopes due to the rainfall shadow (Pociask-Karteczka 2014).

A mean annual discharge of the Upper Vistula River amounts 449 m^3/s (at Zawichost, 1951–1980; Punzet 1991), this is approximately 40 % of the discharge at the river mouth to the Baltic Sea in spite of the draining just 25 % of the watershed (Fig. 1). The Vistula river regime of the middle and lower courses is influenced by the Upper Vistula. There are two flood seasons during the year: one due to the snowmelt (March–April), and another one due to summer rain (July or August, Fig. 11). However, the eastern part of the area is mostly affected by continental type of the climate and the summer flood season is rather not marked or slightly marked. The meteorological conditions in the Upper Vistula Basin in Poland are influenced by polar maritime (west) or continental (east) air masses and for this reason the area's climate is transitional—between oceanic and continental (Niedźwiedź and Obrębska-Starklowa 1991).

The discharge of the River Vistula increases according to the left- and right-bank tributaries confluence. The Carpathian tributaries are predominant in the Vistula

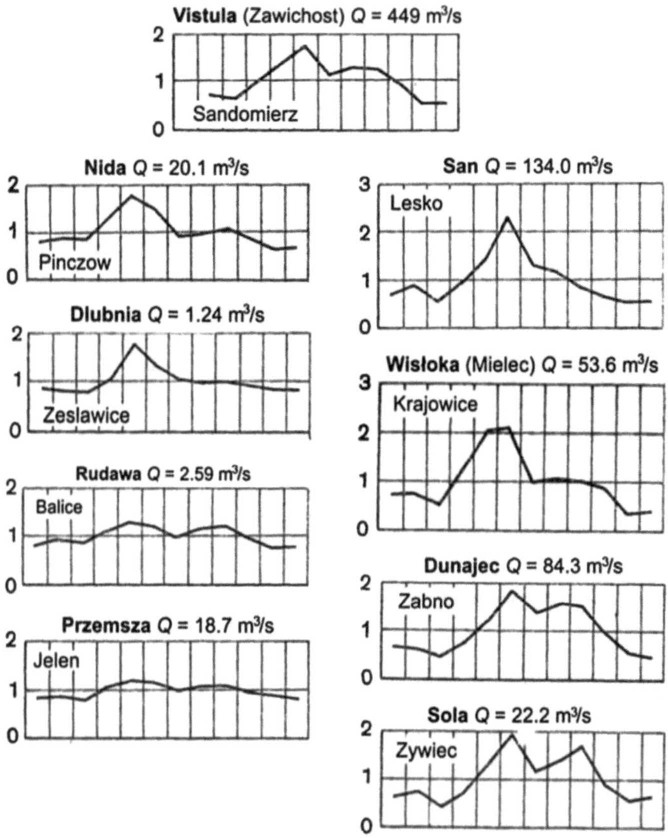

Fig. 11 The monthly Pardé coefficients in the Upper Vistula Basin and the mean annual river discharge (Q) in 1951–1980 (Punzet 1991)

discharge supply (Fig. 3). The river network density in the southern part of the Basin is smaller due to higher permeability of soils and geological formations (limestone, marls). The total mean annual discharge (1951–1980) of some significant Carpathian rivers: Sola, Dunajec, Wisloka, and San (the right-bank Vistula River tributaries) reaches 294.1 m^3/s, i.e. almost 7 times more than the total discharge of upland left-bank Vistula River tributaries: Przemsza, Rudawa, Dlubnia, and Nida (44.0 m^3/s; Punzet 1991).

Steep relief of the Carpathians, relatively small permeability of soil, high density of valleys and field roads favour a very dynamic and rapid hydrological cycle and outflow from the catchments. A runoff coefficient (a runoff/precipitation) in the Carpathians exceeds 60 % and it is the highest in the whole country. An average annual specific runoff is also very high and reaches 50 $dcm^3/s/km^2$ (Lajczak 1996, Fig. 12). A considerable contribution of groundwater in the river runoff is a

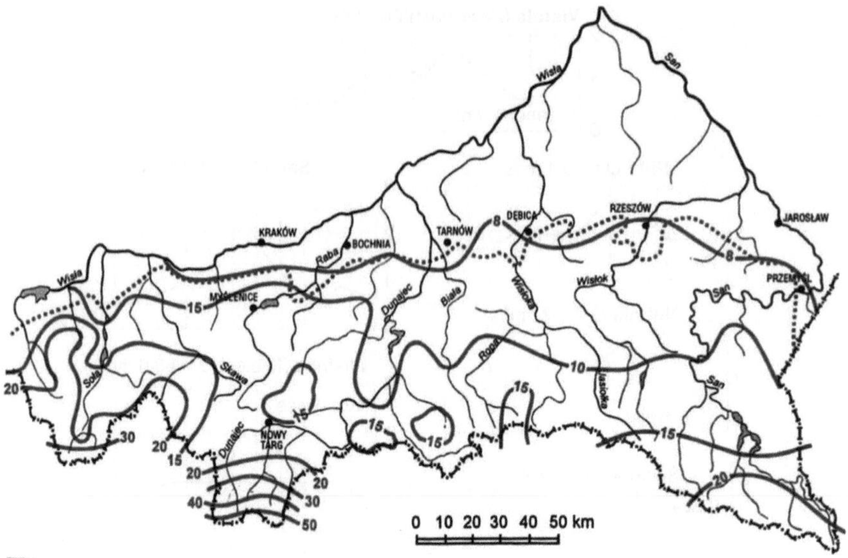

Fig. 12 The average annual specific runoff in the Carpathians (Dynowska 1995)

distinguishing feature of upland rivers. It is related to the rich groundwater aquifer in limestone and marls.

Flood potential of the Carpathian rivers and good water quality contributed to creation of water storage reservoirs. Most of reservoirs were built to retain water during high rainfall events and to prevent or reduce downstream flooding. Over the course of time, reservoirs have acquired some new functions. Finally at present, most of reservoirs are multi-purpose, but yet flood control and hydroelectricity use prevail (Table 2). In some reservoirs water must be stored in order to meet water demand of other regions. Hence significant water withdrawals occur, e.g., from the Sola River to Silesia Region and from the Raba River to Cracow. Water from the reservoir on the Raba River meets approximately 70 % of Cracow water demand.

In the Upper Vistula Basin, one can distinguish three types of landscape with different water cycles: (i) mountainous, (ii) upland and foothills and (iii) lowland (Starkel 1991). The characteristics of the mountain landscape are: quite steep slopes (inclination between 20 and 40° prevails), debris covers, significant denivelation and overall general high altitude above sea level. Hence the mountain area favors a fast surface and subsurface flow, that may lead to formation of floods. High mountains raise above the upper timberline, where highest precipitation occurs. Regionally significant water storage occurs in high mountain postglacial lakes. In medium mountains (Beskid) and foothill region, flysch sediments—a sequence of rhythmically interbedded sandstone and shale/mudstone—do not constitute rich groundwater aquifer, while mantle cover with significant share of clay fraction limits infiltration and considerable slope favours surface flow. Agrarian landscape structure is also an important factor determining the water circulation in the Beskid

Table 2 The largest water storage reservoirs in the Upper Vistula Basin

Name of reservoir	River	Total volume ($\times 10^6$ m^3)	Area (km^2)	Use
Solina	San	472	22	Flood control, hydroelectricity
Czorsztynski	Dunajec	234.5	11	Flood control, hydroelectricity
Roznowski	Dunajec	193	16	Flood control, hydroelectricity, touristic
Swinna Poreba[a]	Skawa	161	10.35	Flood control, hydroelectricity, touristic
Goczalkowicki	Vistula	166.8	32	Water demand, flood control, touristic
Dobczycki	Raba	127	10.7	Water demand, flood control, hydroelectricity
Zywiecki	Sola	94.6	10	Flood control, water demand, hydroelectricity, touristic
Klimkowka	Ropa (Wisloka)	43.5	3.1	Water demand, flood control, hydroelectricity, touristic
Miedzybrodzki	Sola	26.6	3.8	Flood control, water demand, hydroelectricity, touristic

[a]under construction

and Subcarpathian foothills. Land fragmentation, reflected in so-called field patchwork or chessboard pattern, generates favourable conditions for a fast intensive water surface flow, especially in snowmelt season and summer rain period. It also intensifies soil erosion (Woch and Borek 2015). Small river basins with arable land fragmentation are probe to the flash floods which are triggered primarily by intense rainfall.

Uplands are areas of smaller denivelations (50–100 m) and smaller slopes inclination. Carbonate rocks are most extensive there and they are among the most productive aquifers known in the Upper Vistula Basin. Hence there are numerous springs with a high discharge rate, especially karst springs in the Lesser Poland Upland. Loess cover on the widespread plateaus is sensitive to linear erosion. Lowland landscape is typical for the Subcarpathian basins. Post-glacial and fluvial-glacial deposits cover the Miocene loams. This is a transit region for the River Vistula and lower courses of its tributaries.

6 Final Remarks

The Upper Vistula Basin occupies approximately 15 % of Poland. Northern part of the Basin played a very important role in the Middle Ages being the center of the Polish statehood with Wislica and then with Cracow which was the capital of the country from the mid-11th century until 1596. During several centuries of historical development, the region gained a unique ethno-cultural identity.

The regional diversity of natural environment of the Upper Vistula Basin (north—uplands, south—mountains, central part—lowland) caused different economic activity in particular portions of the region. Agriculture and industry prevail in the western, northern and central part of the area. Land cover of the southern, mountainous part remains least affected, with significant share of forest.

The Upper Vistula Basin is of utmost importance in the hydrology of Poland. It embraces the headwater area of the largest river in Poland, i.e. the River Vistula and its tributaries. The water resources (highest precipitation, river runoff) of the area—the water tower of a large part of Poland—are the richest in the whole country. The dynamics of the hydrological cycle in the mountain area is very high due to steep relief and low ground permeability. Flood formation, favoured by intensive surface and subsurface flow at the area with a dense agricultural land fragmentation and field road network, proceeds fast. Hence, the Upper Vistula Basin plays an important role in flood generation in the country.

References

Atlas Hydrograficzny Polski (1986a) Tom II, Zeszyt 2, Metody opracowania i zestawienia liczbowe. In: Stachý J (ed) Wydawnictwa Geologiczne, Warszawa, p 724

Atlas Hydrograficzny Polski (1986b) Tom III, Zeszyt 1, Metody opracowania i zestawienia liczbowe. In: Stachý J (ed) Wydawnictwa Geologiczne, Warszawa, p 131

Atlas Podziału Hydrograficznego Polski (2005) cz. 2, Zestawienia zlewni. In: Czarnecka H (ed) IMGW, Minister Środowiska, Narodowy Fundusz Ochrony Środowiska i Gospodarki Wodnej, Warszawa, p 559

Birek U, Janiec R (1995) Komunikacja. In: Warszyńska J (ed) Karpaty Polskie. Przyroda, człowiek i jego działalność, Uniwersytet Jagielloński, Kraków, pp 301–316

Cebulska M, Szczepanek R, Twardosz R (2013) Rozkład przestrzenny opadów atmosferycznych w dorzeczu górnej Wisły. Opady średnie roczne (1952–1981), IGiGP UJ, Kraków, p 84

Chelmicki W (1991) Położenie, podział i cechy dorzecza. In: Dynowska I, Maciejewski M (eds) Dorzecze górnej Wisły, cz. I, PWN, Warszawa, Krakow pp 15–29

Dynowska I (1995) Wody. In: Warszyńska J (ed) Karpaty Polskie. Przyroda, człowiek i jego działalność, Uniwersytet Jagielloński, Kraków, pp 49–68

Groch J, Kurek W (1995) Turystyka. In: Warszyńska J (ed) Karpaty Polskie. Przyroda, człowiek i jego działalność, Uniwersytet Jagielloński, Kraków, pp 265–299

Hess M (1965) Piętra klimatyczne w polskich Karpatach Zachodnich. Zeszyty Naukowe UJ, Prace Geograficzne 11:267

Jackowski A (1995) Ośrodki kultu religijnego. In: Warszyńska J (ed) Karpaty Polskie. Przyroda, człowiek i jego działalność, Uniwersytet Jagielloński, Kraków, pp 197–206

Kotarba A (1998) Landscape ecology, human impact and extreme erosional events in the Tatra Mountains, Poland. Ambio 27(4):354–357

Lajczak A (1996) Hydrologia. In: Mirek Z, Głowaciński Z, Klimek K, Piękoś-Mirkowa H (eds) Przyroda Tatrzańskiego Parku Narodowego, Tatry i Podtatrze, 3, Tatrzański Park Narodowy, Polskie Towarzystwo Przyjaciół Nauk o Ziemi, Kraków-Zakopane, pp 169–196

Maj M (1995) Kultura ludowa. In: Warszyńska J (ed) Karpaty Polskie. Przyroda, człowiek i jego działalność, Uniwersytet Jagielloński, Kraków, pp 181–196

Niedźwiedzki G, Remin Z (2008) Tropy wielkich teropodów z osadów górnego pliensbachu Gór Świętokrzyskich (Gigantic theropod dinosaur footprints from the upper Pliensbachian of the Holy Cross Mountains, Poland). Przegląd Geologiczny 56(9):823–825

Niedźwiedź T, Obrębska-Starklowa B (1991) Klimat. In: Dynowska I, Maciejewski M (eds) Dorzecze górnej Wisły, cz. I, PWN, Warszawa, Krakow, pp 68–84
Pociask-Karteczka J (2014) Obieg wody w górach wysokich oraz na obszarze młodoglacjalnym – krótkie studium porównawcze. In: Cieśliński R, Jereczek-Korzeniowska K (eds) Problemy badań wody w XX i XXI wieku, Wyd. UG, Gdańsk, pp 308–329
Przyboś K (1995) Dzieje Karpat Polskich. In: Warszyńska J (ed) Karpaty Polskie. Przyroda, człowiek i jego działalność, Uniwersytet Jagielloński, Kraków, pp 147–168
Punzet J (1991) Przepływy charakterystyczne. In: Dynowska I, Maciejewski M (eds) Dorzecze górnej Wisły, cz. I, PWN, Warszawa, Krakow, pp 167–215
Raport z wyników (2011, 2012) Narodowy spis powszechny ludności i mieszkań. GUS, Warszawa, pp. 177
Starkel L (1991) Rzeźba terenu. In: Dynowska I, Maciejewski M (eds) Dorzecze górnej Wisły, cz. I, PWN, Warszawa, Krakow, pp 42–54
Węcławowicz T (1995) Przemiany krajobrazu kulturowego Karpat Polskich. In: Warszyńska J (ed) Karpaty Polskie. Przyroda, człowiek i jego działalność, Uniwersytet Jagielloński, Kraków, pp 169–180
Woch F, Borek R (2015) The role of management of the field-forest boundary in Poland's process of agricultural restructuring. Pap Global Change 22:83–100
Zborowski A (1995) Zróżnicowanie społeczno-gospodarcze Karpat Polskich w świetle urbanizacji. In: Warszyńska J (ed) Karpaty Polskie. Przyroda, człowiek i jego działalność, Uniwersytet Jagielloński, Kraków, pp 317–324

Floods in Mountain Basins

Markus Stoffel, Bartłomiej Wyżga, Tadeusz Niedźwiedź, Virginia Ruiz-Villanueva, Juan Antonio Ballesteros-Cánovas and Zbigniew W. Kundzewicz

Abstract This chapter provides a general introduction to recent research on floods in mountain catchments and reviews state-of-the-art contemporary knowledge on the topic in Poland and Switzerland. The selection of the areas illustrated in this chapter is motivated by the fact that the Swiss Agency for Development and Cooperation (SDC) had funded a research project on floods in the Polish Tatra Mountains and their forelands, to which this book is also dedicated.

Keywords Floods · Flash floods · Mountain environments · Precipitation · Climate change

1 Introduction

Mountain environments cover roughly 25 % of the land surface and are often referred to as 'natural water reservoirs'; this is because a substantial amount of water surplus is usually transported from mountain areas to adjacent lowlands in some of the largest river systems on Earth (Viviroli et al. 2003). Mountain regions cover 52 % of Asia, 36 % of North America, 25 % of Europe, 22 % of South America, 17 % of Australia, and 3 % of Africa, as well as substantial areas of islands including Japan, New Guinea, and New Zealand (Bridges 1990).

M. Stoffel (✉) · V. Ruiz-Villanueva · J.A. Ballesteros-Cánovas
Dendrolab.ch, Institute of Geological Sciences, University of Bern, Bern, Switzerland
e-mail: markus.stoffel@dendrolab.ch

V. Ruiz-Villanueva
e-mail: virginia.ruiz@dendrolab.ch

J.A. Ballesteros-Cánovas
e-mail: juan.ballesteros@dendrolab.ch

M. Stoffel · V. Ruiz-Villanueva · J.A. Ballesteros-Cánovas
Climatic Change and Climate Impacts, Institute for Environmental Sciences, University of Geneva, Geneva, Switzerland

M. Stoffel
Department of Earth Sciences, University of Geneva, Geneva, Switzerland

Z.W. Kundzewicz et al. (eds.), *Flood Risk in the Upper Vistula Basin*,
GeoPlanet: Earth and Planetary Sciences, DOI 10.1007/978-3-319-41923-7_2

Floods in mountain basins are often flashy (Borga et al. 2008, 2014), and therefore differ from most other fluvial floods in that the lead time for warnings is generally very limited (e.g., often much less than two hours). Flash floods usually occur in mountain river catchments draining less than 1000 km^2 (Gaume and Borga 2008; Lumbroso and Gaume 2012). In these environments, direct current meter measurements are often impossible to conduct during flood peaks for safety and technical reasons (Fukami et al. 2008).

One of the greatest difficulties to characterize floods in mountain rivers is that for a given event, several processes coupling between hillslopes and channels may take place concurrently (i.e., debris flows, hyperconcentrated flow, and clear water flow), with different characteristics such as rheology or the number of phases involved (Montgomery and Buffington 1997; Bracken and Croke 2007; Bodoque et al. 2011) Mountain basins often respond rapidly to intense rainfall rates because they have high slopes and a quasi-circular morphology and, as a consequence strong connectivity (Ruiz-Villanueva et al. 2010; Youssef et al. 2011). Likewise, in these basins precipitation has an important orographic component. As a result, precipitation is very variable from a spatio-temporal perspective (Rotunno and Houze 2007). The above determines that mountainous basins are highly prone to have extreme precipitation events, both in terms of total volume and intensity. The resulting floods have a rapid hydrological response, which is characterized by rather "peaky" hydrographs (i.e., hydrographs with short lag times). They reach their flow peaks within a few hours, and thus allow for very little or even no advance warning at all to prevent flood damage (Borga et al. 2007, 2008).

Among the many processes contributing to the generation of floods in mountain rivers, three seem to be key in producing large and/or extreme floods: high-intensity convective rainfall (Gutiérrez et al. 1998; Hicks et al. 2005), extensive deep cyclones generating a few days-long, orographic rainfall (Sturdevant-Rees et al. 2001), and outburst floods resulting from the failure of either natural or artificial

B. Wyżga
Institute of Nature Conservation, Polish Academy of Sciences, al. Mickiewicza 33, 31-120 Kraków, Poland
e-mail: wyzga@iop.krakow.pl

B. Wyżga · T. Niedźwiedź
Faculty of Earth Sciences, University of Silesia, ul. Będzińska 60, 41-200 Sosnowiec, Poland
e-mail: tadeusz.niedzwiedz@us.edu.pl

Z.W. Kundzewicz
Institute for Agricultural and Forest Environment, Polish Academy of Sciences, Poznan, Poland
e-mail: kundzewicz@yahoo.com

Z.W. Kundzewicz
Potsdam Institute for Climate Impact Research, Potsdam, Germany

dams (Cenderelli 2011; Worni et al. 2014; Schwanghart et al. in press). These processes may generate peak discharges greatly exceeding average flood discharges. For instance, glacial-lake outburst floods in Nepal had discharges up to 60 times greater than normal floods generated by snowmelt runoff, glacier melting, and monsoonal precipitation (Cenderelli and Wohl 2003).

In mountain rivers, drainage area and channel gradient, as well as the magnitude and frequency of hillslope failures will not only affect channel morphology, but sediment supply from hillslopes will also influence total sediment load during flood (Lin et al. 2008) as well as the spatial patterns of erosion and deposition along the channel (Cenderelli and Kite 1998; Wohl 2010). Material from mass movement processes (e.g., landslides, debris flows) may not only deliver large amounts of mineral sediments, but also introduce wood debris into the channel corridor. Wood in channels can then favor the creation of temporary debris dams and subsequently produce secondary flood pulses, thereby enhancing erosion, and/or lead to the destruction of infrastructure along the channel (Wohl 2010; Ruiz-Villanueva et al. 2016a, b, c). As a result, floods in mountain rivers often differ from those in lowland environments due to the close coupling between the channel and adjacent hillslopes (Wohl 2010).

Extreme floods can be disastrous, resulting in substantial material losses and/or large numbers of fatalities, if they affect managed valley sections and inhabited valley floors. Examples can be an outburst flood caused by a landslide entering the Vaiont reservoir, Italian Alps, with 2600 deaths (Semenza and Ghirotti 2000), or a flash flood in the Arás basin, Spanish Pyrenees, resulting in 87 fatalities (Gutiérrez et al. 1998). A decisive role in the origin of the disasters must be attributed to inappropriate management decisions, unadjusted to the actual hazard, rather than to the course of natural phenomena. Such was the situation with the Vaiont flood, where water was stored in the reservoir despite the identification of a giant landslide on the valley side above the reservoir (Semenza and Ghirotti 2000) or with the flood on the Arás that destroyed a camp site located on an alluvial fan (Gutiérrez et al. 1998). More frequent are less dramatic situations when flood damages are enhanced by structures that reduce channel conveyance (e.g., Arnaud-Fassetta et al. 2005). Approximately 40 % of the flood-related deaths in Europe between 1950 and 2006 were linked to flash floods (Barredo 2007). The lack of data on flash floods (Marchi et al. 2010), and in particular the lack of accurate discharge estimates (Borga et al. 2014), can often provide an obstacle to improvements in flood forecasting, warning, planning and emergency management.

Less intense floods, especially regularly recurring ones, are key to maintain the function and integrity of aquatic and riparian ecosystems as they allow lateral exchange of water, nutrients and organisms between river channel and the connected floodplain (Junk et al. 1989) and promote recruitment of trees in the riparian areas (Hughes and Rood 2003). However, extreme floods are generally considered as threats for aquatic ecosystems (Wydoski and Wick 2011) that may dramatically though temporarily reduce the abundance of riverine biota such as fish or benthic macroinvertebrates. Because such events cannot be predicted, very few papers compared pre- and post-flood condition of river biocoenosis in mountain rivers.

While the existing studies analyzed direct action of floodwaters on biota (e.g., the effect of the extreme flood of 1997 on the Oder River, Czech Republic, on fish fauna; Lojkasek et al. 2005), no papers have described more prolonged effects of extreme flood events on the physical structure of mountain river habitats.

The Polish-Swiss research project FLORIST has focused on floods in the headwater and foreland reaches of the rivers originating in the Polish Tatra Mountains (Kundzewicz et al. 2014) and has brought together debris-flow, flash flood and mountain river flood experts from both countries. In the following, and in an attempt to set the stage for this book, we briefly summarize a few of the key findings on (flash) flood processes in both countries by reviewing some of the more recent publications on the topic.

2 Physiographic and Meteorological Conditions of Flood Occurrence in the Swiss Alps

In Switzerland, ca. 30 % of the national territory is located between 1000 and 2000 m a.s.l. for which a quick response between precipitation input (often associated with high intensity) and runoff output is the most common process. The sensitivity of these surfaces to produce floods is often exacerbated by the fact that this altitudinal band is also characterized by steep slope gradients and thin soil cover, in addition to the presence of extensive stream networks ensuring high specific discharge and abundant overland flow (Weingartner et al. 2003). Large amounts of precipitation are favored by the uplift of air masses, either in the form of rather localized, convective uplift (i.e. a vertical movement induced by vertical instability in the atmosphere) or cyclonic (also called advective) uplift in a horizontal plane of a warm over a cold air mass (Grebner et al. 2000). Under current climatic conditions prevailing in the European Alps, Schwarb et al. (2001) point to upper intensity rates of up to 100 mm/h for convective rainfalls, but the ongoing increase of mean and extreme temperatures might likely lead to increasing precipitation intensities in the future (Gobiet et al. 2014). In the past, floods induced by convective storms were most frequent during the summer half-year, but recently have started to occur in all seasons as a result of the decreasing snow cover and the reduced buffering effect of snow (Beniston 2005). In the case of advective precipitation events induced by closed cyclonic fields, the areas affected can extend for several 100 km^2 (Grebner et al. 2000). The intensity of advective rainfalls is much smaller than that of convective showers, and typically reaches values of 10–20 mm/h north of the Alpine divide (Grebner et al. 1999).

The last very large flood in the Swiss Alps, Prealpine foreland and plateau occurred in August 2005, causing damage in the order of 3 billion Swiss Francs (DETEC 2008). The circulation pattern triggering the devastating event is referred to as a *Vb situation* and is quite common in this part of Europe. Situations comparable to that of August 2005 occur several times per year, but the large volume of rainfall and the duration of the event over such a large area should be considered as quite unusual. The meteorological event of August 2005 is comparable with the heavy precipitation events of June 1910, July 1977, August 1987 and May 1999. Prolonged and intensive periods of precipitation may also be expected in Switzerland in the future—possibly more frequently than in the past due to global climate change (DETEC 2008).

In terms of process activity in small, Alpine catchments, information is much scarcer and mostly relies on archival records and/or reconstructions. Along with flash floods, debris flows represent one of the most common processes in high mountain areas and are commonly initiated by the mobilization of sediment stored in channels or by shallow landslides, in addition to sudden input of large amounts of water, such as rainstorms, rapid snow melt, rain-on-snow events, or the sudden release of water from glaciers or from dammed lakes. Most commonly, however, debris flows in the Alps have been triggered by high-intensity, short-duration rainstorms or low-intensity, long-duration precipitation events, typically during the summer half-year (Stoffel et al. 2011; 2014a, b; Schneuwly-Bollschweiler and Stoffel 2012; Toreti et al. 2013). Projected changes in mean and extreme temperatures and precipitations are likely to influence the temporal frequency and magnitude of mass wasting in mountain environments (Gobiet et al. 2014). This is especially true for debris flows, where changes in rainfall intensity and duration, in combination with higher temperatures, are thought to lead to enhanced process activity, provided that sediment is not limited and that the occurrence of events is driven primarily by water input above a certain threshold (Stoffel and Huggel 2012; Borga et al. 2014). A warmer climate also results in higher 0 °C isotherms, thus allowing for more precipitation to fall in liquid form even in the uppermost portions of mountain catchments, thereby increasing the area contributing effectively to runoff (Beniston 2005; Stoffel and Beniston 2006). In the case of the Zermatt valley (Valais, Swiss Alps), projections of future precipitation have been realized in the framework of the EU-FP7 ACQWA Project (Beniston and Stoffel 2014). Based on point-based downscaled climate scenarios and for the periods 2001–2050 and 2051–2100, analysis of temperature and rainfall changes above specific thresholds (10–50 mm) as well as the duration of precipitation events (1–3 days) reveals a drying tendency for future summers and more precipitation during the shoulder seasons (Stoffel et al. 2014a, b). At the same time, despite the general decrease in precipitation sums in summer, an increase in the occurrence of heavy (>40 mm) 1-day precipitation events is observed in the region. In conclusion, the drier conditions in future summers and the wetting of springs, falls and early winters are likely to have significant impacts on the behavior of debris flows. Based on the current understanding of the debris-flow systems and their reaction to rainfall inputs, one might expect only slight changes in the overall frequency of events by the mid-21st century, but possibly an increase in the overall magnitude of debris flows due to larger amounts of sediment delivered to the channels and an increase in extreme precipitation events. In the second half of the 21st century, the overall absolute number of days with conditions favorable for the release of debris flows will likely decrease, especially in summer. The anticipated increase of liquid rainfalls during the shoulder seasons (March, April, November, December) is not expected to compensate for the decrease in future heavy summer rainfalls over 2 or 3 days in absolute terms, but magnitudes, in contrast, can be expected to increase in the study area. The volume of entrained debris from the source areas tends to be larger in summer and fall when the active layer of the permafrost bodies is largest and allows for larger volumes of sediment to be mobilized (Lugon and Stoffel 2010), but the situation has been shown to depend also on the stability and climate change-related accelerations of rock-glacier bodies (Stoffel and Huggel 2012). Along with the occurrence of more extreme precipitation events, these rock-glacier instabilities could lead to debris flows without historic precedents in the future.

3 Physiographic and Meteorological Conditions of Flood Occurrence in the Polish Carpathians

The Carpathians cover 6 % of the territory of Poland. They are relatively low mountains in comparison to the Swiss Alps and elevations above 2000 m a.s.l. occur only in the Tatra Mountains shared by Poland and Slovakia (areas of 22.3 and 77.7 %, respectively), with the highest peaks: Rysy (2499 m) in the Polish part and Gerlach (2655 m) in Slovakia. The Polish part of this high-mountain massif represents less than 1 % of the area of the Polish Carpathians. In the remaining part of the Polish Carpathians, almost entirely underlain by flysch, the highest peak reaches 1725 m a.s.l. and relative elevations between the bottoms of intramontane basins and mountain summits vary between 500 and 1200 m. This physiographic setting is clearly reflected in the character of geomorphic hazards. Debris flows occur relatively frequently only in the Tatra Mountains (Kotarba 1992) where, however, they can hardly be considered a threat to man in this uninhabited and undeveloped area. Floods are the hazards most damaging to human property and dangerous to life in the region. Their timing and principal drivers vary along the west-east gradient (Chełmicki et al. 1999). The eastern part of the Polish Carpathians, with lower elevations and more continental climate, is characterized by frequent occurrence of moderate, snow-melt floods in early spring and rare occurrence of large floods caused by summer rainfall. Snow-melt floods are typical of the south-eastern part of the Upper Vistula Basin, and in its Carpathian portion they mainly occur in the foothill parts of the San and Wisłok catchments. Their occurrence depends on the snow-cover duration and the thickness of snow pack at the end of winter. A weak decreasing trend in snow-cover duration and a more significant trend in the snow-pack thickness were observed in the second half of the 20th century (Falarz 2004). Favourable conditions for a snow-melt flood on the Wisłok River were recorded in Rzeszów after the snowy winter 1963/1964 with 52 cm-thick snow pack in March. The last such thick snow pack in Rzeszów (53 cm) was recorded in March 2005. The duration of snow cover at this location during the last warm period 1988-2015 was shorter by 10 days than in the more cooler years 1951–1987. Recently possibility of the occurrence of snow-melt floods has been very low. During the last three winters 2013/2014–2015/2016 with the duration of snow cover between 20 and 28 days, the maximum thickness of snow pack was lower than 10 cm. In higher parts of the Carpathians, snow pack is relatively thick, but snow-melting period extends from April at lower altitudes to early June in the higher parts of the Tatras. As a result, snow-melt floods in these parts of the Carpathians are very rare.

In the western part of the Polish Carpathians, with higher elevations and more oceanic climate, floods are caused by summer rainfall. Summer floods differ in duration and extent according to the triggering rainfall (Starkel 1996). A few days-long rainfall with the total sum of precipitation amounting to a few hundred millimeters, sometimes preceded by a longer period of increased precipitation, that saturates shallow soil, results in floods encompassing the whole or a considerable part of the Polish Carpathians and even the whole Upper Vistula Basin. The four largest floods in the Polish Western Carpathians occurred in July 1934, 1970, 1997 and in May 2010 and were triggered by a few days-long rainfall with average intensity of 8–10 mm h^{-1}. Such precipitation was always caused by cyclones coming along the Vb track from the Mediterranean region to Central-Eastern Europe where they became partly stationary. On the western side of such cyclones, humid air masses are moving from N or NE perpendicularly to the Carpathian chain and produce prolonged, orographically strengthened rainfall. On the northern slopes of the mountains, daily totals of precipitation reach 100–250 mm (Cebulak 1992) and three-days totals amount to 300–500 mm—see Niedźwiedź and Łupikasza (this volume) for more details about these synoptic situations. In the western part of the Upper Vistula Basin (west of the Dunajec), the floods caused by prolonged rainfall are more frequent than in the eastern part (Cebulak 1992).

High-intensity rainfall with 50–150 mm of precipitation in one to a few hours is connected to convective thunderstorms (Cebulak and Niedźwiedź 1998) and results in flash floods of local extent (Bryndal 2014), which however can be highly damaging. Such was a flash flood that occurred in July 2006 in the upper Wisłoka catchment; the intensive rain caused water stage on a tributary to the Wisłoka to increase rapidly by 3.5–4 m and the flood destroyed infrastructure in the valley and caused a few fatalities (Izmaiłow et al. 2006). Several heavy downpours (>8 mm h^{-1}) or even torrential rains (>44 mm h^{-1}) of 0.5–2 h duration were recorded in different parts of the Polish Carpathians (Starkel 2011). Some of them are worthy of note: 7 June 1985 in Szymbark, 25 July 2001 in Maków Podhalański, 14 July 2002 in Muszyna, 26 July 2005 in Baligród, 5 June 2007 in the Western Tatra Mountains, and many others.

Finally, the occurrence of flysch bedrock in most of the Polish Carpathian area causes landslides to be a frequent type of geomorphic hazard that in some cases results in considerable damage to buildings or infrastructure. Landslides are frequently triggered by the same rainfall events which induce major floods in the region, but may also be activated in wet years with relatively high annual sum of precipitation dispersed over several months (Froehlich and Starkel 1995). Such extremely rainy season in the Polish Carpathians was summer 1913 (Starkel 2011).

The high Tatra Mountains feature abundant precipitation, higher than elsewhere in Poland. The highest daily precipitation totals in the Polish Tatra Mountains reach 300 mm (Polish record, overall) observed on 30 June 1973 in Hala Gąsienicowa during a northern cyclonic situation Nc. However, the occurrence of thick and highly porous slope covers and deep, karstic circulation result in relatively low flashiness of runoff from the Tatras (Kundzewicz et al. 2014). For instance, the runoff-irregularity coefficient (the ratio of the highest and the lowest discharge on record) for the Koniówka gauging station on the Czarny Dunajec River equals 453. The flysch Carpathians are typified by low water storativity, and thus runoff from this area is more flashy, despite lower sums of precipitation (Kundzewicz et al. 2014). The most extreme case seems the upper part of the Biała catchment with only shallow, slope aquifers, for which the runoff-irregularity coefficient at the Grybów station amounts to 7500.

4 Recent Advances in Flood Research in the Polish Carpathians

Over the few past decades a change in the seasonality of floods in the Polish Carpathians has occurred (see Ruiz-Villanueva et al. 2016b who demonstrated such a change in rivers in the northern foreland of the Tatra Mountains). Major rain-induced floods, previously typically occurring between June and August, started to occur also in May (2010, 2014) and September (1996, 2007).

Major floods cause substantial modification of channel morphology (Gorczyca et al. 2013; Hajdukiewicz et al. 2016) and result in considerable economic losses. The flood of July 1934 was the largest one in the Dunajec catchment, especially in the northern foreland of the Tatra mountains (Fig. 1), and inundated vast areas in the Upper Vistula Basin. The flood occurred in the final part of a long historical period typified by low forest cover and intense agricultural and pastoral use of hillslopes in the Polish Carpathians. As a result, disastrous character of the flood was largely related to the deposition of huge amounts of coarse bed material in the channels and on valley floors (Fig. 1). In the last decades, large floods originating in the Polish Carpathians occurred in 1997 and 2010, but they were borne in largely reforested mountain catchments and hit the rivers typified by sediment deficit. They resulted in material damage of the order of 1 % of gross national product of the country (Kundzewicz et al. 2012). A considerable proportion of the damage occurred in the valleys of Carpathian tributaries to the Vistula (Biedroń et al. 2011) as a result of the erosional action of the flood flow (Fig. 2); downstream of mountain reaches, the huge masses of floodwaters propagated along the Vistula valley, causing a failure of embankments and inundation of extensive areas at several locations, even 400 km away from the mountains (Kozak et al. 2013).

During the two last decades attention has been paid to the recruitment, transport and deposition of large wood by floods and is role as an agent of flood hazard in Polish Carpathian rivers. Kaczka (1999) was the first who indicated in Polish literature that during floods large numbers of fallen trees are delivered to mountain stream flowing through forested corridors, and the same was next demonstrated with respect to wide mountain rivers (Wyżga and Zawiejska 2005). Wood accumulations causing clogging of bridge cross-sections appeared to be an important

Fig. 1 View of Bystry Stream in its reach within the town of Zakopane just after the catastrophic flood of July 1934. The flood was the largest one on record in this region, with peak discharges of streams flowing from the Tatra Mountains up to three times higher than the largest ones recorded in the second half of the 20th century. Visible are destroyed houses in the riparian area and huge amounts of coarse bed material deposited practically at the level of the developed valley floor. Currently the stream flows in 6 m-wide, stone-lined, artificial channel. The photo no. 1-G-4647-12 of the National Digital Archive of Poland

cause of flood damage during floods (Hajdukiewicz et al. 2016) and research aimed to better understand large wood dynamics in mountain watercourses was carried out using different methods such as wood inventories (Wyżga et al. 2015), numerical modelling (Ruiz-Villanueva et al. 2016c, d, e) or tracking experiments with tagged wood (Wyżga et al. 2016b).

Successful management of flood risk requires proper recognition of the frequencies and magnitudes of flood events, characterization of hydraulic parameters of flood flows and innovative approaches to river management. Small headwater streams are typically ungauged in the Polish Carpathians, but at the same time floods borne in their catchments may provide significant risk to downstream, developed valley reaches because of short concentration times of the flood waves reflected in their flashy nature. Using dendrogeomorphic techniques, Ballesteros-Cánovas et al. (2015) identified 47 flood events that occurred on four streams in the Tatra Mountains over the last 148 years, and discussed synoptic situations leading to their triggering. In turn, Ballesteros-Cánovas et al. (2016) demonstrated that reconstruction of paleodischarges of the floods occurring on these small, headwater streams allows not only to extend existing flow records but also to improve the estimation of flood frequency distributions. Bryndal (2014)

Fig. 2 Damage to infrastructure caused by the 80-year flood of June 2010 on the Biała River. A significant increase in forest cover in the catchment after World War II and widespread in-channel gravel mining conducted in the past decades resulted in sediment starvation of the river and thus the flood action was mainly directed on erosion of the channel boundaries

analysed physiographic parameters of small catchments in the Polish Carpathians that were affected by flash floods and proposed a method allowing for identification of catchments especially prone to the generation of such floods. Wyżga et al. (2016a) analysed variability in the hydraulic importance of channel incision with increasing river size by comparing changes in the frequency of valley floor inundation at gauging stations on the Dunajec River characterizing catchments with an area ranging from 34.5 to 6735 km^2. The analysis indicated that the relative increase in channel capacity and the resultant loss of floodwater storage on valley floors, associated with a given absolute (i.e. expressed in metres) amount of channel incision, will be most severe in the upper river courses where the channels initially had a relatively small capacity. Czech et al. (2016) analysed the impact of river restoration on the conditions for flood flows by comparing hydraulic parameters of flood flows between closely located unmanaged and channelized cross-sections of the Biała River. They indicated that despite a short time since the beginning of the river restoration, it already brings beneficial effects for flood risk management, reducing flow energy and shear forces exerted on the bed and banks of the channel in unmanaged river reaches. Mikuś et al. (2016) analysed geomorphic responses of the multi-thread Czarny Dunajec River to floods of different magnitude and demonstrated that protection of a local road from erosion does not require river channelization but can be attained through flow re-activation in side braids.

Acknowledgments This contribution has been supported by the FLORIST project (Flood risk on the northern foothills of the Tatra Mountains), PSPB no. 153/2010 through a grant from Switzerland through the Swiss Contribution to the enlargement of the European Union.

References

Arnaud-Fassetta G, Cossart E, Fort M (2005) Hydro-geomorphic hazards and impact of man-made structures during the catastrophic flood of June 2000 in the Upper Guil catchment (Queyras, Southern French Alps). Geomorphology 66:41–67

Ballesteros-Cánovas JA, Czajka B, Janecka K, Lempa M, Kaczka RJ, Stoffel M (2015) Flash floods in the Tatra Mountain streams: frequency and triggers. Sci Total Environ 511:639–648

Ballesteros-Cánovas JA, Stoffel M, Spyt B, Janecka K, Kaczka RJ, Lempa M (2016) Paleoflood discharge reconstruction in Tatra Mountain streams. Geomorphology 272C. doi:10.1016/j.geomorph.2015.12.004

Barredo JI (2007) Major flood disasters in Europe: 1950–2005. Nat Hazards 42:125–148

Beniston M (2005) Mountain climates and climatic change: an overview of processes focusing on the European Alps. Pure Appl Geophys 162:1587–1606

Beniston M, Stoffel M (2014) Assessing the impacts of climatic change on mountain water resources. Sci Total Environ 493:1129–1137

Biedroń I, Bogdańska-Warmuz R, Kwiecień M (2011) Szkody i straty powodziowe w dorzeczu Wisły (Flood damages and losses in the Vistula River drainage basin). In: Maciejewski M, Ostojski MS, Walczykiewicz T (eds) Dorzecze Wisły — monografia powodzi maj-czerwiec 2010. Instytut Meteorologii i GospodarkiWodnej, Warszawa, pp 171–182

Bodoque JM, Eguibar MA, Díez-Herrero A, Gutiérrez-Pérez I, Ruiz-Villanueva V (2011) Can the discharge of a hyperconcentrated flow be estimated from paleoflood evidence? Water Resour Res 47:W12535

Borga M, Boscolo P, Zanon F, Sangati M (2007) Hydrometeorological analysis of the August 29, 2003 flash flood in the eastern Italian Alps. J Hydrometeorol 8:1049–1067

Borga M, Gaume E, Creutin JD, Marchi L (2008) Surveying flash flood response: gauging the ungauged extremes. Hydrol Process 22:3883–3885

Borga M, Stoffel M, Marchi L, Marra F, Jakob M (2014) Hydrogeomorphic response to extreme rainfall in headwater systems: flash floods and debris flows. J Hydrol 518:194–205

Bracken LJ, Croke J (2007) The concept of hydrological connectivity and its contribution to understanding runoff-dominated geomorphic systems. Hydrol Proc 21:1749–1763

Bridges EM (1990) World geomorphology. Cambridge University Press, Cambridge, 260 pp

Bryndal T (2014) A method of identification of small Carpathian catchments more prone to flash flood generation. Based on the example of south-eastern part of the Polish Carpathians. Carp J Earth. Environ Sci 9(3):109–122

Cebulak E (1992) Maksymalne opady dobowe w dorzeczu górnej Wisły (Maximum daily precipitation totals in the Upper Vistula Basin). Zesz Nauk UJ, Prace Geogr 90:79–96 (in Polish, with English summary)

Cebulak E, Niedźwiedź T (1998) Ekstremalne zjawiska opadowe w dorzeczu górnej Wisły w latach 1995 i 1996 (Extreme precipitation events in the Upper Vistula Basin in 1995 and 1996). In: Starkel L (ed.), Geomorfologiczny i Sedymentologiczny Zapis Lokalnych Ulew. Dokum Geogr 11:11–30 (in Polish, with English summary)

Cenderelli DA (2011) Floods from natural and artificial dam failures. In: Wohl EE (ed) Inland Flood Hazards: Human, Riparian, and Aquatic Communities. Cambridge Univ. Press, New York, pp 73–103

Cenderelli DA, Kite JS (1998) Geomorphic effects of large debris flows on channel morphology at North Fork Mountain, eastern West Virginia, USA: Earth Surf Proc Landforms 23:1–19

Cenderelli DA, Wohl EE (2003) Flow hydraulics and geomorphic effects of glacial-lake outburst floods in the Mount Everest region. Nepal. Earth Surf Proc Landforms 28:385–407

Chełmicki W, Skąpski R, Soja R (1999) Reżim hydrologiczny rzek karpackich w Polsce (The flow regime of Poland's Carpathian rivers). Folia Geogr 29–30:67–80

Czech W, Radecki-Pawlik A, Wyżga B, Hajdukiewicz H (2016) Modelling the flooding capacity of a Polish Carpathian river: a comparison of constrained and free channel conditions. Geomorphology 272C. doi:10.1016/j.geomorph.2015.09.025

DETEC (Federal Department for the Environment, Transport, Energy and Communications) (2008) The floods of 2005 in Switzerland: synthesis report on the event analysis. DETEC, Berne, 24 pp

Falarz M (2004) Variability and trends in the duration and depth of snow cover in Poland in the 20th century. Intern J Climatol 24:1713–1727

Froehlich W, Starkel L (1995) The response of slope and channel systems to various types of extreme rainfall: a comparison between the temperate zone and humid tropics. Geomorphology 11:337–345

Fukami K, Yamaguchi T, Imamura H, Tashiro Y (2008) Current status of river discharge observation using non-contact current meter for operational use in Japan, World Environmental and Water Resources Congress 2008 Ahupua'a

Gaume E, Borga M (2008) Post-flood field investigations in upland catchments after major flash floods: proposal of a methodology and illustrations. J Flood Risk Manage 1:175–189

Gobiet A, Kotlarski S, Beniston M, Heinrich G, Rajczak J, Stoffel M (2014) 21st century climate change in the European Alps—a review. Sci Total Environ 493:1138–1151

Gorczyca E, Krzemień K, Wrońska-Wałach D, Sobucki M (2013) Channel changes due to extreme rainfalls in the Polish Carpathians. In: Lóczy D (ed) Geomorphological impacts of extreme weather. Case studies from Central and Eastern Europe. Springer, Dordrecht, pp 23–35

Grebner D, Roesch T, Schwarb M (1999) Extreme regional precipitation of varying duration and return period. In: Hydrological Atlas of Switzerland, Plate 2.5, Berne

Grebner D, Aschwanden H, Steinegger U, Zimmermann M (2000) Charakteristik des Hochwassers vom 9. bis 16. Oktober 2000 auf der Alpensüdseite und im Wallis. Wasser, Energie, Luft 11/12:369–377

Gutiérrez F, Gutiérrez M, Sancho C (1998) Geomorphological and sedimentological analysis of a catastrophic flash flood in the Arás drainage basin (Central Pyrenees, Spain). Geomorphology 22:265–283

Hajdukiewicz H, Wyżga B, Mikuś P, Zawiejska J, Radecki-Pawlik A (2016) Impact of a large flood on mountain river habitats, channel morphology, and valley infrastructure. Geomorphology 272C. doi:10.1016/j.geomorph.2015.09.003

Hicks NS, Smith JA, Miller AJ, Nelson PA (2005) Catastrophic flooding from an orographic thunderstorm in the central Appalachians. Water Resour Res 41:W12428

Hughes FMR, Rood SB (2003) Allocation of river flows for restoration of floodplain forest ecosystems: a review of approaches and their applicability in Europe. Environ Manage 32: 12–33

Izmaiłow B, Kamykowska M, Krzemień K (2006) The geomorphological effects of flash floods in mountain river channels. The case of the River Wilsznia (Western Carpathian Mountains). Pr Geogr 116:89–97

Junk JW, Bayley PB, Sparks RE (1989) The flood pulse concept in river flood plain systems. Can Spec Publ Fish Aquat Sci 106:110–121

Kaczka RJ (1999) The role of coarse woody debris in fluvial processes during the flood of the July 1997, Kamienica Łącka Valley, Beskidy Mountains, Poland. Stud Geomorph Carp-Balcan 33:117–130

Kotarba A (1992) High energy geomorphic events in the Polish Tatra Mountains. Geogr Ann 74A:123–131

Kozak J, Ostapowicz K, Bytnerowicz A, Wyżga B (2013) The Carpathian Mountains: Challenges for the Central and Eastern European landmark. In: Kozak J, Ostapowicz K, Bytnerowicz A, Wyżga B (eds) The Carpathians: integrating nature and society towards sustainability. Springer, Berlin, pp 1–11

Kundzewicz ZW, Dobrowolski A, Lorenc H, Niedźwiedź T, Pińskwar I, Kowalczak P (2012) Floods in Poland. In: Kundzewicz ZW (ed) Changes in flood risk in Europe. Intern Assoc Hydrol Sci Spec Publ 10:319–334

Kundzewicz ZW, Stoffel M, Kaczka R, Wyżga B, Niedźwiedź T, Pińskwar I, Ruiz-Villanueva V, Łupikasza E, Czajka B, Ballesteros-Canovas JA, Małarzewski L, Choryński A, Janecka K, Mikuś P (2014) Floods at the northern foothills of the Tatra Mountains—a Polish-Swiss research project. Acta Geophys 62:620–641

Lin GW, Chen H, Hovius N, Horng MJ, Dadson S, Meunier P, Lines M (2008) Effects of earthquake and cyclone sequencing on landsliding and fluvial sediment transfer in a mountain catchment. Earth Surf Proc Landforms 33:1354–1373

Lojkasek B, Lusk S, Halacka K, Luskova V, Drozd P (2005) The impact of the extreme floods in July 1997 on the ichthyocenosis of the Oder catchment area (Czech Republic). Hydrobiologia 548:11–22

Lugon R, Stoffel M (2010) Rock-glacier dynamics and magnitude-frequency relations of debris flows in a high-elevation watershed: Ritigraben, Swiss Alps. Global Planet Change 73:202–210

Lumbroso DM, Gaume E (2012) Reducing the uncertainty in indirect estimates of extreme flash flood discharges. J Hydrol 414–415:16–30

Marchi L, Borga M, Preciso E, Gaume E (2010) Characterisation of selected extreme flash floods in Europe and implications for flood risk management. J Hydrol 394:118–133

Mikuś P, Wyżga B, Radecki-Pawlik A, Zawiejska J, Amirowicz A, Oględcki P (2016) Environment-friendly reduction of flood risk and infrastructure damage in a mountain river: case study of the Czarny Dunajec. Geomorphology 272C. doi:10.1016/j.geomorph.2015.11.003

Montgomery DR, Buffington JM (1997) Channel-reach morphology in mountain drainage basins. Geol Soc Am Bull 109:596–611

Niedźwiedź T, Łupikasza E (2016) Change in circulation patterns. In: Kundzewicz ZW, Stoffel M, Niedźwiedź T, Wyżga B (eds) Flood risk in the Upper Vistula Basin. Springer, Cham (this volume)

Rotunno R, Houze RA (2007) Lessons on orographic precipitation from the Mesoscale Alpine Programme. Quart J Royal Meteorol Soc 133:811–830

Ruiz-Villanueva V, Díez-Herrero A, Stoffel M, Bollschweiler M, Bodoque JM, Ballesteros JA (2010) Dating flash flood events by means of dendrogeomorphic analysis in a small ungauged mountain catchment (Spanish Central System). Geomorphology 118:383–392

Ruiz-Villanueva V, Gurnell A, Piégay H, Marston RA, Stoffel M (2016a) Recent advances quantifying the large-wood cycle in forested river basins: new methods, remaining challenges. Rev Geophys (in review)

Ruiz-Villanueva V, Stoffel M, Wyżga B, Kundzewicz ZW, Czajka B, Niedźwiedź T (2016b) Decadal variability of floods in the northern foreland of the Tatra Mountains. Reg Environ Change 16:603–615

Ruiz-Villanueva V, Wyżga B, Hajdukiewicz H, Stoffel M (2016c) Exploring large wood retention and deposition in contrasting river morphologies linking numerical modelling and field observations. Earth Surf Proc Landforms 41:446–459

Ruiz-Villanueva V, Wyżga B, Mikuś P, Hajdukiewicz H, Stoffel M (2016d) The role of flood hydrograph in the remobilization of large wood in a wide mountain river. J Hydrol. doi:10.1016/j.jhydrol.2016.02.060

Ruiz-Villanueva V, Wyżga B, Zawiejska J, Hajdukiewicz M, Stoffel M (2016e) Factors controlling large wood transport in a mountain river. Geomorphology 272C. doi:10.1016/j.geomorph.2015.04.004

Schneuwly-Bollschweiler M, Stoffel M (2012) Hydrometeorological triggers of periglacial debris flows in the Zermatt Valley (Switzerland) since 1864. J Geophys Res 117:F02033

Schwanghart W, Worni R, Huggel C, Stoffel M, Korup O (in press) 21st century Himalayan hydropower and glacial lake outburst floods. Environ Res Lett

Schwarb M, Daly C, Frei C, Schär C (2001) Mean annual precipitation throughout the European Alps 1971–1990. Hydrological Atlas of Switzerland, Plate, 2.6, Berne

Semenza E, Ghirotti M (2000) History of the 1963 Vaiont slide: the importance of geological factors. Bull Eng Geol Environ 59:87–97

Starkel L (1996) Geomorphic role of extreme rainfalls in the Polish Carpathians. Stud Geomorph Carp-Balcan 30:21–28

Starkel L (2011) Złożoność czasowa i przestrzenna opadów ekstremalnych – ich efekty geomorfologiczne i drogi przeciwdziałania im (Temporal and spatial complexity of extreme rainfalls—their geomorphological effects and ways of counteract them). Landform Anal 5:65–80 (in Polish, with English summary)

Stoffel M, Beniston M (2006) On the incidence of debris flows from the early Little Ice Age to a future greenhouse climate: a case study from the Swiss Alps. Geophys Res Lett 33:L16404

Stoffel M, Huggel C (2012) Effects of climate change on mass movements in mountain environments. Progr Phys Geogr 36:421–439

Stoffel M, Bollschweiler M, Beniston M (2011) Rainfall characteristics for periglacial debris flows in the Swiss Alps: past incidences—potential future evolutions. Climat Change 105:263–280

Stoffel M, Mendlik T, Schneuwly-Bollschweiler M, Gobiet A (2014a) Possible impacts of climate change on debris-flow activity in the Swiss Alps. Climat Change 122:141–155

Stoffel M, Tiranti D, Huggel C (2014b) Climate change impacts on mass movements—case studies from the European Alps. Sci Total Environ 493:1255–1266

Sturdevant-Rees P, Smith JA, Morrison J, Baeck ML (2001) Tropical storms and the flood hydrology of the central Appalachians. Water Resour Res 37:2143–2168

Toreti A, Schneuwly-Bollschweiler M, Stoffel M, Luterbacher J (2013) Atmospheric forcing of debris flows in the southern Swiss Alps. J Appl Meteorol Climatology 52:1554–1560

Viviroli D, Weingartner R, Messerli B (2003) Assessing the hydrological significance of the World's mountains. Mount Res Develop 23(1):32–40

Weingartner R, Barben M, Spreafico M (2003) Floods in mountain areas—an overview based on examples from Switzerland. J Hydrol 282:10–24

Wohl E (2010) Mountain rivers revisited. AGU Water Resources Monograph 19, 576 pp

Worni R, Huggel C, Clague JJ, Schaub Y, Stoffel M (2014) Coupling glacial lake impact, dam breach, and flood processes: a modeling perspective. Geomorphology 224:161–176

Wydoski RS, Wick EJ (2011) Flooding and aquatic ecosystems. In: Wohl EE (ed) Inland flood hazards: human, riparian, and aquatic communities. Cambridge Univ. Press, New York, pp 238–268

Wyżga B, Zawiejska J (2005) Wood storage in a wide mountain river: case study of the Czarny Dunajec, Polish Carpathians. Earth Surf Proc Landforms 30:1475–1494

Wyżga B, Zawiejska J, Mikuś P, Kaczka RJ (2015) Contrasting patterns of wood storage in mountain watercourses narrower and wider than the height of riparian trees. Geomorphology 228:275–285

Wyżga B, Zawiejska J, Radecki-Pawlik A (2016a) Impact of channel incision on the hydraulics of flood flows: examples from Polish Carpathian rivers. Geomorphology 272C. doi:10.1016/j.geomorph.2015.05.017

Wyżga B, Mikuś P, Zawiejska J, Ruiz-Villanueva V, Kaczka RJ, Czech W (2016b) Log transport and deposition in incised, channelized and multi-thread reaches of a wide mountain river: tracking experiment during a 20-year flood. Geomorphology (submitted)

Youssef AM, Pradhan B, Hassan AM (2011) Flash flood risk estimation along the St. Katherine road, southern Sinai, Egypt using GIS based morphometry and satellite imagery. Environ Earth Sci 62:611–623

Anatomy of Flood Risk

Zbigniew W. Kundzewicz and Markus Stoffel

Abstract In this chapter, the notion of flood risk is interpreted and factors influencing flood risk: hazard, exposure, and vulnerability are examined. A holistic perspective on changing flood risk is provided. Economic losses from floods have greatly increased over last decades, principally driven by the growing exposure of assets at risk. It has not been found possible to attribute observed rain-generated peak streamflow trends to anthropogenic climate change, generally. Based on physical reasoning, increases in the frequency and intensity of heavy rainfall should contribute to increase in risk of precipitation-generated local flooding (e.g., flash flooding and urban flooding) in many areas.

Keywords Flood risk · Hazard · Exposure · Vulnerability

Z.W. Kundzewicz (✉)
Institute for Agricultural and Forest Environment, Polish Academy of Sciences, Poznań, Poland
e-mail: kundzewicz@yahoo.com

Z.W. Kundzewicz
Potsdam Institute for Climate Impact Research, Potsdam, Germany

M. Stoffel
Climatic Change and Climate Impacts, Institute for Environmental Sciences, University of Geneva, Geneva, Switzerland
e-mail: markus.stoffel@dendrolab.ch

M. Stoffel
Department of Earth Sciences, University of Geneva, Geneva, Switzerland

M. Stoffel
Dendrolab.ch, Institute of Geological Sciences, University of Bern, Bern, Switzerland

Z.W. Kundzewicz et al. (eds.), *Flood Risk in the Upper Vistula Basin*,
GeoPlanet: Earth and Planetary Sciences, DOI 10.1007/978-3-319-41923-7_3

1 Introduction

Flood is a damaging presence of water in areas which are not normally submerged (e.g., outside of a river bed or of normal confines of a body of water), where damage potential, however, exists. Hence a flood is a combination of two elements —coincidence of the damaging element (abundant water) and of loss potential.

Intense and/or long-lasting rainfall may cause a river (fluvial) flood, a flash flood, an urban flood, a pluvial flood, a groundwater flood, or a sewer flood. In moderate and cold climates, floods can be also caused by snowmelt (sometimes enhanced by rain and/or ice jams). Important to flood generation are antecedent conditions, e.g., saturation of natural storage in the catchment (surface water storage and soil moisture), resulting from earlier precipitation and/or snowmelt. Should storage be limited because groundwater levels are elevated and soil moisture is at maximum capacity, then even moderate amounts of rain can generate a large flood (such as was the case in Poland in May/June 2010). However, the development of a very dry, crusted soil after a prolonged period without rain, can rapidly convert heavy rainfall to runoff (because the runoff coefficient is higher) as well, and would usually result in a flash flood. A combination of factors can contribute to the generation of a large event (e.g., the joint occurrence of rainfall and snowmelt).

Other natural factors that may induce fluvial floods include landslides, reducing conveyance (obstructing river flow), and the sudden failure of inhibiting structures, such as the collapse of landslide dams, ice jams, or glaciers blocking glacial lakes. Catastrophic dike or dam failure (breach or break) can be the direct cause of intense flooding. The extent of inundation can be influenced by blockage of bridges and culverts by debris (Kundzewicz et al. 2012). A special kind of inundations are coastal floods (tidal surges).

Fluvial (river) floods are natural phenomena, manifesting natural spatial and temporal variability of such geophysical variables as river water level and discharge, which, from time to time, may take extremely high values. Yet, besides natural factors, there is a considerable human impact on flooding.

Despite heavy investments into flood defense, floods continue to be an acute problem, causing high material damage and considerable death toll and human suffering in many regions. In last three decades or so, reported annual flood losses (adjusted for inflation) have more than trebled and reach tens of billions of dollars. Globally, the largest flood disasters, killing thousands of people, continue to occur overseas, in particular in Asia. However, despite its economic and social growth and the progress in technology, Europe has not been immune to severe floods. In recent decades, large parts of the continent have been hit by major floods, with tens of fatalities and multi-billion-high (in USD or Euro) material damage. Poland has suffered major floods in last two decades in 1997, 1998, 2001, and 2010, therein the 1997 and 2010 events were really devastating.

As framed by Handmer et al. (2012), adverse impacts of flood events include direct destruction of property, effects on livelihoods, health, and production, as well as indirect effects of these consequences through the wider economy, on a variety of economic sectors including agriculture, transportation, water supply, industry, or tourism. Heavy precipitation and field flooding can delay spring planting, increase soil compaction, and cause crop losses through anoxia and root diseases. However, on the positive side, for ecosystems, flooding can be both a blessing (e.g., advantageous "flood pulse" bringing abundance of water and fertile solids) or a curse, leading to erosion and temporary or even permanent loss of habitat. In some areas, floods are important as a mechanism to replenish wetlands and aquifers.

In the present chapter, we look at how flood risk can be seen from a holistic cradle-to-grave perspective, as a result of a chain of flood-relevant processes and variables, interpreting observational records and using results of simulations with the help of mathematical models. In the following sections, we will review flood risk and its components—flood hazard, exposure, and vulnerability, then changes in flood risk—as well as options for reducing flood risk.

2 Flood Risk and Its Components

The notion of flood risk typically represents an expected value of an undesirable outcome, measuring flood damage, i.e. a product (multiplication) of probability of a flood event and the potential adverse impacts associated with a flood event if it occurs. This aggregate notion combines the probability (likelihood) of various possible flood magnitude classes and their adverse consequences (on lives, livelihoods, health, ecosystems, infrastructure, economic, social and cultural assets, and services) on a numerical scale and multiplying one by the other and summing up to produce a comparative score. Typically, flood risk does not refer to a current problem that must be immediately addressed, but rather it refers to a future problem that will show up in some indefinite and uncertain future time and that can be managed and reduced. Flood risk depends on flood hazard, exposure, and vulnerability (Fig. 1). Flood risk depends on properties of climate, socio-economic, and terrestrial systems. It is influenced by many climatic and non-climatic factors.

2.1 *Flood Hazard*

Flood hazard is a situation that poses a threat to life, health (illness or injury), damage to property, infrastructure, livelihoods, social stability, service provision, and environment. Hazard can be dormant or potential, but once becomes "active", it creates emergency.

Climate is certainly a very important factor driving flood hazard. Fluvial flood hazard is affected by various characteristics of the climatic and atmospheric systems, such as water holding capacity (and water vapor content) of the atmosphere, various of precipitation characteristics—intensity, duration, total amount, timing, phase—rain or snow; spatial and temporal distribution, e.g., via the antecedent

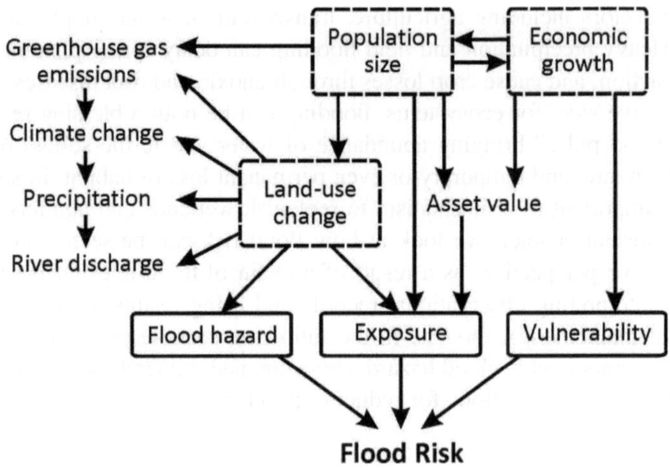

Fig. 1 Anthropogenic drivers of changes in fluvial flood risk. Note that risk reduction activities can reduce the hazard and the potential loss (by affecting exposure and vulnerability). After: Kundzewicz et al. (2014), modified

precipitation index (API), as well as such determinants as large-scale circulation patterns, but also temperature pattern (responsible for such phenomena as soil freezing, snow and ice melt, and ice jam formation). Floods are also affected by drainage basin conditions such as pre-existing water levels in rivers, the snow and ice cover, the soil character and status (permeability, soil moisture content and its vertical distribution), the rate of urbanisation, and the presence of dikes, dams, and reservoirs. River discharge is a result of processes in the drainage basin—from precipitation to runoff.

Floods are intermittent events, possibly of rare recurrence in a given location. However, at some sites extensive inundations are more commonplace, "climatological normal", occurrences, e.g. occurring every spring when the abundant snow cover melts (Kundzewicz 2012).

The characteristics of terrestrial systems strongly influence flood risk. The most pertinent are: catchment size, geology, landscape, topography and soils, the latter often largely modified by human intervention. Human induced impacts on terrestrial systems affecting flood risk include land-use and land-cover changes, influencing the transformation of precipitation into runoff as well as river engineering.

Alterations in catchment surface characteristics, modification of floodplain storage, and of the river network can all modify the physical characteristics of river floods. In urban areas, impermeable surfaces dominate and the value of the runoff coefficient (portion of precipitation that enters a stream) is high, while the water-storage capacity is low, in contrast to rural (and especially forested) areas. Hence, urban and rural catchments of the same size and topography will react differently to the same precipitation input. Peak discharge in an urban area is usually higher, and the time-to-peak is shorter than in a rural area (Kundzewicz et al. 2012). Moreover, processes of urbanization lead to increased occupation of the floodplain, and often inadequate drainage planning (Handmer et al. 2012).

The river stage and thus the risk of flooding also depend on engineered alterations to river courses, depths, and artificial flood containment such as flood dikes. Reservoirs, whether developed specifically for flood protection or being multi-purpose, can substantially reduce short-duration flood waves.

2.2 *Exposure to Floods*

Exposure to floods can be defined as the presence of people, livelihoods, or ecosystems, environmental functions, services, and resources, infrastructure, or economic, social, or cultural assets in places and settings that could be adversely affected by floods (after IPCC 2014).

River floods, jeopardising settlements located in floodplains, have been an ever-present hazard accompanying the mankind since the dawn of civilisation. For millennia, people have been settling in river floodplains in order to till fertile soils, to profit from flat terrain appropriate for settlements, to have easy and safe access to water supply, and to use water for transport. Riparian population occupied higher land to construct dwellings, and used the rest for farming. They benefited from benign floods and their valuable services provided to ecosystems (i.e., wildlife, fisheries, wetlands, and agriculture). In short—people lived much more in harmony with nature. Location of towns proceeded along the banks of lakes and rivers. When towns grew, they sprawled towards river valleys or floodplains, with better soils and better access to navigation routes and merchant tracks. However, such locations were not free of flood risk.

Now, in terms of exposure to flooding, about 800 million people worldwide (i.e. over 11 % of the global population) are currently living in flood-prone areas, and about 70 million (i.e. 1 % of the global population) are, on average, exposed to floods each year (UNISDR 2011).

In some of the countries with high population density, substantial fractions of land surfaces are prone to floods. According to the data compiled by IWR (2011), flood-prone area in the USA, Japan, UK, and the Netherlands, amounts to: 7, 9, 12 and 59 % of total land area, respectively. Population in flood-prone areas in Japan, UK, and the Netherlands is, respectively, 41, 9, and 55 %. Assets in flood plain area in Japan and in the Netherlands are, respectively, 66 and 65 % of the total national assets.

While in the global ranking of countries in terms of absolute GDP exposure to floods, USA comes first, Japan seventh, the Netherlands eleventh and the UK thirteenth, the highest relative share of population and percentage of economy exposed to floods is found in Cambodia, Bangladesh, and Vietnam (Peduzzi et al. 2009; Kundzewicz et al. 2014). In Bangladesh and Cambodia, the relative number of people exposed per year exceeds 14 and 12 %, respectively. In Bangladesh, during the 1998 flood, two-thirds of the country was under water. The same two countries have the highest relative exposure of assets (in relation to the total GDP) per year: in

Cambodia, it exceeds 12 % and in Bangladesh 10 %. Vietnam comes third in the list of countries with the highest relative exposure to floods (both in terms of the number of people exposed to floods, per year, and in terms of absolute numbers and relative proportions, as well as total assets and GDP exposed to floods, per year).

Nevertheless, people like to live near water. A view over water is generally preferred to a view on a flood wall or a dyke. The illusory feeling of safety of riparians protected by structural defences leads to soaring values in exposed areas. This may be further enhanced if people are misinformed or if the risk is played down by local interest groups (e.g., a community that wants to develop and sell real estate, or to promote local tourism, Kron 2012).

2.3 *Vulnerability to Floods*

Among the many existing definitions of vulnerability, let us assume the one after IPCC (2014) where it is defined "the propensity or predisposition to be adversely affected". Vulnerability understood in this way encompasses "a variety of concepts and elements including sensitivity or susceptibility to harm and lack of capacity to cope and adapt." Vulnerability to floods may refer to human health and well-being (human vulnerability), structural integrity (physical vulnerability) or personal wealth (financial vulnerability). In some less developed countries, and in particular in less wealthy areas in these countries, vulnerability can be very high. Even if urbanization issues are universal, they are often at their worst in informal settlements. Hope to overcome poverty drives poor people to migrate to informal settlements in endangered, flood-prone zones around mega-cities in developing countries, which are left uninhabited on purpose, since effective flood protection cannot be assured.

There are multiple measures of losses caused by a particular flood event, ranging from the number of fatalities; number of injured persons; number of flood-related illnesses; number of evacuees; number of relocated persons, or number of affected persons. Material damage indices, on the other hand, include absolute losses or insured losses (in financial terms), but can also go to more detailed specification of losses, e.g., destroyed infrastructure; private dwellings; buildings; commercial and industrial stocks and facilities; transport facilities: railways, bridges, roads; dikes; water supply lines, and telecommunications; on public facilities (e.g., hospitals, schools); losses in cultural heritage, as well as natural resources: water erosion and habitat and soil loss; inundated areas; therein agricultural land, and crop loss. Important, and difficult to estimate, are secondary damages (e.g. lost benefits, foregone opportunities), arising in the complex network of interconnected systems within the "global village", disturbed by flood. All these characteristics compiled above are important in the quantification of flood risk. Indirect damage may be caused by direct damage to physical infrastructure or sources of livelihoods, or because reconstruction pulls resources away from production. Indirect damage costs arise due to the disruption of the flows of goods and services (and therefore economic activity) because of a flood, and are sometimes termed consequential or secondary impacts.

Major floods, with high material damage (in developed and developing countries, alike) and fatalities (predominantly in developing countries), have been caused by the passage of high waters in rivers running through densely populated urban areas. Urban flooding constitutes a risk to life, health, and well-being of the population. It may lead to many adverse effects, such as the disruption of communication (road, rail, air); damage to infrastructure (roads, houses, public utilities, industry); water pollution; ground pollution; and risk to technical and sanitary state of buildings. They may damage urban flora and fauna and enhance the occurrence of mosquitoes (which is especially dangerous in tropical countries where malaria risk is present). Urban areas have shown particularly strong human impact and this also influences flood risk. Costly elements of road infrastructure, such as bridges, culverts, and embankments (for roads and railways) are vulnerable to erosion-caused damage by heavy precipitation and floods.

3 Changes in Flood Risk

A plethora of factors exists that may explain a perceived increase in flood risk, namely flood hazard, exposure, and vulnerability of population and their assets, damage potential (driven by population and its wealth), economic development of flood-prone areas, adaptive capacity, risk awareness and perception (Kundzewicz et al. 2012). Changes in actual flood risk are themselves driven by changes in socio-economic, terrestrial, and climatic systems (Kundzewicz and Schellnhuber 2004). Improved and expanded reporting of disasters also plays a role (and is sometimes referred to as CNN effect).

Since multiple factors control flood risk, changes in any one of these factors lead to changes in flood risk. Changes of some factors may increase the risk, while changes of other factors may decrease risk. Hence, one could envisage a possibility of effective compensation, when risk remains nearly constant despite changes in its components.

Undoubtedly, flood risk has been intensified by humans, who—to use the parlance from mechanics—have contributed to the increase of the load and to the decrease of the resistance of the system. The former can be interpreted as the increase of flood magnitude corresponding to a given design precipitation, while the latter can be understood as amplification of the flood damage potential.

3.1 Changes in Hazard

Flood damage may increase due to several, climatic and non-climatic, reasons. A change in climate could physically alter many of the factors affecting floods, such

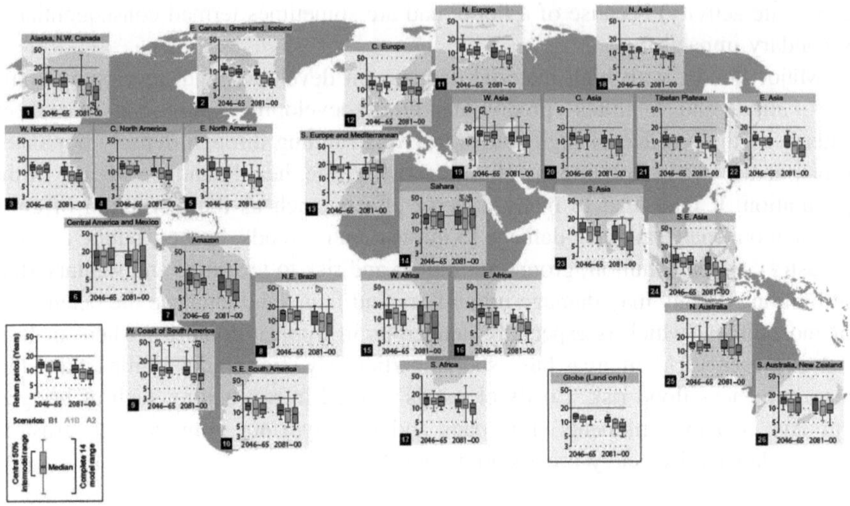

Fig. 2 Projected return period (in years) of late 20th-century 20-year return values of annual maximum 24-h precipitation rates. The bar plots (see legend for more information) show results for regionally averaged projections for two time horizons, 2046 to 2065 and 2081 to 2100, as compared to the late-20th century, and for three different SRES emission scenarios (B1, A1B, A2). Results are based on 14 GCMs contributing to the CMIP3. See Seneviratne et al. (2012) for defined extent of regions. The "Globe" analysis (*inset box*) displays the projected return period (in years) of late 20th-century 20-year return values of annual maximum daily precipitation rates computed using all land grid points. *Source* Seneviratne et al. (2012). Adapted from the analysis of Kharin et al. (2007)

as precipitation (its volume and timing, proportions of precipitation falling as snow and rain), snow cover, antecedent soil moisture content, surface water levels, vegetation, and thus may consequently change the characteristics of floods. The IPCC SREX report (Seneviratne et al. 2012) detected the influence of anthropogenic climate change in some variables affecting floods, including mean precipitation, heavy precipitation, and snowpack, though a direct statistical link between anthropogenic climate change and trends in the magnitude and/or frequency of floods has not been established so far. In climates where seasonal snow storage and melting play a significant role in annual runoff, the hydrological regime is affected by changes in temperature, and there is abundant evidence for changes in the timing (earlier occurrence) of spring peak flows in snowmelt and glacier-fed rivers. However, not all such areas are experiencing changes in the magnitude of peak flow.

Seneviratne et al. (2012) also assessed that it is *likely* that the frequency of heavy precipitation or the proportion of total rainfall from intense events will increase in the 21st century over many areas of the globe, and in particular in the high latitudes and tropical regions, and in winter in the northern mid-latitudes. Figure 2 displays the projected return periods for late-20th-century, '20-year return values' of annual maximum daily precipitation rates in the mid-21st century (on the left of each plot) and in late-21st century (on the right of each plot) under three different emission scenarios SRES B1, A1B and A2. Although there are exceptions, the analyses display an overall tendency for a decrease in return periods—heavy precipitation will become more commonplace. Medians fall below the late-20th 100-year return period for nearly all regions. The 50 % central (inter-quartile) range (colored boxes in each plot) for the end of 21st-century projections generally falls below the late 20th century return period. This analysis illustrates the strong regional dependency of the projected changes, with overall larger changes in the high latitude and tropical regions and larger uncertainty in drier regions. The stronger CO_2 emission scenarios lead to stronger projected increases in the probability (i.e. decreases in the return period) of events considered extreme with respect to the end of 20th century climate, as well as higher percentage increases in the absolute magnitude of heavy precipitation events.

The SREX report (Seneviratne et al. 2012) further assessed that there is *medium confidence* (based on physical reasoning) that projected increases in heavy rainfall would contribute to increases in local flooding in some catchments or regions, and thus these regional variations in projections of changes in heavy precipitation should play an important role in changing flood occurrence at the regional scale.

Climate-driven changes in flood frequency are complex, depending on the generating mechanism, e.g., flood magnitudes are expected to be on the rise where floods are the result of increasing heavy rainfall and flood magnitudes decrease where floods are generated by decreasing spring snowmelt. In some areas, where snowmelt is the principal flood-generating mechanism, the time of the greatest flood risk has shifted from spring to winter. Climate-related changes in future flood frequency are complex, depending on the flood-generating mechanism. Flood magnitudes typically increase with warming if high flows result from heavy rainfall and decrease where they are generated by spring snowmelt. In some places, rapid snowmelt from rain-on-snow events or warm periods in the middle of winter causes a potential flood threat in a warmer world. There has been an increase in westerly weather patterns in Europe in winter; these are very rainy low-pressure systems that often trigger floods (Kron 2012). Similarly, so-called Vb depressions (crossing the northern Mediterranean, where they become saturated with moisture, turning northward east of the Alps and yielding large amounts of rain in Central and Eastern Europe) have been on the rise. Niedźwiedź and Łupikasza (2016) spotted changes in flood-prone circulation patterns of importance for the studied region.

Land-use and land-cover changes as well as engineering developments such as dikes and reservoirs that regulate flow process also affect flood hazard.

River regulations reduced the water storage capacity in wetlands and flood plains. River valleys are now narrowed, often—straightened and shortened, in comparison to the natural situation. River channels are prepared to a faster conveyance of water, between levees, but the system fails acutely if levees break.

Characteristics of terrestrial systems play an essential part in driving flood hazard. Land-use changes (driven by socio-economic factors, e.g., urbanization and deforestation) cause land-cover changes, conditioning transformation of precipitation into river runoff. If catchments change, river discharge and stage (amplitude, frequency statistics, or seasonality) corresponding to the same precipitation input also change. River basins are subject to changes in water storage capacity (e.g., related to the area of floodplains and wetlands), linked with infiltration capacity and runoff coefficient; and changes in portion of impervious area. The net results are a faster and higher peak of the hydrograph (system response to intensive precipitation) and accelerated erosion and transport of water, sediments and other materials in the river. In mountainous areas, development extends to hillslopes which are subject to landslides and debris flows. The problem is increasing with residential area development on hillslopes and deforestation, or road construction.

Over the past decades, a number of developments have clearly caused changes in flood hazard, i.e., in occurrence frequency and in the magnitude of high flows. These changes can be largely attributed to human activities.

3.2 Changes in Exposure

Our natural environment has been changing over the past centuries. Large forested areas have been converted to pastures and farmland and subsequently to built-up areas (with valuable houses, industrial plants, roads, airports).

Economic losses from floods have increased considerably (Handmer et al. 2012), as a result of the very rapid increase in exposure in floodplains fuelled by rapid economic growth, but with large spatial and inter-annual variability (Kundzewicz et al. 2014). Typically, the number of inhabitants in areas that are not flood-safe (human encroaching into harm's way, i.e. floodplains) and ubiquitous economic growth and resultant increase of wealth, hence of damage potential, have been the most essential changes, causing increased flood-related damage in most areas. Problems may grow as people become wealthier and more exposed. Technology in turn also helps populate more "difficult" areas. Many wrong locational decisions have been taken, which cause the flood loss potential to increase.

Global flood disaster losses reported over the last few decades reflect mainly monetized direct damage to assets. Whereas early warning systems can successfully reduce mortality risk through evacuation of the population, crops and infrastructures remain in place, and hence the significant increase in infrastructures has led to a drastic increase in economic risk (UNISDR 2011). Studies that project future flood losses and casualties indicate that, if no adaptation is undertaken, future anthropogenic climate change is likely to lead to increasing flood losses, alongside the increase in exposure linked to ongoing economic development (Handmer et al. 2012).

Today, crops often grow in monocultures, facilitating the use of heavy equipment, which causes soil compaction, reducing soil infiltration capacity, and optimal tillage schemes, which may increase the runoff rate and the flow velocity, and therefore soil erosion (Kron 2012). Within settlements, the anthropogenic increase of ground sealing and of area of impermeable (or less permeable) surfaces, such as house roofs, roads and pedestrian pavements, parking lots, etc. cause increase in runoff and typically increase peak flows of rivers.

The flood damage potential is increasing because of urban concentration world-wide, encroaching of settlements into flood prone areas, and over reliance to the safety provided by flood control works. Anthropogenic pressure causes the tendency to use additional land, that is also the floodplains that attract development due to their flatness, soil fertility and proximity of water. The increase in flood losses is a direct function of the number of people who—by choice or of necessity—settle in flood-prone areas, and a function of the increasing values they possess and the greater susceptibility of these values to water.

Floodplains are—if one disregards flood hazard—well suited for development. They are well suited to leisure and sports facilities, but often development results in large indoor sports centres, amusement areas, restaurants, or shopping facilities, to name a few, that have high flood loss potential (Kron 2012). Municipalities have to allocate land to housing, commerce and industry. If a community is faced with a choice between flood control (i.e. preserving water storage volumes and control of loss potential) and jobs and development, the latter will typically be priority.

Nowadays, properties are more valuable. People have more possessions that are more valuable and more prone to loss or damage, e.g. central heating boilers, freezers, high-tech wash machines, located in the basement, whose value increases. Many homeowners have converted their basements from storage rooms for massive (and inexpensive) items into guest rooms, party cellars, children's playrooms, and home offices, equipped with computers. The basements of commercial buildings often house the central control systems of lifts and air-conditioning facilities, and sometimes even computer centres (Kron 2012). Underground car parkings are more commonplace. Cars are normally easy to move, but they still represent a high loss potential in the event of a flash flood with virtually no warning. People are becoming trapped in cars while driving or in underground garages. If electricity is out, drivers or passengers cannot manually open the door or the window, so that there is a risk of drowning.

3.3 Changes in Vulnerability

Based on data covering the period 1970–2010, flood mortality has decreased globally, despite an increase in exposure. Trends in vulnerability of what is exposed are variable by location, and the global picture is strongly influenced by the vertical development of urbanization in China, where vulnerability to floods has decreased over the past decade. If China is removed from this analysis, the mortality trend remains slightly on the rise (UNISDR 2011).

In the past, people's belongings were generally not susceptible to damage by water. Once the flood had receded, the products could be dried and remained functionable. Now, electrical and electronic machines, appliances, and other devices are highly vulnerable to humidity and to pollution by flood waters—they may suffer total damage when they come in contact with water (Kron 2012).

The change from stove heating to central heating is possibly the key difference. Central heating boilers are installed in the basement, from which they cannot be removed, but so, too, are heating oil tanks, which can cause severe pollution if flooded.

4 Managing and Reducing Flood Risk

The notion of flood risk plays a central role in the European Commission's Directive 2007/60/EC (so called Floods Directive, CEC 2007) on the assessment and management of flood risks. The aim of the Directive is to reduce and manage the risks that floods pose to human life and health, the environment, cultural heritage, economic activity and infrastructure.

To reduce flood risk, one can control the risk-driving factors: hazard, exposure, and vulnerability. The strategy for flood preparedness can be based on the following attitude: protect as far as technically possible and affordable or accommodate, i.e. prepare to "living with floods". If a necessary level of protection cannot be provided and accommodation is not possible, a retreat (relocation) could be a solution.

Increase of flood risk may have been an unwelcome side effect of human activities. In contrast, flood risk reduction is a set of intentional activities. Typically, the dominating risk reduction strategies are: structural flood defences (dams and flood control reservoirs, dikes, bypass channels). Other strategies embrace watershed management, therein enhancing natural storage, zoning, insurance, contingency plans, and building codes. Yet, there is no measure that guarantees a complete protection and absolute safety—a residual risk always remains.

Structural defences are designed based on statistical considerations, so as to withstand river discharge up to a certain magnitude (e.g., a 100-year flood, with a return period of 100 years and probability of exceedance in any given year being 1 %) and may fail if the flood is more extreme (or if the duration of high water is prolonged). Typically, dikes offer adequate protection against small and medium-sized floods, i.e., the number of damaging floods in this range is decreasing. However, even a perfectly maintained dike designed to withstand, for example, a 100-year flood does not, by definition, guarantee absolute protection. Yet, no matter how high a design flood is, there is always a possibility of occurrence of a greater flood. Should one design dikes to withstand a 100-year flood or perhaps a much more robust dike, withstanding a 500-year flood? The latter solution should give a better protection being far more costly. Yet, it may still turn out to be insufficient if a 2000-year flood arrives. Then the dike can be overtopped and/or breached. In this case, losses in a levee-protected landscape are likely higher than they would have been in a natural state (levee-free case). This is so, because existence of a defence is taken as a safety guarantee by the riparian population, so that the considerable wealth is accumulated in apparently protected, but in fact endangered, areas, and damage potential grows. It is simply not possible to build flood protection that caters for extremely rare events, except at the local scale and for particularly high-value areas, such as large cities. Hence, metropolitan and rural areas should be treated differently. Very high safety standards are valid in the low lying country—The Netherlands.

The impacts of changes in flood characteristics are highly dependent on how climate changes in the future, but low confidence generally exists in projections. Planners have to take the higher level of a future design flood into account in their design calculations. The process of adaptation has already begun: several European countries have prescribed the incorporation of a “climate-change” safety factor, adding to the nominal design discharge for new flood-control systems (Madsen et al. 2014).

5 Concluding Remarks

Current studies indicate that increasing exposure of population and assets, and not anthropogenic climate change is responsible for the past increase in flood losses (Kundzewicz et al. 2014). There are indications that exposed population and assets have increased more rapidly than overall population or economic growth.

Assessing the causes of changes in flood hazard is complex as these may be related to both climatic and non-climatic factors, which can be difficult to distinguish in the instrumental record of observations.

Rivers that threaten towns and settlements are not in their natural state. Hence it is difficult to separate changes in the flow data related to climate from ongoing alterations in land use change (including urbanization) and altered river regulation.

There is a potential for floods becoming much higher than ever observed, e.g. if record-high precipitation occurs over a large city (or upstream of a large city) with high (and dynamically growing) damage potential, the consequences could be utterly destructive.

Acknowledgments The authors gratefully acknowledge the financial support of the FLORIST project (Flood risk on the northern foothills of the Tatra Mountains), based on a grant from the Swiss Government through the Swiss Contribution to the enlarged European Union (PSPB No. 153/2010). Useful comments made on the draft of this paper by Professor Bartlomiej Wyzga are also gratefully acknowledged.

References

CEC (Commission of European Communities) (2007) Directive 2007/60/EC of the European Parliament and of the Council of 23 October 2007 on the assessment and management of flood risks. Official J Eur Union L 288:27–34

Handmer J, Honda Y, Kundzewicz ZW, Arnell N, Benito G, Hatfield J, Mohamed IF, Peduzzi P, Wu S, Sherstyukov B, Takahashi K, Yan Z (2012) Changes in impacts of climate extremes: human systems and ecosystems. In: Field CB, Barros V, Stocker TF, Qin D, Dokken DJ, Ebi KL, Mastrandrea MD, Mach KJ, Plattner G-K, Allen SK, Tignor M, Midgley PM (eds) Managing the risks of extreme events and disasters to advance climate change adaptation.

A special report of working groups I and II of the Intergovernmental Panel on Climate Change (IPCC). Cambridge University Press, Cambridge, pp 231–290
IPCC (2014) Annex II: glossary. In: Mach KJ, Planton S, von Stechow C (eds) Climate change 2014: synthesis report. Contribution of working groups I, II and III to the fifth assessment report of the Intergovernmental Panel on Climate Change. Core Writing Team, Pachauri RK, Meyer LA (eds). IPCC, Geneva, Switzerland, pp 117–130
IWR (Institute for Water Resources, US Army Corps of Engineers) (2011) Flood risk management approaches as being practiced in Japan, Netherlands, United Kingdom, and United States. IWR Report No. 2011-R-08, 124 pp
Kharin V, Zwiers FW, Zhang X, Hegerl GC (2007) Changes in temperature and precipitation extremes in the IPCC ensemble of global coupled model simulations. J Clim 20(8):1419–1444
Kron W (2012) Changing flood risk—a reinsurer's viewpoint. Chapter 26 In: Kundzewicz ZW (ed) Changes in flood risk in Europe. IAHS Press, Wallingford
Kundzewicz ZW (ed) (2012) Changes in flood risk in Europe. IAHS Press, Wallingford
Kundzewicz ZW, Schellnhuber H-J (2004) Floods in the IPCC TAR perspective. Nat Hazards 31:111–128
Kundzewicz ZW et al (2012) Flood risk in Europe—setting the stage. In: Kundzewicz ZW (ed) Changes in flood risk in Europe. IAHS Press, Wallingford, pp 11–26
Kundzewicz ZW, Kanae S, Seneviratne SI et al (2014) Flood risk and climate change: global and regional perspectives. Hydrol Sci J 59(1):1–28
Madsen H et al (2014) Review of trend analysis and climate change projections of extreme precipitation and floods in Europe. J Hydrol 519:3634–3650
Niedźwiedź T, Łupikasza E (2016) Change in circulation patterns. In: Kundzewicz ZW, Stoffel M, Niedźwiedź T, Wyżga B (eds) Flood risk in the Upper Vistula Basin. Springer, Cham (this volume)
Peduzzi P, Deichmann U, Maskrey AS et al (2009) Global disaster risk: patterns, trends and drivers, ISDR (2009) global assessment report on disaster risk reduction. United Nations, Geneva, Switzerland, pp 17–57 (chapter 2)
Seneviratne SI, Nicholls N, Easterling D, Goodess CM, Kanae S, Kossin J, Luo Y, Marengo J, McInnes K, Rahimi M, Reichstein M, Sorteberg A, Vera C, Zhang X (2012) Changes in climate extremes and their impacts on the natural physical environment. In: Field CB, Barros V, Stocker TF, Qin D, Dokken DJ, Ebi KL, Mastrandrea MD, Mach KJ, Plattner G-K, Allen SK, Tignor M, Midgley PM (eds) Managing the risks of extreme events and disasters to advance climate change adaptation. A special report of working groups I and II of the Intergovernmental Panel on Climate Change (IPCC). Cambridge University Press, Cambridge, pp 109–230
UNISDR (2011) Revealing risk, redefining development, global assessment report on disaster risk reduction. United Nations International Strategy for Disaster Reduction, Geneva, Switzerland, p 178

Part II
Factors Influencing Flood Risk (80)

Flood Generation Mechanisms and Changes in Principal Drivers

Bartłomiej Wyżga, Zbigniew W. Kundzewicz, Virginia Ruiz-Villanueva and Joanna Zawiejska

Abstract Mechanisms generating floods are reviewed and next discussed with regard to the Upper Vistula Basin. Here, floods typically result from (i) moderate-intensity rain that lasts a few days over a large area and drives large-scale flooding, or (ii) high-intensity, short-lasting convective rain causing local flash floods. Outside the mountain part of the basin, especially in the San River catchment, floods are also caused by intensive snowmelt. Interpretation of climate track in flood generation is presented, based on the analysis of observation records from the last six decades and projections for the future. Catchment and river changes affecting the conditions of flood generation are next considered for the last 130 years. They comprise changes regulating flood runoff (catchment reforestation and dam reservoirs construction), changes reducing floodwater storage and accelerating flood

The original version of this chapter was revised: The text in pages 58 and 59 has been rephrased and properly cited. The erratum to this chapter is available at 10.1007/978-3-319-41923-7_21

B. Wyżga (✉)
Institute of Nature Conservation, Polish Academy of Sciences, al. Mickiewicza 33, 31-120 Kraków, Poland
e-mail: wyzga@iop.krakow.pl

B. Wyżga
Faculty of Earth Sciences, University of Silesia, ul. Będzińska 60, 41-200 Sosnowiec, Poland

Z.W. Kundzewicz
Institute for Agricultural and Forest Environment, Polish Academy of Sciences, ul. Bukowska 19, 60-809 Poznań, Poland
e-mail: kundzewicz@yahoo.com

V. Ruiz-Villanueva
Dendrolab.ch, Institute of Geological Sciences, University of Bern, Baltzerstrasse 1+3, 3012 Bern, Switzerland
e-mail: virginia.ruiz@dendrolab.ch

J. Zawiejska
Institute of Geography, Pedagogical University of Cracow, ul. Podchorążych 2, 30-084 Kraków, Poland
e-mail: zawiejska.joanna@gmail.com

Z.W. Kundzewicz et al. (eds.), *Flood Risk in the Upper Vistula Basin*, GeoPlanet: Earth and Planetary Sciences, DOI 10.1007/978-3-319-41923-7_4

runoff (channel regulation, flood embankments, river incision, and permanent impoundment of the Upper Vistula for navigation purposes), as well as the expansion of riparian forests increasing large wood recruitment to channels during floods.

Keywords Flood hazard · Climate change · Terrestrial drivers · Flood runoff · Upper Vistula Basin

1 Introduction

Changes in flood hazard may be driven by climate changes as well as changes in catchment and river characteristics determining the terrestrial conditions of water runoff (Kundzewicz and Schellnhuber 2004). Potential impacts of climate changes depend largely on flood generation mechanisms, and these mechanisms are subject to review in this chapter, in general terms as well as with respect to flood generation in the Upper Vistula Basin. Interpretation of climate track in flood generation is outlined, based on the analysis of observation records and, albeit highly uncertain, projections for the future. Climate track is typically restricted to the last six decades, for which homogeneous observation records are available. By contrast, changes in terrestrial drivers in the region can be studied over the period exceeding 100 years. We review these drivers, describing catchment and river changes that tended to either regulate or accelerate flood runoff, and changes responsible for the increased flood hazard resulting from large wood delivery to river channels.

2 Climate Track in Flood Generation

There is no doubt that climate is a very important factor driving fluvial flood hazard (cf. Kundzewicz and Stoffel 2016), affected by various characteristics of the climatic/atmospheric systems. Important is the water holding capacity (and the water content) of the atmosphere. Various precipitation characteristics, such as precipitation intensity, duration, total amount, timing, or phase (whether liquid or solid) are virtually essential in shaping flood hazard. Spatial distribution as well as temporal distribution (e.g. measured by the antecedent precipitation index, API) are important (Kundzewicz et al. 2014). Moreover, large-scale circulation patterns as well as temperature patterns (responsible for cold phenomena, such as soil freezing, snow cover, snow and ice melt, and ice jam formation) are of relevance.

The climate change is real and ubiquitous. Significant increases in air temperature have been observed at a range of scales, from local to regional, continental, and global (IPCC 2013). However, changes have not been limited to temperature, but also embraced other variables, of relevance to flood hazard. The effects of climate change on water resources, which vary regionally, largely follow changes in the prime driver—precipitation (additionally, changes in temperature in snow-impacted basins) (Döll et al. 2015).

While observed temperature increases are quite regular, changes in precipitation are not so. Nevertheless, a general large-scale statement can be made that climate variability increases spatially and temporally under climate change, so that wet regions become even wetter and flood hazard is increasing in many locations (Döll et al. 2015). Seneviratne et al. (2012) detected the climate track in some variables affecting floods, including mean precipitation, heavy precipitation, and snow pack, though a direct statistical link between climate change and trends in the magnitude and/or frequency of floods has not been established yet. In climates where seasonal snow storage and melting play a significant role in annual runoff, the hydrological regime is affected by changes in temperature. Changes in the timing (earlier occurrence) of spring peak flows in snowmelt- and glacier-fed rivers have been observed. However, not all such areas are experiencing changes in the magnitude of peak flow (Seneviratne et al. 2012). Intense precipitation has been on the rise and this rhymes with the theoretical Clausius-Clapeyron law, indicating more room for water vapor in the warmer atmosphere (Kundzewicz and Schellnhuber 2004).

Climate-driven changes in flood frequency are complex, depending on whether floods are the result of increasing heavy rainfall (then flood magnitudes are expected to be on the rise) or floods are generated by decreasing spring snowmelt (then flood magnitudes decrease) (cf. Kundzewicz et al. 2014). In some areas, where snowmelt is the principal flood-generating mechanism, the time of the greatest flood risk has shifted from spring to winter (Kundzewicz et al. 2010). A precise understanding of the hydroclimatic characteristics of floods is complicated by a lack of observational data at the spatial and temporal resolution adequate for hydroclimatic research (Kundzewicz et al. 2016). Long time series of good-quality data are not available in many areas. But, even if the data are perfect, it is worthwhile to re-state a tautology: extreme events are rare (Kundzewicz and Schellnhuber 2004). They do not happen frequently, so even where a relatively long time series of instrumental records exists, one still deals with only a small sample of truly extreme and destructive floods (cf. Kundzewicz and Robson 2004).

Climate projections using multi-model ensembles show ubiquitous warming and increases in globally averaged mean water vapor and precipitation over the 21st century (IPCC 2013). Yet, precipitation scenarios show strong regional differences. This is particularly true for precipitation extremes, of relevance for shaping flood hazard. Seneviratne et al. (2012) assessed that it is *likely* that the frequency of heavy precipitation or the proportion of total rainfall from intense events will increase in the 21st century over many areas of the globe. This would translate into an increase of the magnitude and frequency of floods in some regions, but the projections are largely uncertain (Arnell and Gosling 2014; Kundzewicz et al. 2016).

Climate-related changes in future flood frequency are complex, depending on the flood-generating mechanism. Flood magnitudes typically increase with warming if high flows result from heavy rainfall and decrease where they are generated by spring snowmelt. In some places, rapid snowmelt from rain-on-snow events or warm periods in the middle of winter cause a potential flood threat in a warmer world (Kundzewicz et al. 2010).

Clear temperature increases have been detected in the Upper Vistula Basin (cf. Łupikasza et al. 2016). Precipitation trends are more complex and many trends are not statistically significant, as the natural variability is strong. Niedźwiedź and Łupikasza (2016) spotted changes in flood-prone circulation patterns of importance for the studied region. However, the analysis of data from 14 water-gauge stations within the upper Dunajec Basin indicated that it is difficult to find a statistically significant trend in the records of magnitude of annual maximum floods in the region (Ruiz-Villanueva et al. 2016).

Piniewski et al. (2016) examined climatic and hydrological projections for the Upper Vistula Basin for two future time horizons 2021–2050 (near future) and 2071–2100 (far future). They found model agreement about ubiquitous warming on both seasonal and annual scales, and most models (eight out of nine models for the near future and all nine models for the far future) agreed about an increase in projected mean annual precipitation and total runoff. Projected changes in high-streamflow indicator based on the 90th monthly flow percentile (*Q*90) were found to increase in both future horizons, by a few per cent.

Romanowicz et al. (2016) examined projections of future flood hazard changes in two catchments and found disagreements between particular models and between the two locations. They explain these disagreements by climatic variability and the uncertainty of the results.

3 Flood-Generating Mechanisms—An Overview

Floods, understood as destructive abundance of water, are generated by the interaction of various processes. As Nied et al. (2013) indicated, physical controlling factors include "hydrological pre-conditions (e.g. soil saturation, snow cover), meteorological conditions (e.g. amount, intensity, and spatial and temporal distribution of precipitation), runoff generation processes as well as river routing (e.g. superposition of flood waves in the main river and its tributaries)". The combination of these factors and their spatial distribution at a regional scale may be important, as flooding can affect many sites simultaneously, whereas other sites remain unaffected (Merz and Blöschl 2008; Nied et al. 2013). Across Europe, floods have very different characteristics, depending on geography, climate/weather characteristics, and the human occupancy (FloodSite 2007):

- large, slow rising rivers with floods generated by frontal rainfall and/or snowmelt (e.g. the Rhine, the Thames, the Po, and the Loire);
- summer thunderstorm-type events resulting in flash floods with very short warning lead time, often characteristic of the Mediterranean region (e.g. several flash floods in southern France in the 2010s) and/or mountainous regions (e.g. Ore Mountains in August 2002);

- floods caused by heavy precipitation events in urban areas, falling on inadequate sewer systems which are overloaded and may fail owing to inadequate maintenance or insufficient levels of investment (e.g. heavy rainfalls in July 2002 in Germany);
- ice-dammed rivers, generating floods through the back-up of flows spreading onto wide floodplains (e.g. the Vistula River upstreams of Włoclawek dam in January 1982);
- dam failure, representing a very rare type of events, but posing a potential threat (e.g. toxic spill from a mine into the River Tisza in 2000);
- coastal floods, where embankments may be breached or storm surges overwhelm coastal defences (e.g. the North Sea floods in 1953 and 1962).

The two last mechanisms of flood generation are outside the scope of this book.

Nied et al. (2014) summarized three main approaches to describe flood events in terms of their spatio-temporal physical causes as: (i) flood event description, (ii) classification into flood types, and (iii) linkage of flood occurrence to atmospheric circulation patterns. There are many examples following the first approach where detailed descriptions of flood conditions are provided (Kundzewicz et al. 1999; Ulbrich et al. 2003a, b; Ruiz-Villanueva et al. 2012; Blöschl et al. 2013). Nied et al. (2014) described several flood types to assess the second approach. According to their generating process, Merz and Blöschl (2003) distinguished long-rain floods, short-rain floods, flash floods, rain-on-snow floods, and snowmelt floods. Floods can be classified based on storm type, El Niño-Southern Oscillation conditions and decadal-scale climatic variability (Alila and Mtiraoui 2002), or based on precipitation, synoptic weather patterns and snowmelt (Hirschboeck 1987). Finally, the third approach uses statistical or probabilistic links between flood occurrence and atmospheric circulation patterns (e.g. Bárdossy and Filiz 2005; Petrow et al. 2009; Prudhomme and Genevier 2011). Since mechanism generating floods usually depends on the season, the seasonality approach opens the way to studying mixed flood frequency distributions in flood frequency analysis using the information on the seasonality (Sivapalan et al. 2005; Ouarda et al. 2006). In addition to time of the year, the weather circulation patterns associated with floods were used to identify the processes during floods (e.g. Bárdossy and Filiz 2005; Jacobeit et al. 2006; Zehe et al. 2006; Petrow et al. 2009). Most of these studies focused on recognition of flood-triggering circulation patterns to assist in regional flood analyses (Parajka et al. 2010).

Although many studies emphasize the importance of different flood-controlling processes (e.g., Merz and Blöschl 2003; Sivapalan et al. 2005; Bradshaw et al. 2007; McCabe et al. 2007; Parajka et al. 2010; Freudiger et al. 2014; Slater et al. 2015), understanding of regional differences in the process controls of flooding responses is rather limited (Berghuijs et al. 2016). Moreover, the analysis of flood generating processes is important because flood events may reveal aspects of hydrological behaviour that either were unexpected on the basis of weaker responses or highlight anticipated but previously unobserved behaviour (Borga et al. 2010, and references therein). Hence, improved process understanding is a key element for improving the prediction and interpretation of flood trends, especially under environmental change (Kundzewicz et al. 2014; Hall et al. 2014).

4 Flood Generating Mechanisms in the Upper Vistula Basin

Floods that encompass the largest parts or even the entire Upper Vistula Basin and cause most damage are generated by a few days-long rainfall with average intensity of 8–10 mm h^{-1} and the total sum of precipitation of a few hundred millimetres. Such precipitation has always been linked with cyclones moving along the Vb track (van Bebber 1891) from the Mediterranean region to Central-Eastern Europe and inducing southward movement of air masses on their western side. Depending on the location of the cyclone centre, high precipitation totals may occur (i) on both sides of the Upper Vistula (as in 2001; Sasim and Mierkiewicz 2002), (ii) on the northern slopes of the Western Carpathians and the Sudetes where precipitation is additionally enhanced by orographic effect (as in 1997; Niedźwiedź 1999), or (iii) only in the southernmost parts of the Polish Carpathians (in 2014). Local, heavy downpours, sometimes with intensities typical of torrential rains, with duration of 0.5–2 h and sums of precipitation up to 150 mm, occur in the upland (Cebulak and Niedźwiedź 1998) and the Carpathian parts of the Upper Vistula Basin (Niedźwiedź 1999), and in the lowland Sandomierz Basin (Starkel 2011). They result in local flash floods, especially if linked with specific physiographic catchment features, such as a high proportion of arable lands in the total catchment area, low soil-infiltration rates, and high density of drainage network (Bryndal 2014). Finally, snowmelt floods occur in the catchments with relatively small variation in altitude in which snow pack thawing occurs at the same time catchment-wide. Such floods are thus typical of the left-side tributaries to the Upper Vistula as well as of foothill and foreland reaches of its Carpathian tributaries, although flood magnitudes are usually lower than those of the floods generated by prolonged summer rainfall. Different situation occurs in the catchments of the lower San and its tributaries where snow pack can retain considerable amounts of water as a result of more severe weather conditions during winters. Here, a sudden increase in temperature after long winter conditions with considerable amounts of snowfall may lead to a catastrophic flood as in March 1924 (Punzet 1984). Moreover, the high variability of peak stages of spring floods recorded along the lower San indicates frequent occurrence of ice jams that may locally increase water levels (Punzet 1984).

5 Changes in Terrestrial Drivers of Floods

Over the 20th century catchments and channels of the rivers in the Upper Vistula Basin experienced considerable changes resulting in altered conditions for runoff formation and the passage of flood waves. Below we discuss three groups of such changes with different impacts on flood conditions and the related flood hazard and risk.

5.1 Factors Regulating Flood Runoff

5.1.1 Reforestation of Catchments

Deforestation of catchments in the Polish Carpathians progressed until the early 19th century, and subsequently the forest cover in the region remained at the low level until the 1930s (Kozak 2009). The scarce forest cover facilitated rapid flood runoff and the formation of high, flashy flood waves. The widespread occurrence of braided rivers (Wyżga et al. 2015) and the formation of massive, loosely packed gravels in their channels (Wyżga 1993a; Wyżga et al. 2012) represent morphological and sedimentary records of the rapid flood runoff from the Polish Carpathian catchments during the 19th and an early part of the 20th century. Reconstructions of paleodischarges in the Tatra Mountains streams by means of dendrochronological dating of tree scars and hydraulic modelling evidence higher flood magnitudes in the first half of the 20th century than in its second half (Ballesteros-Cánovas et al. 2016).

In the 20th century and especially in its second half, a considerable increase in forest cover occurred in Polish Carpathian catchments (Kozak 2009). The largest increase typified the eastern part of the Polish Carpathians that was dramatically depopulated in the mid-1940s. For instance, in the upper Wisłoka catchment, the forest cover increased from 26 % in 1900 to 30 % in 1938 and to 67 % in 1995 (Lach and Wyżga 2002). Less intense land use changes occurred in the western part of the Polish Carpathians (Kozak 2009) that remained densely populated and where the changes were conditioned by socio-economic changes (reduced profits from agricultural production, the emergence of new sources of income for the previously purely agricultural community—Kukulak 1994). For instance, in the upper part of the Dunajec catchment, the forest cover increased from 27 % in 1901 to 42 % in 2000 (Wyżga et al. 2012). The increase in forest cover was associated with a reduction in the area of arable land (Munteanu et al. 2014), the scale of which might have been greater than that of the increase in forest cover. In the upper part of the Dunajec catchment, the area of arable land decreased from 42 % in 1901 to 17.5 % in 2000, partly in favour of meadows. The decline in agricultural practices was especially important in the eastern part of the mountains as it has led to the abandonment and overgrowth with bushes of cart tracks that ceased to function as pathways for rapid evacuation of water and sediment from the hillslopes (Lach and Wyżga 2002).

The land use changes led to decreasing flashiness of flood runoff that was clearly reflected in a shift from the formation of loosely packed gravels to normally and closely packed gravels and the onset of the formation of bed armour in channels (Wyżga 1993a; Wyżga et al. 2012), and in the appearance of a tendency of rivers to

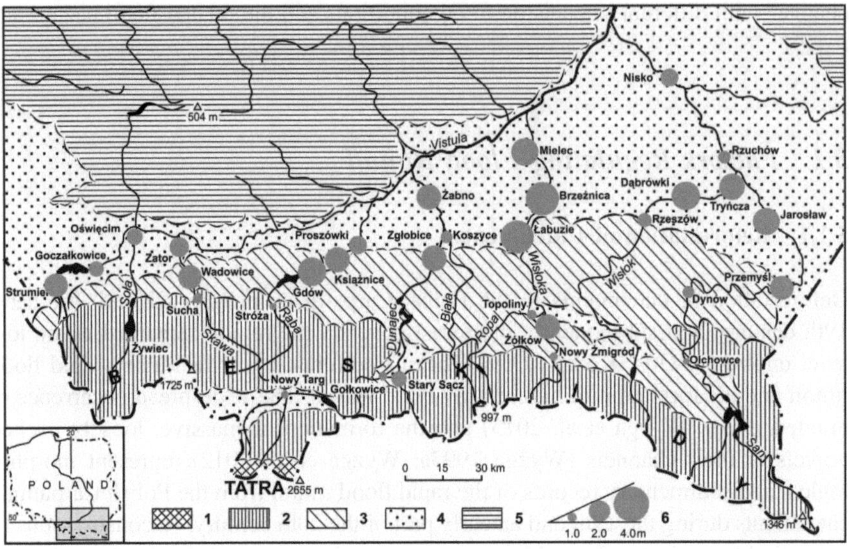

Fig. 1 Dimensions of channel incision of Carpathian tributaries to the Vistula during the 20th century inferred from the lowering of minimum annual water stage at gauging stations on the rivers. *1* high mountains; *2* mountains of intermediate and low height; *3* foothills; *4* intramontane and submontane depressions; *5* uplands; *6* lowering of minimum annual stage at water-gauge stations during the 20th century. After Wyżga (2008), modified, reproduced with permission of Elsevier

meander (Wyżga 2001a; Wyżga et al. 2015). They must have been also reflected in reduced peak flows of flood waves (Ballesteros-Cánovas et al. 2016). Analysis of changes in flood runoff reveals a linkage between the scale of land use change and a reduction in mean annual discharges (Wyżga 2008). At the Wadowice station on the Skawa River draining the western part of the Polish Carpathians (Fig. 1), mean annual flood ($Q_{2.33}$) in the years 1956–2000 was 8 % smaller than that recorded in the years 1921–1955, whereas at the Łabuzie station on the Wisłoka River draining the eastern part of the mountains (Fig. 1) the respective reduction in mean annual flood amounted to 31 %.

5.1.2 Construction of Dam Reservoirs

After the catastrophic flood of 1934, dam reservoirs started to be constructed on Polish Carpathian rivers, with the first two completed (Porąbka Dam on the Soła River—1936; Rożnów Dam on the Dunajec River—1941) within a few years after the works had been initiated. To date, 14 dam reservoirs (including the almost finished Świnna Poręba Dam on the Skawa River) have been constructed in the Upper Vistula Basin, with a total flood capacity of 413 million m^3 during summer (Walczykiewicz and Rataj 2011). If compared with the flow of the Vistula River at the Zawichost gauging station characterizing its whole upper basin, the capacity corresponds to 11 days-long runoff with mean annual discharge or to 14 h of runoff with the peak discharge of 1 % probability of exceedance. Such a capacity must be considered modest as the hydrographs of large floods at the station may last up to 10 days. However, the efficiency of particular dam reservoirs in the reduction of peak flows of flood waves is highly diverse, depending on the location of a given reservoir, its function(s), inflow to and the work of a reservoir during particular flood waves. For instance, during the flood of 2010, small reservoirs located on left-side tributaries to the Upper Vistula reduced the peak flow by 8–75 %. At the same time, large reservoirs located on the Carpathian tributaries to the Upper Vistula exhibited the reduction rate between 11 and 59 % (Walczykiewicz and Rataj 2011) and both these extreme values refer to the same river—the Dunajec; the Rożnów-Czchów reservoirs located in the middle river course could reduce the peak flood flow by only 11 %, despite 59 % reduction of the flood peak by the Czorsztyn-Sromowce reservoirs located in the upper river course.

A few factors limit the efficiency of the dam reservoirs in the Upper Vistula Basin in reducing flood peaks. First, apart from protection from flooding, the dams play also other roles, mainly hydropower production or water supply, which require the maintenance of relatively high water level in the reservoirs. Second, reliable forecasts of rainfall and thus also of water inflow to the reservoirs are available shortly before an event, which limits the time span required to considerably increase their flood capacities. Third, a full flood capacity of the reservoirs is typically kept during summer months, whereas in the remaining part of a year higher water level is maintained. Therefore, floods occurring outside the "usual period" (e.g. in mid-May 2010) are likely to coincide with reservoir capacities lower than the maximal ones. Fourth, dam reservoirs disrupt the continuity of bed material transport in the rivers, inducing or strengthening the tendency to channel incision in the downstream reaches. The resultant loss of floodwater storage in floodplain areas partly counteracts the effect of dam reservoirs on flood peaks. In several countries, river reaches downstream of dam reservoirs are fed with gravel to prevent sediment starvation of the rivers (Kondolf 1997), but so far this solution has not been implemented in Poland.

5.2 *Factors Accelerating Flood Runoff*

5.2.1 Construction of Flood Embankments and Channel Regulation

Channelization works in the Upper Vistula River began in the mid-19th-century and were continued to the second half of the 20th century (Łajczak 1995). In the late 19th century channelization works were also initiated in the lower courses of Carpathian tributaries to the Vistula (Szumański 1986; Zawiejska and Wyżga 2010). In both cases the works consisted in channel narrowing and straightening through meander cutoffs. In the first half of the 20th century channelization works encompassed also middle courses of Carpathian tributaries to the Vistula, whereas in the second half of the century they were conducted in the middle and upper courses of the rivers (Wyżga 2001a; Zawiejska and Wyżga 2010). In these river reaches, the works resulted in considerable channel narrowing and replacement of the multi-thread channel by a single, regulated channel. As a result of the works, mean velocity at given discharges increased (Wyżga 1993b) and bed degradation was initiated leading to deep channel incision over several decades (Łajczak 1995; Wyżga 2001a, 2008).

In the 1890s–1900s flood embankments were constructed along the Upper Vistula and lower courses of its main tributaries (Łajczak 1995; Zawiejska and Wyżga 2010). Over the 20th century, flood embankments were also constructed along short reaches of the middle and upper courses of Carpathian tributaries to the Vistula. Since their construction, many reaches of the embankments have been modified by increasing the height or straightening some sections, especially along the Upper Vistula where several river meanders were cut off during the past century. A reduction in floodplain width caused by the construction of embankments increased water stages attained at particular flood discharges and this effect must have been particularly important on the Upper Vistula where the ratio of channel width to inter-embankment zone width was especially high. A combination of channelization works and flood embankment construction increased the frequency of high flood stages and discharges resulting in non-stationarity of flow record—according to Punzet (1972), flood flows from the periods before and after 1920 represent different populations, although at particular water-gauge stations the change might have occurred at somewhat different time. The non-stationarity of the record of maximum annual water stages can be shown on the example of the Żabno station located in the lowest course of the Dunajec River. Since around 1930 peak stages of large floods have started to reach about 1 m higher than previously, and this effect has been persistent despite the operation of the large Rożnów reservoir (located upstream) since the early 1940s and continuing channel incision at the gauge cross-section (apparent in the progressive lowering of minimum annual stages) (Fig. 2). The timing of the change at the station allows to attribute it to the channel narrowing and straightening in the course of channelization works conducted in the lower and middle course of the Dunajec in the first three decades of the 20th century rather than to the construction of embankments along the lower river course in the 1890s–1900s (Zawiejska and Wyżga 2010).

Still another effect of the channelization works, especially channel straightening through meander cutoffs, was a shortening of the travel time of flood waves. For instance, the travel time of flood waves on the Vistula River between its tributaries the Skawa and the Raba was shortened by half from 44 h at the beginning of the 20th century to 22 h in the 1980s (Punzet 1985). This has increased synchronicity of the flood waves on the Vistula and on its mountain tributaries, especially the Dunajec, with the resultant increased peak discharges of the flood waves in the downstream course of the Vistula.

The increase in water stages of embanked rivers, resulting from floodplain constriction by embankments and the increase in flood discharges induced by

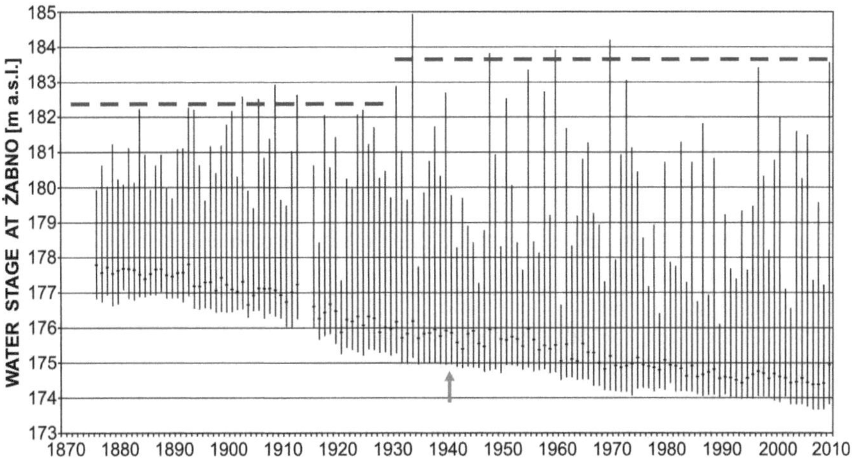

Fig. 2 Amplitude of water stages (*vertical lines*) and mean annual stages (*points*) at the Żabno gauging station in the lowest course of the Dunajec River since 1876. *Dashed lines in the left and right parts* of the diagram indicate approximate maximum stage of the large floods occurring before and after channelization of the river in its lower and middle courses. The *arrow* indicates the onset of the operation of the Rożnów dam reservoir in the middle river course

channel change in the upstream reaches, facilitated failure or overtopping of the embankments during large floods. The embankments along the Upper Vistula and the lower courses of its Carpathian tributaries within the Sandomierz Basin were breached/overtopped during floods in 1903, 1925, 1934, 1960, 1970, 1997, and 2010. The failure of the embankments resulted in storage of up to 400 million m^3 of water (in 1934) outside the embankments, and this prevented or slowed down the increase in peak discharges of the flood waves in the downstream reaches of the Vistula. However, it was also the reason for enormous flood damage as the inundated area outside the embankments reached up to 1000 km^2 (in 1934).

5.2.2 Channel Incision

Rivers draining the Polish Carpathians incised by 0.5–3.8 m over the 20th century (Fig. 1). In many river reaches, incision resulted in the dissection of the whole thickness of alluvium on the valley floor and transformation of the alluvial channels into bedrock ones (Hajdukiewicz and Wyżga 2013). Incision has been caused by the channelization-induced increase in transport capacity of the rivers caused by their channelization, the concomitant decrease in sediment supply to the channels and in-channel gravel mining (Wyżga 2008). The term *channel incision* describes bed degradation or channel deepening that leads to the increase in channel capacity (Wyżga et al. 2016). Absolute (expressed in metres) amounts of channel incision in Polish Carpathian rivers are typically greater in their lower and middle courses than in the upper course (Fig. 1). However, Wyżga et al. (2016) examined incision-caused changes in channel conveyance and the frequency of valley-floor inundation at water-gauge cross-sections along the Dunajec River and found the increasing impact of incision with decreasing river size. They concluded that as channel dimensions decrease and channel slope increases in the upstream direction, the influence of a given absolute amount of river incision on channel conveyance and on the resultant loss of floodwater storage on floodplains will also increase toward the upper reaches of a river network.

Rivers draining the eastern part of the Polish Carpathians exhibit somewhat greater absolute amounts of incision than rivers draining the western part of the mountains (Fig. 1). Also in this case a different impact of incision on the loss of floodwater storage in floodplain areas was demonstrated by Wyżga (2001b) who compared changes in the percentage of total flood flows conveyed on the valley floor between the Łabuzie gauge cross-section on the Wisłoka River and the Wadowice gauge cross-section on the Skawa River. Despite similar absolute amounts of incision recorded at both gauging stations during analysed periods, in the Skawa a reduction in the amount of floodwater conveyed on the valley floor was considerably greater, especially at low-frequency, high-magnitude flood flows.

The impact of channel incision on the flood hazard to downstream river reaches was analysed by comparing inflow and outflow peak discharges of flood waves passing the modified river reaches (Wyżga 1997). The analysis focused on relatively long river reaches, along which the catchment area increases relatively little; such reaches thus provide good conditions for flood-wave transformation, but the recorded changes in flood flows cannot be attributed to changed inflow from tributaries. Analysis of annual maximum discharges, peak discharges of ten largest flood waves and of all flood waves from particular decades consistently indicated a shift toward higher values of flood discharges recorded at the downstream end of such reaches with the progress in channel incision (Wyżga 1997). This can be illustrated by a comparison of the magnitude of floods of given recurrence intervals at the Łabuzie and Brzeźnica stations on the Wisłoka (Fig. 1), estimated for the record periods 1921–1955 and 1956–2000 typified, respectively, by small and high degree of channel incision (Table 1). As a result of reforestation of the montane part of the Wisłoka catchment, flood magnitudes recorded at both stations were lower in the second period than in the first one. However, the stations differed in the scale of the reduction, which amounted to 31 % at the upstream located Łabuzie station and to 19 % at the downstream located Brzeźnica station, and the difference can be ascribed to changed conditions of flood-wave transformation between the stations. In the years 1921–1955 all considered index floods ($Q_{1.5}$–Q_{20}) slightly decreased in magnitude in the Łabuzie-Brzeźnica reach, with mean annual flood ($Q_{2.33}$) decreasing by 4 %. After 1955, peak discharges of flood waves increased between the stations, and the increase varied between 5 % for a 20-year flood and 14 % for mean annual flood (Table 1).

The changes in flood flows recorded downstream of the river reaches modified by channel incision can be explained by increased concentration of flood flows in

Table 1 Flood flows (in $m^3\ s^{-1}$) of given recurrence intervals at the Łabuzie and Brzeźnica stations on the Wisłoka River estimated for the record periods 1921–1955 and 1956–2000

	Łabuzie		Brzeźnica	
	1921–1955	1956–2000	1921–1955	1956–2000
$Q_{1.5}$	385	272	370	295
$Q_{2.33}$	525	360	505	410
Q_5	720	575	690	630
Q_{10}	880	745	850	800
Q_{20}	1035	910	990	960

After Wyżga (2008), reproduced with permission of Elsevier
Q_x denotes discharge of given recurrence interval

the deepened channels. It reduced floodplain retention of floodwater and increased relative smoothness (ratio of water depth to the height of protrusion of bed-material particles to the flow) of flood flows, resulting in decreased attenuation and faster propagation of flood waves.

5.2.3 Loss of Channel Storage in the Impounded Reach of the Upper Vistula River

Between 1949 and 2003, six barrages were constructed on the Upper Vistula with the aim to increase water depth in the river for navigation purposes. Their construction changed the river into a sequence of shallow water reservoirs on the distance of 86 km. Because the river was impounded for navigation, the barrages maintain the same water level except at high flows. This distinguishes these shallow reservoirs from typical dam reservoirs, from which a considerable proportion of stored water can be released before the flood in order to increase their flood capacity.

Channel storage is the volume of water that can be temporarily retained in a channel up to the bankfull stage (Fig. 3). In the shallow reservoirs formed upstream of the barrages, a considerable proportion of channel storage is exhausted as a result of the permanent river impoundment (Fig. 3). When discussing a possible

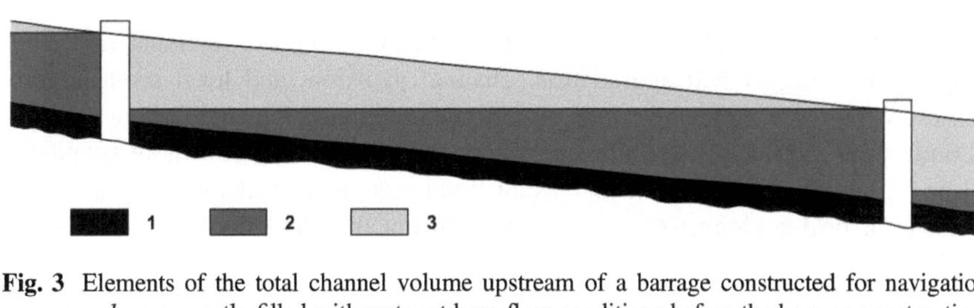

Fig. 3 Elements of the total channel volume upstream of a barrage constructed for navigation purposes: *1* permanently filled with water at base-flow conditions before the barrage construction, *2* permanently filled with water after the barrage construction, *3* filled with water during floods. Available channel storage of floodwater encompasses parts 2 and 3 before, while only part 3 after the barrage construction

construction of a new barrage downstream, Wyżga et al. (2014) indicated that its construction would permanently exhaust 2.4 million m^3 of channel storage (189 m^3 per 1 m of river length, on average), i.e., 57 % of the volume hitherto available along the planned reservoir. In turn, the amount of channel storage that was already lost as a result of the construction of the six barrages can be estimated at ~12 million m^3 (with an average of 139 m^3/1 m of river length). This lost storage volume is rather large as it corresponds to one-fifth of the flood capacity of the large dam reservoir constructed in past years at Świnna Poręba on the Skawa River.

5.3 Expansion of Riparian Forests Increasing Large Wood Recruitment to River Channels

Topographic maps of the Polish Carpathians prepared in the late 18th to the late 19th centuries indicate a lack or only a scarce occurrence of riparian forests in the river valleys (Wyżga et al. 2012, 2015). This situation was typical of European mountain areas of that time (Kondolf et al. 2002; Rinaldi et al. 2013) and reflected high intensity of grazing or cultivation of riparian areas (preventing development of riparian forests) combined with high dynamics of the rivers draining highly deforested catchments, that facilitated rapid turnover of their active zones. During the 20th century riparian forests developed in the valleys of Polish Carpathian rivers (Wyżga et al. 2012). For instance, in the middle course of the Czarny Dunajec, the riparian area in the late 19th century was devoid of forest; at the mid-20th century forest expanded to ca. 40 % of the reach length, whereas at present it grows along the entire reach (Wyżga 2007). The expansion of riparian forests was stimulated by a reduction in agricultural and pastoral activities in the riparian area and by river narrowing caused by reduced river dynamics and channelization of many river sections.

Currently, large amounts of woody material are recruited during floods from the forested channel banks and islands to the channels of Polish Carpathian watercourses and may cause flood damage if deposited at vulnerable sites such as bridges or urban reaches. Negative impacts of wood debris during floods are mainly related to the clogging of narrow river sections and bridges, which causes a quick succession of backwater effects as a result of the reduction of cross-sectional area. Such clogging can trigger bed aggradation, channel avulsion, and local scouring processes, which can ultimately lead to floodplain inundation and bridge collapse (Comiti et al. 2012). As a result, the nearby area may be flooded more frequently (Ruiz-Villanueva et al. 2013), increasing flood risk (Ruiz-Villanueva et al. 2014). Impacts of bridge clogging were observed during the large flood of 2010 on the Biała River when deposition of wood debris under a bridge (Fig. 4) directed flood flow onto the adjacent valley floor, resulting in flood damage to the nearby houses and infrastructure (Hajdukiewicz et al. 2016).

Fig. 4 Partial clogging of the bridge at Jankowa on the Biała River with large woody debris trapped on the bridge pillar during the flood of June 2010. The bridge clogging directed the flood flow onto the adjacent valley floor, causing flood damage to the nearby houses and infrastructure

As demonstrated by Mikuś et al. (2016, in this volume), large wood is particularly mobile in larger rivers where it can be transported long distances along single-thread, channelized and incised reaches, which are currently common in the rivers (Wyżga et al. 2015). As a result, during floods substantial amounts of wood can be delivered to and deposited at vulnerable sites located within or downstream of such reaches irrespective of preventive measures (e.g. riparian forest clearing) undertaken in the vicinity of these sites.

6 Concluding Remarks

Floods in the Upper Vistula Basin are generated by three principal mechanisms: (i) a few days-long rainfall of moderate intensity, (ii) short-duration, high-intensity rainfall, and (iii) snowmelt, locally linked with ice jam formation. During the last century, changes in climatic and terrestrial drivers of floods have affected the flood hazard in the region, although both the changes themselves and their impacts on flood hazard are better documented for terrestrial drivers. An increase in temperature, already recorded over the past century and forecast to continue during the 21st century, tends to reduce the magnitudes of snowmelt floods and to increase the frequency of occurrence of local flash floods generated by high-intensity, convective rainfall. However, the ongoing climate changes may have no significant influence on the frequencies and magnitudes of floods caused by the rainfall generated by widespread cyclones as the Upper Vistula Basin is located between the Southern Europe, where precipitation totals are forecast to decrease, and Northern Europe, where they are expected to increase (European Commission 2005).

The increase in forest cover in the Polish Carpathians had an apparent regulating effect on the flood runoff from the mountain catchments, and the construction of dam reservoirs allowed for some reduction in peak discharges of flood waves, although multiple functions of the reservoirs limit their efficiency in reducing flood discharges. In turn, widespread channel regulation and river incision tended to reduce floodwater retention in floodplain areas and to accelerate the passage of flood waves, with the resultant increase in flood hazard in downstream river reaches. Construction of flood embankments considerably limited the lateral extent of active floodplains, stimulating enhanced management of the river valleys. However, as water stages of flood flows on the embanked rivers increased, flood embankments—especially in river reaches within the Sandomierz Basin—have been repeatedly breached/overtopped, resulting in the inundation of managed areas outside the embankments and considerable material damage. Shallow water reservoirs constructed for navigation purposes on the Upper Vistula substantially reduced channel storage of floodwater in that river because of the specific mode of functioning of these reservoirs. Finally, increased delivery of large wood to river channels following the 20th-century expansion of riparian forests in the valleys of Polish Carpathian rivers currently makes wood jamming during floods an important factor of flood hazard in vulnerable river reaches/cross-sections.

Analyses of long-term records of flood discharges at numerous gauging stations indicated no systematic increase in the magnitudes of floods on the rivers of the region (Soja 2002; Ruiz-Villanueva et al. 2016), and this can be explained by the counteracting operation of different drivers. However, this seemingly optimistic conclusion can be reformulated into a less optimistic one. The lack of common and significant decreasing trends in flood magnitudes indicates a lost opportunity for permanent reduction of flood hazard on the rivers in the region, that could have been achieved as a result of the reforestation of catchments (cf. Wyżga 1997) and the construction of dam reservoirs. Moreover, it should be remembered that the lack of increase in flood hazard over the few past decades does not translate to the same situation in flood risk because of the increased exposure and vulnerability of the valley communities to floods, caused by the enhanced management of the valley floors and the increased wealth of the communities.

Acknowledgments This work was supported by the project FLORIST (Flood risk on the northern foothills of the Tatra Mountains) supported by a grant from the Swiss Government through the Swiss Contribution to the enlarged European Union (PSPB No. 153/2010). The study makes use of hydrological data provided by the Institute of Meteorology and Water Management—State Research Institute (IMGW-PIB).

References

Alila Y, Mtiraoui A (2002) Implications of heterogeneous flood-frequency distributions on traditional stream-discharge prediction techniques. Hydrol Proc 16:1065–1084

Arnell NW, Gosling SN (2014) The impacts of climate change on river flood risk at the global scale. Clim Change 134:387–401

Ballesteros-Cánovas JA, Stoffel M, Spyt B, Janecka K, Kaczka RJ, Lempa M (2016) Paleoflood discharge reconstruction in Tatra Mountain streams. Geomorphology 272C. doi:10.1016/j.geomorph.2015.12.004

Bárdossy A, Filiz F (2005) Identification of flood producing atmospheric circulation patterns. J Hydrol 313:48–57

Berghuijs WR, Woods RA, Hutton CJ, Sivapalan M (2016) Dominant flood generating mechanisms across the United States. Geophys Res Lett 43:4382–4390

Blöschl G, Nester T, Komma J, Parajka J, Perdigão RAP (2013) The June 2013 flood in the Upper Danube basin, and comparisons with the 2002, 1954 and 1899 floods. Hydrol Earth Syst Sci Discuss 10:9533–9573

Borga M, Anagnostou EN, Blöschl G, Creutin JD (2010) Flash floods: observations and analysis of hydrometeorological controls. J Hydrol 394:1–3

Bradshaw CJA, Sodhi NS, Peh KS-H, Brook BW (2007) Global evidence that deforestation amplifies flood risk and severity in the developing world. Global Change Biol 13:2379–2395

Bryndal T (2014) A method of identification of small Carpathian catchments more prone to flash flood generation. Based on the example of south-eastern part of the Polish Carpathians. Carp J Earth Environ Sci 9(3):109–122

Cebulak E, Niedźwiedź T (1998) Ekstremalne zjawiska opadowe w dorzeczu górnej Wisły w latach 1995 i 1996 (Extreme precipitation events in the Upper Vistula Basin in 1995 and 1996). In: Starkel L (ed) Geomorfologiczny i Sedymentologiczny Zapis Lokalnych Ulew. Dokum Geogr 11:11–30 (in Polish, with English summary)

Comiti F, D'Agostino V, Moser M, Lenzi MA, Bettella F, Dell'Agnese A, Rigon E, Gius S, Mazzorana B (2012) Preventing wood-related hazards in mountain basins: From wood load estimation to designing retention structures. In: 12th Congress Interpraevent 2012—Grenoble/France, pp 651–662

Döll P, Jiménez-Cisneros B, Oki T, Arnell NW, Benito G, Cogley JG, Jiang T, Kundzewicz ZW, Mwakalila S, Nishijima A (2015) Integrating risks of climate change into water management. Hydrol Sci J 60:4–13

European Commission (2005) Climate change and the European water dimension. EU Report No. 21553. European Commission—Joint Research Centre, Ispra, Italy

FloodSite (2007) Evaluating flood damages: guidance and recommendations on principles and methods. http://www.floodsite.net/html/partner_area/project_docs/t09_06_01_flood_damage_guidelines_d9_1_v2_2_p44.pdf

Freudiger D, Kohn I, Stahl K, Weiler M (2014) Large-scale analysis of changing frequencies of rain-on-snow events with flood-generation potential. Hydrol Earth Syst Sci 18:2695–2709

Hajdukiewicz H, Wyżga B (2013) Degradacja rzek wielonurtowych polskich Karpat w XX wieku (Twentieth-century degradation of multi-thread rivers in the Polish Carpathians). In: Wyżga B (ed) Stan środowiska rzek południowej Polski – znaczenie środowiskowe, degradacja i możliwości rewitalizacji rzek wielonurtowych. Instytut Ochrony Przyrody PAN, Kraków, pp 33–58

Hajdukiewicz H, Wyżga B, Mikuś P, Zawiejska J, Radecki-Pawlik A (2016) Impact of a large flood on mountain river habitats, channel morphology, and valley infrastructure. Geomorphology 272C. doi:10.1016/j.geomorph.2015.09.003

Hall J, Arheimer B, Borga M, Brázdil R, Claps P, Kiss A, Kjeldsen TR, Kriaučiūnienė J, Kundzewicz ZW, Lang M, Llasat MC, Macdonald N, McIntyre N, Mediero L, Merz B, Merz R, Molnar P, Montanari A, Neuhold C, Parajka J, Perdigão RAP, Plavcová L, Rogger M,

Salinas JL, Sauquet E, Schär C, Szolgay J, Viglione A, Blöschl G (2014) Understanding flood regime changes in Europe: a state-of-the-art assessment. Hydrol Earth Syst Sci 18:2735–2772
Hirschboeck KK (1987) Catastrophic flooding and atmospheric circulation anomalies. In: Mayer L, Nash DB (eds) Catastrophic flooding. Allen & Unwin, Boston, pp 23–56
IPCC (Intergovernmental Panel on Climate Change) (2013) Summary for policymakers. In: Stocker TF, Qin D, Plattner G-K, Tignor M, Allen SK, Boschung J, Nauels A, Xia Y, Bex V, Midgley PM (eds) Climate change 2013: the physical science basis. Contribution of working group I to the fifth assessment report of the intergovernmental panel on climate change. Cambridge University Press, Cambridge
Jacobeit J, Philipp A, Nonnenmacher M (2006) Atmospheric circulation dynamics linked with prominent discharge events in Central Europe. Hydrol Sci J 51:946–965
Kondolf GM (1997) Hungry water: effects of dams and gravel mining on river channels. Environ Manage 21:533–551
Kondolf GM, Piégay H, Landon N (2002) Channel response to increased and decreased bedload supply from land use change: contrasts between two catchments. Geomorphology 45:35–51
Kozak J (2009) Forest cover changes and their drivers in the Polish Carpathian Mountains since 1880. In: Nagendra H, Southworth J (eds) Reforesting landscapes: linking patterns and process. Springer, Dordrecht, pp 253–273
Kukulak J (1994) Antropogeniczne przemiany w środowisku przyrodniczym Podhala w latach 1931–1988 (Anthropogenic changes in natural environment of Podhale in the years 1931–1988). In: Górz B (ed) Studia nad przemianami Podhala. Pr Monogr WSP w Krakowie, vol 172, pp 265–285
Kundzewicz ZW, Robson AJ (2004) Change detection in river flow records—a review of the methodology. Hydrol Sci J 49:7–19
Kundzewicz ZW, Schellnhuber HJ (2004) Floods in the IPCC TAR perspective. Nat Hazards 31:111–128
Kundzewicz ZW, Stoffel M (2016) Anatomy of flood risk. In: Kundzewicz ZW, Stoffel M, Niedźwiedź T, Wyżga B (eds) Flood risk in the Upper Vistula Basin. Springer (in this volume)
Kundzewicz ZW, Szamałek K, Kowalczak P (1999) The great flood of 1997 in Poland. Hydrol Sci J 44:855–870
Kundzewicz ZW, Lugeri N, Dankers R, Hirabayashi Y, Döll P, Pińskwar I, Dysarz T, Hochrainer S, Matczak P (2010) Assessing river flood risk and adaptation in Europe—review of projections for the future. Mitig Adapt Strat Glob Change 15:641–656
Kundzewicz ZW, Kanae S, Seneviratne SI, Handmer J, Nicholls N, Peduzzi P, Mechler R, Bouwer LM, Arnell N, Mach K, Muir-Wood R, Brakenridge GR, Kron W, Benito G, Honda Y, Takahashi K, Sherstyukov B (2014) Flood risk and climate change: global and regional perspectives. Hydrol Sci J 59:1–28
Kundzewicz ZW, Krysanova V, Dankers R, Hirabayashi Y, Kanae S, Hattermann FF, Huang S, Milly PCD, Stoffel M, Driessen PPJ, Matczak P, Quevauviller P, Schellnhuber H-J (2016) Differences in flood hazard projections in Europe—their causes and consequences for decision making. Hydrol Sci J (in review)
Lach J, Wyżga B (2002) Channel incision and flow increase of the upper Wisłoka River, southern Poland, subsequent to the reafforestation of its catchment. Earth Surf Proc Landforms 27: 445–462
Łajczak A (1995) The impact of river regulation, 1850–1990, on the channel and floodplain of the Upper Vistula River, southern Poland. In: Hickin EJ (ed) River geomorphology. Wiley, Chichester, pp 209–233
Łupikasza E, Niedźwiedź T, Pińskwar I, Ruiz-Villanueva V, Kundzewicz ZW (2016) Observed changes in temperature and precipitation and relationship between them in the Upper Vistula Basin. In: Kundzewicz ZW, Stoffel M, Niedźwiedź T, Wyżga B (eds) Flood risk in the Upper Vistula Basin. Springer (in this volume)
McCabe GJ, Clark MP, Hay LE (2007) Rain-on-snow events in the western United States. Bull Am Meteorol Soc 88:319–328
Merz R, Blöschl G (2003) A process typology of regional floods. Water Resour Res 39:1340

Merz R, Blöschl G (2008) Flood frequency hydrology—1. Temporal, spatial and causal expansion of information. Water Resour Res 44:W08432

Mikuś P, Wyżga B, Ruiz-Villanueva V, Zawiejska J, Kaczka RJ, Stoffel M (2016) Methods to assess large wood dynamics and the associated flood hazard in Polish Carpathian watercourses of different size. In: Kundzewicz ZW, Stoffel M, Niedźwiedź T, Wyżga B (eds) Flood risk in the Upper Vistula Basin. Springer (in this volume)

Munteanu C, Kuemmerle T, Boltiziar M, Butsic V, Gimmi U, Halada L, Kaim D, Király G, Konkoly-Gyuró E, Kozak J, Lieskovský J, Mojses M, Müller M, Ostafin K, Ostapowicz K, Shandra O, Štych P, Walker S, Radeloff VC (2014) Forest and agricultural land change in the Carpathian region—a meta-analysis of long-term patterns and drivers of change. Land Use Policy 38:685–697

Nied M, Hundecha Y, Merz B (2013) Flood-initiating catchment conditions: a spatio-temporal analysis of large-scale soil moisture patterns in the Elbe River basin. Hydrol Earth Syst Sci 17:1401–1414

Nied M, Pardowitz T, Nissen K, Ulbrich U, Hundecha Y, Merz B (2014) On the relationship between hydro-meteorological patterns and flood types. J Hydrol 519:3249–3262

Niedźwiedź T (1999) Rainfall characteristics in southern Poland during the severe flooding event of July 1997. Stud Geomorph Carp-Balcan 33:5–25

Niedźwiedź T, Łupikasza E (2016) Change in atmospheric circulation patterns. In: Kundzewicz ZW, Stoffel M, Niedźwiedź T, Wyżga B (eds) Flood risk in the Upper Vistula Basin. Springer (in this volume)

Ouarda TBMJ, Cunderlik JM, St-Hilaire A, Barbet M, Bruneau P, Bobeé B (2006) Data-based comparison of seasonality-based regional flood frequency methods. J Hydrol 330:329–339

Parajka J, Kohnová S, Bálint G, Barbuc M, Borga M, Claps P, Cheval S, Dumitrescu A, Gaume E, Hlavčova K, Merz R, Pfaundler M, Stancalie G, Szolgay J, Blöschl G (2010) Seasonal characteristics of flood regimes across the Alpine-Carpathian range. J Hydrol 394:78–89

Petrow T, Zimmer J, Merz B (2009) Changes in the flood hazard in Germany through changing frequency and persistence of circulation patterns. Nat Hazards Earth Syst Sci 9:1409–1423

Piniewski M, Szcześniak M, Mezghani A, Kundzewicz ZW (2016) Hydroclimatic projections for the Upper Vistula Basin. In: Kundzewicz ZW, Stoffel M, Niedźwiedź T, Wyżga B (eds) Flood risk in the Upper Vistula Basin. Springer (in this volume)

Prudhomme C, Genevier M (2011) Can atmospheric circulation be linked to flooding in Europe? Hydrol Proc 25:1180–1990

Punzet J (1972) Tendencje zmian przebiegu wielkich wód w dorzeczu górnej Wisły (Tendency of variations in the course of high waters in the Upper Vistula basin). Wiad Służ Hydrol Meteorol 8(3–4):29–41

Punzet J (1984) Maksymalne przepływy Sanu — badania weryfikacyjne (The highest discharges of the San—a control study). Gosp Wodna 1:7–11

Punzet J (1985) Wezbrania Wisły w obrębie Krakowa – dawniej i dziś (Floods on the Vistula River in the Cracow region—in the past and today) Gosp Wodna 8:190–193

Rinaldi M, Wyżga B, Dufour S, Bertoldi W, Gurnell A (2013) River processes and implications for fluvial ecogeomorphology: a European perspective. In: Shroder J, Butler D, Hupp CR (eds) Treatise on geomorphology. Vol 12, Ecogeomorphology. Academic Press, San Diego, pp 37–52

Romanowicz RJ, Kundzewicz ZW, Meresa HK, Stoffel M, Krysanova V, Doroszkiewicz J (2016) Projections of changes in flood hazard in two headwater catchments of the Vistula in the context of European-scale studies. In: Kundzewicz ZW, Stoffel M, Niedźwiedź T, Wyżga B (eds) Flood risk in the Upper Vistula Basin. Springer (in this volume)

Ruiz-Villanueva V, Borga M, Zoccatelli D, Marchi L, Gaume E, Ehret U (2012) Extreme flood response to short-duration convective rainfall in south-west Germany. Hydrol Earth Syst Sci 16:1543–1559

Ruiz-Villanueva V, Bodoque JM, Díez-Herrero A, Eguibar MA, Pardo-Igúzquiza E (2013) Reconstruction of a flash flood with large wood transport and its influence on hazard patterns in an ungauged mountain basin. Hydrol Proc 27:3424–3437

Ruiz-Villanueva V, Bodoque JM, Díez-Herrero A, Bladé E (2014) Large-wood transport as significant influence on flood risk in a mountain village. Nat Hazards 74:967–987

Ruiz-Villanueva V, Stoffel M, Wyżga B, Kundzewicz ZW, Czajka B, Niedźwiedź T (2016) Decadal variability of floods in the northern foreland of the Tatra Mountains. Reg Environ Change 16:603–615

Sasim M, Mierkiewicz M (2002) Powódź w dorzeczu środkowej i dolnej Wisły w lipcu 2001 (The flood of July 2001 in the Middle and Lower Vistula River catchment. Gosp Wodna 1:10–16 (in Polish, with English summary)

Seneviratne SI, Nicholls N, Easterling D, Goodess CM, Kanae S, Kossin J, Luo Y, Marengo J, McInnes K, Rahimi M, Reichstein M, Sorteberg A, Vera C, Zhang X (2012) Changes in climate extremes and their impacts on the natural physical environment. In: Field CB, Barros V, Stocker TF, Qin D, Dokken DJ, Ebi KL, Mastrandrea MD, Mach KJ, Plattner G-K, Allen SK, Tignor M, Midgley PM (eds) Managing the risks of extreme events and disasters to advance climate change adaptation. A special report of working groups I and II of the intergovernmental panel on climate change (IPCC). Cambridge University Press, Cambridge, pp 109–230

Sivapalan M, Blöschl G, Merz R, Gutknecht D (2005) Linking flood frequency to long-term water balance: incorporating effects of seasonality. Water Resour Res 41:W06012

Slater LJ, Singer MB, Kirchner JW (2015) Hydrologic versus geomorphic drivers of trends in flood hazard. Geophys Res Lett 42:370–376

Soja R (2002) Hydrologiczne aspekty antropopresji w polskich Karpatach (Hydrological aspects of anthropopression in the Polish Carpathians). Pr Geogr IGiPZ PAN 186:1–130

Starkel L (2011) Złożoność czasowa i przestrzenna opadów ekstremalnych — ich efekty geomorfologiczne i drogi przeciwdziałania im (Temporal and spatial complexity of extreme rainfalls—their geomorphological effects and ways of counteract them). Landform Analysis 15:65–80 (in Polish, with English summary)

Szumański A (1986) Postglacjalna ewolucja i mechanizm transformacji dna doliny Dolnego Sanu (Late Glacial and Holocene evolution and mechanism of transformation of the floor of the lower San valley). Geologia 12(1):5–92 (in Polish, with English summary)

Ulbrich U, Brücher T, Fink AF, Leckebusch GC, Krüger A, Pinto JG (2003a) The central European floods of August 2002: part 1—rainfall periods and flood development. Weather 58:371–377

Ulbrich U, Brücher T, Fink AF, Leckebusch GC, Krüger A, Pinto JG (2003b) The central European floods of August 2002: part 2—synoptic causes and considerations with respect to climate change. Weather 58:434–442

Van Bebber WJ (1891) Die Zugstrassen der barometrischer Minima. Meteorol Z 8:361–366

Walczykiewicz T, Rataj C (2011) Analiza pracy zbiorników retencyjnych (Analysis of the operation of dam reservoirs). In: Maciejewski M, Ostojski MS, Walczykiewicz T (eds) Dorzecze Wisły — monografia powodzi maj-czerwiec 2010. Instytut Meteorologii i Gospodarki Wodnej, Warszawa, pp 121–148

Wyżga B (1993a) Present-day changes in the hydrologic regime of the Raba River (Carpathians, Poland) as inferred from facies pattern and channel geometry. In: Marzo M, Puigdefabregas C (eds) Alluvial Sedimentation. International Association Sedimentology Special Publication, vol 17, pp 305–316

Wyżga B (1993b) River response to channel regulation: case study of the Raba River, Polish Carpathians. Earth Surf Proc Landforms 18:541–556

Wyżga B (1997) Methods for studying the response of flood flows to channel change. J Hydrol 198:271–288

Wyżga B (2001a) A geomorphologist's criticism of the engineering approach to channelization of gravel-bed rivers: case study of the Raba River, Polish Carpathians. Environ Manage 28: 341–358

Wyżga B (2001b) Impact of the channelization-induced incision of the Skawa and Wisłoka Rivers, southern Poland, on the conditions of overbank deposition. Regul Rivers Res Manage 17: 85–100

Wyżga B (2007) Gruby rumosz drzewny: depozycja w rzece górskiej, postrzeganie i wykorzystanie do rewitalizacji cieków górskich (Large woody debris: storage in a mountain river, perception and use in the restoration of mountain watercourses). Instytut Ochrony Przyrody PAN, Kraków

Wyżga B (2008) A review on channel incision in the Polish Carpathian rivers during the 20th century. In: Habersack H, Piégay H, Rinaldi M (eds) Gravel-bed rivers VI: from process understanding to river restoration. Dev Earth Surf Proc 11, Elsevier, Amsterdam, pp 525–555

Wyżga B, Zawiejska J, Radecki-Pawlik A, Hajdukiewicz H (2012) Environmental change, hydromorphological reference conditions and the restoration of Polish Carpathian rivers. Earth Surf Proc Landforms 37:1213–1226

Wyżga B, Radecki-Pawlik A, Hajdukiewicz H, Przebięda M (2014) O celowości budowy stopnia Niepołomice na Wiśle (On purposefulness of the construction of the Niepołomice barrage on the Vistula River). Gosp Wodna 7:261–268

Wyżga B, Zawiejska J, Hajdukiewicz H (2015) Multi-thread rivers in the Polish Carpathians: occurrence, decline and possibilities of restoration. Quat Int. doi:10.1016/j.quaint.2015.05.015

Wyżga B, Zawiejska J, Radecki-Pawlik A (2016) Impact of channel incision on the hydraulics of flood flows: examples from Polish Carpathian rivers. Geomorphology 272C. doi:10.1016/j.geomorph.2015.05.017

Zawiejska J, Wyżga B (2010) Twentieth-century channel change on the Dunajec River, southern Poland: patterns, causes and controls. Geomorphology 117:234–246

Zehe E, Singh AK, Bárdossy A (2006) Modelling of monsoon rainfall for a mesoscale catchment in North-West India I: assessment of objective circulation patterns. Hydrol Earth Syst Sci 10:797–806

Methods to Assess Large Wood Dynamics and the Associated Flood Hazard in Polish Carpathian Watercourses of Different Size

Paweł Mikuś, Bartłomiej Wyżga, Virginia Ruiz-Villanueva, Joanna Zawiejska, Ryszard J. Kaczka and Markus Stoffel

Abstract Applicability, advantages and limitations of a range of methods applied to determine large wood dynamics in Kamienica Stream and the Czarny Dunajec River, Polish Carpathians, are discussed. Results of a 6-year-long monitoring suggest an increased rate of wood recruitment to Kamienica Stream caused by recent bark beetle infestation of the spruce forests in the valley. However, both monitoring of wood transport and wood inventories indicate that the mobility of large wood in the stream is low and can increase only during major floods. Thus flood hazard to downstream valley reaches potentially resulting from the consid-

P. Mikuś (✉) · B. Wyżga
Institute of Nature Conservation, Polish Academy of Sciences,
al. Mickiewicza 33, 31-120 Kraków, Poland
e-mail: mikus@iop.krakow.pl

B. Wyżga
e-mail: wyzga@iop.krakow.pl

P. Mikuś · B. Wyżga · R.J. Kaczka
Faculty of Earth Sciences, University of Silesia, ul. Będzińska 60, 41-200 Sosnowiec, Poland
e-mail: kaczka@wnoz.us.edu.pl

V. Ruiz-Villanueva · M. Stoffel
Dendrolab.ch, Institute of Geological Sciences, University of Bern,
Baltzerstrasse 1+3, 3012 Bern, Switzerland
e-mail: virginia.ruiz@dendrolab.ch

M. Stoffel
e-mail: markus.stoffel@dendrolab.ch

V. Ruiz-Villanueva · M. Stoffel
Climatic Change and Climate Impacts, Institute for Environmental Sciences,
University of Geneva, Boulevard Carl Vogt 66, 1205 Geneva, Switzerland

M. Stoffel
Department of Earth Sciences, rue des Maraîchers 13, 1205 Geneva, Switzerland

J. Zawiejska
Institute of Geography, Pedagogical University of Cracow, ul. Podchorążych 2,
30-084 Kraków, Poland
e-mail: zawiejska.joanna@gmail.com

Z.W. Kundzewicz et al. (eds.), *Flood Risk in the Upper Vistula Basin*,
GeoPlanet: Earth and Planetary Sciences, DOI 10.1007/978-3-319-41923-7_5

erable amounts of large wood stored in the upper stream reach is limited. In the Czarny Dunajec, wood inventories, a tracking experiment with logs tagged with radio transmitters, and numerical modelling indicated high potential for wood transport in the narrow river reaches formed by channelization or channel incision, and high potential for wood deposition in the wide, multi-thread channel. Vegetative regeneration of living willow wood considerably reduces its remobilization by subsequent floods. Efficient transport of large wood along narrow river reaches implicates that during floods substantial amounts of wood may be delivered from distant sources to the channel sections located downstream of the narrow reaches. Wide, multi-thread reaches operate as natural wood traps, considerably limiting further transfer of wood to vulnerable sites/reaches.

Keywords Large wood dynamics · Flood hazard · Wood inventory · Wood monitoring · Numerical modelling · Wood tracking

1 Introduction

Over the last decades, large wood (i.e. trees and shrubs fallen to river channels) has received increasing interest from scientists who recognize its significance as a functional component of fluvial ecosystems (Gurnell 2013; Wohl 2013; Le Lay et al. 2013). When large wood is being analysed, it is possible to conceptualize its recruitment/mobilization, transport and storage in rivers, which together constitute the river's wood budget (Benda and Sias 2003).

Individual pieces of wood and their aggregates (jams) create obstructions that can substantially increase the frictional resistance to flow (Shields and Gippel 1995; Curran and Wohl 2003). These obstructions modify flow velocity (Gippel 1995; Davidson and Eaton 2013) and increase local storage of sediment and organic matter around the wood (Faustini and Jones 2003). However, when large quantities of wood are mobilized (usually during floods), the resultant flow obstructions, particularly at critical sections such as bridges, can increase the magnitude, frequency, and duration of overbank flows (Ruiz-Villanueva et al. 2013, 2014c; Lucía et al. 2015). Therefore, our understanding of large wood dynamics and estimation of the related potential hazards requires in situ observations and measurements, which still remain very limited (MacVicar et al. 2009). Only a few field studies analysed wood dynamics using different approaches such as surveys from either field data or aerial imagery (Latterell and Naiman 2007; Wohl and Goode 2008), video cameras or time-lapse photography (Moulin and Piégay 2004; Bertoldi et al. 2014; Kramer and Wohl 2014) or tracer methods (Lassettre and Kondolf 2012; Schenk et al. 2014; Ravazzolo et al. 2015). Most of these studies were carried out in small streams, whereas much less is known about wood transport in large rivers where large wood dynamics and associated effects on fluvial processes might be very different (MacVicar and Piégay 2012). Modelling (physical and numerical) is a powerful tool to analyse large wood dynamics; however, few flume experiments have been performed so far with the aim to characterise wood transport and deposition (Braudrick and Grant 2001; Welber et al. 2013; Bertoldi et al. 2014). As to numerical models, a few attempts have recently been made to analyse large wood phenomena in rivers (Ruiz-Villanueva et al. 2014a).

The type and accuracy of the information about large wood strongly depend on the approach and its spatial and temporal resolution (Gurnell 2013). In the case of large wood-related hazards, it is particularly important to gather accurate information as effective management of large wood requires an in-depth understanding of the stability of individual wood pieces and jams within channels and on floodplains, and physical and ecological effects created by wood (Wohl et al. 2016). Therefore, the recognition of potential impacts of large wood on flood hazard and its relevance for flood risk in mountain watercourses is still very challenging (Mazzorana and Fuchs 2010; Comiti et al. 2012; Kundzewicz et al. 2014; Ruiz-Villanueva et al. 2014b). Successful achievement of this goal may require a combination and adjustment of different techniques to river morphology and hydrological conditions as well as the duration of the study and the accuracy of the desired results.

The aim of this chapter is (i) to discuss different approaches to gathering data about large wood dynamics based on the study in two watercourses of different size in the Polish Carpathians, (ii) to indicate the advantages and limitations of particular techniques, (iii) to assess large wood dynamics in the two types of watercourses, and (iv) to infer about the large wood-related flood hazard in the watercourses under consideration and other watercourses of similar size in the region.

We summarize below investigations conducted in the second- to fourth-order reaches of Kamienica Stream within the Gorce Mountains National Park and in the middle course of the fifth-order Czarny Dunajec River. Recent bark beetle infestation and the resultant dieback of riparian forest along the headwater reach of Kamienica Stream could have considerably increased the delivery of fallen trees to the channel, and as the wood is not removed from the stream under the national park regulations, a question arises as to whether it can threat downstream, developed parts of the valley. Pronounced development of riparian forest along the Czarny Dunajec occurred in the 20th century (Wyżga et al. 2012); as a result, substantial quantities of large wood are currently recruited to the river during floods (Wyżga and Zawiejska 2005, 2010). At the same time, considerable variability in channel morphology in the middle course of the Czarny Dunajec should be reflected in marked differences in large wood dynamics and the associated flood hazard between particular river reaches.

2 Field Setting

Kamienica Stream drains mountains of medium height (with the highest peak in the catchment at 1311 m a.s.l.) in the Outer Western Carpathians (Fig. 1). In its upper course, the stream flows through a narrow, mostly V-shaped valley with 90 % of

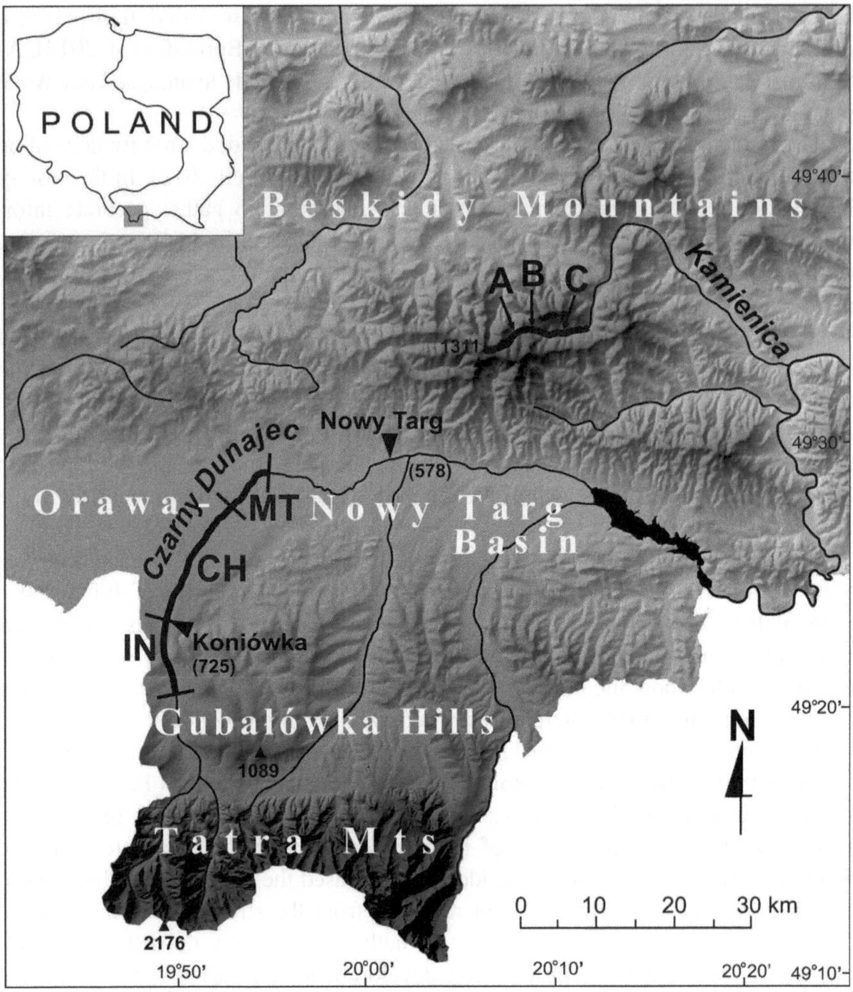

Fig. 1 Location of the studies of large wood dynamics in the watercourses of the northern foreland of the Tatra Mountains. Studied reaches of the Czarny Dunajec River and Kamienica Stream are indicated by *thick lines*. Reaches of the Czarny Dunajec: *IN* incised; *CH* channelized; *MT* multi-thread. *A*, *B* and *C* indicate reaches of Kamienica Stream with tagged riparian trees

the area covered by forest. Since the establishment of the Gorce Mountains National Park in 1980, this part of the stream has been almost free of human impact on the channel and the riparian zone. Over the last few years bark beetle infestation resulted in widespread dieback of spruce forest in the valley, that might have increased wood delivery to the channel and its transport to downstream, developed valley reaches. The study was conducted in second- to fourth-order stream reaches located between 0.3 and 9.1 km from the stream source. Along these reaches, the catchment area increases from 0.1 to 14.9 km^2 and channel slope decreases from

0.1 to 0.035 m m^{-1}. The second-order stream reach has an average channel width of 5.9 m and exhibits a step–pool channel pattern, with numerous steps developed on wood dams (Kaczka 2003). In the third- and fourth-order reaches, the channel is 10.3 m wide on average, with step–pool pattern in narrower parts of the stream and riffle–and–pool pattern in its wider parts. Along the most part of the study reaches, the banks of Kamienica Stream are overgrown with the upper subalpine forest composed of spruce (*Picea abies*), whereas the subalpine forest in the lowermost part of the study area is composed of spruce, beech (*Fagus sylvatica*) and fir (*Abies alba*).

The Czarny Dunajec River rises in the high-mountain Tatra massif, with the highest elevation in the catchment at 2176 m a.s.l. In the Tatra Mountains foreland, it flows through the Gubałówka Hills and the Orawa-Nowy Targ Basin to the confluence with the Biały Dunajec River at an altitude of 578 m (Fig. 1). Investigations of large wood dynamics were carried out in the middle river course, where the fifth-order Czarny Dunajec flows on a large glaciofluvial-alluvial fan and receives no significant tributaries. During the past few decades, some parts of the middle river course were considerably modified by channelization or gravel-mining-induced channel incision, whereas the other parts avoided significant human disturbances and remained unmanaged (Zawiejska and Wyżga 2010; Wyżga et al. 2012). As a result, currently the river's morphology varies considerably (Wyżga and Zawiejska 2005, 2010) with relatively narrow, single-thread, incised or channelized reaches and a wide, multi-thread reach (Fig. 1). Despite considerable variation in river widths, the middle course of the Czarny Dunajec—with its average width of 52 m and the maximum height of riparian trees amounting to 18 m—can be considered a large channel with respect to in-stream wood (Gurnell et al. 2002; Wyżga et al. 2015). The river flows through a forested corridor, with the riparian forest consisting of alder (*Alnus incana*) and a few willow species (*Salix eleagnos*, *S. purpurea*, *S. fragilis* and *S. alba*).

3 Investigations of Large Wood Dynamics in Kamienica Stream

3.1 Monitoring of the Recruitment and Transport of Large Wood in a Mountain Stream

The relatively small size of Kamienica Stream and the abundance of wood dams in its channel allowed us to expect relatively short distances of large wood transport; moreover, the relatively small channel size facilitated searching for displaced wood pieces. This is why we used metal plates as tags in the monitoring of wood dynamics in the stream. In October 2009, 429 trees growing along three reaches of the upper course of Kamienica Stream (Fig. 1) were tagged with numbered metal plates and their position was recorded with a Trimble GeoXT GPS receiver with

0.5 m precision. Different metals were used in particular study reaches: aluminium in reach A located at the distance of 2500–2950 m from the stream source, copper in reach B (4000–4450 m from the stream source), and steel in reach C (7850–8300 m from the stream source). The different metals were chosen so as to allow later use of a metal detector to distinguish between trees delivered to the stream in different reaches or to find a tagged tree in case the plate is impossible to reach (e.g. buried). In each reach, we tagged all trees with a diameter >10 cm and closest to the channel either on the floodplain or on the valley side; in fact, their distance from the channel margin ranged from 0.5 to 12 m. The positions of standing and fallen trees were monitored a few times per year, especially after flood or strong windstorm events in the region, to determine the causes and timing of the delivery of trees to the stream and their transport distance during particular flood events.

No significant floods occurred on Kamienica Stream during the six years of monitoring. Over that time, 32 trees (7.5 % of the tagged sample) were recruited to the stream, with 17 (53 %) delivered as a result of windthrow of living specimens, 8 (25 %) first died as a result of bark beetle infestation and were next thrown by wind, 5 (16 %) were undercut by bank erosion, and 2 (6 %) fell because of snow overload (Fig. 2). Two high-intensity weather events occurred over the monitoring period. At the very beginning of that period, an intense snowfall occurred, (ca. 80 cm of snow over 2 days), resulting in the recruitment of 2 trees from snow overload. In summer 2010 a strong windstorm occurred and the associated rainfall resulted in a moderate flood; during this event 8 trees growing on steep, undercut stream banks were toppled in the lowermost reach A. Although bark beetle infestation has not been a direct cause of the recruitment of spruce trees, trees weakened or killed by the infestation were very susceptible to breakage by wind, and this type of wood recruitment was characteristic of spruce growing along the stream. If such trees fell perpendicular to the channel, they were often broken into several smaller fragments. Such relatively small logs could have been transported even by small floods over distances of a few tens of metres before they were braced against other, more stable wood pieces. By contrast, living trees were frequently anchored on both stream banks after they fell, resulting in the formation of complete dams (Fig. 3). Living coniferous or deciduous trees were recruited to the stream as a result of windthrow, bank erosion or snow overload.

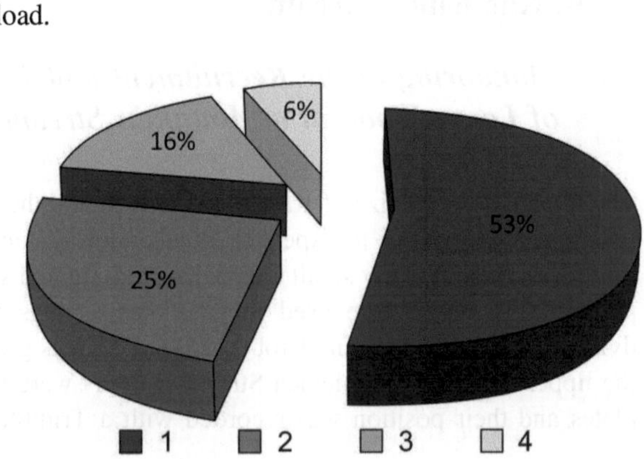

Fig. 2 Percentage of tagged trees recruited to Kamienica Stream between 2009 and 2015 as a result of particular factors: *1* windthrow of living trees; *2* tree dieback followed by windthrow; *3* bank erosion; *4* overloading with snow

Fig. 3 Spruce tree fallen to the channel of Kamienica Stream in a monitored reach. The tree was tagged with a numbered aluminium plate (shown by a *white arrow*) when growing close to the channel bank to allow its identification after displacement in the stream. The fallen tree forms a complete dam

Based on the number on the metal plate fixed to a tree, it is possible to determine the length of tree displacement from its initial position on a channel bank or its earlier position within the stream (Fig. 3). As the study period lacked major flood events, the majority of fallen trees were not transported and some were moved short distances, which did not exceed 100 m.

3.2 *Wood Inventory in Kamienica Stream*

Further information about wood dynamics in Kamienica Stream was obtained from the inventory of large wood performed in September–October 2012. It encompassed the whole lengths of second- and third-order stream reaches and the upper part of the fourth-order reach; in total, seventy 100-m channel segments were surveyed (Fig. 1). In each segment, channel slope was measured with a level and average channel width was determined from measurements at the beginning, middle, and end of the segment. Wood deposits were classified as logs or log jams, i.e. the aggregates of at least three logs. We measured the volume of logs, either isolated or aggregated into jams, which was next converted to wood mass by multiplying by the wood mass estimate of 500 kg m^{-3} to allow comparability of results with those from the Czarny Dunajec River. The orientation of individual logs and log jams in relation to the channel axis was also measured. Finally, a degree of wood decay was classified into one of four categories (1—fresh, bark adheres tightly; 2—loose bark; 3—no bark, wood hard; 4—no bark, wood soft) according to the approach of Lienkaemper and Swanson (1987).

In the surveyed part of the stream, channel width varied between 4 and 20.9 m. With the average height of mature trees on the channel banks amounting to 24 m, this part of the stream represents a watercourse of small- to medium-width (Wyżga et al. 2015). The total number of 2275 wood deposits was recorded, with particular stream segments retaining from 10 to 58 deposits, 32.5 on average. Of this total, 87 % was represented by individual logs and 12.3 % by log jams (28.5 logs and 4 log jams in a segment, on average). Mean mass of wood deposits was calculated as 167 kg and varied among the stream segments between 47 and 369 kg. Particular stream segments stored from 0.47 to 14.2 t of wood, with the average of 5.4 t (Wyżga et al. 2015). Notably, both the number and the mass of wood deposits in a segment were unrelated to channel width, and this was reflected in a lack of dependence of total wood storage (i.e. the total mass of wood in a channel segment) on channel width. Only the number of jams in a segment increased by 1 with each 3-m increase in channel width; however, with the small proportion of jams in the total number of wood deposits, this relationship was not reflected in an increase in the number of wood deposits with increasing channel width. The average value of specific wood storage (i.e. the amount of wood per unit channel area) for the surveyed segments of Kamienica amounted to 69.7 t ha^{-1}. However, the generally invariable total amounts of wood in stream segments were reflected in a nonlinear decrease of specific wood storage with increasing channel width. According to the estimated reciprocal regression model, the values of specific wood storage decreased from ca. 140 t ha^{-1} in the narrowest channel segments to ca. 25 t ha^{-1} in the widest segments (Wyżga et al. 2015).

Comparison of the patterns of large wood orientation recorded in the second- and fourth-order reaches of Kamienica indicates a change in the predominant wood orientation with increasing stream size (Fig. 4). In the second-order reach, a near-perpendicular orientation predominated, with a mode at 60°. A single mode indicating a predominant alignment of fallen trees towards the right stream bank and some deviation from the fully perpendicular orientation of wood pieces may be attributed to the predominant wind direction from NW rather than to the reorientation of wood pieces by stream current in the headwater reach. The fourth-order reach was typified by a bimodal orientation of wood pieces, with the longitudinal alignment slightly predominating over the orientation towards the right bank. Here, a proportion of wood pieces were apparently reoriented by stream current, whereas perpendicular alignment typified some other pieces, especially those longer than the channel width that formed complete and active dams. The patterns of large wood orientation in the stream were typified by a greater proportion of perpendicular to near-perpendicular alignment of wood pieces than those recorded shortly after the large flood of 1997 (Kaczka 1999).

In the second-order reach of Kamienica, 16 % of wood pieces were relatively fresh, representing class 1 and class 2 of wood decay, whereas as much as 84 % of

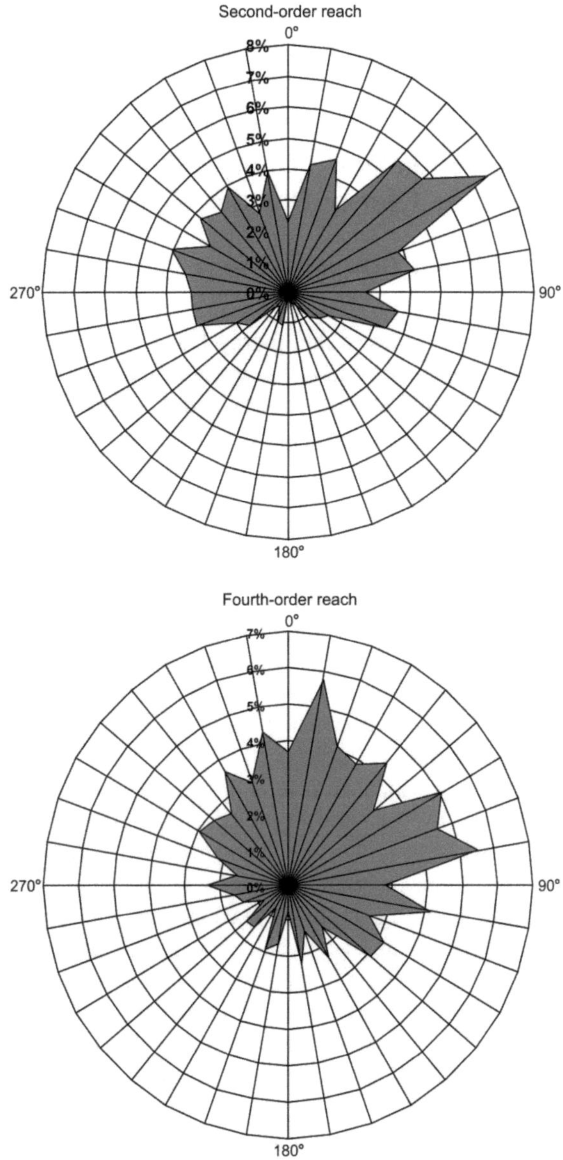

Fig. 4 Orientation of wood pieces in relation to the channel axis in the second-order reach (*upper diagram*) and the fourth-order reach (*lower diagram*) of Kamienica Stream

pieces exhibited an advanced state of decay linked with classes 3 and 4. In the fourth-order reach, classes 1 and 2 constituted 31 % of all wood pieces, while 69 % were classified as representing class 3 and class 4. These results differed considerably from those recorded after the large flood in 1997 (Kaczka 1999). In the second-order reach, 56 % of wood pieces were then associated with classes 1 and 2, whereas 44 % were in an advanced state of decay (classes 3 and 4). In the fourth-order reach, 65 % of pieces were classified as representing class 1 and class 2, while only 35 % exhibited an advanced state of decay (classes 3 and 4).

4 Investigations of Large Wood Dynamics in the Czarny Dunajec River

4.1 Wood Inventories in a Wide Mountain River

Wood inventories in the Czarny Dunajec were carried out in 2001 and 2011, and were completed by less systematic observations performed after the 20-year flood of May 2014. During the inventories wood deposits were classified into three categories: logs, wood jams, or whole shrubs and trees. Wood jams occurring in the river consisted not only of logs but they were heterogeneous mixtures of logs, branches, root boles, and twigs, sometimes with a considerable addition of fine organic matter and inorganic sediment (see Fig. 6a below). As the volume of wood contained in wood jams and shrubs and trees cannot be easily measured, we used the method developed by Thévenet et al. (1998) that involves measuring the volume of wood deposits and multiplying it by wood mass estimates for particular types of wood deposits (50, 100 and 500 kg m^{-3} for shrubs and trees, jams, and logs, respectively). Wood deposits were recorded in 100-m river segments, for which river width was determined based on the measurements performed at each end of every segment.

Below we indicate results of the inventory performed after the 7-year flood of July 2001 (Wyżga and Zawiejska 2005, 2010; Wyżga et al. 2015). Wood deposits were recorded in 89 river segments distributed along 17.2-km long section of the Czarny Dunajec with three geomorphological styles: incised, channelized and multi-thread (Fig. 1). In the section, river width varied between 18.5 and 148.5 m. The variation in river width and geomorphological river style was reflected in substantial variation in the number and the mean mass of wood deposits. Particular river segments stored between 0 and 154 wood deposits, with an average of 19.7 deposits in a segment. Logs constituted 21.3 %, shrubs and trees 37.6 %, and jams 41.1 % of the deposits. Mean mass of wood deposits stored in particular segments equalled 233 kg; it ranged from 0 (in a few segments that lacked large wood) to 733 kg. However, the mean mass differed considerably among particular types of deposits and amounted to 23 kg for logs, 182 kg for shrubs and trees, and 389 kg for wood jams. Both the number and the mean mass of wood deposits in a river segment increased linearly with increasing river width and the relationships were highly significant statistically.

Total wood storage in a 100-m segment of the Czarny Dunajec varied between 0 and 40.7 t, with an average of 4.6 t. The values of total wood storage were directly related to river width; on average, they increased by 1 t with each 5-m increase in river width. Despite similar values of total wood storage in the Czarny Dunajec

River and Kamienica Stream, the substantially greater width of the river was reflected in markedly lower values of specific wood storage—7.9 t ha^{-1} on average. Specific wood storage varied between 0 (in the segments lacking wood deposits) to 33 t ha^{-1} and was directly related to river width; on average, it increased by 1 t ha^{-1} with each 6-m increase in river width. The values of total and specific wood storage differed markedly between relatively narrow, single-thread river segments with a heavily regulated or incised channel and wider segments with a partly constrained or unmanaged, multi-thread channel. Average values of total wood storage amounted to 0.53 t in the former and 10 t in the latter, whereas the respective values of specific wood storage equalled 1.6 and 10.6 t ha^{-1}. Moreover, the two types of river segments differed considerably in the pattern of large wood distribution (Wyżga and Zawiejska 2010). Narrow, single-thread river segments were typified by random wood distribution, reflected in a lack of dependence of total and specific wood storage on channel width. By contrast, in wider river segments the values of total and specific wood storage were directly related to river width. These different patterns of wood distribution in the two types of river segments are illustrated with the results of the wood inventory carried out in 2011 (Fig. 5). Even though the absolute amounts of wood recorded in 2011 were smaller than in 2001, perhaps reflecting the lower magnitude of the preceding flood, the spatial patterns of wood distribution in the river were the same as those found 10 years before.

One of the key findings of the inventory from 2001 was a marked difference in the state of wood preservation and the predominant types of wood deposits between different parts of the wide, unmanaged river reach (Wyżga and Zawiejska 2005). Close to the end of the narrow, channelized reach, most wood was highly disintegrated and abraded and structured into jams. At a larger distance from the end of the channelized reach, many shrubs and trees were well preserved, with unbroken crowns and intact leaves and bark, and shrubs and trees constituted a considerable proportion of the total mass of wood. The same differences in the appearance of large wood were also observed after the 20-year flood of May 2014 (Fig. 6).

4.2 *Observations of the Vegetative Regeneration of Living Wood*

Capability of some tree species to re-sprout after deposition in a river considerably reduces their potential for remobilization during subsequent flood events. Below an altitude of ca. 800–900 m, riparian forests of Polish Carpathian rivers, including the Czarny Dunajec, are composed of alder and Salicaceae (willows, in lower altitudes also poplars). Alder exhibits relatively little capability to re-sprout, but willows can vigorously regenerate vegetatively after being deposited on relatively moist surfaces (Fig. 7) (Moggridge and Gurnell 2009). A study carried out in a multi-thread reach in the middle course of the Czarny Dunajec indicated that a considerable proportion of willow shrubs and trees deposited on gravel bars sprout and root to

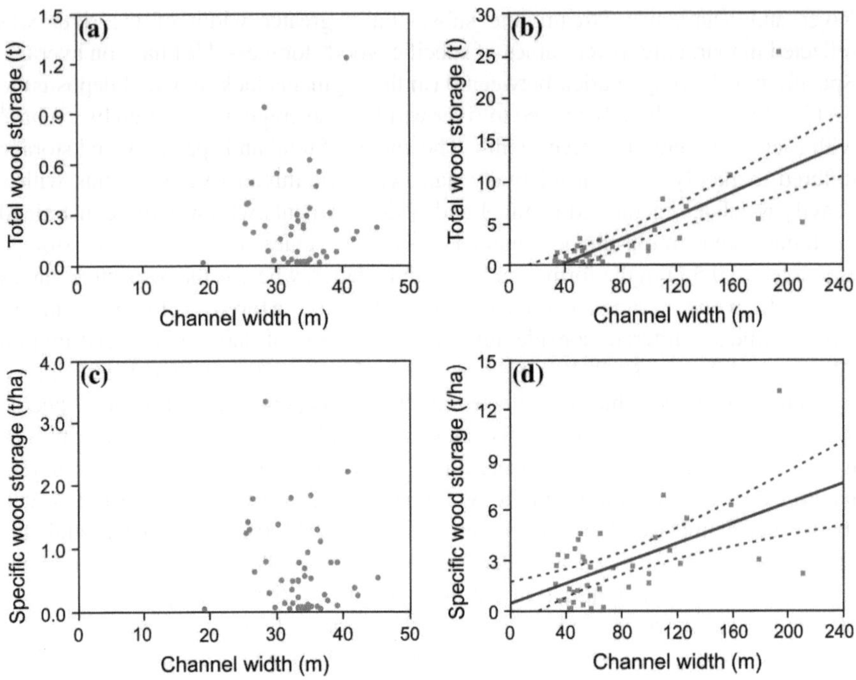

Fig. 5 Scatter plots and estimated regression relationships between total or specific wood storage and river width in heavily channelized and incised segments of the Czarny Dunajec (**a**, **c**) as well as partly reinforced and unmanaged river segments (**b**, **d**). The diagrams present results of the wood inventory performed in the Czarny Dunajec in 2011. *No regression lines* are indicated for the relationships which are not significant

the bar surface, developing into pioneer islands (Mikuś et al. 2013). The elevation of these pioneer islands ranged from 0.2 to 1 m above the water surface in the nearest low-flow channel, indicating that pioneer islands in the multi-thread reach can develop in a wide range of relative elevation. This can be partly attributed to the high ability of deposited willow shrubs and dense shoots sprouting from them to effectively trap fine sediment (Mikuś et al. 2013; cf. Gurnell and Petts 2006) that subsequently provides moisture for the developing willow shrubs. Moreover, some proportion of wood pieces contained in jams deposited at the head of older, building and established islands sprout leading to the upstream growth of the islands (Wyżga and Zawiejska 2010; Mikuś et al. 2013; cf. also Kaczka et al. 2008). The development of shrubs from the living willow driftwood effectively binds all wood pieces contained in the jams formed at island heads, preventing their remobilization during subsequent floods.

Notably, pioneer islands do not form in narrow, single-thread reaches of the Czarny Dunajec with a regulated or incised channel. Wider parts of the incised reach and narrower parts of the multi-thread reach with bar-braided morphology

Fig. 6 Large wood deposited in the upstream (**a**) and middle (**b**) parts of the wide, multi-thread reach of the Czarny Dunajec, ca. 700 and 1600 m below the narrow, channelized river reach, during a 20-year flood in May 2014. Wood deposited close to the channelized reach is highly disintegrated and aggregated into jams (**a**), whereas farther downstream many shrubs and trees are well preserved, with unbroken crowns (**b**)

Fig. 7 4-year pioneer island formed by dense shoots sprouted from a living willow shrub deposited on a gravel bar of the Czarny Dunajec by the flood of 2001. The shrub from which the pioneer island developed is still visible during the cold part of a year but disappears from the riverscape during vegetation season

support only scarce, pioneer and building islands (Mikuś et al. 2013) which are thus only transient traps for the large wood delivered from upstream.

4.3 Tracking Experiment with Radio Transmitter-Tagged Logs

Relatively large distances of wood displacement in wide rivers (Van der Nat et al. 2003; Schenk et al. 2014; Ravazzolo et al. 2015) and large channel area of such rivers underlay a decision to use radio telemetry in a tracking experiment intended to study large wood dynamics in the Czarny Dunajec. The single-frequency, microprocessor-controlled pulsed radio transmitters LPI 2800, produced by Wildlife Materials (USA), were purchased for the experiment. They are water resistant and emit signal in the band of 150–151 MHz (with the frequency of particular devices differing by 0.01 MHz) via 25 cm-long external antenna. We also prepared 30 alder logs of 3 m length and a diameter between 16 and 24 cm. When a weather forecast in mid-May 2014 indicated a possible flood on mountain rivers in the region, the transmitters were activated, mounted and sealed (with spray polyurethane foam) in the previously prepared holes in the logs (Fig. 8a). During the rising limb of the flood, the tagged logs (10 at each place) were put into the river at

Fig. 8 **a** Installation of a radio transmitter in an alder log prepared for the tracking experiment. **b** Searching for the tagged logs after the flood of May 2014 using a TRX 1000WR signal receiver and a Yagi antenna

three locations: (i) at the beginning of the incised river reach, (ii) close to the beginning of the channelized reach, and (iii) 1 km upstream from the beginning of the wide, multi-thread reach. After the flood we systematically searched for the logs on about 50 km of the river length (to the Czorsztyn Reservoir on the Dunajec River—Fig. 1) using a TRX-1000WR signal receiver and a handheld Yagi antenna (Fig. 8b). In total, twenty-four logs were found, eight from each set, such that recovery rate was 80 %. The log found farthest downstream was deposited 7.7 km from the place of its delivery and 19.6 km from the uppermost delivery site. With unknown fate of the lacking logs, only the recovered logs were taken into account in the analysis.

Considerable differences in the length of displacement were found between the logs put into the river at the three locations (Wyżga et al. in press). Logs delivered at the beginning of the incised reach were displaced between 3.7 and 14.6 km, 11.4 km on average. Logs delivered close to the beginning of the channelized reach were displaced by 0.03–9.0 km, with the mean displacement by 6.4 km. Finally, logs delivered 1 km upstream from the beginning of the multi-thread reach travelled from 1.2 to 7.6 km, with an average of 2.6 km. Only one out of the 8 recovered logs delivered to the river at the beginning of the 4.6-km long incised reach was retained in this reach. Out of the 15 logs that entered the 4.6-km long channelized reach, 5 were deposited in it. Finally, 18 logs entered the 4-km long multi-thread reach and 17 were deposited there, mostly in the uppermost part of the reach.

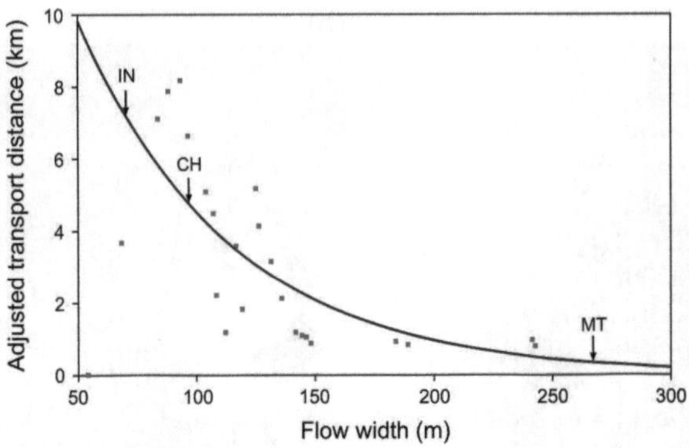

Fig. 9 Relationship between adjusted transport distance of the logs tagged with radio transmitters and average flow width over that distance at peak discharge of the 20-year flood of May 2014. Adjusted transport distance is a product of the distance to a given river cross-section with deposited log and a proportion of logs that were transported on that distance. *Arrows* indicate average flow widths at peak discharge of the flood in the incised (*IN*), channelized (*CH*), and multi-thread (*MT*) reaches of the Czarny Dunajec

The three river reaches differed markedly in the flow width at the flood peak; on average, it amounted to 70 m in the incised reach, 97 m in the channelized reach and 267 m in the multi-thread reach (Fig. 9). Average flow width between a site of log delivery to the river and particular cross-sections where consecutive tagged logs were deposited was found to be a highly significant predictor of the length of transport distance. Figure 9 presents a nonlinear regression model estimated for the so-called adjusted transport distance; this parameter is calculated as a product of the transport distance and a proportion of logs that are transported on that distance. Average flow widths along the whole distance of log travel to particular depositional cross-sections, described by the regression model, can be compared with average flow widths in particular reaches of the Czarny Dunajec (shown by arrows). The comparison indicates that at the flow width of the 20-year flood, typical of the incised reach, logs can be transported long distances along the river, whereas the length of log displacement dramatically diminishes as the flood flow expands to the width typical of the multi-thread reach.

4.4 Numerical Modelling of Large Wood Dynamics in Contrasting River Morphologies

Numerical modelling provides another approach to understanding processes, testing hypotheses and running scenarios, and thus it is a potentially powerful tool for analyzing large wood dynamics. We applied the 2D hydrodynamic model developed by Ruiz-Villanueva et al. (2014a) to two reaches of the Czarny Dunajec with different morphology to analyze various aspects of large wood dynamics in the wide mountain river. The model is based on the finite volume method with a second-order Roe Scheme coupled to a Lagrangian model for the simulation of individual pieces of wood. The incipient motion of logs, considered as cylinders, is determined by the balance of forces acting on the centre of mass of each log. Once the log is put in motion, two possible transport mechanisms are implemented: sliding on the river bed or floating. In all cases, translation and rotation are considered depending on the velocity field. When the piece of wood is sliding on the river bed, friction is the main control factor of movement, and thus its velocity may be very different from the flow velocity. If the log is floating, its velocity is assumed to be the same as the flow velocity, unless turbulence is considered. Turbulent fluctuations of velocity affect wood motion, introducing a random component into the movement of logs. This model fully couples the hydrodynamics with the large wood transport. The presence of wood adds a new shear stress (produced by the drag force of the logs) term in the Saint Venant equations. In addition, the velocity and movement of logs can also be modified when they interact with the river boundaries or between logs themselves (Ruiz-Villanueva et al. 2014a, b, c).

The numerical model needs initial and boundary conditions for wood. The initial position of each log (x, y coordinates of the mass centre and angle with respect to the flow), its length, diameter and wood density for the initial time step should be provided. Moreover, inlet boundary conditions (i.e. logs entering the river reach) can also be assigned to the simulation domain boundaries, specifying a number of wood pieces per minute and their characteristics. Based on detailed knowledge of the fluvial corridor, riparian vegetation, and wood availability, ranges of the main characteristics of logs need to be established: maximum and minimum lengths, diameters and density of wood. Stochastic variations of these parameters together with position and angle are used to characterize the wood entering the domain.

This model was applied to the Czarny Dunajec with the aim to analyze factors controlling wood transport and retention such as wood size, flow conditions and river morphology (Ruiz-Villanueva et al. 2016a, b). Analysis was based on a multi-run scenario simulation in order to include the complexity and stochastic variability of large wood dynamics in the deterministic model. Large wood dynamics was simulated in two river reaches of similar length: single-thread, channelized reach with a relatively narrow and deep channel and multi-thread, unmanaged reach with a wide and shallow channel. Data from the Koniówka stream gauging station was used to characterize the inlet flow and design several flood scenarios, and the available rating curve was used for roughness (Manning's n) calibration. In total, more than 200 scenarios have been designed under steady and unsteady flow conditions, and using different sizes of logs and different wood inputs.

Results showed that a larger proportion of all logs introduced upstream of the study reach is transported downstream the single-thread, channelized reach than downstream the multi-thread reach, and, accordingly, a larger proportion of all introduced logs is deposited in the latter. In the single-thread channel, log length was found to predominantly control large wood transport, whereas in the wider, multi-thread channel, log diameter was more important (Ruiz-Villanueva et al. 2016b). This agrees with the findings of other researchers who observed the same pattern in the field and in flume experiments (Welber et al. 2013; Bertoldi et al. 2014). In each reach, the elevation of deposition of simulated logs above the low-flow water surface was found to significantly differ between various flood magnitudes, generally increasing with increasing flood discharges. At each flood magnitude considered, it was also significantly higher in the single-thread channel than in the multi-thread channel (Fig. 10). The preferential sites for large wood to be deposited at different flood magnitudes were identified by means of depositional probability. Results showed that the preferential sites of wood deposition vary with flood magnitude and that the probability of deposition is significantly controlled by the relative elevation of the different geomorphic units in relation to the water level (Ruiz-Villanueva et al. 2016a).

Simulating unsteady flow conditions, Ruiz-Villanueva et al. (2016c) analyzed the influence of a flood hydrograph (in terms of peak discharge, time to peak, and total flood duration) on the transport of wood. Results revealed a lag between the beginning of a flood and large wood remobilization, which is related to the flood responsible for the initial wood deposition. Furthermore, the peak in large wood transport is generally reached before the flood peak, and wood transport decreases close to or slightly after the hydrograph peak. During the falling flood limb, large wood transport is likely negligible, unless an additional supply of wood is provided to the river as wood has already been subject to the same or larger discharges during the rising limb. This results in a hysteresis between discharge and large wood transport, which was also recorded using video monitoring in the Ain River (MacVicar and Piégay 2012).

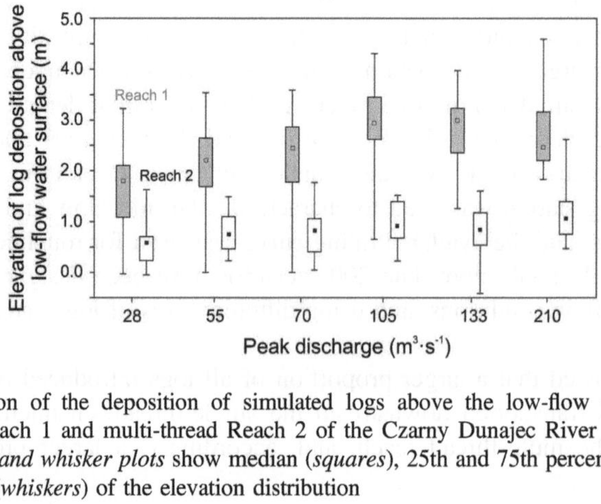

Fig. 10 Elevation of the deposition of simulated logs above the low-flow water surface in single-thread Reach 1 and multi-thread Reach 2 of the Czarny Dunajec River at different peak discharges. *Box and whisker plots* show median (*squares*), 25th and 75th percentiles (*boxes*) and extreme values (*whiskers*) of the elevation distribution

5 Discussion and Conclusions

5.1 *Suitability of the Used Methods for Examination of Large Wood Dynamics in the Watercourses*

Tagging riparian trees along Kamienica Stream and their subsequent monitoring appeared to be highly informative about large wood dynamics in the stream. This approach allowed us to approximately date their delivery to the channel, to infer about its cause, and to determine the lengths of displacement of fallen trees during particular flood events. Tagging trees with metal plates was highly cost-effective; the cost of metal plates was, on average, two orders of magnitude lower than that of radio transmitters. This enabled us to tag a relatively large number of riparian trees, with the total length of the monitored stream reaches of 1.35 km. Moreover, the tags are typified by long functionality; during six years of the monitoring, no tag was lost or destroyed. The lack of major floods precluded verification of the tag effectiveness in case of long-distance displacement of fallen trees. In turn, tagging logs with radio transmitters provided useful information about large wood dynamics in the Czarny Dunajec during a major flood event. With relatively long distances of log displacement in the river during a single flood event and the large river area, gaining the information would not be possible without the use of radio telemetry. However, after the tracking experiment we did not find 6 out of 30 devices and 3 others could not be retrieved as the logs were retained underwater. This means 30 % total loss of the devices and indicates relatively high costs of tracking large wood dynamics in wide rivers.

Large wood inventories in the stream and in the river were associated with similar surveying effort; the number of wood deposits recorded in Kamienica Stream exceeded that in the Czarny Dunajec by 30 %, but with the substantially larger area of the river, both morphometric and wood surveys were more effort- and time-consuming. The inventories in both watercourses revealed different spatial patterns of large wood distribution, with an increase in channel width reflected in decreasing specific wood storage in the stream and increasing in the river. These different patterns of wood distribution were attributed to different mechanisms of large wood retention in watercourses narrower and wider than the height of riparian trees (Wyżga et al. 2015).

Wood inventories provided detailed information about the amounts, location and character of large wood in the river and some relations of these parameters to morphometric river characteristics at discrete time periods. Wood and morphometric surveys were easy to do and inexpensive. However, they recorded a static image of large wood, without information on the time of its recruitment and deposition in the river (for instance, the inventory in 2001 was performed after two flood waves from June and July 2001 with the same peak discharges), whereas the length of wood transport could be only indirectly inferred from the state of wood preservation. As the inventories were time-consuming, their spatial scale was limited, and the obtained results may not be easily extrapolated to other river reaches with a specific combination of morphometric, hydraulic, channel boundary and riparian vegetation conditions.

Observations of the vegetative regeneration of wood were a valuable source of information about the fate of living wood deposited in wider river reaches, indicating that a considerable proportion of such wood becomes rooted to the bar surface and subsequently buried, acting as propagules from which new willow/alder shrubs and trees develop (Mikuś et al. 2013). However, obtaining quantitative information about the amount of deposited wood that becomes permanently immobile as a result of vegetative regeneration would require systematic classification of all wood deposits into dead and living wood (Gurnell et al. 2000) during an inventory performed after a flood and verification of the fate of the living wood (that can die as a result of unfavourable moisture conditions, be remobilized before it is buried, or be eroded together with the underlying bar) in the subsequent years. As vegetative regeneration of living wood typically occurs in the widest reaches of the Czarny Dunajec with the largest amounts of deposited wood, this would be nearly equivalent with repetition of the wood inventory in each of a few subsequent years.

The tracking experiment with logs tagged with radio transmitters indicated that log displacement distance differed between reaches into which they had been delivered and that it depended on the flow width at the flood peak. The use of logs with the same dimensions that were consistently placed in the river at very similar parts of the flood hydrograph (Wyżga et al. in press) was an unquestionable advantage of the experiment; it allowed us to consider differences in the length of log displacement as reflecting different morphological and hydraulic conditions in particular river reaches. Disadvantages comprised (i) relatively high cost of the radio transmitters that limited the number of devices used in the experiment, (ii) limited representativeness of the logs for the range of large wood types recruited to and deposited in the river, and (iii) limited number of places where the logs could be safely delivered to the river at flood conditions (this is why the lowermost delivery site was located 1 km upstream from the beginning of the multi-thread reach).

Numerical modelling of large wood phenomena in channelized and multi-thread reaches of the Czarny Dunajec confirmed different potential of these river morphologies for the transport and deposition of wood previously demonstrated by wood inventories, indicated log length as a major control on wood dynamics in the channelized reach and log diameter in the multi-thread reach, revealed the dependence of the vertical and horizontal location of wood deposition on flood magnitude and a complex temporal pattern of wood remobilization during the passage of flood waves. The definite advantage of this source of information was that it provided the opportunity to examine (i) wood dynamics during floods of various magnitudes, (ii) temporal and spatial complexity of wood mobilization, transport and deposition, and (iii) the influence of changes in wood parameters on wood dynamics. The most important weakness seems related to the limited representation by the simulated logs of the natural complexity of large wood delivered to the Czarny Dunajec, with whole shrubs and trees representing more than one-third of the total number of wood deposits recorded in the river.

5.2 *Large Wood Dynamics in the Mountain Stream and the Mountain River*

Based on the rate of delivery of riparian trees to Kamienica Stream during six years of monitoring, the turnover period of riparian trees is estimated at 80 years, provided that the same rate is maintained in the long term. However, the monitoring period lacked major floods such as the one that occurred in 1997 when plenty of large wood was recruited to the stream from cutbanks and valley-side landslides (Kaczka 1999). The occurrence of such a flood in the monitoring period would result in even faster turnover of riparian trees. At the same time, the riparian area supports trees with ages up to ~160 years. This indicates that bark beetle infestation considerably increased tree delivery to the stream and accelerated the turnover of the riparian forest.

The obtained evidence indicates that mobility of large wood is very low in the second-order reach of the stream and greater but still limited in the fourth-order reach. In the former, the very low mobility is indicated by the predominance of near-perpendicular orientation of wood pieces, the occurrence of numerous wood dams (Kaczka 2003, 2009), and the prevalence of highly decayed wood. In the latter, considerable mobility of large wood must have typified the flood of 1997 as evidenced by the failure of numerous wood dams and the occurrence of longitudinal wood orientation (Kaczka 1999). Kamienica Stream is ungauged but anecdotal evidence from local farmers indicated the flood as occurring once in a few tens of years. Since that event the mobility of wood in the fourth-order reach of the stream must have been substantially lower as indicated by the formation of the second mode of wood orientation with near-perpendicular alignment of pieces, considerably increased proportion of the wood in advanced state of decay, and the maximum displacement by 100 m of tagged fallen trees during the six years of monitoring.

By contrast, the fifth-order Czarny Dunajec is typified by relatively large, although spatially highly diverse wood mobility. The tracking experiment and the numerical modelling indicated high potential for long-distance transport of large wood in single-thread river reaches, and this was also implicated by the high degree of mechanical disintegration of the wood deposited by floods immediately below the narrow, channelized reach. The high potential for wood transport is typical of rivers with channel width equal to or somewhat larger than the riparian tree height, in which flows are fast and deep and a general lack of major roughness elements such as bars or islands prevents wood retention (Wyżga et al. 2015). It also agrees with the observation by Seo and Nakamura (2009) that fluvial export of wood is maximised in the rivers with channel width somewhat larger than wood piece length and with relatively high stream power.

In turn, the wide, multi-thread river reach exhibits much less potential for the transport of wood and much higher potential for its retention. This reflects lower flow depths and unit stream power facilitating wood deposition on gravel bars and on the floodplain, and the abundance of retention features such as islands, against which wood can be braced. Rapid flow expansion immediately downstream of the end of the narrow, channelized reach causes a large proportion of wood pieces transported from the upstream reaches to be deposited on short distances in the wide, multi-thread reach. Much of the remaining wood delivered from the upstream reaches is trapped in wood jams that form at the head of islands (Wyżga and Zawiejska 2010) and along concave banks of channel bends (Mikuś et al. 2016). Vigorous bank erosion occurring in the multi-thread reach during floods results in the delivery of substantial amounts of large wood to the river; however, it mostly consists of whole shrubs and trees with complex, three-dimensional structure, many of which are stranded on the nearest channel bars (Wyżga and Zawiejska 2005; cf. Welber et al. 2013). Willows constitute a large proportion of shrubs and trees deposited in the reach and often root to the bar surface, which limits the potential for their subsequent entrainment even if hydraulic conditions in the reach facilitate wood remobilization (Ruiz-Villanueva et al. 2016c).

5.3 Implication of Large Wood Dynamics in the Mountain Watercourses of Different Size to Flood Hazard

Bark beetle infestation that resulted in the widespread dieback of spruce forest in the Kamienica valley must have increased the rate of tree delivery to the stream. Despite this greater recruitment of large wood to the channel, little potential for wood export to the downstream, developed valley reaches exists, except during large, rare floods. However, a large proportion of wood pieces stored in the stream are in the advanced state of decay and will be subjected to rapid mechanical disintegration if entrained by a large flood. This allows us to conclude that large wood retained in the studied reach of the Kamienica valley within the national park does not represent a significant flood hazard to downstream valley reaches. However, the location of Kamienica within the national park precludes a simple generalization of the above conclusion for other mountain streams in the region. On the one hand, removal of wood from streams that is a common practice in the Polish Carpathians greatly reduces the amounts of wood that might be exported to downstream valley reaches during floods. On the other hand, wood dams—abundant in Kamienica channel—facilitate dissipation of stream energy and trap a proportion of wood pieces mobilized by floods; a general lack of such dams in the mountain streams outside nature-protected areas will thus facilitate flushing out of wood recruited to these streams during floods.

The reaches of the Czarny Dunajec and other mountain rivers narrowed as a result of channelization or channel incision provide conditions for efficient transfer of large wood over relatively long distances. As a result, during floods channel sections located downstream of such reaches may be supplied with substantial amounts of wood from distant locations irrespective of protection measures (e.g. bank reinforcement, riparian forest clearing) undertaken in the vicinity of vulnerable sites such as bridges or urban reaches. In turn, wide, multi-thread reaches with their high potential to efficiently retain wood delivered from the upstream operate as natural wood traps that can considerably and efficiently limit further transfer of wood to vulnerable sites/reaches (Wyżga and Zawiejska 2010). Preservation or restoration of such reaches is thus crucial not only because of their high environmental value but also for the enhanced management of flood hazard related to presence of large wood in mountain rivers (Mikuś et al. 2016).

Acknowledgments This work was supported by the project FLORIST (Flood risk on the northern foothills of the Tatra Mountains) supported by a grant from the Swiss Government through the Swiss Contribution to the enlarged European Union (PSPB No. 153/2010). We thank Dieter Rickenmann for his critical comments on the manuscript.

References

Benda LE, Sias JC (2003) A quantitative framework for evaluating the mass balance of in-stream organic debris. Forest Ecol Manage 172:1–16

Bertoldi W, Welber M, Mao L, Zanella S, Comiti F (2014) A flume experiment on wood storage and remobilization in braided river systems. Earth Surf Proc Landforms 39:804–813

Braudrick CA, Grant GE (2001) Transport and deposition of large woody debris in streams: a flume experiment. Geomorphology 41:263–283

Comiti F, D'Agostino V, Moser M, Lenzi MA, Bettella F, Dell'Agnese A, Rigon E, Gius S, Mazzorana B (2012) Preventing wood-related hazards in mountain basins: from wood load estimation to designing retention structures. In: 12th congress Interpraevent 2012—Grenoble/France, pp 651–662

Curran JH, Wohl EE (2003) Large woody debris and flow resistance in step-pool channels, Cascade Range, Washington. Geomorphology 51:141–157

Davidson SL, Eaton BC (2013) Modeling channel morphodynamic response to variations in large wood, Implications for stream rehabilitation in degraded watersheds. Geomorphology 202:59–73

Faustini JM, Jones JA (2003) Influence of large woody debris on channel morphology and dynamics in steep, boulder-rich mountain streams, western Cascades, Oregon. Geomorphology 51:187–205

Gippel C (1995) Environmental hydraulics of large woody debris in streams and rivers. J Environ Eng 121:388–395

Gurnell AM (2013) Wood in fluvial systems. In: Shroder J, Wohl E (eds) Treatise on geomorphology. Vol 9, Treatise on fluvial geomorphology. Academic Press, San Diego, pp 163–188

Gurnell A, Petts G (2006) Trees as riparian engineers: the Tagliamento River, Italy. Earth Surf Proc Landforms 31:1558–1574

Gurnell AM, Petts GE, Hannah DM, Smith BPG, Edwards PJ, Kollmann J, Ward JV, Tockner K (2000) Wood storage within the active zone of a large European gravel-bed river. Geomorphology 34:55–72

Gurnell AM, Piégay H, Swanson FJ, Gregory SV (2002) Large wood and fluvial processes. Freshwater Biol 47:601–619

Kaczka RJ (1999) The role of coarse woody debris in fluvial processes during the flood of the July 1997, Kamienica Łącka Valley, Beskidy Mountains, Poland. Stud Geomorph Carp-Balcan 33:117–130

Kaczka R (2003) The coarse woody debris dams in mountain streams of central Europe, structure and distribution. Stud Geomorph Carp-Balcan 37:111–127

Kaczka RJ (2009) Dynamics of large woody debris and wood dams in mountain Kamienica Stream, Polish Carpathians. In: Kaczka RJ, Malik I, Owczarek P, Gärtner H, Helle G, Heinrich I (eds) Tree rings in archaeology, climatology and ecology 7. Association for Tree Ring Research, Potsdam, Germany, pp 171–175

Kaczka RJ, Wyżga B, Zawiejska J (2008) Tree-ring study of the island formation and riparian forest along a gravel-bed river in the Polish Carpathians. In: Elferts D, Brumelis G, Gartner H, Helle G, Schleser G (eds) Tree rings in archaeology, climatology and ecology 6. Association for Tree Ring Research, Potsdam, Germany, pp 106–110

Kramer N, Wohl E (2014) Estimating fluvial wood discharge using time-lapse photography with varying sampling intervals. Earth Surf Proc Landforms 39:844–852

Kundzewicz ZW, Stoffel M, Kaczka RJ, Wyżga B, Niedźwiedź T, Pińskwar I, Ruiz-Villanueva V, Łupikasza E, Czajka B, Ballesteros-Canovas J, Małarzewski Ł, Choryński A, Janecka K, Mikuś P (2014) Floods at the northern foothills of the Tatra Mountains—a Polish-Swiss research project. Acta Geophys 62:620–641

Lassettre NS, Kondolf GM (2012) Large woody debris in urban stream channels, redefining the problem. River Res Applic 28:1477–1487

Latterell JJ, Naiman RJ (2007) Sources and dynamics of large logs in a temperate forest river. Ecol Applic 17:1127–1141

Le Lay YF, Moulin B, Piégay H (2013) Wood entrance, transfer, deposition and effects on fluvial forms and processes: problem statements and challenging issues. In: Shroder J, Butler D, Hupp CR (eds) Treatise on geomorphology. Vol 12, Ecogeomorphology. Academic Press, San Diego, pp 20–36

Lienkaemper GW, Swanson FJ (1987) Dynamics of large woody debris in streams in old-growth Douglas-fir forests. Can J For Res 17:150–156

Lucía A, Comiti F, Borga M, Cavalli M, Marchi L (2015) Dynamics of large wood during a flash flood in two mountain catchments. Nat Hazards Earth Syst Sci 3:1643–1680

MacVicar BJ, Piégay H (2012) Implementation and validation of video monitoring for wood budgeting in a wandering piedmont river, the Ain River (France). Earth Surf Proc Landforms 37:1272–1289

MacVicar BJ, Piégay H, Henderson A, Comiti F, Oberlin C, Pecorari E (2009) Quantifying the temporal dynamics of wood in large rivers: field trials of wood surveying, dating, tracking, and monitoring techniques. Earth Surf Proc Landforms 34:2031–2046

Mazzorana B, Fuchs S (2010) Fuzzy formative scenario analysis for woody material transport related risks in mountain torrents. Environ Model Softw 25:1208–1224

Mikuś P, Wyżga B, Kaczka RJ, Walusiak E, Zawiejska J (2013) Islands in a European mountain river: linkages with large wood deposition, flood flows and plant diversity. Geomorphology 202:115–127

Mikuś P, Wyżga B, Radecki-Pawlik A, Zawiejska J, Amirowicz A, Oględcki P (2016) Environment-friendly reduction of flood risk and infrastructure damage in a mountain river: case study of the Czarny Dunajec. Geomorphology 272C. doi:10.1016/j.geomorph.2015.11.003

Moggridge HL, Gurnell AM (2009) Controls on the sexual and asexual regeneration of Salicaceae along a highly dynamic, braided river system. Aquat Sci 71:305–317

Moulin B, Piégay H (2004) Characteristics and temporal variability of large woody debris trapped in a reservoir on the River Rhone (Rhone): implications for river basin management. River Res Applic 20:79–97

Ravazzolo D, Mao L, Picco L, Lenzi MA (2015) Tracking log displacement during floods in the Tagliamento River using RFID and GPS tracker devices. Geomorphology 228:226–233

Ruiz-Villanueva V, Bodoque JM, Díez-Herrero A, Eguibar MA, Pardo-Igúzquiza E (2013) Reconstruction of a flash flood with large wood transport and its influence on hazard patterns in an ungauged mountain basin. Hydrol Proc 27:3424–3437

Ruiz-Villanueva V, Bladé E, Sánchez-Juny M, Marti-Cardona B, Díez-Herrero A, Bodoque JM (2014a) Two-dimensional numerical modelling of wood transport. J Hydroinform 16:1077–1096

Ruiz-Villanueva V, Bodoque JM, Díez-Herrero A, Bladé E (2014b) Large-wood transport as significant influence on flood risk in a mountain village. Nat Hazards 74:967–987

Ruiz-Villanueva V, Castellet EB, Díez-Herrero A, Bodoque JM, Sánchez-Juny M (2014c) Two-dimensional modelling of large wood transport during flash floods. Earth Surf Proc Landforms 39:438–449

Ruiz-Villanueva V, Wyżga B, Hajdukiewicz H, Stoffel M (2016a) Exploring large wood retention and deposition in contrasting river morphologies linking numerical modelling and field observations. Earth Surf Proc Landforms 41:446–459

Ruiz-Villanueva V, Wyżga B, Zawiejska J, Hajdukiewicz M, Stoffel M (2016b) Factors controlling large wood transport in a mountain river. Geomorphology 272C. doi:10.1016/j.geomorph.2015.04.004

Ruiz-Villanueva V, Wyżga B, Mikuś P, Hajdukiewicz H, Stoffel M (2016c) The role of flood hydrograph in the remobilization of large wood in a wide mountain river. J Hydrol. doi:10.1016/j.jhydrol.2016.02.060

Schenk ER, Moulin B, Hupp CR, Richter JM (2014) Large wood budget and transport dynamics on a large river using radio telemetry. Earth Surf Proc Landforms 39:487–498

Seo JI, Nakamura F (2009) Scale-dependent controls upon the fluvial export of large wood from river catchments. Earth Surf. Proc Landforms 34:786–800

Shields FD, Gippel CJ (1995) Prediction of effects of woody debris removal on flow resistance. J Hydraul Eng 121:341–354

Thévenet A, Citterio A, Piégay H (1998) A new methodology for the assessment of large woody debris accumulations on highly modified rivers (example of two French piedmont rivers). Regul Rivers Res Manag 14:467–483

Van der Nat D, Tockner K, Edwards PJ, Ward JV (2003) Large wood dynamics of complex Alpine river flood plains. J North Am Benthol Soc 22:35–50

Welber M, Bertoldi W, Tubino M (2013) Wood dispersal in braided streams: Results from physical modeling. Water Resour Res 49:7388–7400

Wohl E (2013) Floodplains and wood. Earth Sci Rev 123:194–212

Wohl E, Goode JR (2008) Wood dynamics in headwater streams of the Colorado Rocky Mountains. Water Resour Res 44:W09429

Wohl E, Bledsoe PB, Fausch KD, Kramer K, Bestgen KR, Gooseff MN (2016) Management of large wood in streams: an overview and proposed framework for hazard evaluation. J Am Water Res Assoc 52:315–335

Wyżga B, Zawiejska J (2005) Wood storage in a wide mountain river: case study of the Czarny Dunajec, Polish Carpathians. Earth Surf Proc Landforms 30:1475–1494

Wyżga B, Zawiejska J (2010) Large wood storage in channelized and unmanaged sections of the Czarny Dunajec River, Polish Carpathans: implications for the restoration of mountain rivers. Folia Geogr Ser Geogr-Phys 41:5–34

Wyżga B, Zawiejska J, Radecki-Pawlik A, Hajdukiewicz H (2012) Environmental change, hydromorphological reference conditions and the restoration of Polish Carpathian rivers. Earth Surf Proc Landforms 37:1213–1226

Wyżga B, Zawiejska J, Mikuś P, Kaczka RJ (2015) Contrasting patterns of wood storage in mountain watercourses narrower and wider than the height of riparian trees. Geomorphology 228:275–285

Wyżga B, Mikuś P, Zawiejska J, Ruiz-Villanueva V, Kaczka RJ, Czech W (in press) Log transport and deposition in incised, channelized and multi-thread reaches of a wide mountain river: tracking experiment during a 20-year flood. Geomorphology, submitted

Zawiejska J, Wyżga B (2010) Twentieth-century channel change on the Dunajec River, southern Poland: patterns, causes and controls. Geomorphology 117:234–246

Large Wood Transport, Deposition and Remobilization during Floods in the Czarny Dunajec River: Outcomes from Numerical Modelling

Virginia Ruiz-Villanueva, Bartłomiej Wyżga, Joanna Zawiejska, Paweł Mikuś, Hanna Hajdukiewicz, Maciej Hajdukiewicz and Markus Stoffel

Abstract The knowledge of large wood (LW) dynamics regarding factors controlling wood transport, preferential depositional sites and the timing and duration of wood flux, is crucial for the maintenance of a good ecological status of rivers and for the management of wood-related potential hazards. Besides field surveys, tracking experiments or physical modelling, numerical models represent an alternative and complementary approach to explore LW dynamics, to test hypotheses

V. Ruiz-Villanueva (✉) · M. Stoffel
Dendrolab.ch, Institute of Geological Sciences, University of Bern,
Baltzerstrasse 1+3, 3012 Bern, Switzerland
e-mail: virginia.ruiz@dendrolab.ch

M. Stoffel
e-mail: markus.stoffel@dendrolab.ch

B. Wyżga · P. Mikuś · H. Hajdukiewicz
Institute of Nature Conservation, Polish Academy of Sciences,
al. Mickiewicza 33, 31-120 Kraków, Poland
e-mail: wyzga@iop.krakow.pl

P. Mikuś
e-mail: mikus@iop.krakow.pl

H. Hajdukiewicz
e-mail: hanahaj@gmail.com

B. Wyżga · P. Mikuś
Faculty of Earth Sciences, University of Silesia, ul. Będzińska 60, 41-200 Sosnowiec, Poland

J. Zawiejska
Institute of Geography, Pedagogical University of Cracow,
ul. Podchorążych 2, 30-084 Kraków, Poland
e-mail: zawiejska.joanna@gmail.com

M. Hajdukiewicz
Department of Environmental Engineering, Geomatics and Energetics, Kielce University of Technology, Al. Tysiąclecia Państwa Polskiego 7, 25-314 Kielce, Poland
e-mail: mhajdukiewicz@gmail.com

Z.W. Kundzewicz et al. (eds.), *Flood Risk in the Upper Vistula Basin*,
GeoPlanet: Earth and Planetary Sciences, DOI 10.1007/978-3-319-41923-7_6

and to run scenarios. We used a 2D numerical model which simulates the transport of large wood together with flow dynamics. The model is able to predict and simulate wood transport and deposition, and reproduces interactions between wood, channel bed, floodplain surface and infrastructures. We applied this model combined with direct field observations to explore main factors controlling large wood dynamics in the Czarny Dunajec River in Poland. We simulated different types of logs under different flood magnitude scenarios (steady and unsteady flow conditions) to analyse wood transport, deposition and remobilization in two contrasting river morphologies. We summarized in this chapter the main outcomes from this work. Results illustrate that a wide range of quantitative information about LW transport and deposition can be obtained from the use of numerical modelling together with the proper assessment of inlet and boundary conditions and validation based on field data.

Keywords In-stream wood · Woody debris · Large wood dynamics · Iber-wood 2D model · Poland

1 Introduction

In forested mountain streams, the interaction between riparian vegetation and geomorphic processes is amplified by high stream power, high sediment transport rates and abundant wood delivery to the channels (Badoux et al. 2015; Rickenmann et al. 2015). Besides other parameters influencing wood distribution along the channels (e.g., recruitment processes, forest stand and age or forest and river management), geomorphology is a major control on the transport and distribution of large wood in rivers (Wohl and Cadol 2011). In addition, wood dynamics is also controlled by flow patterns (Gurnell 2012). Flood frequency and magnitude are significant factors influencing the distribution of large wood in rivers (Moulin et al. 2011). However, the knowledge about transport, deposition and remobilization processes of large wood within fluvial corridors is still limited because of a lack of accurate field data (MacVicar et al. 2009; MacVicar and Piégay 2012). In spite of that, researchers have tried to understand the hydraulic conditions at which wood movement occurs, proposing empirical and conceptual models relating river morphology and hydraulic parameters to the frequency and size of wood (Benda and Sias 2001), examining the influence of wood on flow resistance (Wilcox and Wohl 2006; Allen and Smith 2012), or running physical modelling experiments (Braudrick and Grant 2000, 2001; Bocchiola et al. 2002; Haga et al. 2002; Schmocker and Hager 2011; Welber et al. 2013; Bertoldi et al. 2014). These studies have demonstrated that wood transport may be more complicated than sediment transport (Wohl and Cadol 2011; MacVicar and Piégay 2012), and therefore more knowledge is still needed.

A variety of techniques have been employed to measure wood retention: detailed mapping of the locations of wood pieces (Elosegi et al. 1999), remote sensing

(Lassettre et al. 2008; Bertoldi et al. 2013), wood tagging and tracking (MacVicar et al. 2009) and physical experiments (Bertoldi et al. 2014). This chapter shows another approach to analysing wood dynamics by combining numerical modelling and field measurements. A wide range of quantitative information about wood transport and deposition can be obtained from the use of numerical modelling together with the proper assessment of boundary conditions and validation based on field data (Ruiz-Villanueva et al. 2016a, b, c). In this work we combine the numerical simulation of wood transport and deposition with field observations to analyse wood dynamics along the mountainous Czarny Dunajec River in southern Poland. This river has been studied in detail in regard to wood storage, which gives an excellent opportunity to verify and compare field observations and modelling results. As numerical modelling was run here in a multi-run mode, the results can be analysed in a probabilistic manner. This chapter summarizes main outcomes from the combination of numerical modelling with information obtained from field surveys and tracking of wood in the river. The goals were to determine main factors controlling wood transport, depositional patterns of large wood that may result from the occurrence of floods of different magnitude, and the impacts of flood sequencing as well as to analyse relationships with flow regime and discharge.

2 Study Site: The Czarny Dunajec River

The Czarny Dunajec (Fig. 1) is a gravel-bed river mostly draining the Inner Western Carpathians; it originates at about 1500 m a.s.l. in the high-mountain Tatra massif, with the highest peak in the catchment at 2176 m a.s.l. Hydrological regime of the river is determined by the high-mountain part of the catchment and is typified by low winter flows and floods occurring between May and August as a result of prolonged frontal rains, sometimes superimposed on snow-melt runoff. Mean annual discharge of the river is 4.4 $m^3\ s^{-1}$ at Koniówka, in the middle course of the river (catchment area of 134 km^2) and 8.8 $m^3\ s^{-1}$ at Nowy Targ (432 km^2), close to the confluence of the Czarny Dunajec and the Biały Dunajec rivers.

As a result of spatially variable human impacts over the past few decades (Zawiejska and Wyżga 2010), the river varies highly in width and morphology. This allows distinction of two different reaches: a single-thread, partially channelized reach 1 and an unmanaged, multi-thread reach 2 (Fig. 1). The total length of the study reaches is 5.5 km. In reach 1 the river has relatively small, uniform width, and a few drop structures reduce slope locally. One or both channel banks are reinforced with gabions or rip rap. In reach 2 the width of the active river zone amounts to 116 m on average, but it varies considerably from 60 m at the upstream end of the reach, where islands are small and scarce, to about 180 m near its downstream end where islands consequently become more important (Mikuś et al. 2013). The substantial differences in river width between the reaches are reflected in markedly different unit stream power of flood flows and in higher average flow

Fig. 1 Location of the Czarny Dunajec River (*in red*) and the location and appearance of the study reaches. Flow direction in the photos is towards the camera for reach 1 and out of the camera for reach 2

depth in the narrower reach 1. Moreover, the differences in channel management and river morphology underlie differences in the intensity of LW recruitment and the availability of LW retention sites between the reaches (Wyżga and Zawiejska 2005, 2010).

The river banks in both reaches as well as the forested islands in reach 2 are overgrown with forest stands composed of alder and willow species, with predominating young, shrubby forms of *Alnus incana*, *Salix eleagnos, S. purpurea* and *S. fragilis*, less frequent stands of older *A. incana* trees and occasional *S. alba* trees (Mikuś et al. 2013). With riparian tree height reaching 18 m, the study reaches represent large channels with respect to in-stream wood (Wyżga et al. 2015; cf. Gurnell et al. 2002; Wohl 2013).

3 Methods

The wood transport module of 2D hydrodynamic model *Iber* has been tested in real rivers in previous works (Ruiz-Villanueva et al. 2014a, b), and thus only a brief description is provided here.

Wood incipient motion is considered performing a balance of forces (i.e. the gravitational force acting in a downstream direction; the friction force in the direction opposite to flow; and the drag force, also acting in the flow direction) acting on each single piece of wood (assuming logs as cylinders). Some of the parameters involved in the governing equations are: wood density, angle of the log relative to flow, log length, log diameter, friction coefficient between the wood and the river bed and the drag coefficient of the wood in water. The method couples the flow variables calculated with the hydrodynamic module to update the position and velocity of the wood logs at every time step. The movement of logs includes two possible transport mechanisms (i.e. floating or sliding) based on wood density; and both translation and rotation, the latter reflecting the fact that one end of a piece of wood is moving faster than the other end (based on flow velocity field). Interactions between logs and the channel configuration and among logs themselves are also taken into account in the model. The influence of wood on hydrodynamics can be, in general terms, expressed as a reduction of the average velocity and local elevation of the water surface profile (Gippel 1995); it is solved in the model similarly to the effect of roughness—by including an additional shear stress term in the 2D Saint Venant equations. Therefore, flow conditions exert an influence on the logs but also the presence of logs affects the flow. In order to incorporate wood transport into the model, initial wood and boundary conditions need to be established.

The model works with a non-structured mesh of elements which may have 3 or 4 sides. To obtain this non-structured mesh, the detailed geometry of the entire reach was produced with available digital elevation models (DEM) and a topographical survey aimed at improving the DEM in those sections where its accuracy was insufficient (i.e. critical sections such as bridges, dikes or bends). As a result, we obtained detailed (1 m spatial accuracy) geometry of the studied reaches.

Data from the Koniówka stream gauging station was used to characterise the inlet flow. This station is located 6.2 km upstream of the study reaches. Data was used first for the calculation of flood discharges of a given probability/recurrence interval (for running different inlet discharge scenarios) before the available rating curve was used for roughness (Manning's n) calibration. Roughness coefficient was obtained from the delineation, both in the channel and the flooding areas, of homogeneous land units in terms of their roughness (roughness homogeneous units; RHU) and by using in situ measurements of bed material size in selected channel transects (Wyżga et al. 2012, Zawiejska et al. 2015). Each RHU delimited in the field was digitized in ArcGIS and assigned a possible range of roughness values, following the criteria of Chow (1959) and applying different empirical equations (Meyer-Peter and Müller 1948; Bray 1979; Strickler 1923) in transects. Different discharge ranges were run to calibrate the obtained values of Manning roughness coefficient for high and low flows and to estimate assumed errors.

Wood characteristics used in the simulations were based on the riparian vegetation in the Czarny Dunajec (Table 1). In the model runs, we varied log length (L_w), log diameter (D_w) and wood density (ρ_w). Wood density was assigned between 0.4 and 0.7 g cm^{-3}, except for type 1. In order to simulate the lack of buoyancy for type 1 of wood, we assumed very high wood density (0.85–0.95 g cm^{-3}). In addition, a sensitivity analysis of this factor was also carried out (Ruiz-Villanueva et al. 2016c).

Significant floods occurred recently in the Czarny Dunajec River, namely in 2001, 2010, and 2014. Field observations regarding wood transport and deposition were made after these floods, with published results for the flood of 2001 (Wyżga and Zawiejska 2005, 2010). The data obtained during these post-flood surveys allow validation, interpretation and discussion of model results. In addition, during the most recent flood in May 2014 (peak discharge of 130 m^3 s^{-1} with a 20-year recurrence interval at the Koniówka gauging station), 30 logs tagged with radio transmitters (length: 3 m; diameter: ca. 20 cm) were placed into the river just before the flood peak. Tracking of these logs allowed further analysis of wood transport during this relatively high-magnitude flood. The flood magnitude was high enough for the flow to overtop the reach 1, and activate all low-flow channels and inundate gravel bars and islands in reach 2.

The analysis of the factors controlling wood transport was based on the transport ratio. The transport ratio, Tr, is defined as the ratio between the outlet and inlet number of logs (Tr = pieces transported downstream the studied reach/total inlet logs) and it is inverse to the wood retention capacity: Rc = deposited logs/inlet logs; ($Tr = 1 - Rc$).

4 Results and Discussion

4.1 Large Wood Transport

One of the first aspects we explored was the dependence of wood transport on log volume. We combined diameters and lengths (always within a reliable range based on vegetation and wood in the Czarny Dunajec) resulting in different piece volumes, and simulated them under different flow conditions (different steady discharges). We observed that the number of pieces in transport (defined here by the transport ratio) strongly decreased with increasing piece volume (Fig. 2a, b). However, the scatter of points in Fig. 2 shows that this relationship is not linear but can be better explained by a power function. A significant relationship (p-value < 0.01) was identified between the quantities of transported wood and the volume of the pieces in motion. The scatter in the diagrams is because different discharges resulted in different transport ratios for the same piece volume. According to our results, the scatter is higher in reach 1, especially for larger piece volumes (Fig. 2a). In addition, we

Table 1 Wood characteristics used in the simulations and based on the riparian vegetation in the Czarny Dunajec

Type of simulated logs	Length (m)	Diameter (m)	Wood density (gr cm^{-3})
1. Large alders (*Alnus incana*) and mature willows (*Salix eleagnos*)	10–18	0.3–0.8	0.85–0.95
2. Very large willows (*Salix fragilis* and *S. alba*)	10–15	0.15–0.3	0.4–0.7
3. Young willows (*S. purpurea* and *S. eleagnos*) and alder (*A. incana*)	3–10	0.1–0.2	
4. Branches and small pieces	1–3	0.05–0.1	

observed that the two reaches also differ in the mean and maximum transport ratios. The mean transport ratio is lower in reach 2 (the wide, multi-thread channel; mean transport ratio: 0.26, SD: 0.16) than in reach 1 (the single-thread channel; mean transport ratio: 0.43, SD: 0.20). The maximum transport ratio in reach 2 is 0.57, while for reach 1 the highest value is 1, meaning that all input logs are transported downstream the reach. These values are in the range observed by other researchers. Cadol and Wohl (2010) reported mobility rates ranging between 0.8 and 0.59 in tropical headwater streams. Wohl and Goode (2008) found average mobility ranging from 0.16 to 0.23 in streams in the Rocky Mountains, while Schenk et al. (2014) reported mobility of 0.41 in a large low-gradient river. Recently Iroumé et al. (2015) reported ratios of wood mobility up to 0.28 in headwater streams in the Andes.

Moreover, the lower transport capacity of reach 2 is indicated by the fact that pieces larger than 3 m^3 in volume are not mobilized even under very high-flow conditions, whereas in reach 1 very large pieces (>6 m^3) are still in transport (Fig. 2). The results clearly show that logs of similar dimensions behave differently in both reaches as a consequence of the differences in channel geometry and hydrodynamics (flow energy and therefore transport capacity). This was confirmed during field surveys when a majority of the large wood deposited immediately downstream of the single-thread channel was observed to be highly disintegrated and abraded, indicating that the wood pieces were transported long distances (Wyżga and Zawiejska 2005).

Differences in the transport ratios show differences in the transport capacity between river reaches with different geomorphic configurations. Braided or wide multi-thread channels (reach 2) show a higher retention capacity than narrower single-thread channels (reach 1), as observed and reported by other researchers in similar fluvial environments. In the Tagliamento River in Italy, wood storage was observed to differ significantly between different geomorphic configurations (island-braided and bar-braided reaches; van der Nat et al. 2003; Bertoldi et al. 2013), whereas in the Piave River (Italy), Pecorari (2008) reported higher storage in braided as compared to wandering reaches. Also during flume experiments the crucial role of local-scale morphology in wood dispersal was observed (Welber et al. 2013).

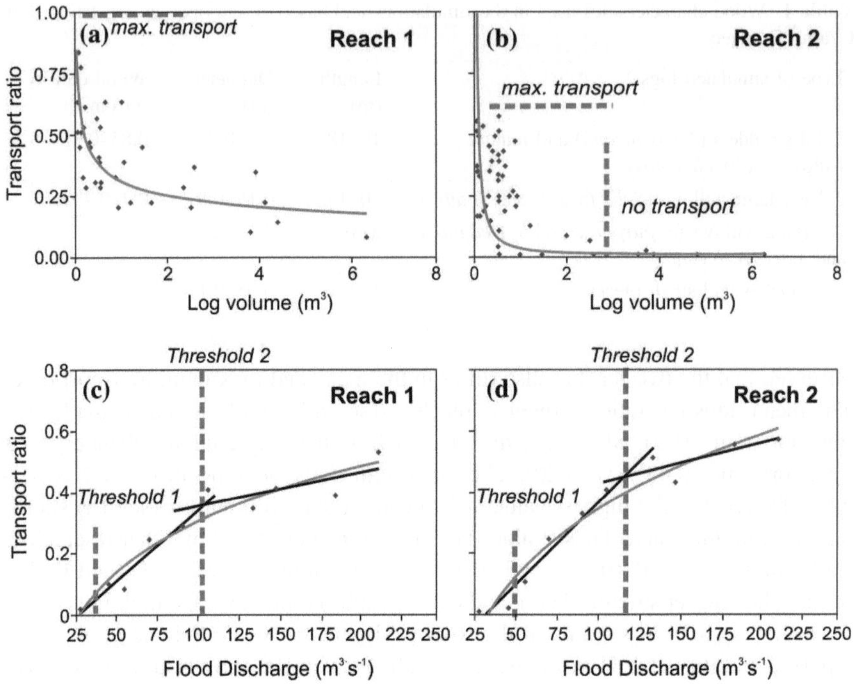

Fig. 2 Scatter plots and estimated regression relationships between large-wood transport ratio and piece volume for (**a**) the single-thread reach 1 and (**b**) the multi-thread reach 2, and between wood transport ratio and simulated flood discharges in (**c**) the single-thread reach 1 and (**d**) the multi-thread reach 2. Wood piece diameter is 0.23 m; wood piece length 12.5 m; wood density 0.56 g cm^{-3}

Besides the type of wood and river morphology, wood transport is controlled by hydrodynamics. We observed variation in wood transport ratios for the same type of logs transported under different flood magnitudes (Fig. 2c, d). The threshold discharge below which wood transport is negligible is different in each reach. In reach 1, discharges ≥ 28 m^3 s^{-1} (very frequent floods) have the capacity to transport wood, while in the multi-thread reach 2 discharges above 45 m^3 s^{-1} are needed for wood transport. In the narrow reach 1, a flow of 28 m^3 s^{-1} has enough energy and water depth to mobilize wood, whereas in the much wider reach 2 the energy and depth of the same flow are not sufficient for wood entrainment. Besides this, transport ratio increases with discharge in both reaches until it reaches an upper threshold, and then decreases and/or increases much more slowly (Fig. 2). This upper threshold is also different for each reach because is related to the bankfull discharge. For the single-thread reach 1, wood transport ratio increases rapidly with discharge until the bankfull stage is reached. At floods of higher magnitude, water begins to inundate the floodplain and a proportion of wood pieces are directed onto the channel banks where flow is shallow and slow, so that most logs will be deposited. In reach 2, once all channels are flooded, the flow over the bars and floodplains will become shallow and slow, so that wood entering these areas cannot be transported and is deposited. However, a higher discharge is required for such situation to occur in reach 2. Similar nonlinear relationships between wood transport and discharge were observed in other rivers, such as the Erlenbach and the Ain, after floods of different magnitude (MacVicar and Piégay 2012; Turowski et al. 2013).

Previous studies demonstrated that the mobility of large wood is a function of the relation between the length of pieces and channel width (Gurnell 2003). Ratios of average piece size to channel dimensions (i.e. the ratios of average piece length to channel width and of average piece diameter to flow depth), may be used to characterise the likelihood of wood mobility (Lienkaemper and Swanson 1987; Braudrick and Grant 2001). Differences in geomorphology between the studied reaches translate to differences in the hydrodynamic context, such as deeper flow, higher velocity, and higher stream power in reach 1. These differences allow the transport of thicker pieces in reach 1, whereas in reach 2 logs with a diameter similar to water depth are not moved at all (Fig. 3). Figure 3 shows that in reach 2 the slope of the linear relationship between the transport ratio and log diameter is steeper and, therefore, log diameter plays a more important role in wood transport than in reach 1 (Fig. 3b, d).

In reach 1, very short pieces (1 m long) are easily transported by the simulated flood (105 $m^3\ s^{-1}$). However, for longer pieces (up to 17 m) the transport ratio decreases significantly (Fig. 3a). This relationship is not the same for reach 2 (Fig. 3c) where longer pieces (between 8 and 16 m) are more readily transported than shorter pieces (up to 8 m), whereas for very long logs (from 16 up to 22 m) the transport ratio is reduced. Longer wood pieces are generally less likely to be deposited in this reach, although a more detailed inspection of results for this scenario reveals a more complicated pattern. Shorter pieces (up to 8 m) are more easily deposited than longer pieces (between 8 and 16 m) as the flow readily introduces shorter pieces to shallow channel areas, whereas longer pieces, with greater momentum, tend to be transported along the thalweg. In turn, very long logs (from 16 m up to 22 m) are also easily deposited in reach 2 as their length facilitates their anchoring on the margins of particular braids.

Flume experiments also suggested that for braided rivers the strongest control on piece stability is wood diameter (Braudrick and Grant 2000; Welber et al. 2013). As we observed in our study and described by Ruiz-Villanueva et al. (2016c), the main factor controlling wood transport in the single-thread channel is wood piece length (Fig. 3a). On the contrary, in the multi-thread reach the main factor controlling wood transport is piece diameter (Fig. 3d).

4.2 Large Wood Deposition

Geomorphic units which are more likely to retain LW within rivers have been previously reported (Piégay et al. 1999; Gurnell et al. 2000b; Montgomery et al.

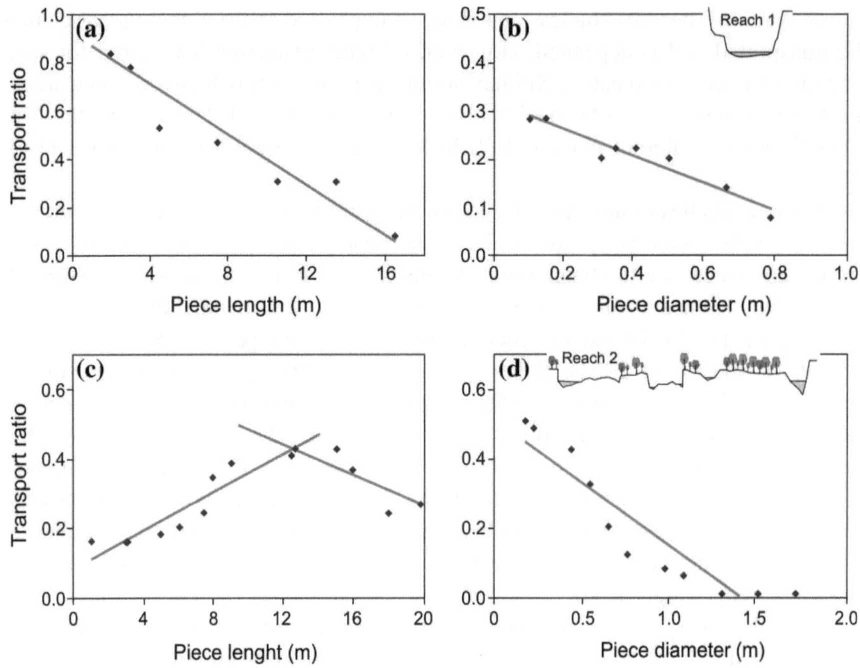

Fig. 3 **a**, **c** Relationships between wood transport ratio and the piece length for **a** the single-thread reach 1 and **c** the multi-thread reach 2. The results are shown for the simulation of the 10-year flood (105 $m^3\ s^{-1}$) and wood pieces with mean diameter of 0.2 m but different lengths (Lw) ranging from 1 to 20 m. **b**, **d** Relationships between wood transport ratio and the piece diameter for **b** the single-thread reach 1 and **d** the multi-thread reach 2. The results are shown for the simulation of the 10-year flood (105 $m^3\ s^{-1}$) and wood pieces with mean length of 12.5 m but different diameters (Dw) ranging from 0.1 to 0.8 m. Wood density is 0.56 g cm^{-3} in all cases

2003). According to our findings, LW is more likely to be deposited along the main channel, on point bars and in the forested areas adjacent to the main channel during ordinary floods (>10-year recurrence interval) in single-thread channel configurations. In multi-thread reaches, ordinary floods will tend to deposit LW on bars as well as vegetated and forested islands (Fig. 4). These preferential sites for LW deposition were identified based on the probability of deposition computed by the ensemble of the results from the multi-run model approach. Figure 4 demonstrates that for each flood scenario and for each river geomorphic configuration, different values and different spatial distributions were obtained in terms of depositional probability.

Figure 5 shows that in the case of single-thread reach 1 and for high-frequency floods, the preferential sites for LW deposition are the main channel, bars and the forested areas adjacent to the main channel. As flood magnitude increases, the probability for LW to be deposited in the main channel of reach 1 decreases. LW is likely to be transported downstream of this reach or to be deposited in the areas covered by mature forest along the floodplain. During very extreme floods, water

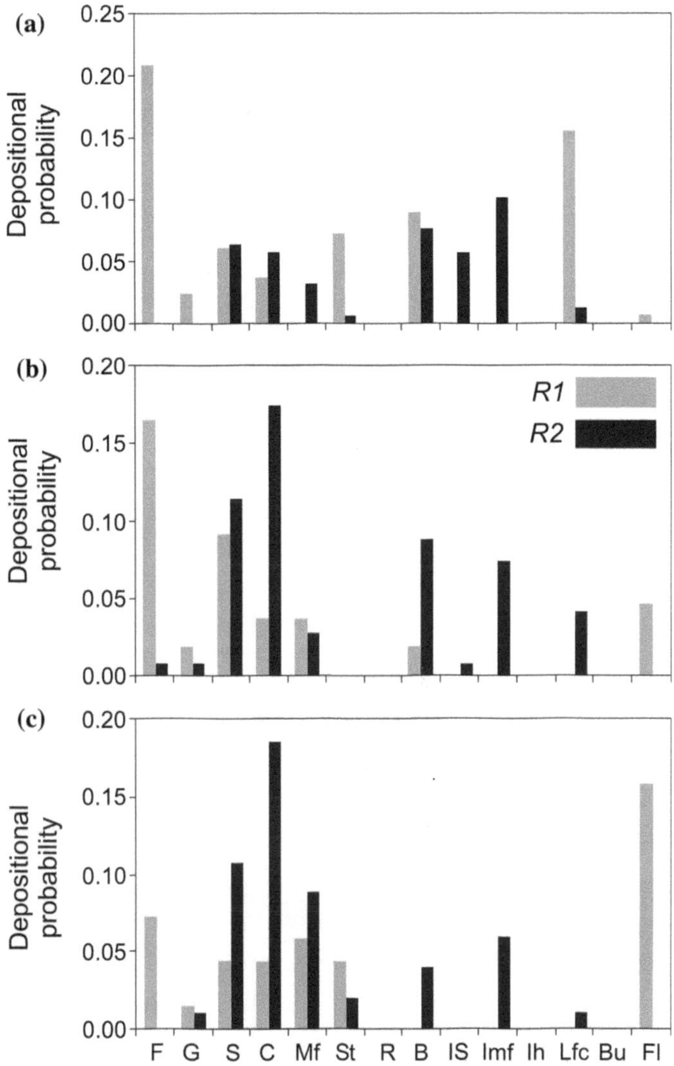

Fig. 4 **a**, **b**, **c** Depositional probability values for different RHU: Forest (F), gravelly and sandy surfaces within the floodplain (G), shrubs (S), meadows/cultivated (C), mature forest (Mf), road (R), scattered trees (St), gravel bars without vegetation (B), vegetated island (Is), forested island (Imf), island with shrubs (Ih), low-flow channel (Lfc), floodplain (Fl); and for different flood scenarios: **a** frequent floods; **b** ordinary floods; **c** very extreme floods. R1 is reach 1 and R2 is reach 2

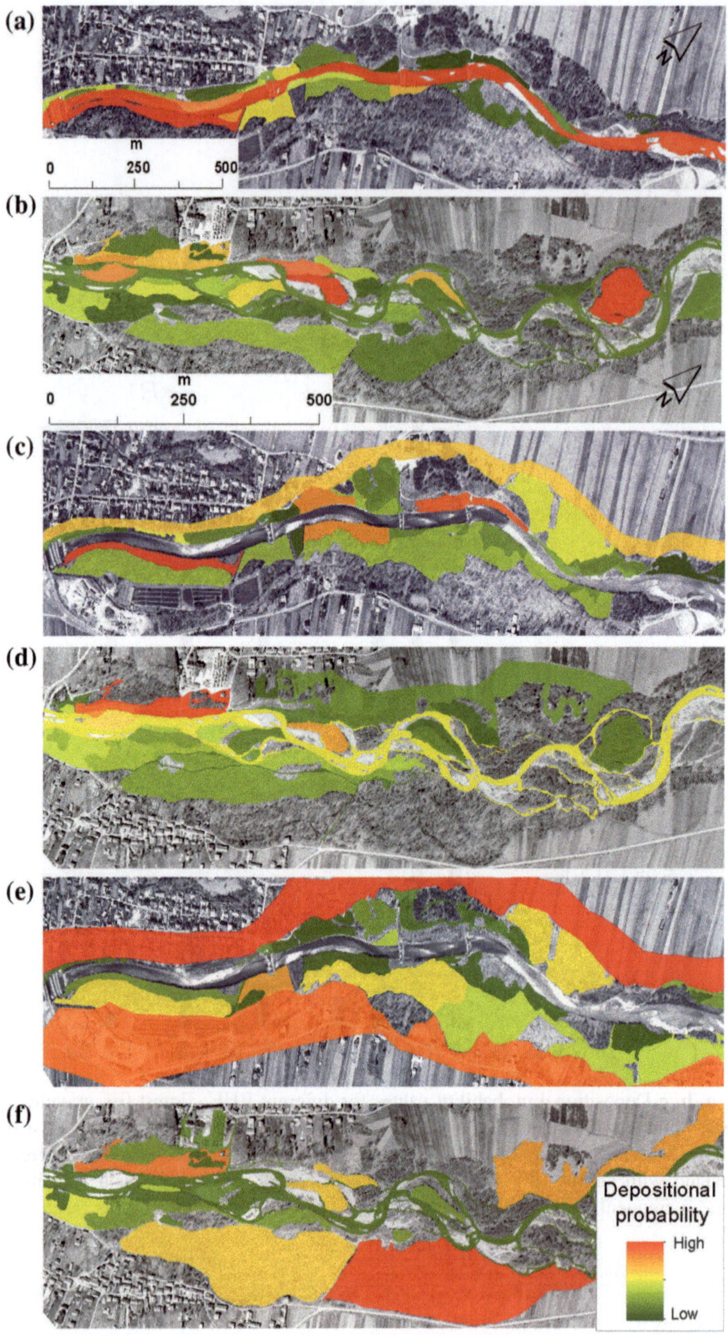

Fig. 5 Maps of wood depositional probability values for different RHU at frequent flood (1.2-year flood with a peak discharge of 28 $m^3\ s^{-1}$) in Reach 1 (**a**) and Reach 2 (**b**), at ordinary flood (10-year flood; 105 $m^3\ s^{-1}$) in Reach 1 (**c**) and Reach 2 (**d**), and at very extreme flood (80-year flood; 210 $m^3\ s^{-1}$) in Reach 1 (**e**) and Reach 2 (**f**). Flow direction is from left to right

inundates the floodplain and a proportion of wood pieces are directed onto the channel banks where the flow is slow and shallow.

In the case of the multi-thread reach 2, during frequent floods bars, vegetated islands, and forested islands are the preferential sites for LW deposition. LW generally is not deposited on the floodplain during frequent floods. In the case of ordinary floods, LW is preferentially deposited on forested islands, in mature forest and crops on the floodplain, and along the margins of braids. During very extreme floods ($Q = 210\ m^3\ s^{-1}$), LW is retained far away from the main channel and the active river zone, in areas covered by meadows and crops as well as mature forest.

Correlation coefficients between LW depositional probability and individual controlling variables (tested by the Spearman rank correlation test) indicate that LW depositional probability is related to log dimensions (length and diameter), wood density, Manning roughness coefficient of depositional sites and flow velocity. However, only log diameter and surface roughness influence the probability of LW deposition in both reaches and under different flood conditions. Flow velocity has an influence on wood deposition only in the case of low flood flows and in reach 1 (Table 2).

The dependence of depositional probability on multiple controlling variables was verified by means of generalized multiple regression analysis (linear regression models that allow for variables with non-normal distribution of residuals). The obtained models explain between 12 and 44 % of the variation in LW depositional

Table 2 Results of the Spearman rank correlation analysis between wood depositional probability and controlling variables at the sites with LW deposits in reach 1 and reach 2 and for very frequent and very extreme flood scenarios. After Ruiz-Villanueva et al. (2016a), reproduced with permission of Wiley.

Flood magnitude	Very frequent flood		Very extreme flood	
Independent variable	Spearman correlation coefficient	*p*-value	Spearman correlation coefficient	*p*-value
Reach 1				
Wood density ($g\ cm^{-3}$)	0.156	0.043	*−0.031*	0.797
Piece diameter (m)	0.283	0.007	*−0.229*	0.057
Piece length (m)	0.147	0.057	*−0.180*	0.137
Manning roughness coefficient ($m^{1/2}\ s^{-1}$)	0.638	<0.001	*−0.219*	0.068
Flow velocity ($m\ s^{-1}$)	*	*	0.011	0.927
Reach 2				
Wood density ($g\ cm^{-3}$)	*−0.160*	0.044	*−0.358*	<0.001
Piece diameter (m)	*−0.283*	<0.001	*−0.458*	<0.001
Piece length (m)	*−0.324*	<0.001	*−0.488*	<0.001
Manning roughness coefficient ($m^{1/2}\ s^{-1}$)	0.258	0.001	*−0.468*	<0.001
Flow velocity ($m\ s^{-1}$)	*−0.336*	<0.001	0.088	0.380

Negative correlation is highlighted in italics. Water depth was not included among the independent variables because similar to flow velocity*, water depth equalled zero at most deposition locations probability, and point to the decisive role of either log length or log diameter, and surface roughness. At the same time, however, we observe that the direction of the relationship differs between low- and high-magnitude floods. For low flood flows, depositional probability is higher in the areas with higher roughness within the flooded area. In the case of very extreme floods in reach 1, the LW depositional probability is negatively related to both Manning coefficient and log diameter. The depositional probability is thus higher in the areas with low surface roughness. The negative relation between depositional probability and log diameter can be explained by the fact that with the mean length of simulated pieces at 12.5 m, even thin logs can interact easily with and be anchored on the margins of the narrow, single-thread channel. By contrast, in the case of shorter pieces, interactions with the channel margins are less likely and the ratio between log diameter and water depth is high enough to allow their downstream transport, even if the logs have large diameters.

In the case of low flood flows ($Q = 28\ m^3\ s^{-1}$) in multi-thread reach 2, the variation of depositional probability was also higher in the areas with higher roughness within the flooded area; however, a much lower proportion of the total variance can be explained by this variable.

In the scenario describing a very extreme flood ($Q = 210\ m^3\ s^{-1}$) in reach 2, the variation in depositional probability is negatively related to surface roughness. This observation again reflects the fact that at the extreme discharge LW is mostly deposited in the areas with lower values of roughness coefficient (i.e. on the floodplain). With equal roughness of depositional surfaces, logs of larger diameter are less likely to be deposited in the reach. This may reflect the fact that thicker logs tend to be transported by the flow along deeper areas with the faster current as a result of their greater momentum.

Therefore, not only roughness controls wood deposition. There has been little information in the literature on the elevation at which LW is deposited in rivers (Gurnell et al. 2000a; Bertoldi et al. 2013). Based on the results of the model runs, we indicate that the relative elevation of LW deposits differs between floods of different magnitude and we also show that LW is not always deposited in those geomorphic units where the highest roughness values occur (as it occurs at low-magnitude floods). We tested the relative elevation of LW depositional sites with respect to (i) the low-flow water surface and (ii) the lower river bank. These relative elevations were analyzed for each reach and for different peak discharges (Ruiz-Villanueva et al. 2016a). For all flood magnitudes, the average elevation of LW deposits above the low-flow water surface is higher in the single-thread reach 1 (with its narrow and deep channel) than in the multi-thread reach 2 (with its wide and shallower channel). In both reaches, the relative elevation of LW deposits changes significantly with changing flood magnitude. In a similar way, elevation of LW deposits relative to the lower river bank also depends significantly on flood magnitude. However, if the relative elevation of LW deposits is considered, differences between the two reaches apparently increase with increasing flood magnitude, but become statistically significant once a certain flood magnitude is attained. Overall, these results demonstrate that LW deposition is strongly controlled by water depth (Ruiz-Villanueva et al. 2016a).

4.3 Large Wood Remobilization

In the previous sections we assumed that wood was entering the studied reaches from upstream, but previously deposited wood can be mobilized during subsequent floods (Haga et al. 2002; Wohl and Goode 2008). As observed by MacVicar and Piégay (2012), two consecutive floods with similar peak discharges and volumes mobilized different quantities of wood; as a result of antecedent flood effects, the second flood transported significantly less wood. We analysed wood remobilization using the output from a simulated flood (i.e., the resultant spatial distribution of wood) as an input for the next flood scenario without the supply of wood recruited from upstream. Results of these runs showed that the number of remobilized wood pieces is smaller in reach 1 than in reach 2 (Fig. 6) and remobilized logs travelled larger distance in reach 1 than in reach 2 (Fig.7).

One of the reasons for the greater ratio of wood remobilization from reach 2 was the magnitude of the preceding flood—consequently; in this reach the flood deposited wood pieces mostly within the active river zone. If the previous flood was greater, more pieces might be deposited on the floodplain in reach 2 and the disparity in wood remobilization ratio between the reaches might be reduced, eliminated, or maybe reversed, depending on the magnitude of the antecedent flood. We thus emphasize the importance of the magnitude of the preceding flood in predisposing the potential of deposited wood for remobilization during a subsequent flood.

4.4 Large Wood Dynamics Under Unsteady Flow Conditions

A significant feature of wood dynamics is its temporal dimension, and it is represented by simulating the entire flood hydrograph (i.e., unsteady flows). To design the hydrographs, we used the Dimensionless Unit Hydrograph method of the Soil Conservation Service (SCS 1972). This method uses the ratio between the discharge at one time step and the peak discharge ($Q \cdot Qp^{-1}$) and the ratio between time and the time to peak ($t \cdot tp^{-1}$) to build the final hydrograph. Subsequently we modified the resultant common hydrographs (i.e. the hydrographs with the most common shape in the river) to obtain flashy and flattened flood waves with the same flow volume. We defined the common flood scenario ($Q \cdot Qp^{-1} = 1$ and $t \cdot tp^{-1} = 1$) where the flood peak is reached 5 h after the beginning of the flood wave, and the total flood duration equals 25 h. In the flattened flood wave scenario ($Q \cdot Qp^{-1} = 0.5$ and $t \cdot tp^{-1} = 2$), floods have a total duration of 35 h and the peak is

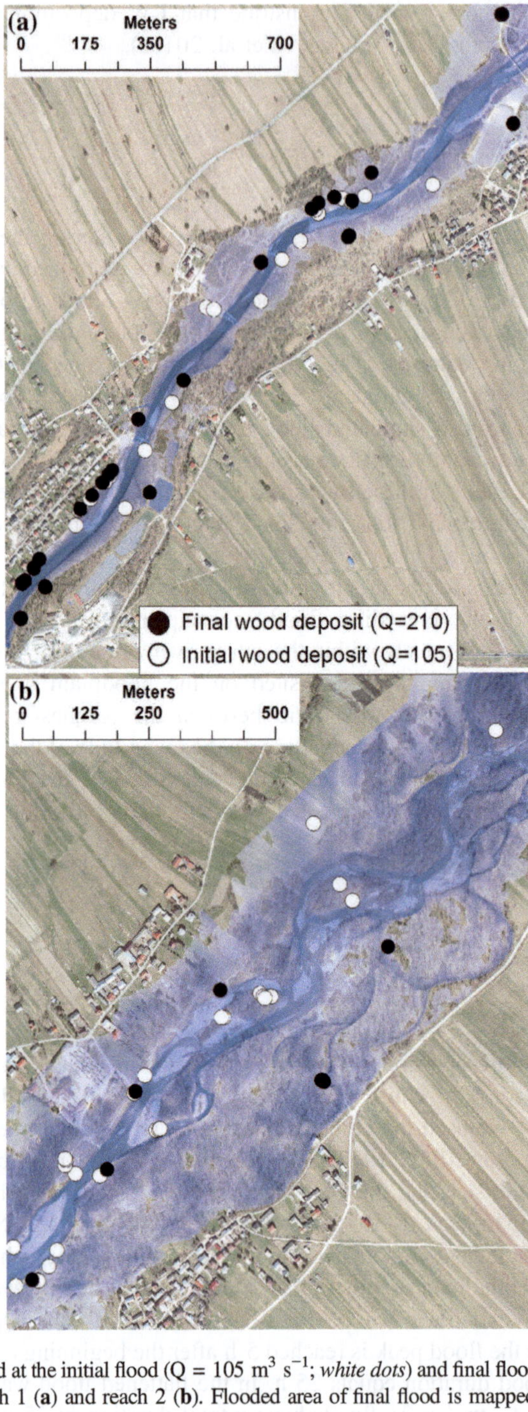

Fig. 6 Logs stored at the initial flood (Q = 105 m^3 s^{-1}; *white dots*) and final flood (Q = 210 m^3 s^{-1}; *black dots*) in reach 1 (**a**) and reach 2 (**b**). Flooded area of final flood is mapped in *blue*

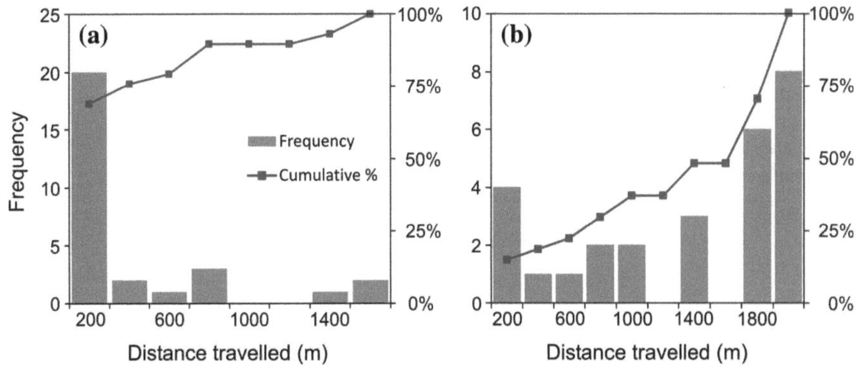

Fig. 7 Distribution of travelled distance (i.e. distances between the locations of remobilization and deposition of the remobilized logs, including those which exited the modelling domain) for Reach 1 (**a**) and Reach 2 (**b**)

reached after 10 h, whereas in the flashy flood scenario ($Q \cdot Qp^{-1} = 1.5$ and $t \cdot tp^{-1} = 0.5$) floods last for 17 h and the peak is passing after 3 h. In all scenarios, the shape of the hydrographs reflects the shape of flood waves occurring in the river, with a rising limb being steeper than the falling limb. In this set of modelling tests, we also analysed wood remobilization, using the result of the simulated discharge of 28 $m^3 \cdot s^{-1}$ (1.2-year flood) as an initial condition. Therefore, we assumed that wood recruitment is not occurring upstream from the study reaches or laterally (due to bank erosion), but that only the initially deposited wood is being remobilized.

Results of the simulated hydrographs showed that large wood is transported downstream during the rise of the hydrograph, and we observed a time lag between the beginning of the flood and wood motion (Fig. 8). The peak of the wood flux is systematically reached before the flood peak in all scenarios and in both reaches (Fig. 8 shows just three cases). This observation is in agreement with Schenk et al. (2014) who state that most wood is mobilized at the very beginning of the flood in the Roanoke River. This result was confirmed by Ravazzolo et al. (2015) who, in addition, observed that most of their tagged logs in the Tagliamento River were deposited right at the flood peak.

Our findings also confirmed that wood transport decreases near or slightly after hydrograph peaks. This suggests that the mobilization of in-stream wood is likely negligible during the falling limb of the hydrograph as wood has already been subject to the same or larger discharges during the flood rise. All these observations might be of great interest to manage potential wood-related hazards during floods.

Common and flashy hydrographs with higher peak discharges mobilized more logs than the equivalent flattened floods. A 5-year flood with the common hydrograph shape and the flattened equivalent of a 25-year flood have similar peak discharges (78 $m^3 \cdot s^{-1}$ and 73 $m^3 \cdot s^{-1}$, respectively) but a slightly higher percentage of mobilized wood is associated with the latter, despite its slightly lower peak discharge. On the other hand, a 25-year flood with a common hydrograph is

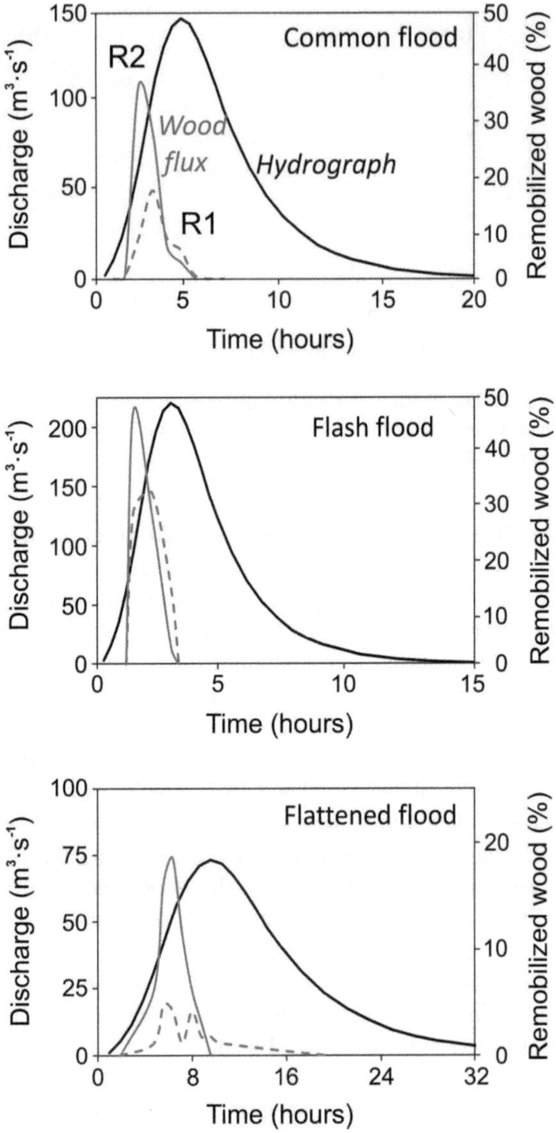

Fig. 8 Flood hydrographs (*black solid lines*) and temporal variation in the mobilization of initial wood pieces (in %) in reach 1 (R1; *dashed grey lines*) and reach 2 (R2; *solid grey lines*)

able to transport a smaller number of pieces downstream from the reaches than the flashy equivalent of the 5-year flood, even though the former has a higher peak discharge. Therefore, the hydrograph shape and the duration of the flood also significantly influence the mobilization of in-stream wood and its duration.

While the number of mobilized logs was positively related to flow peak, the length of wood mobilization period is negatively related to peak discharge, but positively to the duration of the flood, particularly to the time to peak. As we observed when we simulated large wood transport under steady-flow conditions, the relationships between wood transport and the time to peak and between wood transport and the peak discharge of the flood are not linear (Ruiz-Villanueva et al. 2016b). The highest wood flux occurs during flashy floods which generally have higher peak discharges, and the lowest flux is associated with the equivalent flattened flood scenarios (Fig. 9). The common hydrograph scenarios represent an intermediate situation (Fig. 9).

The transport of wood commences during the rising limb and ceases before the peak of the flood wave in the flash flood scenario, whereas it extends into an early phase of flood recession in both reaches in the flattened flood scenario (Fig. 9). Here, the largest pieces are moved during the flood peak, whereas in the case of the flash flood with a higher discharge, the largest pieces are already transported before the peak (Fig. 9). When both reaches are compared, we observe that floods in reach 2 move pieces of smaller sizes than in reach 1. This is because of differences in water depth and flow velocity between the reaches.

Wood transport in rivers is a nonlinear process, and the relation between the proportion of mobilized wood and discharge during (simulated) floods is complex (Ruiz-Villanueva et al. 2016b). In-stream wood starts to move as soon as a certain discharge is reached, and in the case of previously deposited wood, this threshold is clearly related to the magnitude of the antecedent flood. In both cases, the number

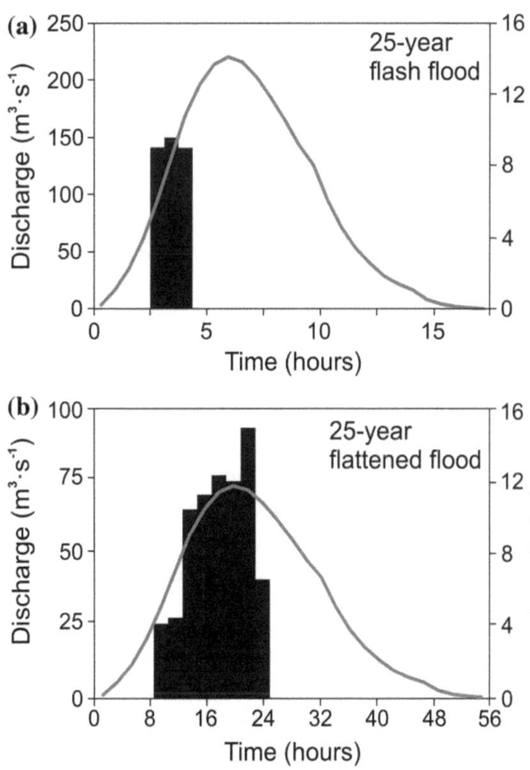

Fig. 9 Flood hydrographs (*grey line*) and the length of mobilized wood pieces (*black histogram*) in reach 1 during the flashy (**a**) and flattened (**b**) equivalents of the 25-year flood

of mobilized pieces increases with increasing discharge until the maximum amount of transported wood is reached. After this point, and even if discharge is still increasing, the number of wood pieces in motion will decrease and eventually the motion will cease completely. We attribute this phenomenon to the hysteresis (Ruiz-Villanueva et al. 2016b) in mobilization of wood pieces with discharge as previously proposed by Marcus et al. (2011) and MacVicar and Piégay (2012).

5 Concluding Remarks

The results of this study help to understand the complex relationships between floods and wood motion, despite the fact that several relevant processes that occur at the field scale—such as sediment transport and bank erosion—could not be considered in the modelling approach. Moreover, still only very limited information is available on the actual mechanics of wood recruitment and transport within streams, such that we still do not know how the timing of individual tree fall, mass recruitment, jam formation or jam breakup will correspond to flow hydrographs (Wohl et al. 2011). Despite these limitations and possible shortcomings, we remain convinced that the results from this study can be generalized and therefore extrapolated to other rivers with similar characteristics.

After decades of modelling sediment transport the knowledge is far from complete, so there is still a long way to travel and several challenges to be met to fully understand wood transport, even more for combining both processes. Recent advances in computer technology have increased our capabilities for developing more detailed, computationally intensive models. Similarly, advances in GIS technologies and field-measurement equipment and techniques have provided extensive and highly accurate data, and allowed for modelling, analysis, and monitoring of wood transport over various spatial and temporal scales. An integrated framework, combining several of those techniques would help improve the theoretical and numerical models of wood transport and allow for performing more extensive testing and validation of these models. Thus, it would allow for comprehensive understanding of the transport processes, their influence on hydrodynamics and geomorphology and the spatial and temporal environmental variability.

The knowledge of wood dynamics and the outcomes regarding timing and duration of wood load transport and the influence of flood hydrograph presented in this paper are crucial for developing adaptive management of the potential hazards of LW to human communities and infrastructure, both in the Czarny Dunajec valley (cf. Kundzewicz et al. 2014) and elsewhere.

Acknowledgments This work was supported by the project FLORIST (Flood risk on the northern foothills of the Tatra Mountains; PSPB no. 153/2010). Authors acknowledge the revision made by Dr. Walter Bertoldi which improved the early version of this chapter.

References

Allen JB, Smith DL (2012) Characterizing the impact of geometric simplification on large woody debris using CFD. J Hydraul Eng 1:1–14

Badoux A, Böckli M, Rickenmann D, Rickli C, Ruiz-Villanueva V, Zurbrügg S (2015) Large wood transported during the exceptional flood event of 24 July 2014 in the Emme catchment (Switzerland). In: Picco L, Lenzi MA, Bertoldi W et al (eds) Proceedings of the third international conference 'Wood in world rivers 3', Padova, Italy, pp 109–111

Benda LE, Sias JC (2001) A quantitative framework for evaluating the mass balance of in-stream organic debris. Forest Ecol Manag 172:1–16

Bertoldi W, Gurnell AM, Welber M (2013) Wood recruitment and retention: the fate of eroded trees on a braided river explored using a combination of field and remotely-sensed data sources. Geomorphology 180:146–155

Bertoldi W, Welber M, Mao L, Zanella S, Comiti F (2014) A flume experiment on wood storage and remobilization in braided river systems. Earth Surf Proc Landforms 39:804–813

Bocchiola D, Catalano F, Menduni G, Passoni G (2002) An analytical-numerical approach to the hydraulics of floating debris in river channels. J Hydrol 269:65–78

Braudrick CA, Grant GE (2000) When do logs move in rivers? Water Resour Res 36:571–583

Braudrick CA, Grant GE (2001) Transport and deposition of large woody debris in streams: a flume experiment. Geomorphology 41:263–283

Bray DI (1979) Estimating average velocity in gravel-bed rivers. J Hydr Div 105:110–1122

Cadol D, Wohl E (2010) Wood retention and transport in tropical, headwater streams, La Selva Biological Station, Costa Rica. Geomorphology 123:62–73

Chow VT (1959) Open-channel hydraulics. McGraw-Hill, New York

Elosegi A, Díez JR, Pozo J (1999) Abundance, characteristics and movement of woody debris in four Basque streams. Arch Hydrobiol 147:455–471

Gippel CJ (1995) Environmental hydraulics of large woody debris in streams and rivers. J Environ Eng 121:388–395

Gurnell AM (2003) Wood storage and mobility. In: Gregory SV, Boyer KL, Gurnell AM (eds) The ecology and management of wood in world rivers. Am Fish Soc Symp 37:75–91

Gurnell AM (2012) Wood and river landscapes. Nat Geosci 5:93–94

Gurnell AM, Petts GE, Hannah DM, Smith BPG, Edwards PJ, Kollmann J, Ward JV, Tockner K (2000a) Wood storage within the active zone of a large European gravel-bed river. Geomorphology 34:55–72

Gurnell AM, Petts GE, Harris N, Ward JV, Tockner K, Edwards PJ, Kollmann J (2000b) Large wood retention in river channels: the case of the flume Tagliamento, Italy. Earth Surf Proc Landforms 25:255–275

Gurnell AM, Piégay H, Swanson FJ, Gregory SV (2002) Large wood and fluvial processes. Freshwater Biol 47:601–619

Haga H, Kumagai T, Otsuki K, Ogawa S (2002) Transport and retention of coarse woody debris in mountain streams: an in situ field experiment and a field survey of coarse woody debris distribution. Water Resour Res 38:11–26

Iroumé A, Mao L, Andreoli A, Ulloa H, Ardiles M (2015) Large wood mobility processes in low-order Chilean river channels. Geomorphology 228:681–693

Kundzewicz ZW, Stoffel M, Kaczka RJ, Wyżga B, Niedźwiedź T, Pińskwar I, Ruiz-Villanueva V, Łupikasza E, Czajka B, Ballesteros-Canovas J, Małarzewski Ł, Choryński A, Janecka K, Mikuś P (2014) Floods at the northern foothills of the Tatra Mountains—A Polish-Swiss Research Project. Acta Geophys 62:620–641

Lassettre NS, Piégay H, Dufour S, Rollet A (2008) Decadal changes in distribution and frequency of wood in a free meandering river, the Ain River, France. Earth Surf Proc Landforms 33:1098–1112

Lienkaemper GW, Swanson FJ (1987) Dynamics of large woody debris in streams in old-growth Douglas-fir forests. Can J Forest Res 17:150–156

MacVicar BJ, Piégay H (2012) Implementation and validation of video monitoring for wood budgeting in a wandering piedmont river, the Ain River (France). Earth Surf Proc Landforms 37:1272–1289

MacVicar BJ, Piégay H, Henderson A, Comiti F, Oberlin C, Pecorari E (2009) Quantifying the temporal dynamics of wood in large rivers: field trials of wood surveying, dating, tracking, and monitoring techniques. Earth Surf Proc Landforms 34:2031–2046

Marcus WA, Rasmussen J, Fonstad MA (2011) Response of the fluvial wood system to fire and floods in Northern Yellowstone. Ann Assoc Am Geogr 101:21–44

Meyer-Peter E, Müller R (1948) Formulas for bed-load transport. Proceedings of the 2nd meeting of the international association for hydraulic structures research, pp 39–64

Mikuś P, Wyżga B, Kaczka RJ, Walusiak E, Zawiejska J (2013) Islands in a European mountain river: linkages with large wood deposition, flood flows and plant diversity. Geomorphology 202:115–127

Montgomery DR, Collins BD, Buffington JM, Timothy BA (2003) Geomorphic effects of wood in rivers. In: Gregory SV, Boyer KL, Gurnell AM (eds) The ecology and management of wood in world rivers. American Fisheries Society Symposium 37:21–47

Moulin B, Schenk ER, Hupp CR (2011) Distribution and characterization of in-channel large wood in relation to geomorphic patterns on a low-gradient river. Earth Surf Proc Landforms 36:1137–1151

Pecorari E (2008) Il materiale legnoso in corsi d'acqua a canali intrecciati: volumi, mobilità, degradazione ed influenza morfologica. PhD thesis, University of Padova, Italy

Piégay H, Thévenet A, Citterio A (1999) Input, storage and distribution of large woody debris along a mountain river continuum, the Drôme River, France. Catena 35:19–39

Ravazzolo D, Mao L, Picco L, Lenzi MA (2015) Tracking log displacement during floods in the Tagliamento River using RFID and GPS tracker devices. Geomorphology 228:226–233

Rickenmann D, Waldner P, Usbeck T, Köchil D, Sutter F, Rickli C, Badoux A (2015) Large wood transport during the 2005 flood events in Switzerland. In: Picco L, Lenzi MA, Bertoldi W et al (eds) Proceedings of the third international conference 'Wood in world rivers 3', Padova, Italy, pp 124–126

Ruiz-Villanueva V, Bladé E, Díez-Herrero A, Bodoque JM, Sánchez-Juny M (2014a) Two-dimensional modelling of large wood transport during flash floods. Earth Surf Proc Landforms 39:438–449

Ruiz-Villanueva V, Bladé E, Sánchez-Juny M, Marti-Cardona B, Díez-Herrero A, Bodoque JM (2014b) Two-dimensional numerical modelling of wood transport. J Hydroinform 16:1077–1096

Ruiz-Villanueva V, Wyżga B, Hajdukiewicz H, Stoffel M (2016a) Exploring large wood retention and deposition in contrasting river morphologies linking numerical modelling and field observations. Earth Surf Proc Landforms 41:446–459

Ruiz-Villanueva V, Wyżga B, Mikuś P, Hajdukiewicz H, Stoffel M (2016b) The role of flood hydrograph in the remobilization of large wood in a wide mountain river. J Hydrol. doi:10.1016/j.jhydrol.2016.02.060

Ruiz-Villanueva V, Wyżga B, Zawiejska J, Hajdukiewicz M, Stoffel M (2016c) Factors controlling large wood transport in a mountain river. Geomorphology 272C. doi:10.1016/j.geomorph.2015.04.004

Schenk ER, Moulin B, Hupp CR, Richter JM (2014) Large wood budget and transport dynamics on a large river using radio telemetry. Earth Surf Proc Landforms 39:487–498

Schmocker L, Hager WH (2011) Probability of drift blockage at bridge decks. J Hydraul Eng 137:480–492

SCS (1972) Soil conservation service. National engineering handbook, section 4. U.S. Department of Agriculture, Washington, DC

Strickler A (1923) Beiträge zur Frage der Geschwindigkeitsformel und der Rauhigkeitszahlen für Ströme, Kanäle und geschlossene Leitungen. Mitteilungen des Amtes für Wasserwirtschaft 16, Bern

Turowski JM, Böckli M, Rickenmann D, Beer A (2013) Field measurements of the energy delivered to the channel bed by moving bed load and links to bedrock erosion. J Geophys Res Earth Surf 118:2438–2450

Van der Nat D, Tockner K, Edwards PJ, Ward JV, Gurnell AM (2003) Habitat change in braided flood plains (Tagliamento, NE-Italy). Freshwater Biol 48:1799–1812

Welber M, Bertoldi W, Tubino M (2013) Wood dispersal in braided streams: results from physical modeling. Water Resour Res 49:7388–7400

Wilcox AC, Wohl E (2006) Flow resistance dynamics in step-pool stream channels: 1. Large woody debris and controls on total resistance. Water Resour Res 42:1–16

Wohl E, Goode JR (2008) Wood dynamics in headwater streams of the Colorado Rocky Mountains. Water Resour Res 44:W09429

Wohl E (2013) Floodplains and wood. Earth-Sci Rev 123:194–212

Wohl E, Cadol D (2011) Neighborhood matters: patterns and controls on wood distribution in old-growth forest streams of the Colorado Front Range, USA. Geomorphology 125:132–146

Wohl E, Polvi LE, Cadol D (2011) Wood distribution along streams draining old-growth floodplain forests in Congaree National Park, South Carolina, USA. Geomorphology 126:108–120

Wyżga B, Zawiejska J (2005) Wood storage in a wide mountain river: case study of the Czarny Dunajec, Polish Carpathians. Earth Surf Proc Landforms 30:1475–1494

Wyżga B, Zawiejska J (2010) Large wood storage in channelized and unmanaged sections of the Czarny Dunajec River, Polish Carpathians: implications for the restoration of mountain rivers. Folia Geogr Ser Geogr-Phys 41:5–34

Wyżga B, Oglęcki P, Radecki-Pawlik A, Skalski T, Zawiejska J (2012) Hydromorphological complexity as a driver of the diversity of benthic invertebrate communities in the Czarny Dunajec River, Polish Carpathians. Hydrobiologia 696:29–46

Wyżga B, Zawiejska J, Mikuś P, Kaczka RJ (2015) Contrasting patterns of wood storage in mountain watercourses narrower and wider than the height of riparian trees. Geomorphology 228:275–285

Zawiejska J, Wyżga B (2010) Twentieth-century channel change on the Dunajec River, southern Poland: patterns, causes and controls. Geomorphology 117:234–246

Zawiejska J, Wyżga B, Radecki-Pawlik A (2015) Variation in surface bed material along a mountain river modified by gravel extraction and channelization, the Czarny Dunajec, Polish Carpathians. Geomorphology 231:353–366

Modelling Hydraulic Parameters of Flood Flows for a Polish Carpathian River Subjected to Variable Human Impacts

Artur Radecki-Pawlik, Bartłomiej Wyżga, Wiktoria Czech, Paweł Mikuś, Joanna Zawiejska and Virginia Ruiz-Villanueva

Abstract Channelization and channel incision have considerably modified channel morphology of the Czarny Dunajec River, and now it varies from a single-thread, incised or regulated channel to an unmanaged, multi-thread channel. Effects of these distinct channel morphologies on the conditions for flood flows were investigated in a study of 25 cross-sections from the middle river course. Cross-sectional morphology, channel slope and roughness were used as input data for the 1D steady-flow hydraulic modelling performed for discharges with recurrence interval between 1.5 and 50 years. Adjustment of roughness coefficients to obtain the agreement between simulated and observed peak levels of the 2014 flood allowed calibration of the model for particular cross-sections. As a result of differences in flow widths, cross-sectional flow areas and channel slope, flood flows in the three river reaches differ in unit stream power and bed shear stress, with the highest values of the parameters recorded in the incised reach, intermediate values in the

A. Radecki-Pawlik (✉) · W. Czech
Department of Hydraulics Engineering and Geotechnique,
University of Agriculture, al. Mickiewicza 24/28, 30-059 Kraków, Poland
e-mail: rmradeck@cyf-kr.edu.pl

W. Czech
e-mail: vviki7@o2.pl

B. Wyżga · P. Mikuś
Institute of Nature Conservation, Polish Academy of Sciences,
al. Mickiewicza 33, 31-120 Kraków, Poland
e-mail: wyzga@iop.krakow.pl

P. Mikuś
e-mail: mikus@iop.krakow.pl

J. Zawiejska
Institute of Geography, Pedagogical University of Cracow,
ul. Podchorążych 2, 30-084 Kraków, Poland
e-mail: zawiejska.joanna@gmail.com

V. Ruiz-Villanueva
Dendrolab.ch, Institute of Geological Sciences, University of Bern,
Baltzerstrasse 1+3, 3012 Bern, Switzerland
e-mail: virginia.ruiz@dendrolab.ch

Z.W. Kundzewicz et al. (eds.), *Flood Risk in the Upper Vistula Basin*,
GeoPlanet: Earth and Planetary Sciences, DOI 10.1007/978-3-319-41923-7_7

channelized reach and the lowest values in the multi-thread reach. The recognised differences in the flow power and in tractive forces exerted on the flow boundary underlie and explain different evolutionary tendencies of particular river reaches during the past decades. Stabilization of river banks in channelized reaches induces a progressive increase in floodplain elevation; sedimentation in the analysed channelized cross-section of the Czarny Dunajec have reduced its initial flow conveyance by half and elevated water stages at given flood discharges by about 0.7 m during 30 years since the river channelization.

Keywords Hydraulic modelling · River channelization · Channel incision · Roughness calibration · Overbank deposition · Flow conveyance

1 Introduction

In the 20th century and especially in its second half, the morphology of Polish Carpathian rivers was considerably modified as a result of diverse human impacts. The rivers were channelized along most of their courses, with the change resulting in the replacement of wide and shallow, sinuous or multi-thread channels by single, narrow and straight ones (Wyżga 2001a; Wyżga et al. 2015). In the second half of the century, in-channel gravel mining was a common practice, resulting in a deficit of sediment available for fluvial transport (Rinaldi et al. 2005; Wyżga et al. 2010). Increased transport capacity of the channelized rivers and the deficit of bed material, caused by gravel mining and a reduction in sediment supply to the channels accompanying an increase in forest cover in the catchments (Lach and Wyżga 2002), induced intense bed degradation leading to deep channel incision (Wyżga 2008; Wyżga et al. 2016).

Variation in river morphology throughout stream networks is central to the problems of catchment management because of its influence on overlying flow and sediment routing, physical habitats and channel–floodplain interactions (Stewardson 2005; Wyrick et al. 2014). Intense human impacts on rivers, such as channelization or channel incision, may lead to adverse effects on physical habitat conditions and river biocoenoses (Wyżga et al. 2011) but also modify flood conditions in rivers (Czech et al. 2016). River straightening and narrowing in the course of channelization increases flow velocities and unit stream power associated with given discharges (Wyżga 1993, 2001a), inducing bed degradation and the resultant increased concentration of flows in the deepened channels (Wyżga 2001b; Wyżga et al. 2016). Flow concentration in incised channels considerably reduces flood-water storage in floodplain areas and the role of floodplains as sediment sinks during floods (Wyżga 2001b).

Different river morphologies may variously influence propagation of flood waves (e.g. Burkham 1976; Wyżga 1996) and result in different values of flow energy and tractive forces at similar discharges (Czech et al. 2016), with differing effects on channel stability and trends of channel adjustments. Establishing how hydraulic conditions during floods differ between neighbouring river reaches of different morphology is thus a key question in flood risk (Kundzewicz et al. 2014) and river restoration assessment (Czech et al. 2016).

One-dimensional or two-dimensional numerical models can be used to analyse hydraulic conditions during floods (Horritt and Bates 2002) and they are powerful tools to understand the effects of human disturbances in rivers on these conditions. We applied 1D modelling to determine hydraulic conditions of flood flows in a number of cross-sections of the Czarny Dunajec River, Polish Carpathians. During the second half of the 20th century, this river was subjected to considerable though spatially varied modifications caused by human impacts; in consequence, currently it runs through neighbouring modified single-thread reaches with an incised or regulated channel, and an undisturbed reach with multi-thread morphology. We hypothesize that these diverse channel morphologies result in different hydraulic conditions of flood flows that variously affect channel stability and the conveyance and storage of floodwater.

2 Study Area

The Czarny Dunajec (Fig. 1) is one of two headwaters of the Dunajec, the second largest river in the Polish Carpathians. The Czarny Dunajec rises at about 1500 m a.s.l. in the high-mountain Tatra massif. In the foreland of the Tatras, the river flows 38 km to the confluence with the Biały Dunajec River in the middle part of the Orawa-Nowy Targ Basin. The Tatra part of the catchment determines the hydrological regime of the river, which is characterized by low winter flows and the occurrence of floods between May and August. It also feeds the Czarny Dunajec with coarse material that composes a non-cohesive alluvial plain formed by the river in the Tatra Mountains foreland.

In the second half of the 20th century, the river was affected by spatially variable human disturbances (Wyżga et al. 2009, 2011; Zawiejska et al. 2015), which changed its formerly uniform multi-thread channel pattern (Zawiejska and Wyżga 2010; Wyżga et al. 2015). During the 1950s–1960s gravel was intensely mined from the river bed at several locations, and in the following decades larger cobbles were commonly extracted from the channel (Wyżga et al. 2010). These activities induced rapid bed degradation that has resulted in up to 3.5 m of channel incision and transformation from alluvial to bedrock boundary conditions over most of the river length within the Gubałówka Hills (Zawiejska and Wyżga 2010). A 7-km-long reach in the middle river course was progressively channelized between the 1960s and 1990s, which resulted in the replacement of the former wide, multi-thread channel by a narrow, nearly straight, single-thread channel with reinforced banks and slope reduced by 0.7–2.1-m-high concrete drop structures. Along 4 km downstream of the channelized reach, the river remains unmanaged, with channel pattern varying from bar-braided to heavily island-braided. The active river zone in this reach is from 60 to 180 m wide, and the river bed has been

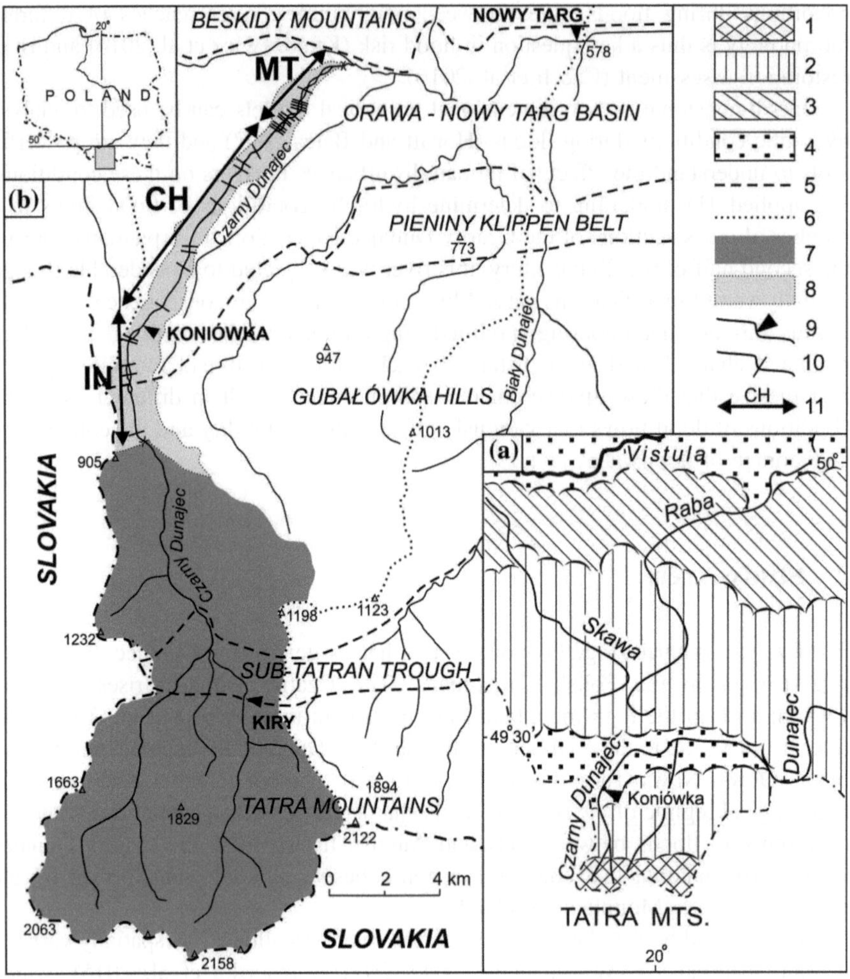

Fig. 1 **a** Location of the Czarny Dunajec River in relation to physiogeographic regions of southern Poland. **b** Drainage network of the Czarny Dunajec catchment and detailed setting of the investigated reaches and cross-sections of the river. *1* high mountains; *2* mountains of intermediate and low height; *3* foothills; *4* intramontane and submontane depressions; *5* boundaries of physiogeographic units; *6* boundary of the Czarny Dunajec catchment; *7* the Czarny Dunajec catchment to the beginning of the study section; *8* catchment area increment along the study section; *9* water-gauge stations; *10* river cross-sections investigated; *11* river reaches with different morphology/channel management: *IN* incised; *CH* channelized; *MT* multi-thread

vertically stable or slowly aggrading in recent decades (Zawiejska and Wyżga 2010). Further downstream, the river flows in a narrow, regulated channel, but its slope is not reduced by hydraulic drop structures.

This study was performed in the middle course of the Czarny Dunajec with a deeply incised channel in the upper part, a regulated channel in the middle part, and an unmanaged, multi-thread channel in the lower part. Along its middle course, the river flows on a large, glaciofluvial–alluvial fan; as a result of this hydrographic setting, the river receives no tributaries and the catchment area increases relatively little (Fig. 1). As flood discharges are here only little affected by water inflow from the catchment, it provides an opportunity to verify how these diverse morphological conditions of the river are reflected in hydraulic conditions of flood flows.

3 Study Methods

3.1 Estimation of Flood Discharges of Given Recurrence Interval

The study analyses 25 river cross-sections from the middle course of the Czarny Dunajec: 5 from the incised reach, 9 from the channelized reach, and 11 from the multi-thread reach (Fig. 1). With very low sinuosity (sinuosity index, SI = 1.05) and a high width/depth ratio of its channel (37.5 on average), the Czarny Dunajec in its middle course forms a typical high-flow system, in which flood waves are transmitted without a significant change in hydrograph shape and changes in peak discharges reflect water addition by tributaries (Burkham 1976; Wyżga 1996). Fitting the General Extreme Value function (GEV) to the flood frequency distributions for the same record period (1966–2010) at the Kiry, Koniówka and Nowy Targ gauging stations (Fig. 1), we calculated discharge quantiles of given recurrence intervals at the stations. Discharges were calculated for the following recurrence intervals and the associated probabilities of exceedance (in parentheses): 1.5 years (67 %), 2 years (50 %), 2.5 years (40 %), 3.33 years (30 %), 4 years (25 %), 5 years (20 %), 10 years (10 %), 20 years (5 %), 25 years (4 %), 33 years (3 %), and 50 years (2 %). Differences in catchment area between particular cross-sections and the Koniówka gauge cross-section were measured on a 1:10000 scale map, and flood discharges of given recurrence intervals in the cross-sections were calculated from their values at the three gauging stations and the determined gradients of their increase with increasing catchment area between the gauging stations.

3.2 Field-Derived Data

Geometry of the studied cross-sections was surveyed with a level during base-flow conditions in summer 2014. The bed-material grain size was determined separately for channel bars and each low-flow channel; details of the sampling procedures were described by Wyżga et al. (in press). We then used the empirical equation of Strickler (1923) to calculate Manning roughness coefficient from the median grain size of channel bar and low-flow channel sediments. In the case of vegetated parts of the river cross-sections with different types of vegetation cover/land use, Manning roughness coefficient was assigned according to the criteria of Chow (1959). We used different approaches to estimate the slope of flood water in freely developed cross-sections and in those located between concrete drop structures in the channelized reach. In the former, the flood-water slope was approximated by base-flow water slope measured on a distance of 150 m upstream and downstream of the analysed cross-sections. The distance of 300 m corresponds in the middle course of the Czarny Dunajec to ca. 6 channel widths (cf. Wyżga and Zawiejska 2005) representing an average spacing of riffle–pool sequences in rivers (Leopold et al. 1964). In the cross-sections between drop structures, water slope at the flow of 1.5-year frequency was considered to be the same as the base-flow water slope measured between points located immediately below the end sill of an upstream structure and on the crest of the downstream structure. At the flow of 50-year frequency, the weirs are considered to be fully drawn down and thus the water slope can be approximated by the slope measured between the crests of the upstream and downstream drop structures. The water-surface slope at intermediate flood discharges was calculated from the two above-mentioned slopes proportionately to the differences between a given flood frequency and the frequencies of 1.5 and 50 years.

3.3 Hydraulic Modelling and Data Analysis

Modelling of hydraulic conditions of flood flows in the cross-sections was carried out with 1D, steady-flow HEC-RAS model (USACE 2010), independently for each cross-section. For each cross-section, the model was calibrated with the level of the 20-year flood of May 2014, determined on the basis of high-water marks such as trash line. To calibrate the model, Manning's roughness coefficients were modified so as to adjust the calculated water level to the observed level. The following parameters calculated by the model were analysed in this study: energy grade slope, flow width, cross-sectional area of flow, mean flow depth, mean flow velocity, Froude number, bed shear stress and unit stream power.

Average values of particular hydraulic parameters characterizing the investigated cross-sections from the incised, channelized and multi-thread reaches of the Czarny Dunajec at flood discharges of given recurrence intervals were compared and statistical significance of the differences among them was determined with a Kruskal-Wallis test. Patterns of hydraulic parameters at flood flows of given frequency, apparent among all the studied cross-sections, were determined by means of principal component analysis.

3.4 Analysis of Changes in Channel Conveyance

Plans of the channelization scheme of the middle Czarny Dunajec have been preserved only for its downstream part constructed in the 1980s–1990s. We used the channelization designs and river appearance on aerial photos taken in 1983, when the scheme was under construction, and in 1994, a few years after its construction, to reconstruct the river geometry and roughness conditions existing shortly after the scheme implementation in 1985 at the location of one of the channelized cross-sections surveyed in 2014. This allowed us to compare an initial and the contemporary channel geometry and conveyeance of the channelization scheme. Sedimentological and morphological observations of the development and preservation of natural levees in the channelized and multi-thread river reaches were used to infer about the role of natural levee sediments in the long-term changes of channel conveyeance in channelized and freely developing river reaches.

4 Results

4.1 Differences in Hydraulic Parameters Between Incised, Channelized and Multi-thread Cross-Sections

Average values of flood discharges in the channelized reach are larger by 5 % than those in the upstream incised reach and smaller by 3 % than in the downstream multi-thread reach. These small differences in flood discharges allow us to compare hydraulic parameters of flood flows between the three reaches. Incised, channelized, and multi-thread cross-sections differ significantly in energy grade slope and the hydraulic parameters characterizing geometry of flood flows (Fig. 2; Table 1). Energy grade slope of the flows in incised cross-sections equals about 0.012 and is significantly steeper than that typifying the flows in channelized and multi-thread cross-sections ($\sim$0.007). Mean flow depth in the multi-thread cross-sections is about twice smaller than in the incised and channelized cross-sections. In turn, flow width in the multi-thread cross-sections is 3.5–4 times larger than in the incised and channelized cross-sections. As a result of such differences, cross-sectional flow area in the multi-thread cross-sections is about twice as large as in the channelized cross-sections and even more than in the incised ones (Fig. 2).

Parameters characterizing hydrodynamics of flood flows also differ significantly between the three river reaches (Table 1); however, the pattern of these differences varies among the hydraulic parameters (Fig. 2). Mean flow velocity in multi-thread cross-sections is lower by about half than that in incised cross-sections and by slightly less than half in comparison with that in channelized cross-sections. Mean bed shear stress in the multi-thread cross-sections is also lower by half than that in the channelized cross-sections, but more than three times lower than that in the incised cross-sections. Unit stream power in the multi-thread cross-sections is lower

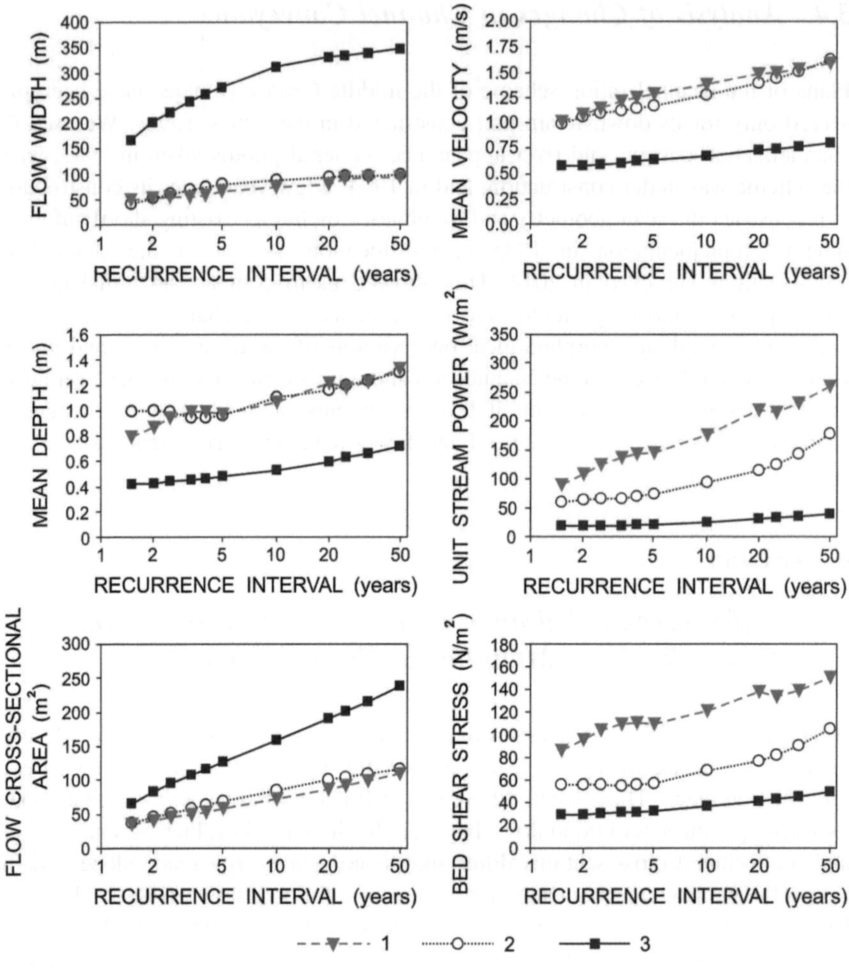

Fig. 2 Selected hydraulic parameters of the flood flows of given recurrence interval averaged for the investigated cross-sections from the incised, channelized, and multi-thread reaches of the Czarny Dunajec. *1–3* means of the parameters for: *1* incised, *2* channelized, and *3* multi-thread cross-sections

than in the channelized cross-sections from 3.3 times at a 1.5-year flow to 4.5 times at a 50-year flood, whereas it is lower than the respective values in the incised cross-sections from 5.0 to 6.5 times at these flood frequencies. Froude number, which varies with flow velocity and depth, presents the most complicated pattern; while the lowest values of the parameter typify the multi-thread cross-sections, the highest values are recorded in the incised cross-sections at low flood flows, but in the channelized cross-sections at high flood flows.

Table 1 Average values of the analysed hydraulic parameters characterizing the flows of given recurrence interval in the incised (IN), channelized (CH), and multi-thread (MT) cross-sections of the Czarny Dunajec River, and results of a Kruskal-Wallis test for the significance of difference between the averages for the three cross-section types (statistically significant differences are indicated in bold)

Recurrence interval (years)	Flow width (m)				Mean depth (m)				Flow area (m^2)			
	IN	CH	MT	*p* value	IN	CH	MT	*p* value	IN	CH	MT	*p* value
2	53.0	51.9	202.0	**0.0001**	0.87	1.00	0.43	**0.0001**	41.1	45.1	83.0	**0.0002**
5	65.0	79.0	272.5	**0.0001**	0.97	0.96	0.48	**0.0001**	58.1	69.1	126.3	**0.0001**
10	74.9	86.8	311.0	**0.0001**	1.06	1.10	0.53	**0.0001**	71.9	85.1	158.4	**0.0001**
20	78.6	93.4	331.3	**0.0001**	1.22	1.16	0.60	**0.0004**	86.1	100.0	190.9	**0.0001**
50	94.3	98.1	348.3	**0.0001**	1.33	1.30	0.71	**0.0008**	109.6	116.3	237.5	**0.0001**
Recurrence interval (years)	Mean velocity ($m\ s^{-1}$)				Unit stream power ($W\ m^{-2}$)				Bed shear stress ($N\ m^{-2}$)			
	IN	CH	MT	*p* value	IN	CH	MT	*p* value	IN	CH	MT	*p* value
2	1.08	1.05	0.58	**0.0002**	108.7	63.2	17.5	**0.0002**	96.1	55.6	29.5	**0.0003**
5	1.28	1.17	0.63	**0.0001**	146.2	73.0	21.0	**0.0001**	109.4	57.2	32.8	**0.0005**
10	1.38	1.27	0.67	**0.0001**	176.5	94.4	24.8	**0.0001**	121.0	68.2	36.4	**0.0008**
20	1.49	1.39	0.72	**0.0001**	218.4	114.3	30.2	**0.0001**	138.3	77.2	41.2	**0.0003**
50	1.59	1.63	0.80	**0.0001**	259.7	178.4	39.8	**0.0001**	150.7	105.3	49.0	**0.0004**

4.2 Patterns of the Variation in Hydraulic Parameters Among the Cross-Sections at Different Flood Frequencies

Principal component analysis indicated that patterns of the variation in hydraulic parameters among the studied cross-sections vary with the frequency of flood discharges. In the case of a 2-year flood, the two first components of the PCA explained 99.2 % of the total variation in the hydraulic parameters, with 83.7 % of the variation accounted for by the first component and 15.5 % by the second one. The first component correlated positively with unit stream power ($r = 0.76$), mean bed shear stress ($r = 0.72$), mean flow velocity ($r = 0.72$), and mean flow depth ($r = 0.72$), while negatively with flow width ($r = -0.98$), and cross-sectional flow area ($r = -0.92$). The second component correlated positively with energy grade slope ($r = 0.72$), mean bed shear stress ($r = 0.68$), and unit stream power ($r = 0.65$) (Fig. 3b). The positions of multi-thread cross-sections were clearly differentiated along the first ordination axis from those of incised and channelized cross-sections, with the former typified by negative scores and almost all the latter by positive scores of the component (Fig. 3a). The multi-thread cross-sections and the channelized cross-sections were typified by the same range of negative and positive scores of the second component, whereas the incised cross-sections exhibited only positive scores of the component (Fig. 3a).

The PCA performed for 50-year flood conditions revealed a somewhat different pattern. In that case, the degree of explanation of the total variance by the two first components of the PCA was nearly the same—98.4 %—but the first component explained 79.4 % of the total variation in hydraulic parameters and the second one 19.0 %. The first ordination axis correlated positively with unit stream power ($r = 0.88$), mean bed shear stress ($r = 0.85$), mean flow velocity ($r = 0.79$), and mean flow depth ($r = 0.79$). It also correlated negatively with flow width ($r = -0.92$), and cross-sectional flow area ($r = -0.86$). The second ordination axis correlated positively with mean bed shear stress ($r = 0.49$), and unit stream power ($r = 0.48$) (Fig. 3d). Multi-thread cross-sections, with negative scores of the first component, now had their positions even more clearly differentiated from the positions of the two other cross-section types along the first ordination axis. Channelized cross-sections were typified by lower and incised cross-sections by somewhat higher positive scores of the first component, but these cross-section types exhibited considerable overlap of their positions along the first ordination axis (Fig. 3c). The cross-sections from the three river reaches were relatively little differentiated along the second component; the exception was the cross-section from the uppermost, bedrock part of the incised reach, which distinguished by its high positive score of the component (Fig. 3c).

Because differences along the first component of the PCA explain most of the total variation in hydraulic conditions among the studied cross-sections, we illustrate them by comparing exemplary incised, channelized, and multi-thread cross-sections, with the first two typified by positive scores and the third one by

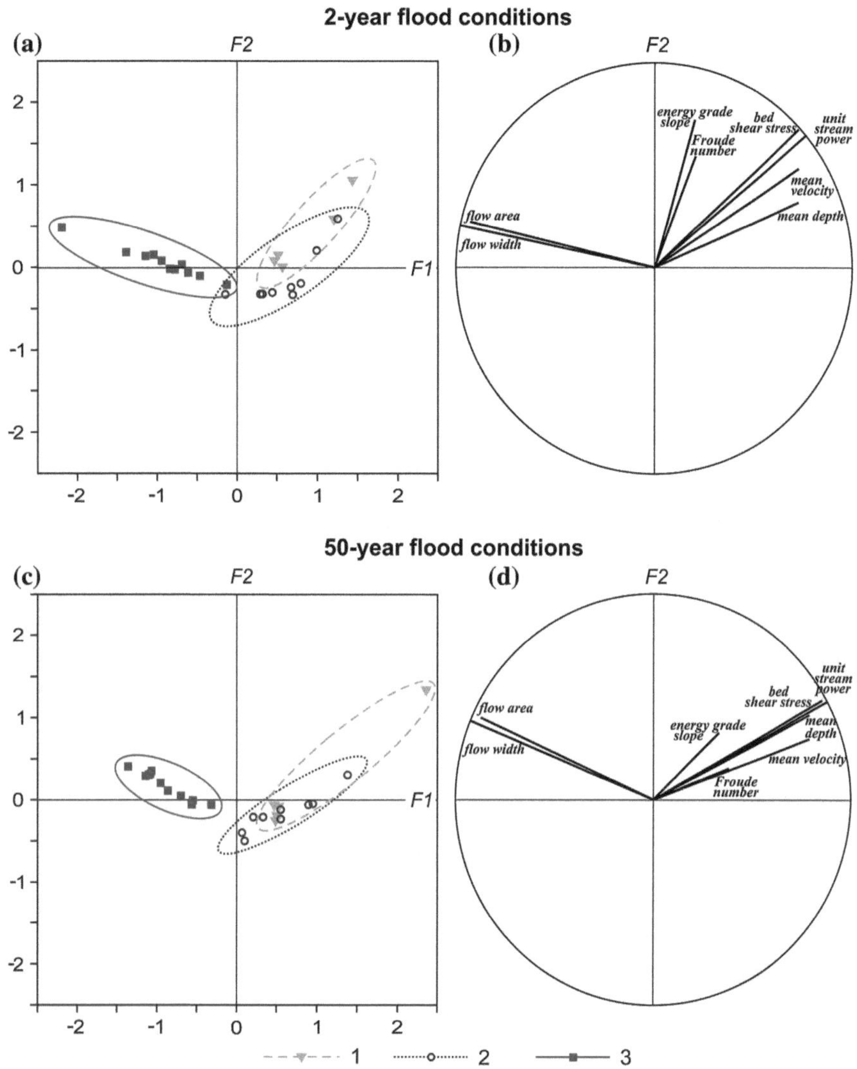

Fig. 3 Results from principal component analysis for the hydraulic parameters associated with the flow of 2-year frequency (*upper diagrams*) and 50-year frequency (*lower diagrams*) in the incised, channelized, and multi-thread cross-sections of the Czarny Dunajec River. **a**, **c** positions of the cross-sections on the first factorial plane, with envelopes of the positions of cross-sections from particular reaches; **b**, **d** correlation circles identifying the positions of hydraulic parameters on the first two PCA axes. *1* incised cross-sections; *2* channelized cross-sections; *3* multi-thread cross-sections

a negative score of the component (Figs. 3 and 4). In each of the cross-sections, water levels associated with 2-year and 50-year flood discharges are indicated (Fig. 4). In multi-thread cross-section 15, the channel is wide and the floodplain also has a considerable extent. Consequently, width and cross-sectional area of the flood flows in the cross-section are large. In incised cross-section 5, the channel is relatively deep and has moderate width, whereas the highly elevated floodplain is narrow. Here, flow width corresponds to only 31 % of that in the multi-thread cross-section at the 2-year flood and to 29 % at the 50-year flood, whereas the respective values of cross-sectional flow area amount to 63 and 57 % of that in the multi-thread cross-section. In channelized cross-section 9, the channel is narrow but the floodplain has moderate width. At the 2-year flood, the flow width corresponds to 30 % and the cross-sectional flow area to 64 % of their values in the multi-thread cross-section, whereas at the 50-year flood the respective values of the parameters equal 34 and 67 % of their values in the multi-thread cross-section (Fig. 4). These differences in flow geometry are reflected in marked discrepancies in dynamic characteristics of flood flows between the cross-sections. Mean velocity in the incised cross-section exceeds that in the multi-thread cross-section by 49 % at the 2-year flood and by 61 % at the 50-year flood; in the channelized cross-section it is larger than in the multi-thread cross-section by 51 % at the 2-year flood and by 43 % at the 50-year flood. Mean bed shear stress in the incised cross-section exceeds that in the multi-thread cross-section 2.8 times at both flood frequencies; in the channelized cross-section it is 1.7 times larger than in the multi-thread cross-section at the 2-year flood and 2.6 times larger at the 50-year flood. Finally, unit stream power in the incised cross-section exceeds that in the multi-thread cross-section from 4 to 4.5 times, whereas in the channelized cross-section it is 2.6 times larger at the 2-year flood and 3.7 times at the 50-year flood (Fig. 4).

4.3 Differences Between Initial and Finally Used Roughness Values Indicated by the Calibration of Modelled Water Stages

Observations of the peak stage of the 20-year flood of May 2014 in the study cross-sections enabled us to calibrate roughness coefficients used in the hydraulic modelling of flood flows. In single-thread, incised or channelized cross-sections, calculated water levels were consistently higher than those observed during the flood, and a reduction of roughness coefficients was necessary to adjust the calculated water levels. In order to obtain the agreement between the calculated and observed water levels, maximum water depth at the 20-year flood in the cross-sections had to be reduced by 1–5 %. In these cross-sections, a large proportion of the total flow was conveyed in the channel zone and thus we reduced roughness coefficients for this part of the river cross-sections to adjust the modelled water levels. In most cases, the Manning roughness coefficient determined on the basis of granulometry analysis of the surface bed material had to be reduced by 10 % and for one cross-section, 20 % reduction was necessary.

For most multi-thread cross-sections, water levels calculated on the basis of initially obtained roughness coefficients agreed with the observed peak stage of the

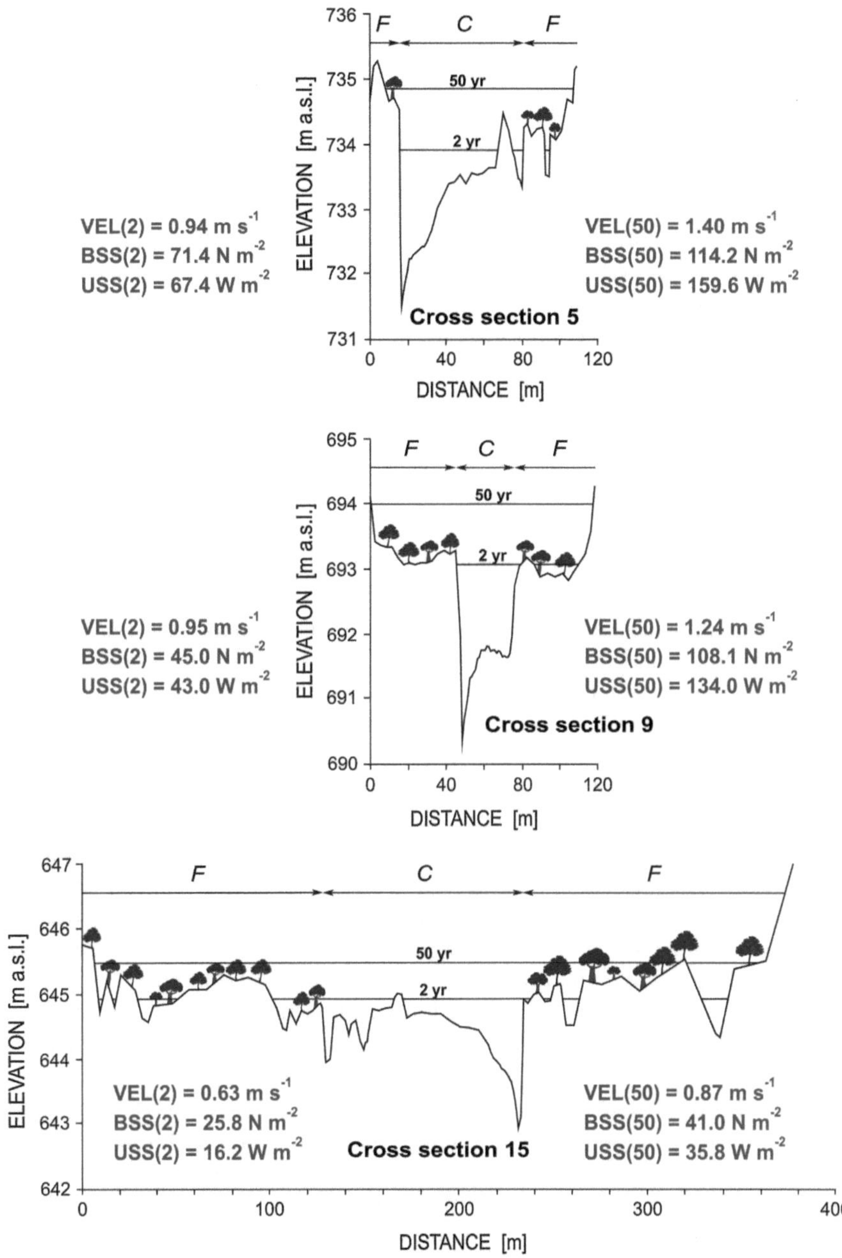

Fig. 4 Extent of inundation and the values of mean flow velocity (*VEL*), mean bed shear stress (*BSS*), and unit stream power (*USS*) at 2-year (*left values*) and 50-year flood discharges (*right values*) in cross-section 5 from the incised reach of the Czarny Dunajec, in cross-section 9 from the channelized reach, and in cross-section 15 from the multi-thread reach. Morphological zones of the cross-sections: *C* channel; *F* floodplain

flood. However, in a few cross-sections water levels calculated for the 20-year flood were lower than the observed ones. These cross-sections are located at a distance of a few hundred metres from the beginning of the multi-thread reach, where the river morphology changes from bar-braided to island-braided and where especially intense deposition of large wood occurred during the flood (Fig. 5). To obtain the agreement between the calculated and observed water levels, maximum water depth at the 20-year flood had to be increased by 2–3 %. Most of the width of these cross-sections was located on the floodplain, and thus the adjustment of the calculated water levels was attained by increasing the Manning's roughness coefficients for vegetated parts of the cross-sections by 15–20 % (but always within the range indicated by Chow (1959) for a given type of vegetation cover).

4.4 *Changes in the Conveyance of the Channelization Scheme*

The channelization scheme of the middle Czarny Dunajec, with a regulated channel partitioned by concrete drop structures and with flood embankment(s) on one or both sides of artificially constricted floodplain, was constructed in the 1960s–1990s. The construction plans for its downstream part built in the 1980s–1990s document the scheme to have been designed to convey a flow of at least 330 $m^3\ s^{-1}$ that was then considered a 100-year flood in the middle river course (based on a short record period with a few larger floods in the 1970s at the Koniówka gauging station). Reconstruction of the channel geometry formed in the course of the channelization at the location of the current cross-section 11 indicated that in 1985 the river flowed in a flat-bottomed channel bounded on both sides by narrow, gravelly floodplains with smooth bulldozed surface; on the right side, the floodplain was bordered by a flood embankment (Fig. 6). The channel conveyed a 2-year flood, whereas the whole channelized cross-section had a capacity of 433 $m^3\ s^{-1}$ equal to $\sim$1000-year flood according to the current flood frequency estimates (Table 2).

Over the next 30 years the river formed a sinuous thalweg, undercutting opposite banks between successive pairs of drop structures. As a result, the cross-section became triangular and a side bar was deposited along its left bank (Fig. 6), considerably reducing channel capacity to the elevation of the former channel banks. At the same time, prominent natural levees were formed along the channel margins, elevating bank edges and considerably reducing the cross-sectional area of floodplain flows, especially on the right-side floodplain (Fig. 6). Finally, a dense, alder-willow forest developed on the floodplains (Fig. 7); the forest, together with butterbur understorey developing during a vegetation season radically increased hydraulic roughness of the floodplains. As a result of all these changes, the whole channelized cross-section now can convey a flow of 194 $m^3\ s^{-1}$ with a recurrence interval of 72 years (Table 2). This means that the conveyance of the channelization scheme was reduced by more than half in terms of

Fig. 5 Wood jams deposited during the flood of May 2014 on a channel bar in the island-braided part of the multi-thread reach of the Czarny Dunajec

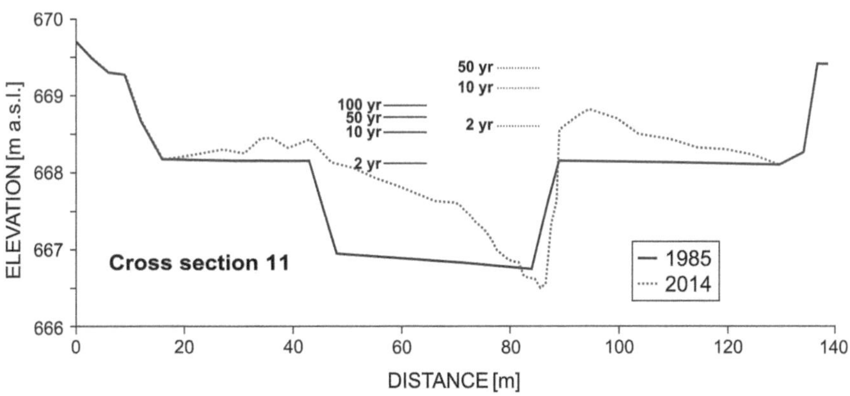

Fig. 6 Channelized cross-section 11 in 1985, shortly after the construction of the channelization scheme, and in 2014. Water levels associated in 1985 and 2014 with flood discharges of given recurrence intervals are indicated

discharge and about 14 times in terms of flood frequency. Water stages associated with the discharges lower than 194 m^3 s^{-1} increased from 0.48 to 0.66 m (Table 2). Higher discharges will overtop the flood embankment and its height would have to be increased by 0.87 m to allow the channelized cross-section to accommodate the flow of 433 m^3 s^{-1} equal to the initial conveyance of the channelization scheme at the analysed cross-section (Table 2).

Table 2 Changes in water stage of the Czarny Dunajec associated with the discharges of given recurrence intervals in channelized cross-section 11 after the construction of the channelization scheme in 1985 and in 2014, and the current vertical distance of the water stages from the crest of the flood embankment

Recurrence interval (years)	Discharge (m^3/s)	Stage in 1985 (m a.s.l.)	Stage in 2014 (m a.s.l.)	Stage increase between 1985 and 2014 (m)	Vertical distance to the embankment crest in 2014 (m)
1.5	36.0	667.95	668.47	0.52	0.94
2	46.0	668.12	668.60	0.48	0.81
3.33	62.5	668.29	668.78	0.49	0.63
5	76.9	668.39	668.91	0.52	0.50
10	102.8	668.52	669.09	0.57	0.32
20	133.0	668.63	669.24	0.61	0.17
33	158.3	668.69	669.31	0.62	0.10
50	182.2	668.72	669.36	0.64	0.05
72	194.0	668.76	**669.41**	0.66	0.00
100	227.8	668.87	*669.56*	*0.69*	*−0.15*
340	330.0	669.16	*669.95*	*0.79*	*−0.53*
~1000	433.0	**669.41**	*670.28*	*0.87*	*−0.87*

The elevation of the embankment crest at the cross-section is indicated in bold. Values in italics refer to water stages higher than the embankment crest, that would be attained if the cross-section were bordered by a flood embankment higher than the current one

4.5 *Development of Natural Levees in Channelized and Unmanaged River Reaches*

During 30 years of the functioning of the channelization scheme, natural levee sediments were deposited along the channel; in the analysed cross-section 11, they reach 0.3 m in thickness on the left channel side and 0.7 m on the right side (Fig. 6). The cross-section is located within a gentle channel bend and the formation of natural levees was more intense along the outer channel bank. The intensity of natural levee formation depends on flood magnitude determining the depth of floodplain inundation and the intensity of suspended sediment transport, and up to a half of the total thickness of the natural levees might have been deposited during the 20-year flood in May 2014 (Fig. 7). Bank reinforcement in the channelized reach prevents erosion of the natural levees and thus they are permanent fluvial forms enduringly reducing the conveyance of the channelization scheme.

Similar thickness of the deposition of natural levee sediments during the flood of May 2014 (up to 20–30 cm) was also observed along the outer channel banks in the multi-thread river reach (Fig. 8). However, here the natural levee deposits could have been soon completely eroded in the course of outer bank retreat and floodplain reworking by the laterally migrating channel (Fig. 9), especially where the

Fig. 7 Sandy sediments of a natural levee deposited in channelized cross-section 11 during the flood of May 2014, visible above a *thin, dark layer*. 0.5-m scale on the rod. Alder-willow forest grows on the floodplain in the channelized river reach

retreating bank was not reinforced with the roots of larger trees. Such was the situation in multi-thread cross-section 21 where natural levee deposits formed by the flood of May 2014 were eroded by the laterally migrating channel during 1.5 year after the flood (Fig. 9). In this way a hydraulic impact of the formation of the natural levee (markedly lower than in the channelized reach because of much larger lateral extent of the floodplain in the multi-thread reach) was removed in a relatively short period after the major flood.

5 Discussion

5.1 Hydraulic Conditions in the River Reaches of Different Morphology

Hydraulic conditions in rivers are usually complex flows (i.e., unsteady, non-uniform flows of high Reynolds number in a complex geometry), but they can be analysed with use of 1D or 2D modelling (Horritt and Bates 2002). Whilst 1D approach is often considered a gross simplification of the flow field (Knight and Shiono 1996), one can justify the approach by assuming that the approximations involved in treating out-of-bank flow as one-dimensional are small compared to

Fig. 8 Sandy sediments of a natural levee deposited near to the outer channel bank in multi-thread cross-section 21 during the flood of May 2014. The sediments were eroded by the laterally migrating channel during 1.5 year after the flood

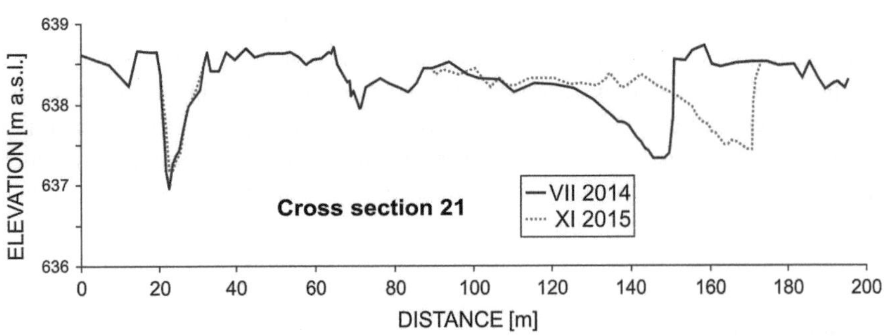

Fig. 9 Multi-thread cross-section 21 in July 2014, shortly after the 20-year flood of May 2014, and in November 2015. Visible lateral shift of the right, main braid

other uncertainties (Ali and Goodwin 2002). Moreover, in-channel flow varies much more in the downstream direction than in the cross-stream or vertical directions. In this situation, it is common (Knight and Shiono 1996) to treat flow using the one-dimensional Saint Venant equations (Bates et al. 2014). In addition, two-dimensional hydraulic modelling requires detailed knowledge of the channel and floodplain geometry (Bates et al. 1992) but it was hardly available for a longer river stretch shortly after the major flood that had considerably changed the river geometry in the incised and multi-thread reaches (cf. Mikuś et al. 2016). As demonstrated by Horritt and Bates (2002) for a 60 km reach of the River Severn, UK, one-dimensional and two-dimensional models performed equally well. This suggests that although the assumptions about flow conditions required for 1D hydraulic modelling (Jowett and Duncan 2012) may not be fully met for the multi-thread reach or for overbank flooding, the additional energy losses can be compensated for using a calibrated friction coefficient (this is discussed more in detail in Sect. 5.2). Importantly, the approach allowed us to obtain the detailed river geometry in the studied cross-sections and to extend the study to a 15 km-long river stretch.

A previous reconstruction of unit stream power at the peak flow of a 7-year flood in 2001 indicated considerable variation of the parameter along the river (Wyżga and Zawiejska 2005). This study has demonstrated that marked differences in river morphology and the type of channel management between particular reaches of the Czarny Dunajec are reflected in significant differences of the hydraulic conditions of flood flows. Flood flows conveyed in the incised and channelized reaches are typified by markedly smaller width and cross-sectional area, and markedly greater depth than those in the multi-thread reach. This difference in flow geometry is reflected in substantially higher average flow velocity in the incised and channelized reaches that in the multi-thread reach. With similar flow depths in the incised and channelized reaches, flood flows in the incised reach are typified by steeper energy grade slope than in the channelized reach; as a result, values of mean bed shear stress and unit stream power are the highest in the incised reach, intermediate in the channelized reach, and the smallest in the multi-thread reach. The recognised differences in the energy of flood flows and in tractive forces exerted on the flow boundary underlie and explain different evolutionary tendencies of particular river reaches during the past decades, with rapid channel incision occurring in the upper part of the studied river stretch and slow aggradation in the multi-thread reach (Zawiejska and Wyżga 2010). They also explain marked differences in the size of bed material between the three river reaches (Zawiejska et al. 2015) and in the potential for either transport or deposition of large wood between river reaches with narrow and wide channel morphologies (Wyżga and Zawiejska 2005; Ruiz-Villanueva et al. 2016a, b; Wyżga et al., in press).

The study confirmed usefulness of principal component analysis to reveal how the patterns of differences in hydraulic conditions among the studied river cross-sections and reaches vary with the frequency of flood discharges (cf. Czech et al. 2016). The analysis indicated that differences in the patterns of hydraulic conditions among the three reaches are most pronounced at low flood discharges when the flows are conveyed within the channels of different cross-sectional geometry and slope. At high flood discharges, differences in hydraulic conditions among the reaches slightly diminish as a result of greater similarities in cross-sectional flow geometry and energy grade slope between the incised and channelized reaches that also tend to reduce differences in flow energy and tractive forces between the reaches. However, the high flood discharges accentuate differences in hydraulic conditions among particular cross-sections of a given reach, with most spectacular differences between four cross-sections from the alluvial part of the incised reach and a cross-section from the bedrock part of this reach.

5.2 Roughness Calibration as a Way to Integrate Various Effects on the Accuracy of Flood-Flow Modelling

In the context of flood modelling in a river reach, Manning's n roughness coefficient is an important source of uncertainty, among others such as errors in the input data (i.e., hydrological data and quantile estimation), model uncertainty and error in observed data (Bates et al. 2014). Therefore, when roughness parameter is being calibrated, we are also making allowance for other physical and non-physical effects (i.e., model structural error and energy losses or flow processes which are not adequately reproduced by the model). The different morphologies analysed in this work—incised, regulated and multi-thread channels—show different sources of frictional losses, which must be incorporated into the 1D hydraulic resistance term, using Manning roughness. We could constrain from field data only certain components of the hydraulic resistance, such as bed-material grain size (based on the granulometry analysis) and drag due to vegetation (based on Chow 1959). However, our model results using the initial estimated roughness values showed that water level simulated for the peak discharge of a 20-year flood of May 2014 was higher than the observed levels in single-thread, incised and channelized cross-sections, but equal to or lower than that observed in multi-thread cross-sections. To attain the agreement between the simulated and the observed water levels, it was thus necessary to reduce roughness coefficients in the single-thread cross-sections and to increase them in some multi-thread cross-sections. A few factors might have contributed to these differences in the direction of necessary correction of roughness coefficients between single-thread and multi-thread river cross-sections. First, the size of surface bed material measured at base-flow conditions seems poorly representative of the conditions existing at high flood flows when bedload is vigorously transported by the river. Flume experiments of Dietrich et al. (1989) indicated that intense transport of sand and gravel in a gravel-bed channel is associated with bed smoothing by sand grains deposited between gravel particles; this reduces the height of roughness elements on the riverbed and a lower portion of the total flow energy is used to overcome the resistance to flow resulting from bed roughness. Parker et al. (2008) suggested that the adjustment of bed-surface size distribution may not require a change in the grain size of bed material but is attributed to a transient layer of bedload. This factor should reduce bed resistance to flow at both single- and multi-thread river morphologies, although its effect may be stronger in a single-thread channel where more spatially concentrated bedload layer tends to be thicker. According to the equation of Strickler (1923) used in this study for calculation of bed roughness, a 10 % reduction in roughness coefficient would require 45 % reduction in the median grain size of the riverbed if the fining of bed surface at flood conditions were the only reason for the discrepancy between the simulated and observed water levels.

Large wood deposits can considerably increase resistance to flow (Dudley et al. 1998) as they induce local backwater effects and flow non-uniformity. The inventories of large wood performed in the Czarny Dunajec after former floods indicated that the multi-thread river reach retains 1–2 orders of magnitude greater amounts of wood per unit river area than the single-thread reaches (Wyżga and Zawiejska 2005, 2010). Considerable amounts of wood were deposited in the multi-thread reach also during the flood of May 2014 and they must have increased resistance to flow in the vicinity of multi-thread cross-sections. However, as the large wood was mostly deposited on channel bars and the floodplain, its presence was not reflected in the grain size of bed material within low-flow channels, measured at base-flow conditions.

Finally, the use of post-flood channel and floodplain geometry in the assessment of hydraulic parameters at the peak flow of major floods may be also a source of error as changes in river geometry, including those in channel width, may continue after the flood peak (cf. Surian et al. 2016). In the Czarny Dunajec, the post-flood survey in multi-thread cross-sections might have overestimated the channel width and capacity existing at the peak flow of the 2014 flood, hence resulting in too low simulated flood levels in comparison with the observed ones. In turn, if the deposition of natural levee sediments in channelized cross-sections continued during the flood recession, it might have reduced flow capacity of the cross-sections in relation to that existing at the flood peak, hence leading to the overestimation of simulated flood levels in comparison with the observed ones.

The calibration of roughness coefficients most likely integrated the above mentioned effects that variously affected the accuracy of modelling of peak water levels in river cross-sections of different morphology.

5.3 *Influence of Overbank Deposition on Channel Conveyance*

Rivers that can actively migrate within their floodplains tend to maintain similar flow capacity of their channels through continual erosion of the higher floodplain along concave banks and the formation of a new floodplain at lower elevation along convex banks. A part of this process is the formation and erosion of natural levees —ridges composed of relatively coarse overbank deposits that form along channels and attain the height inversely proportional to the rate of channel migration (Hudson 2005). As illustrated by the multi-thread reach of the Czarny Dunajec, vertically stable mountain rivers tend to rework their floodplains rapidly, and this prevents the formation of thick cover of overbank deposits and prominent natural levees and an excessive increase in channel capacity. The situation changes with artificial stabilization of river banks in the course of channel regulation. Elimination of channel migration causes that the elevation of floodplain surface progressively increases with each successive episode of floodplain inundation and natural levees rapidly grow in height. If the river is embanked, as in the channelized reach of the Czarny Dunajec, deposition of floodplain sediments will lead to a progressive reduction of the channelization-scheme conveyance and increased hazard of overtopping of flood embankments during major floods. The case study of the Czarny Dunajec presented in this paper demonstrates that to remain effective, a channelization scheme requires costly and time-consuming unsustainable maintenance, including periodic clearance of floodplain vegetation (to keep low surface roughness of the floodplain) and the removal of overbank deposits after major floods.

6 Concluding Remarks

This study has demonstrated distinct patterns of hydraulic parameters of flood flows in the incised, channelized and multi-thread reaches of the mountainous Czarny Dunajec River. Differences in the cross-sectional geometry and the energy grade slope between the reaches are reflected in significant differences in flow energy and tractive forces. More than half a century after the period of intense in-channel gravel mining that disturbed equilibrium conditions and initiated rapid channel incision in the upstream part of the middle river course (Zawiejska et al. 2015), even relatively high flood flows are conveyed within the deep channel and the resultant high shear forces prevent the accumulation of bed material and bed aggradation. A hydrological consequence of the channel incision is a dramatic reduction of floodplain storage of floodwater (cf. Wyżga et al. 2016) and the rapid passage of flood waves through the unmanaged reach that would otherwise be very suitable for such storage because of the lack of human investment on the river banks. In the channelized reach, the reduction of channel gradient by drop structures reduces flow energy and tractive forces in comparison with the incised reach. However, because of the floodplain constriction by flood embankment(s), flow velocities are nearly as high as in the incised reach, also resulting in the rapid evacuation of floodwater from the reach. Only in the multi-thread reach, the wide and shallow active channel with complex geometry and the extensive floodplain and islands enable slow velocities of flood flows and provide suitable conditions for the channel and floodplain storage of floodwater. Currently, incised and regulated channel types predominate in Polish Carpathian rivers (Wyżga 2001a; Wyżga et al. 2016) and thus their hydrological impacts on flood hazard at the regional scale must be considered highly detrimental.

Stabilization of river banks by channelization structures prevents floodplain reworking by the laterally migrating channel and initiates a progressive growth in floodplain elevation. In the rivers that are embanked or bounded by relatively low, managed terraces this process leads to a reduction in the conveyance of their channelization schemes or floodplains and to increased flood hazard to the areas located outside the flood embankments or on the low terraces. This flood hazard may suddenly manifest during large floods that can transport large amounts of sediment and deposit them along regulated channels, resulting in the overtopping and failure of flood embankments and catastrophic inundation of the areas outside the embankments (cf. Lenar-Matyas et al. 2013). The recognition of the negative impact of overbank sedimentation along regulated channels leads to two conclusions. Wherever bank stabilization is not necessary to eliminate the erosional hazard to valuable managed riparian areas, channel regulation should be avoided and floodplain reworking by the laterally migrating river should be allowed. Where channel regulation and bank stabilization are unavoidable as the protection of settlements and infrastructure that cannot be achieved otherwise, they should be followed by maintenance operations allowing to keep low floodplain roughness and periodically remove overbank sediments that reduce the channelization-scheme conveyance.

Acknowledgments This study was completed within the scope of the Research Project DEC-2013/09/B/ST10/00056 financed by the National Science Centre of Poland.

References

Ali S, Goodwin P (2002) The predictive ability of 1D models to simulate floodplain processes. In: Falconer RA, Lin B, Harris EL et al. (eds) Hydroinformatics 2002. Proceedings of the fifth international conference on hydroinformatics. Volume one: model development and data management. IWA Publishing, London, pp 247–252

Bates PD, Anderson MG, Baird L, Walling DE, Simm D (1992) Modelling floodplain flows using a two-dimensional finite element model. Earth Surf Proc Landforms 17:575–588

Bates PD, Pappenberger F, Romanowicz RJ (2014) Uncertainty in flood inundation modelling. In: Beven K, Hall J (eds) Applied uncertainty analysis for flood risk management. Imperial College Press, London, pp 232–269

Burkham DE (1976) Effects of changes in an alluvial channel on the timing, magnitude, and transformation of flood waves, Southeastern Arizona. US Geol Surv Prof Paper 655-K:1–25

Chow VT (1959) Open-channel hydraulics. McGraw-Hill, New York

Czech W, Radecki-Pawlik A, Wyżga B, Hajdukiewicz H (2016) Modelling the flooding capacity of a Polish Carpathian river: a comparison of constrained and free channel conditions. Geomorphology 272C. doi:10.1016/j.geomorph.2015.09.025

Dietrich WE, Kirchner JW, Ikeda H, Iseya F (1989) Sediment supply and the development of the coarse surface layer in gravel-bedded rivers. Nature 340:215–217

Dudley SJ, Fischenich JC, Abt SR (1998) Effect of woody debris entrapment on flow resistance. J Am Water Resour Assoc 34:1189–1197

Horritt MS, Bates PD (2002) Evaluation of 1D and 2D numerical models for predicting river flood inundation. J Hydrol 268:87–99

Hudson PF (2005) Natural Levees. In: Trimble S (ed) Encyclopedia of water science. Taylor and Francis, New York, pp 1–4

Jowett IG, Duncan MJ (2012) Effectiveness of 1D and 2D hydraulic models for instream habitat analysis in a braided river. Ecol Eng 48:92–100

Knight DW, Shiono K (1996) River channel and floodplain hydraulics. In: Anderson MG, Walling DE, Bates PD (eds) Floodplain Processes. Wiley, Chichester, pp 139–182

Kundzewicz ZW, Stoffel M, Kaczka RJ, Wyżga B, Niedźwiedź T, Pińskwar I, Ruiz-Villanueva V, Łupikasza E, Czajka B, Ballesteros-Canovas J, Małarzewski Ł, Choryński A, Janecka K, Mikuś P (2014) Floods at the northern foothills of the Tatra Mountains—a Polish-Swiss research project. Acta Geophys 62:620–641

Lach J, Wyżga B (2002) Channel incision and flow increase of the upper Wisłoka River, southern Poland, subsequent to the reafforestation of its catchment. Earth Surf Proc Landforms 27: 445–462

Lenar-Matyas A, Łapuszek M, Ratomski J (2013) Określenie wpływu sedymentacji rumowiska na przepustowość międzywala (Determination of debris sedimentation and plant cover influence on inter-embankment zone capacity). Gosp Wodna 8:303–305

Leopold LB, Wolman MG, Miller JP (1964) Fluvial processes in geomorphology. Freeman, New York

Mikuś P, Wyżga B, Radecki-Pawlik A, Zawiejska J, Amirowicz A, Oglęcki P (2016) Environment-friendly reduction of flood risk and infrastructure damage in a mountain river: case study of the Czarny Dunajec. Geomorphology 272C. doi:10.1016/j.geomorph.2015.11.003

Parker G, Hassan MA, Wilcock P (2008) Adjustment of the bed surface size distribution of gravel-bed rivers in response to cycled hydrographs. In: Habersack H, Piégay H, Rinaldi M (eds) Gravel-bed rivers VI: from process understanding to river restoration. Elsevier, Amsterdam, pp 241–289

Rinaldi M, Wyżga B, Surian N (2005) Sediment mining in alluvial channels: physical effects and management perspectives. River Res Applic 21:805–828

Ruiz-Villanueva V, Wyżga B, Hajdukiewicz H, Stoffel M (2016a) Exploring large wood retention and deposition in contrasting river morphologies linking numerical modelling and field observations. Earth Surf Proc Landforms 41:446–459

Ruiz-Villanueva V, Wyżga B, Zawiejska J, Hajdukiewicz M, Stoffel M (2016b) Factors controlling large wood transport in a mountain river. Geomorphology 272C. doi:10.1016/j.geomorph.2015.04.004

Stewardson M (2005) Hydraulic geometry of stream reaches. J Hydrol 306:97–111

Strickler A (1923) Beiträge zur Frage der Geschwindigkeitsformel und der Rauhigkeitszahlen fur Ströme. Kanäle und Geschlossene Leitungen, Bern

Surian N, Righini M, Lucía A, Nardi L, Amponsah W, Benvenuti M, Borga M, Cavalli M, Marchi L, Rinaldi M, Viero A (2016) Channel response to extreme floods: insights on controlling factors from six mountain rivers in northern Apennines, Italy. Geomorphology 272C. doi:10.1016/j.geomorph.2016.02.002

USACE (2010) HEC-RAS river analysis system version 4.1, user's manual. US Army Corps of Engineers Hydrologic Engineering Center, Davis, CA

Wyrick JR, Senter AE, Pasternack GB (2014) Revealing the natural complexity of fluvial morphology through 2D hydrodynamic delineation of river landforms. Geomorphology 210:14–22

Wyżga B (1993) River response to channel regulation: case study of the Raba River, Carpathians, Poland. Earth Surf Proc Landforms 18:541–556

Wyżga B (1996) Changes in the magnitude and transformation of flood waves subsequent to the channelization of the Raba River, Polish Carpathians. Earth Surf Proc Landforms 21:749–763

Wyżga B (2001a) A geomorphologist's criticism of the engineering approach to channelization of gravel-bed rivers: case study of the Raba River, Polish Carpathians. Env Manage 28:341–358

Wyżga B (2001b) Impact of the channelization-induced incision of the Skawa and Wisłoka Rivers, southern Poland, on the conditions of overbank deposition. Regul Rivers Res Manag 17:85–100

Wyżga B (2008) A review on channel incision in the Polish Carpathian rivers during the 20th century. In: Habersack H, Piégay H, Rinaldi M (eds) Gravel-bed rivers VI: From process understanding to river restoration. Dev Earth Surf Proc 11, Elsevier, Amsterdam, 525–555

Wyżga B, Zawiejska J (2005) Wood storage in a wide mountain river: case study of the Czarny Dunajec, Polish Carpathians. Earth Surf Proc Landforms 30:1475–1494

Wyżga B, Zawiejska J (2010) Large wood storage in channelized and unmanaged sections of the Czarny Dunajec River, Polish Carpathans: implications for the restoration of mountain rivers. Folia Geogr Ser Geogr-Phys 41:5–34

Wyżga B, Amirowicz A, Radecki-Pawlik A, Zawiejska J (2009) Hydromorphological conditions, potential fish habitats and the fish community in a mountain river subjected to variable human impacts, the Czarny Dunajec, Polish Carpathians. River Res Applic 25:517–536

Wyżga B, Hajdukiewicz H, Radecki-Pawlik A, Zawiejska J (2010) Eksploatacja osadów z koryt rzek górskich – skutki środowiskowe i procedury oceny (Exploitation of sediments from mountains river beds—environmental impact and evaluation procedures). Gosp Wodna 6:243–249

Wyżga B, Oglęcki P, Radecki-Pawlik A, Zawiejska J (2011) Diversity of macroinvertebrate communities as a reflection of habitat heterogeneity in a mountain river subjected to spatially variable human impacts. In: Simon A, Bennett SJ, Castro JM (eds) Stream restoration in dynamic fluvial systems: scientific approaches, analyses and tools. American Geophysical Union, Washington, pp 189–207

Wyżga B, Zawiejska J, Hajdukiewicz H (2015) Multi-thread rivers in the Polish Carpathians: occurrence, decline and possibilities of restoration. Quatern Int. doi:10.1016/j.quaint.2015.05.015

Wyżga B, Zawiejska J, Radecki-Pawlik A (2016) Impact of channel incision on the hydraulics of flood flows: examples from Polish Carpathian rivers. Geomorphology 272C. doi:10.1016/j.geomorph.2015.05.017

Wyżga B, Mikuś P, Zawiejska J, Ruiz-Villanueva V, Kaczka RJ, Czech W (in press) Log transport and deposition in incised, channelized, and multithread reaches of a wide mountain river: tracking experiment during a 20-year flood. Geomorphology

Zawiejska J, Wyżga B (2010) Twentieth-century channel change on the Dunajec River, southern Poland: patterns, causes and controls. Geomorphology 117:234–246

Zawiejska J, Wyżga B, Radecki-Pawlik A (2015) Variation in surface bed material along a mountain river modified by gravel extraction and channelization, the Czarny Dunajec, Polish Carpathians. Geomorphology 231:353–366

Part III
Deciphering Changes in Observation Records

Observed Changes in Air Temperature and Precipitation and Relationship between them, in the Upper Vistula Basin

Ewa Łupikasza, Tadeusz Niedźwiedź, Iwona Pinskwar, Virginia Ruiz-Villanueva and Zbigniew W. Kundzewicz

Abstract The chapter presents changes in air temperature and precipitation in the Upper Vistula Basin. Data from 18 meteorological stations covering the 1951–2015 period was used to investigate variability and trends in air temperature, precipitation and linkages between them. Air temperature in the Upper Vistula Basin was significantly rising during the research period. Distinct warming on annual scale started in the first half of the eighties. Spring and summer air temperatures have been significantly increasing; winter air temperature trends were much weaker while no significant changes were found in autumn. Upward trends in air temperature within the Upper Vistula Basin were also reflected in the frequency of thermally characteristic days which was significantly changing during the research period. Strong downward trends were found in the frequency of winter days (Tavg $\leq$ 0 °C) while trends in warm characteristic days were positive. Most observed changes in precipitation were not statistically significant at the level of 0.05. This indicates that significant increase in air temperature is not currently accompanied by significant increase in precipitation thus changes in precipitation are not directly related to changes in air temperature and they possibly vary in time. Relations between

E. Łupikasza (✉) · T. Niedźwiedź
Faculty of Earth Sciences, Department of Climatology, University of Silesia, ul. Będzińska 60, 41-200 Sosnowiec, Poland
e-mail: ewa.lupikasza@us.edu.pl

T. Niedźwiedź
e-mail: tadeusz.niedzwiedz@us.edu.pl

I. Pinskwar · Z.W. Kundzewicz
Institute for Agricultural and Forest Environment, Polish Academy of Sciences, ul. Bukowska 19, 60-809 Poznań, Poland
e-mail: iwona.pinskwar@isrl.poznan.pl

Z.W. Kundzewicz
e-mail: kundzewicz@yahoo.com

V. Ruiz-Villanueva
Dendrolab.ch, Institute of Geological Sciences, University of Bern, Baltzerstrasse 1+3, 3012 Bern, Switzerland
e-mail: virginia.ruiz@dendrolab.ch

Z.W. Kundzewicz et al. (eds.), *Flood Risk in the Upper Vistula Basin*,
GeoPlanet: Earth and Planetary Sciences, DOI 10.1007/978-3-319-41923-7_8

precipitation and air temperature are not straightforward. The results indicate that strong increase in air temperature is rather accompanied by decrease in precipitation frequency and amount. However, this applies only to overall precipitation totals and not to extreme events which are random and can occur unexpectedly. Flood precipitation can also occur during drier periods, as in the last decade of the research period. It however must be mentioned that for nearly all stations mean precipitation totals from the warmer period of 1991–2013 (1991–2011) were higher than in previous period 1961–1990. Moreover the study revealed that increasing, statistically significant, trend (from 0.15 to 0.24 °C/decade) in mean annual air temperature likely impacted changes in heavy precipitation. The links between air temperature ranges (0–10, 10–20 °C and above 20 °C) and precipitation ranges (0–10 mm, 10–20 mm, etc. to above 50 mm) for two periods examined at nine stations in the Upper Vistula Basin revealed that when air temperature exceeded 20 °C more intense precipitation was observed in the second warmer period 1991–2013.

Keywords Air temperature • Precipitation • Trends • Climate change • The Upper Vistula Basin

1 Introduction

The Upper Vistula River Basin (about 51,000 km^2) is located in the southern and south-eastern part of Poland (Fig. 1). It is the most differentiated part of the country with regard to natural environment, with elevation varying between 140 m in Sandomierz Basin up to 2499 m in the Polish part of the Tatra Mountains and to 2655 m in Slovakia (Gerlach). This is the highest range of the whole Carpathian Mountains.

Four climatic zones were distinguished within the Upper Vistula Basin on the basis of annual course of air temperature and precipitation (Niedźwiedź and Obrębska-Starklowa 1991; Fig. 1). The mountain climate region (A) encompasses the highest part of the Carpathian Mountains above 900 m, with mean annual temperature lower than 7 °C and precipitation above 900 mm. Hess (1965) counted this region among the fifth vertical climatic belt categorized on the base of vertical profile of mean annual air temperature. The moderately cool climate with annual temperature below 4 °C extends to 1150 m. The cool belt (temperature between 4 and 2 °C) reaches the upper tree line (1520 m). The very cool subalpine belt with mean annual air temperature between 2 and 0 °C is located between 1520 and 1850 m. The moderately cold alpine belt stretches to the climatic snow line (2200 m) which is marked by location of the isotherm of −2 °C. Above this line the cold belt reaching the highest summits of the Tatra Mountains is located. The lowest part of the Carpathians (250–900 m) is encompassed by the temperate-warm climatic region (B) with the mean annual temperature between 7 and 8 °C and precipitation between 700 and 900 mm. The warmest and moderately dry climate (region C) embraces the Oświęcim-Sandomierz Basin. The upland climatic region (D) is cooler and more humid comparing to region C. In the Holy-Cross Mountains (611 m) near Kielce mean annual temperature drops to about

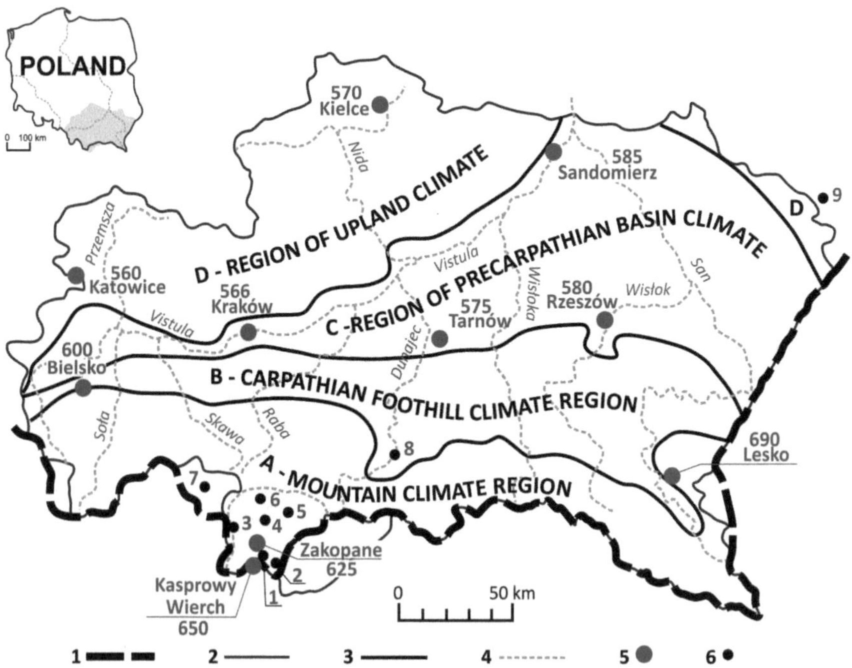

Fig. 1 Location of the studied area in Poland and location of the meteorological stations in the area of the Upper Vistula Basin. Climatic regions after Niedźwiedź and Obrębska-Starklowa (1991). *Borders 1*—cross-country borders, *2*—the Upper Vistula Basin, *3*—climatic regions; *4*—rivers, *5*—main meteorological stations with WMO numbers, *6*—other meteorological stations (*1*–Hala Gąsienicowa, *2*–Morskie Oko, *3*–Witów, *4*–Poronin, *5*–Białka Tatrzańska, *6*–Szaflary, *7*–Jabłonka, *8*–Nowy Sącz, *9*–Tomaszów Lubelski)

6 °C and precipitation increases to 900 mm. Climatological characteristics of air temperature and precipitation in the research period for selected 18 stations in the Upper Vistula Basin are gathered in Table 1. In the research period mean annual air temperature at majority of stations varied between c.a. 7.5 and 8.5 °C with an exception of Zakopane and Kasprowy Wierch where air temperature was much lower as a result of vertical decrease (Table 1).

Despite clear vertical decrease in air temperature the Tatra Mountains climate is characterized by abundant precipitation. Annual precipitation totals exceed 1500–1750 mm above the altitude of 1400 m a.s.l. In the research period maximum 10-day precipitation totals were higher than 100 mm at many stations and exceeded the threshold of 200 mm at three stations. The country-wise peak of daily precipitation total amounting to 300 mm was recorded on 30 June 1973 at Hala Gąsienicowa station (1520 m a.s.l.) (Niedźwiedź 2003). Precipitation extremes typically occur during warm part of the year, from May to September, with maximum in July. Several-day high rainfalls have not only caused severe flash floods and landslides in small catchments, but also catastrophic large-scale floods in the

Table 1 The list of stations with their altitudes (Hs), the research periods and mean value of indices for available periods

Station	Altitude Hs (m a.s.l.)	Available period	Mean annual precipitation totals (mm)	Mean maximum 10-day precipitation totals (mm)	Mean maximum monthly precipitation totals (mm)	Annual mean temperature (°C)
Kasprowy Wierch	1991	1951–2013 1951–2015	1752 1750	213	307	−0.6 −0.5
Hala Gąsienicowa	1520	1951–2011	1696	242	349	–
Morskie Oko	1408	1954–2011	1583	205	301	–
Zakopane	857	1951–2013 1951–2015	1126 1129	160	236	5.4 5.5
Witów	835	1954–2011	1033	145	212	–
Poronin	773	1954–2011	992	151	214	–
Białka Tatrzańska	700	1951–2011	846	114	166	–
Szaflary	655	1951–2011	857	116	170	–
Jabłonka	615	1955–2011	746	94	243	–
Bielsko-Biała Aleksandrowice	398	1951–2013 1951–2015	990 987	144	166	8.1 8.2
Lesko	386	1954–2013 1954–2015	806 811	107	162	7.5 7.6
Katowice	317	1951–2013 1951–2015	719 715	89	134	8.2 8.3
Nowy Sącz	292	1954–2013	733	109	349	8.3
Tomaszów Lubelski	273	1951–2013	634	84	124	–
Kielce-Suków	268	1951–2013 1951–2015	630 630	88	129	7.5 7.6
Kraków-Balice	237	1951–2015	670			8.2
Sandomierz	217	1951–2013 1951–2015	567 566	84	121	8.0 8.1
Tarnów	209	1951–2013 1951–2015	710 712	106	148	8.6 8.7
Rzeszów-Jasionka	200	1952–2013 1951–2015	636 631	86	127	– 8.1

Upper Vistula Basin, and further downstream. The most severe floods occurred in July 1903, 1934, 1960, 1970, 1997, 2001, and 2010. The flood in July 1934 was one of the most pronounced deluges in the 20th century in the Vistula Basin, causing death of 55 people, while the flood in May and June 2010 caused economic losses up to 3.2 billion US dollars (Chorynski et al. 2012).

According to the Clausius–Clapeyron law, higher temperatures are associated with higher saturated water vapour pressure in the atmosphere. Hence, warming can lead to occurrence of more intense precipitation as well as to extend the periods of precipitation occurrence. Although studies (Niedzwiedz et al. 2015) on intense precipitation at two stations located on the northern foothills of the Tatra Mountains (Kasprowy Wierch and Zakopane) indicated no statistically significant trends, upward trends in maximum daily, maximum five-day and warm half-year precipitation totals as well as in the frequency of days with precipitation ≥ 50 mm were found at both stations.

Since flood risk is likely to depend on current climate change manifesting itself in significant climate warming, this chapter discusses changes in selected characteristic of air temperature and precipitation in the Upper Vistula Basin. The first part of the chapter concerns changes in air temperature including monthly seasonal and annual means. Moreover, we also investigated variability in the frequency of selected thermally characteristic days which were recognized as direct indicators of current climate change. Regional air temperature changes are discussed on the background of global changes. Next we explored changes in monthly, seasonal and annual precipitation including extremes. We paid special attention to warm half-year and summer season which are the most flood parts of the year in the Upper Vistula Basin. Further part of the chapter compares precipitation climatology in the Tatra Mountains and Podhale Basin with the remaining part of the Upper Vistula Basin and seeks relations between precipitation and air temperature. We used different sets of the stations, depending on analysis. Majority of analyses use 10 stations located in the Upper Vistula Basin. These stations represent different parts of the research area including the highest altitudes (Kasprowy Wierch, Zakopane) and the foothills of the Carpathian Mountains (Bielsko, Lesko). The remaining six stations represent the areas of the Upper Vistula Basin located further to the north (Tarnów, Kraków, Rzeszów, Katowice, Sandomierz, Kielce) as shown in Fig. 1. The data from these stations cover the period 1951–2015. The further nine stations used in the chapter are listed in Table 1. Summing up we used the data from 19 stations covering various periods to explore changes in air temperature and precipitation.

2 Changes in Air Temperature of the Upper Vistula Basin in the Light of Global Temperature Trends

The enhanced violence of extreme weather events is currently being attributed to contemporary climate change, manifesting itself in significant air temperature increase on both regional and global scales (Trepińska 1983, 1997; Limanówka

1997; Trepińska et al. 1997; Fortuniak et al. 2001; Kożuchowski and Żmudzka 2001; Wibig and Głowicki 2002; Cebulak and Limanówka 2007; Bielec-Bakowska and Lupikasza 2009; Bielec-Bąkowska and Piotrowicz 2013). A recent report on global climate state (NOAA 2016) announced that three warmest years of the last 136 years (1880–2015) were recorded at the end of this period. In 2015, the average global air temperature across land and ocean was warmer by +0.90 °C comparing to the 1910–2000 average. The annual air temperature in the previous year, 2014, exceeded the long-term average by +0.74 °C and was the second warmest in the series. The mean air temperature increase in Poland was up to +0.8 °C in the second half of the 20th century (cf. Fortuniak et al. 2001; Kożuchowski and Żmudzka 2001; Degirmendžić et al. 2004).

This chapter uses average, maximum and minimum air temperature from 10 stations, location of which is shown in Fig. 1. Long-term course of air temperature in the Upper Vistula Basin exhibits a lot of natural, inter-annual and intra-annual variability but general trend relates to global changes. This is illustrated in Fig. 2 showing standardized global annual air temperature and analogous temperature in the Upper Vistula Basin in the 1951–2015 period. Fluctuations and no clear tendency dominated before 1975. Since the middle of the 1970s evident increase was recognized. Average annual air temperature for 1988–2015 was higher by +0.9 °C than for previous long-term period (1951–1987). The increase was restrained at the turn of the centuries but the distinctive maximum of the annual air temperature which occurred in 2014 and 2015 (+8.8 °C which is higher by 2.5 °C than the average from 1951–1975) may suggest renewal of the warming rate both within the research area and globally. Finally two periods can be distinguished on the basis of long-term course of annual air temperature—the cooler temperatures covered the 1951–1987 period and a distinct increase of annual temperature encompassed the 1988–2015 period.

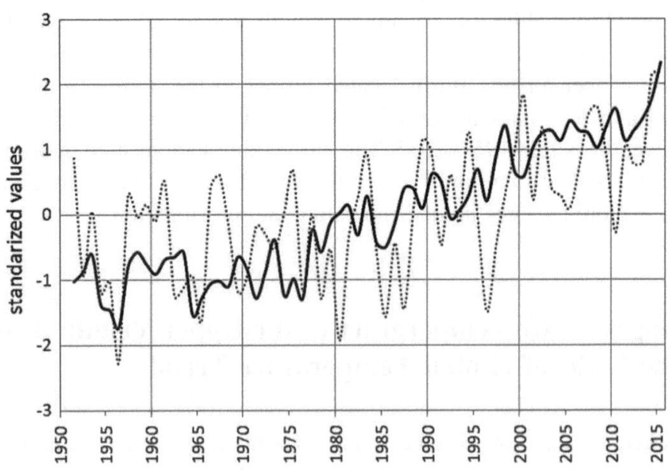

Fig. 2 Standardized values of annual air temperature for the Upper Vistula Basin (*spot line*) and global air temperature (*solid line*) in the 1951–2015 period

3 Annual, Seasonal and Monthly Air Temperature

Spatial and seasonal distribution of air temperature in the Upper Vistula Basin is influenced by various factors of which topography is the most crucial local one. Climatological characteristics of air temperature in the research period are presented in Table 2. The annual range of monthly air temperature within the Upper Vistula Basin reached almost 27 °C. The extreme monthly values varied from −8.3 °C in February at Kasprowy Wierch station to 18.6 °C in July at Tarnów and Sandomierz located in the lowest altitudinal belt. Although the lowest average monthly temperature fell in February (see Kasprowy Wierch in Table 2), it is January which is the coolest month at majority of stations. In that month air temperature changed in the vertical profile from −1.7 °C in Lesko to −8.0 at Kasprowy Wierch (Table 2). At most stations monthly temperatures exceeded 10 °C from May to September with maximum in July varying between 14.9 and 18.6 °C, depending on the station. The period between May and September is the most flood-prone season in the Upper Vistula Basin. At the Kasprowy Wierch high mountain station, the annual range of temperature spread between −8.3 °C (February) and 7.7 °C (August).

The absolute range of air temperature at particular stations within the Upper Vistula Basin ranged from more than 53 °C at Kasprowy Wierch to more than 70 °C in Rzeszów and Kielce (Table 3). The highest ever daily temperature occurred on 8 August 2013 in Tarnów while the lowest one was recorded on 28 February 1963 in Rzeszów located on the borderline between Sandomierz Basin and the Carpathian Foothills. The occurrence of the lowest daily temperature minimums at hollows and basins of the Carpathian Mountains is due to the slide down of cold heavy air along the slopes and its stagnation in a concave topographic form.

The long-term course of mean seasonal temperature averaged for the region indicates various seasonal patterns with clear increase in spring and summer temperatures since the beginning of the 1980s. Similar tendency was also seen in the long-term course of autumn temperature but since the beginning of the 1990s. Winter air temperature was dominated by vivid fluctuations with no clear tendency (Fig. 3). In all seasons, the highest air temperatures were recorded in the last decade of the 20th century or in the 21st century (9.0 °C in spring 2007, 18.4 °C in summer 2015, 9.7 °C in autumn 2006 and 1.4 °C in winter 2006/2007). These features can clearly be seen in Fig. 3 showing standardized seasonal air temperatures calculated as arithmetical averages from ten stations. Only in spring there was a continuous increase in air temperature throughout the entire period as shown by fitted 5th order polynomial. However, in the last years, the most rapid increase was found in summer temperature. For example eight of ten highest summer temperatures were recorded in the 21st century. On the other hand, vividly cooler summers occurred in the WMO normal period of 1961–1990. All the lowest seasonal temperatures fell before 1980: 3.8 °C in spring 1955, 13.9 °C in summer 1978, 5.6 °C in autumn 1956, −8.4 °C in winter 1962/1963.

Long-term course of air temperature discussed above is well reflected in the results of trend analysis showing distinct, statistically significant changes between

Table 2 Average monthly, seasonal and annual air temperature in the Upper Vistula Basin in the period 1951–2014

Season	12 650	12 625	12 600	12 690	12 566	12 575	12 580	12 585	12 560	12 570
	Kasprowy Wierch	Zakopane	Bielsko-Biała	Lesko	Kraków Balice	Tarnów	Rzeszów	Sandomierz	Katowice	Kielce
Jan	−8.0	−4.2	−1.7	−2.7	−2.6	−2.0	−2.9	−3.1	−2.1	−3.1
Feb	−8.3	−3.4	−0.9	−1.7	−1.4	−0.9	−1.9	−1.9	−1.0	−2.2
Mar	−6.3	−0.1	2.8	2.1	2.6	3.1	2.3	2.2	2.8	1.6
Apr	−2.4	5.1	8.0	7.6	8.3	8.8	8.2	8.4	8.3	7.5
May	2.4	10.1	12.7	12.5	13.4	13.7	13.4	13.7	13.2	12.9
Jun	5.7	13.2	15.8	15.5	16.6	16.9	16.7	16.9	16.4	16.2
Jul	7.6	14.9	17.6	17.2	18.3	18.6	18.4	18.6	18.1	17.8
Aug	7.7	14.4	17.2	16.6	17.7	18.0	17.7	18.0	17.5	17.2
Sep	4.3	10.5	13.4	12.7	13.4	13.8	13.4	13.5	13.3	12.8
Oct	1.1	6.2	9.1	8.4	8.5	9.2	8.5	8.3	8.7	7.9
Nov	−3.3	1.4	4.1	3.5	3.4	4.3	3.5	3.1	3.7	2.9
Dec	−6.5	−2.6	0.1	−0.9	−0.7	0.0	−0.8	−1.1	−0.4	−1.1
MAM	−2.1	5.0	7.8	7.4	8.1	8.5	7.9	8.1	8.1	7.4
JJA	7.0	14.2	16.9	16.4	17.5	17.9	17.6	17.8	17.3	17.1
SON	0.7	6.0	8.9	8.2	8.4	9.1	8.5	8.3	8.6	7.8
DJF	−7.6	−3.4	−0.8	−1.8	−1.6	−0.9	−1.9	−2.0	−1.1	−2.2
Year	−0.5	5.5	8.2	7.6	8.2	8.7	8.1	8.1	8.3	7.6

Table 3 Absolute extremes of daily air temperature at selected stations in the Upper Vistula Basin (1951–2015)

WMO No	Station name	Hs (m a.s.l.)	Tmax (°C)	Date	Tmin (°C)	Date	Temperature (range K)
12 650	Kasprowy Wierch	1991	23.0	5 July 1957	−30.2	17 January 1963	53.2
12 625	Zakopane	857	32.8	8 August 2013	−34.1	1 February 1956	66.9
12 600	Bielsko-Biała	398	36.4	8 August 2013	−29.6	9 February 1956	66.0
12 690	Lesko	420	33.8	7 August 1956	−32.5	28 February 1963	66.3
12 566	Kraków-Balice	237	37.3	8 August 2013	−29.9	13 January 1987	67.2
12 575	Tarnów	209	**37.9**	**8 August 2013**	−30.3	20 January 1963	68.2
12 580	Rzeszów	200	36.5	15 August 1952	**−35.8**	**28 February 1963**	72.3
12 585	Sandomierz	217	37.1	15 August 1952	−28.6	9 February 1956	65.7
12 560	Katowice	284	37.2	8 August 2013	−30.0	9 February 1956	67.2
12 570	Kielce	260	36.4	9 August 2013	−33.9	8 January 1987	70.3

Bolded values are the highest and the lowest air temperatures

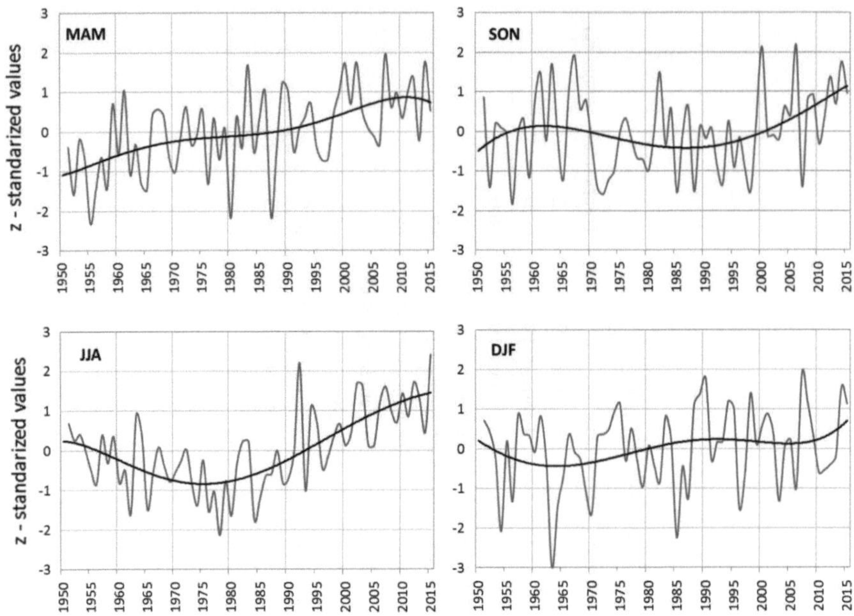

Fig. 3 Standardized values of seasonal air temperature for the Tatra Mountains in the 1951–2015 period (*grey line*—observed temperature, *black line*—fitted 5th order polynomial)

March and August (Table 4). Significant increase in annual air temperature of the rate of +0.13 °C per 10 years to +0.29 °C per 10 years, depending on station, was found within the entire research area. On a seasonal scale, the most significant was increase in summer and spring air temperature, particularly in August and July. No significant changes were found in autumn and winter months, however majority of air temperature trends were positive despite those for October and December at Lesko station. It is also worth to notice that magnitudes of insignificant trends in January and February air temperatures were much higher comparing to the monthly trends in air temperature from September to December. If the rate and pattern of current climate change continues to the future, trends in January and February air temperature will likely become statistically significant.

Significant increase in air temperature is also well reflected in the frequency of thermally characteristic days which are considered as direct indicators of current climate change. We analysed five characteristic days selected on the basis of average, maximum and minimum air temperature for eight stations representing topographically different parts of the Upper Vistula region. Winter and summer days were defined as days with average daily mean air temperature ≤ 0 and ≥ 15 °C, respectively. Daily maximum air temperature was used to select ice days (Tmax $\leq$ 0 °C), hot days (Tmax $\geq$ 25 °C) and very hot days (Tmax $\geq$ 30 °C). Long-term course of thermally characteristic days frequency at particular stations is presented in Figs. 4, 5, 6, 7 and Fig. 8. For the most part of the Upper Vistula Basin

Table 4 Trends (change per 10 years) in monthly, seasonal and annual mean air temperature in the Upper Vistula Basin in the 1951–2015 period

Season	12 650 Kasprowy Wierch	12 625 Zakopane	12 600 Bielsko-Biała	12 690 Lesko	12 566 Kraków Balice	12 575 Tarnów	12 580 Rzeszów	12 585 Sandomierz	12 560 Katowice	12 570 Kielce
Jan	+0.30	+0.42*	+0.34	+0.30	+0.36	+0.38	+0.44*	+0.38	+0.36	+0.33
Feb	+0.20	+0.35	+0.37	+0.24	+0.38	+0.40	+0.47	+0.41	+0.40	+0.38
Mar	+0.13	+0.34*	+0.37*	+0.28	+0.39*	+0.44**	+0.51**	+0.48**	+0.41*	+0.40*
Apr	+0.24*	+0.34**	+0.36**	+0.26*	+0.31**	+0.34**	+0.35***	+0.32**	+0.38***	+0.28**
May	+0.35**	+0.34**	+0.37***	+0.23*	+0.33***	+0.38***	+0.35***	+0.28**	+0.35***	+0.26*
Jun	+0.16	+0.20*	+0.22**	+0.08	+0.16*	+0.18*	+0.18*	+0.06	+0.16*	+0.02
Jul	+0.33**	+0.32***	+0.38***	+0.29**	+0.34***	+0.36***	+0.36***	+0.30**	+0.35***	+0.26*
Aug	+0.33***	+0.34***	+0.36***	+0.31***	+0.33***	+0.31***	+0.36***	+0.27**	+0.33***	+0.26**
Sep	+0.03	+0.11	+0.06	+0.04	+0.11	+0.04	+0.14	+0.08	+0.10	+0.06
Oct	+0.03	+0.12	+0.07	−0.01	+0.09	+0.04	+0.07	+0.03	+0.07	+0.00
Nov	+0.20	+0.21	+0.21	+0.14	+0.10	+0.14	+0.23	+0.16	+0.18	+0.09
Dec	+0.11	+0.10	+0.08	−0.03	+0.03	+0.04	+0.08	+0.09	+0.09	+0.03
MAM	+0.24**	+0.34***	+0.37***	+0.26**	+0.34***	+0.38***	+0.40***	+0.36***	+0.38***	+0.31***
JJA	+0.28***	+0.29***	+0.32***	+0.23***	+0.28***	+0.28***	+0.30***	+0.21**	+0.28***	+0.18**
SON	+0.09	+0.14	+0.11	+0.06	+0.10	+0.07	+0.15*	+0.09	+0.12	+0.05
DJF	+0.21*	+0.31*	+0.27	+0.15	+0.26	+0.29	+0.33*	+0.30*	+0.29*	+0.25
Year	+0.20***	+0.27***	+0.27***	+0.13*	+0.24***	+0.25***	+0.29***	+0.24***	+0.26***	+0.20***

Trends calculated with linear regression of least squares method, statistical significance tested with t-test, significance levels: $^{*}\alpha \leq 0.05$, $^{**}\alpha \leq 0.01$, $^{***}\alpha \leq 0.001$

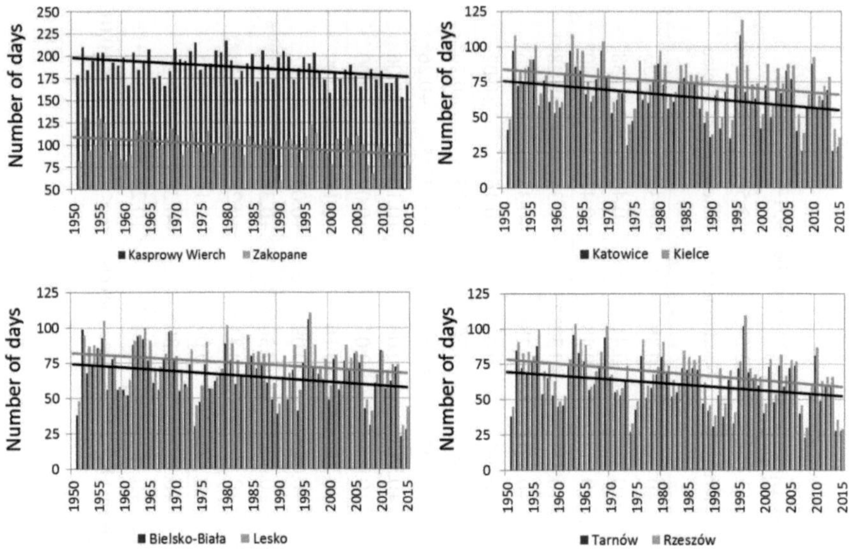

Fig. 4 Annual number of winter days (Tmean $\leq$ 0 °C) at selected stations in the Upper Vistula Basin in the 1951–2015 period (*bars*—observed number of days, *lines*—fitted linear trend)

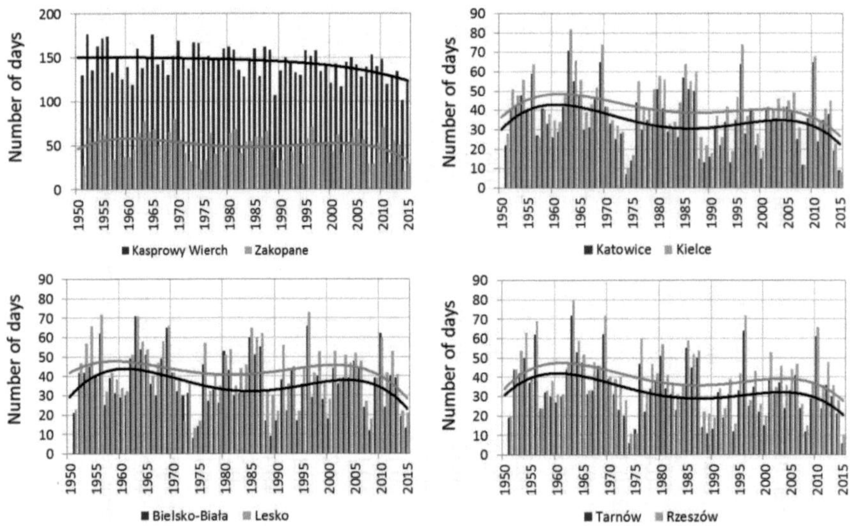

Fig. 5 Annual number of ice days (Tmax $\leq$ 0 °C) at selected stations in the Upper Vistula Basin in the 1951–2015 period (*bars*—observed number of days, *lines*—fitted 4th order polynomial)

winter days occurred during 18–21 % of days a year (102–119 days). At more elevated Zakopane and Kasprowy Wierch stations such days constituted 27 and 51 % of days a year, respectively. The 1996 year was the most exceptional over the

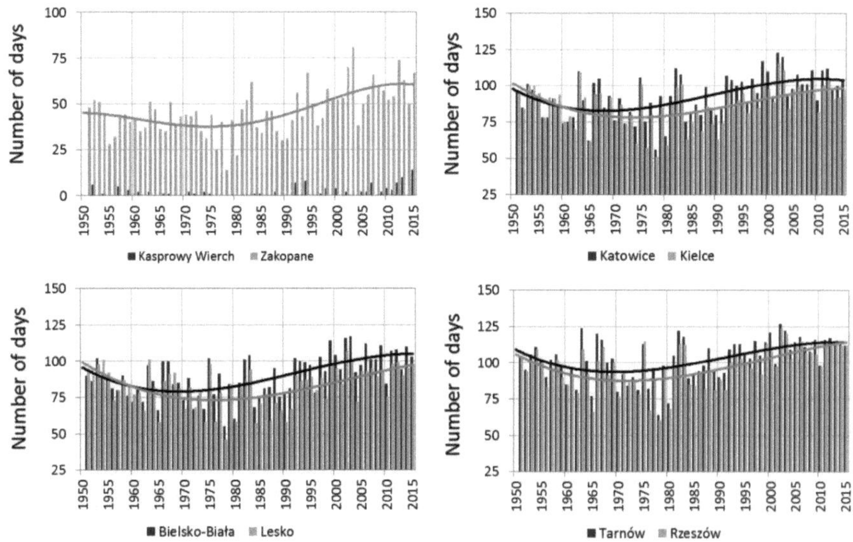

Fig. 6 Annual number of summer days (Tmean ≥ 15 °C) at selected stations in the Upper Vistula Basin in the 1951–2015 period (*bars*—observed number of days, *lines*—fitted 3rd [Zakopane] and 4th order polynomial)

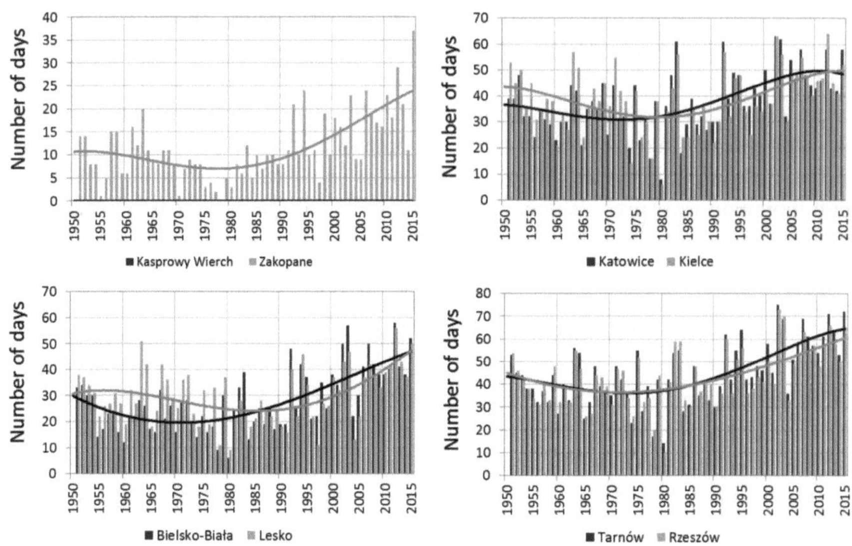

Fig. 7 Annual number of hot days (Tmax ≥ 25 °C) at selected stations in the Upper Vistula Basin in the 1951–2015 period (*bars*—observed number of days, *lines*—fitted 3rd [Zakopane] and 4th order polynomial)

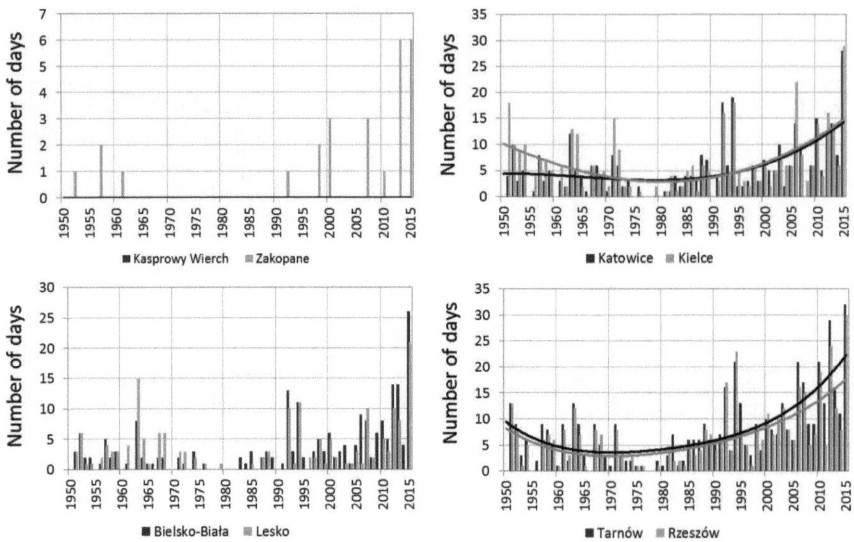

Fig. 8 Annual number of very hot days (Tmax $\geq$ 30 °C) at selected stations in the Upper Vistula Basin in the 1951–2015 period (hot days were not noted at the Kasprowy Wierch station in the research period) (*bars*—observed number of days, *lines*—fitted 4th order polynomial)

entire research period due to the highest number of winter days at majority of meteorological stations. At the Kasprowy Wierch station the frequency of winter days reached its maximum in 1980 when more than 59 % of days (217 days) had mean temperature below freezing point. The frequency of winter days was significantly decreasing over the research period at majority of stations with a rate of about −3 days per 10 years. We found the most significant downward trend at the high mountain Kasprowy Wierch station ($\alpha \leq 0.001$). In the last two years of the research period (2014, 2015) the number of winter days was very low.

In the Upper Vistula Basin, ice days (Fig. 5) constituted about 10 % of days a year; their frequency was distinctly higher at Kasprowy Wierch station (40 % of days). At majority of stations, the highest number of ice days was found in 1963, except for Zakopane and Kasprowy Wierch where the maximum occurred in the 1950s (82 days in 1956 and 117 days in 1952, respectively). At majority of stations ice days were very rare in the first part of the 1970s but the index reached its minimum value in 2015. The fraction of ice days was diminishing in the Upper Vistula Basin over the research period, however the trends were weaker than those for winter days, being significant at only two of eight stations (Tarnów: −2 days per 10 years and Kasprowy Wierch: −3 days per 10 years). In the recent years clear decrease in the number of ice days can be seen since the first pentad of the 21st century with the only exception of the 2010 flood year.

Summer days with average temperature $\geq$ 15 °C (Fig. 6) constituted on average 22–35 % of days per year including the Zakopane station. At the Kasprowy peak, their frequency dropped to only 4 % of days a year. The year with the highest percentage of summer days was 2002 and, at some stations, 2003 (Zakopane, Bielsko-Biała). Hot days (Fig. 7) did not occur at all at Kasprowy Wierch station in the research period while at other stations they constituted from 15 % (Lesko) to 21 % (Tarnów) of days in a year. In the Tatra Mountains, the upper limit of the hot days occurrence is placed at the level of about 1800 m a.s.l. The record-breaking frequency of hot days occurred in 2002 (Tarnów, Rzeszów and Katowice) or in 2012 (Bielsko-Biała, Lesko, Kielce). Zakopane experienced the highest number of hot days in 2015.

In the Upper Vistula Basin, very hot days were rather rare constituting on average from 2 % (Zakopane) to 9 % (Tarnów) of days a year. Such days did not occur at all at Kasprowy Wierch (Fig. 8) because their occurrence is limited to the altitude of about 1200 m. In the research period it was possible to point out several years with clearly high frequency of very hot days including 1963, 1992, 1994, 2003, 2006, 2007, 2010, 2012, 2013. However, their record breaking frequency was noted in 2015 when daily maximum temperature exceeded the threshold of 30 °C during 21 days (Lesko) to 32 days (Tarnów), depending on station. Even in Zakopane where average frequency of very hot days equalled 0.4 day, 6 such days were noted in 2015. The long-term courses of warm temperature indices were less spatially variable over the research period compared with cold temperature indices. Actually, at every station two periods with opposite trend direction can be distinguished. Decrease in all warm temperature indices covered the first three decades of the research period (1951–1980). The minimum values of warm day indices fell at the turn of the 1970s and 1980s and then there was their gradual increase. Linear trends calculated over the entire research period were statistically significant at majority of station excluding Lesko and Tarnów. Magnitude of statistically significant trends (at least $\alpha \leq 0.5$) varied from 0.5 days (Kasprowy Wierch) to 3.8 days (Bielsko-Biała) per 10 years for summer days, from 1.9 days (Zakopane) to 3.9 days (Tarnów) per 10 years for hot days and from 1.1 day (Bielsko-Biała) to 1.9 day (Tarnów) per 10 years for very hot days. Trends in both air temperature and thermally characteristic days indicate statistically significant (often at $\alpha < 0.001$) increase in air temperature which in future is likely to manifest itself as increase in the frequency of extreme weather events.

4 Changes in Precipitation of the Upper Vistula Basin

The occurrence of precipitation and its amount strongly depend on altitude and atmospheric circulation which is also triggering temporal precipitation variability. In the Upper Vistula Basin, linkage between precipitation amount and altitude is clearly seen both in annual, seasonal and monthly precipitation totals which at Kasprowy Wierch station (1750.4 mm) are twice or even more as high as at other stations (Table 5). In summer, the region received about 40 % (from 35 % at Kasprowy Wierch to 43 % in Zakopane) while in winter about 16 % of annual total (from 13 % in Zakopane to 20 % at Kasprowy Wierch). Spring and autumn precipitation totals were almost equal. Such an annual course of precipitation is typical in moderate continental climate. Monthly extremes fell in July (maximum) and February (minimum) (Table 5).

The most recent flood year (2010) had the highest ever annual precipitation total at nine stations in the large region (1646 mm in Zakopane, 1562 mm in Poronin, 1518 mm in Witów, 1215 mm in Tarnów, 1154 mm in Nowy Sącz, 1150 mm in Szaflary, 1104 mm in Lesko, 1029 mm in Tomaszów Lubelski, 964 mm in Jabłonka). At other nine stations precipitation in 2010 was counted between the highest values in the series: 2nd highest record in Hala Gąsienicowa: 2359 mm (in 2001: 2626 mm), 2nd in Bielsko-Biała: 1478 mm (in 1966: 1505), 2nd in Rzeszów: 966 mm (in 1966: 1009 mm), 3rd in Morskie Oko: 2256 mm (in 1974: 2290 mm; in 2001: 2271 mm), 3rd in Katowice: 965 mm (in 1974: 1011 mm; in 1977: 973 mm), 6th in Sandomierz 720 mm (in 1966: 860 mm), 7th in Białka Tatrzańska: 1015 mm (in 1970: 1128 mm; in 2001 1062 mm), 10th in Kasprowy Wierch: 2117 mm (in 2001: 2600 mm), and 11th in Kielce: 744 mm (in 1966: 1000 mm; 2001: 918 mm).

The highest daily precipitation totals recorded in the Upper Vistula Basin were, in some cases, only slightly lower than mean monthly totals (Cebulak 1992, 1997; Ustrnul and Czekierda 2009). Daily precipitation exceeded 100 mm at least once at six out of 10 stations. The highest daily total at Kasprowy Wierch reached 232.0 mm but it did not beat the record of 300 mm measured on 30 June 1973 at Hala Gąsienicowa. Daily maximum precipitation totals for each month and each station are presented in Table 6. Flood precipitation in the Tatra Mountains is usually noted at days with air advection from north-east under an influence of low pressure system. Such a situation is usually linked to stationary low with its centre located over Ukraine. Additionally, resulting long-lasting precipitation is enhanced by orographic effect. Sometimes extreme precipitation totals are a result of torrential rainfall of a short duration and local extent as it happened in Sandomierz (140.5 mm) at night 25/26 of July 2011 and had damaging effect. This flood-causing rainfall was linked to the convective cell which flew from the south-eastern sector and stopped to give exceptional 3-h shower of an unexpected intensity. Variability of flood conditions in the Southern Poland and the Tatra Mountains was presented in several latest publications (cf. Kundzewicz et al. 2012; Niedźwiedź et al. 2015; Ruiz-Villanueva et al. 2016).

Changes in precipitation within the Upper Vistula Basin were earlier discussed in a few general publications (Cebulak 1997; Degirmendžić et al. 2004; Łupikasza 2010) including those based on the longest time series for Cracow (Trepińska 1969, 1983, 1997; Twardosz 1999, Niedźwiedź et al. 2009). In order to assess long-term variability in precipitation characteristics during the last 65 years (1951–2015), we investigated its annual, summer and warm half-year totals as well as maximum daily precipitation in these seasons. We also analysed the frequencies of days with high precipitation (≥ 30 and ≥ 50 mm) which can be used as indirect index of flood-prone conditions. Since we did not identify any monotonic increases in the chronological series of precipitation indices we decided to fit a polynomial of the 5th order to the data in order to identify the periods of precipitation abundance and

Table 5 Mean monthly, annual and seasonal precipitation totals (in mm) for selected stations in the Upper Vistula Basin (1951–2015)

Season	12 650	12 625	12 600	12 690	12 566	12 575	12 580	12 585	12 560	12 570
	Kasprowy Wierch	Zakopane	Bielsko-Biała	Lesko	Kraków Balice	Tarnów	Rzeszów	Sandomierz	Katowice	Kielce
Jan	118.9	47.7	46.2	36.3	36.4	36.5	31.8	27.7	41.2	40.3
Feb	106.6	46.3	45.4	34.7	30.7	31.8	27.9	24.2	36.3	32.9
Mar	118.3	58.6	54.0	41.6	36.1	37.7	33.5	29.6	42.9	37.7
Apr	132.0	79.6	70.4	55.9	47.1	50.3	44.6	39.4	47.8	39.8
May	174.6	130.0	117.3	91.8	78.1	81.5	74.9	63.6	77.4	63.0
Jun	218.1	169.2	141.9	106.1	90.9	101.6	83.2	74.9	86.5	75.1
Jul	218.5	179.4	142.3	121.9	90.6	108.0	90.8	90.0	102.0	92.4
Aug	180.1	136.6	110.3	91.7	79.7	78.3	69.0	62.9	78.8	71.0
Sep	132.2	97.9	89.0	78.9	56.3	61.2	56.7	50.1	60.9	51.2
Oct	107.2	70.7	60.1	55.3	44.4	44.6	43.8	39.5	46.6	39.5
Nov	120.9	61.2	59.9	46.7	42.6	41.5	38.6	34.1	48.7	44.0
Dec	123.1	51.3	50.2	47.3	37.1	38.3	36.6	30.6	45.6	43.3
MAM	424.9	268.2	241.7	189.3	161.2	169.6	153.0	132.6	168.1	140.5
JJA	616.7	485.2	394.5	319.6	261.2	288.0	243.0	227.7	267.2	238.6
SON	360.3	229.8	209.0	180.9	143.3	147.3	139.2	123.7	156.3	134.7
DJF	348.5	145.3	141.8	118.3	104.2	106.6	96.2	82.5	123.1	116.5
Year	1750.4	1128.6	987.0	811.0	669.8	711.6	631.4	566.5	714.7	630.2

Table 6 Maximum daily precipitation totals (in mm) for selected stations in the Upper Vistula Basin (1951–2015)

Season	12 650	12 625	12 600	12 690	12 566	12 575	12 580	12 585	12 560	12 570
	Kasprowy Wierch	Zakopane	Bielsko-Biała	Lesko	Kraków Balice	Tarnów	Rzeszów	Sandomierz	Katowice	Kielce
Jan	65.3	31.4	31.6	31.0	20.0	22.7	23.8	45.6	24.5	28.1
Feb	76.4	44.3	26.4	26.0	17.4	28.4	28.1	27.6	21.2	17.4
Mar	53.2	59.8	37.6	37.0	24.0	30.2	24.5	20.9	28.2	21.5
Apr	70.5	56.7	53.3	45.4	44.8	53.8	41.2	36.0	81.6	42.1
May	130.1	104.2	162.7	71.4	87.4	81.0	62.2	47.2	62.1	59.0
Jun	232.0	138.7	88.2	53.6	71.9	71.9	65.2	49.5	74.1	76.5
Jul	166.1	138.7	132.1	68.0	74.4	110.8	59.2	140.5	58.4	155.2
Aug	118.7	90.4	147.4	66.4	62.4	65.2	61.8	59.5	61.6	66.3
Sep	100.5	77.6	91.5	80.7	64.0	83.6	45.7	51.2	47.2	44.5
Oct	63.7	73.1	46.4	41.5	51.4	32.8	55.7	37.4	40.6	28.2
Nov	66.4	48.4	54.1	48.1	33.9	33.8	25.0	24.9	27.4	26.8
Dec	59.4	33.1	36.6	32.0	29.8	34.2	24.3	22.9	28.6	26.7
MAM	130.1	104.2	162.7	71.4	87.4	81.0	62.2	47.2	81.6	59.0
JJA	232.0	138.7	147.4	68.0	74.4	110.8	65.2	140.5	74.1	155.2
SON	100.5	77.6	91.5	80.7	64.0	83.6	55.7	51.2	47.2	44.5
DJF	76.4	44.3	36.6	32.0	29.8	34.2	28.1	45.6	28.6	28.1
Year	232.0	138.7	162.7	80.7	87.4	110.8	65.2	140.5	81.6	155.2

deficit. The patterns of long-term precipitation variability, shown in Figs. 9, 10, 11, 12, 13 and 14, allow identification of wetter and drier periods which in case of all indices were quite synchronic. The most individual features were found in the

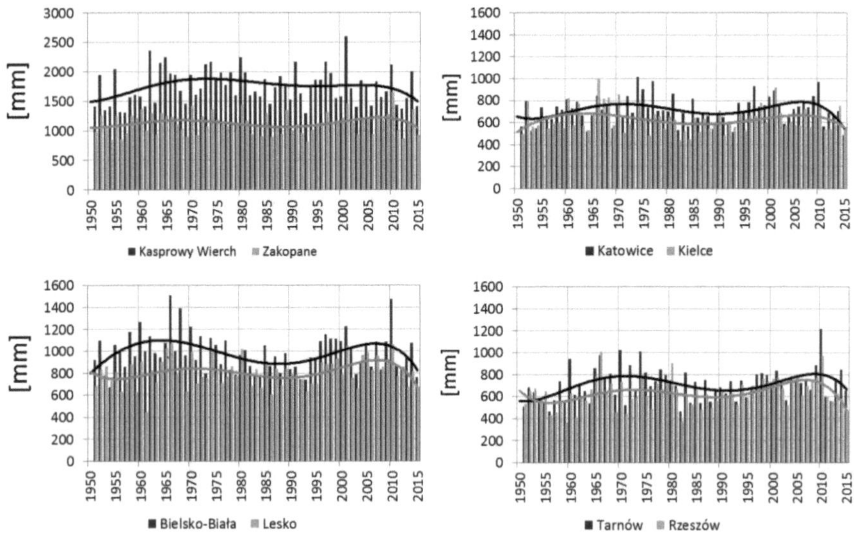

Fig. 9 Annual precipitation totals at selected stations in the Upper Vistula Basin in the period 1951–2015 (*bars*—observed number of days, *lines*—fitted 5th order polynomial)

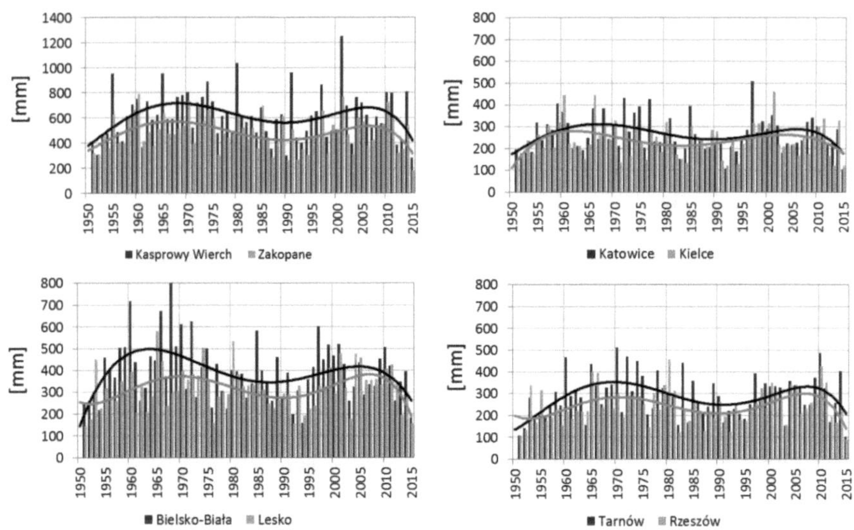

Fig. 10 Summer precipitation totals at selected stations in the Upper Vistula Basin in the period 1951–2015 (*bars*—observed number of days, *lines*—fitted 5th order polynomial)

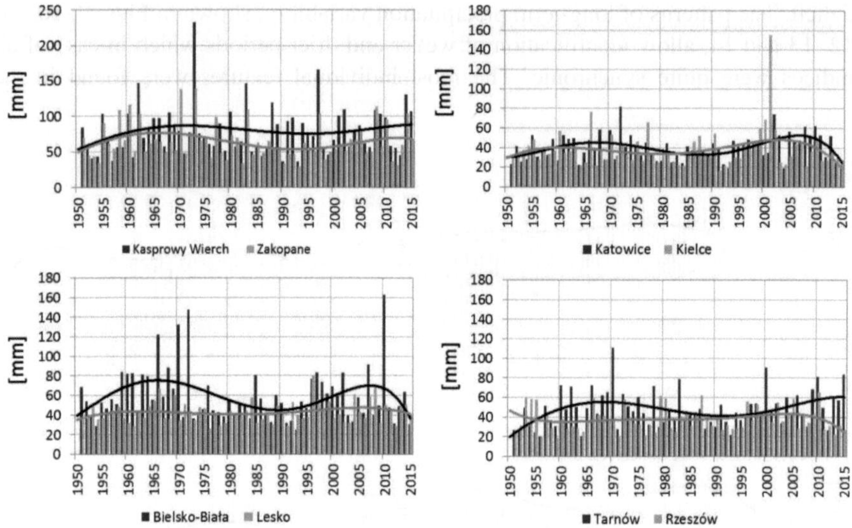

Fig. 11 Maximum daily summer precipitation totals at selected stations in the Upper Vistula Basin in the period 1951–2015 (*bars*—observed number of days, *lines*—fitted 5th order polynomial)

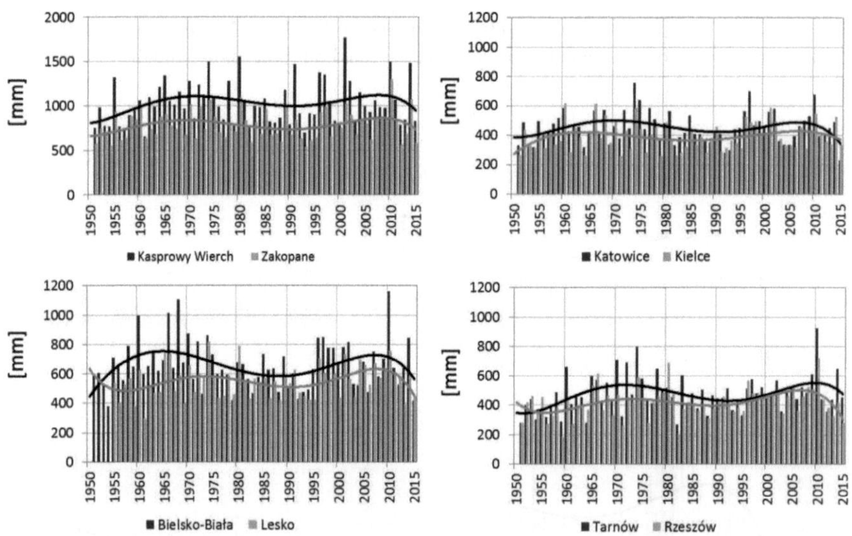

Fig. 12 Warm half-year precipitation totals at selected stations in the Upper Vistula Basin in the period 1951–2015 (*bars*—observed number of days, *lines*—fitted 5th order polynomial)

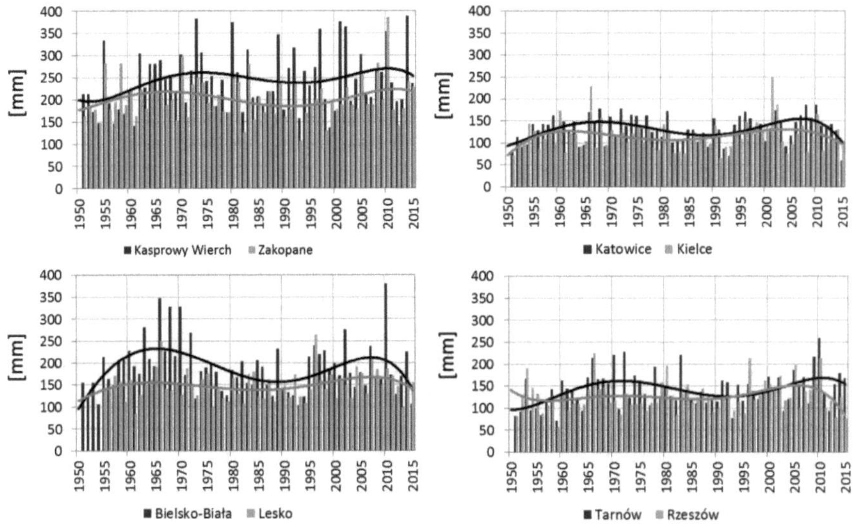

Fig. 13 Maximum daily summer half-year precipitation totals in the Upper Vistula Basin in the 1951–2015 period (*bars*—observed number of days, *lines*—fitted 5th order polynomial)

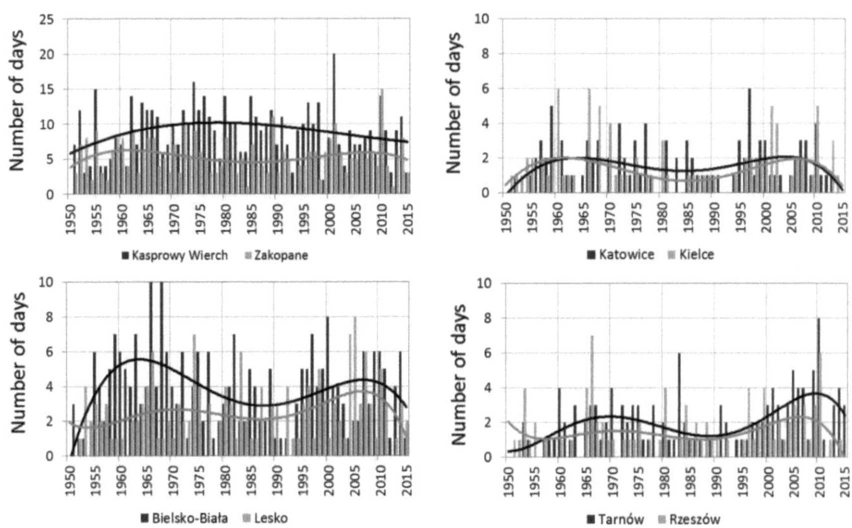

Fig. 14 Annual number of days with precipitation ≥ 30 mm at selected stations in the Upper Vistula Basin in the period 1951–2015 (*bars*—observed number of days, *lines*—fitted 5th order polynomial)

pattern for Kasprowy Wierch. Annual precipitation totals which ranged between 566.5 mm (Sandomierz) to 1750.4 mm (Kasprowy Wierch) were lower than average in the first decade of the research period covering the 1950s and between

1980 and 1995. The wettest years were identified to occur in 1960s and at the turn of the centuries. The third pentad of the 21st century, just after the exceptionally floody year 2010, exhibits clear decrease in annual precipitation totals (Fig. 9). The year 2015 was recognized as one of the driest in the series. Relatively dry 1950s and 1980s and relatively wet 1960s and the turn of the centuries were also identified, however with small shifts, in the long-term course of other indices.

In summer the Upper Vistula Basin received on average 334 mm (from 616.7 mm at Kasprowy Wierch to 227.7 mm in Sandomierz). Average summer precipitation totals from wet decades were by about 20 % higher than averages from dry period. In fact, this disproportion was not very distinct but noticeable (Fig. 10). Summer might be expected as the season of the highest maximum daily precipitation totals. Indeed, at six of ten meteorological stations, daily precipitation records were recorded in summer, but at other stations absolute maximums were noted at other seasons. At all stations, summer precipitation totals have been decreasing since 2010. Such a tendency was also spotted in the long-term course of summer daily maximums but not at every station. For example at Kasprowy Wierch there was an increase in the variability of daily maximums from year to year which finally resulted in weak increasing tendency since the beginning of the 21th century (Fig. 12). In the Upper Vistula Basin, significant increase in summer air temperature was not accompanied by any clear tendency in precipitation.

Long-term course of warm half-year precipitation totals and daily maximums did not differ from general pattern identified (Figs. 12 and 13). It could be mentioned, however, that this half-year received from 59 % (1030.7 mm) to 69 % (783.8 mm) of the annual precipitation total. In the wettest years warm half-year precipitation constituted from 149 % (Lesko) to 195 % (Tarnów) of the average value for 1951–2015. In contrast, precipitation totals in the driest year were as low as 43 % (Rzeszów) to 69 % (Zakopane) of the average value for 1951–2015. Changes in precipitation totals seem not to be directly related to changes in air temperature and do strongly vary in time. However, in some periods, for example in the first and the last decade of the research period, an increasing tendency in air temperature was accompanied by a decreasing tendency in precipitation. This suggests that the periods with lower temperature are typically wetter than the periods of high temperature. This applies only to overall aggregated precipitation totals and not to short intervals and extreme events. This is indicated by high maximum daily precipitation totals which were registered in the summers and warm half-years of the last decade of the research period (Figs. 11 and 13).

There is strong spatial and temporal variability in precipitation, so that trends are usually insignificant and can change in a short time. Trends in monthly, seasonal and annual precipitation totals in the Upper Vistula region are shown in Table 7. These trends were calculated with linear regression of least squares method and its significance was tested with nonparametric Mann-Kendall test. Most trends were not statistically significant. Noticeable changes were identified in a few cases only. Furthermore trend directions also varied depending on the season and the station. Even trends in annual totals were of various signs. March and September were the only months with increases noted at every station while in August opposite,

Table 7 Trends (mm/10 years) in monthly, seasonal and annual precipitation totals in the Upper Vistula Basin in the 1951–2015 period

Season	12 650	12 625	12 600	12 690	12 566	12 575	12 580	12 585	12 560	12 570
	Kasprowy Wierch	Zakopane	Bielsko-Biała	Lesko	Kraków Balice	Tarnów	Rzeszów	Sandomierz	Katowice	Kielce
Jan	−0.4	+0.5	+0.7	+1.7	+2.2	+1.1	+1.8	−0.6	+2.7	−0.8
Feb	−5.5	+0.7	−1.6	+0.3	−0.9	+0.4	+0.7	−2.5**	0.0	−2.0
Mar	+0.6	+2.9	+1.7	+2.2	+1.7	+1.8	+2.9*	+0.1	+2.7	+1.1
Apr	−3.2	−1.9	−2.9	+1.1	−0.6	+0.8	+1.1	−0.3	−1.4	−1.6
May	+5.0	+8.5	+3.1	+5.7	+1.6	+4.5	+3.3	+1.9	−0.1	+3.3*
Jun	−5.0	−6.4	−2.0	−2.6	−0.9	+2.3	+1.1	−3.8	−1.3	−2.5
Jul	+8.0	+4.7	−2.0	+4.6	−0.4	+3.5	+4.2	+1.7	−1.4	+1.7
Aug	−2.9	−4.5	−5.7	−1.0	−2.2	−2.3	−1.3	−2.2	−0.4	−0.2
Sep	+9.7	+5.6	+5.2	+5.0	+3.4	+6.0*	+4.2	+2.4	+2.0	+1.6
Oct	+3.1	+1.9	−0.1	+1.2	+0.5	+0.1	+1.0	+1.2	+0.9	+0.5
Nov	−1.1	−1.2	−1.3	−0.7	+0.1	−1.3	−0.8	−2.1	+1.1	−0.5
Dec	−6.5	−3.5*	−2.3	−3.7*	−1.1	−1.9	−1.6	−2.9*	+0.1	−3.1
MAM	+2.3	+9.4	+1.9	+9.0*	+2.6	+7.1	+7.3**	+1.7	+1.2	+2.9
JJA	+0.1	−6.3	−9.7	+1.0	−3.5	+3.5	+4.0	−4.3	−3.0	−1.0
SON	+11.6*	+6.2	+3.8	+5.6	+4.0	+4.8	+4.4	+1.4	+4.0	+1.5
DJF	−9.6	−1.2	−2.1	−1.4	+1.1	+0.5	+1.7	−5.2*	+3.9	−5.0
Year	+1.7	+7.1	−7.2	+11.1	+3.3	+14.9	+16.5*	−7.2	+4.9	−2.4

Statistical significance: $*\alpha \leq 0.05$, $**\alpha \leq 0.01$, $***\alpha \leq 0.001$

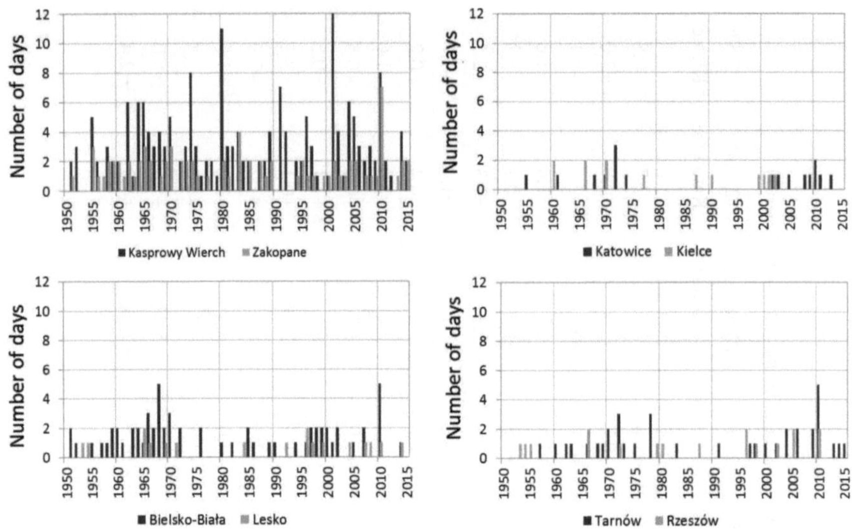

Fig. 15 Annual number of days with precipitation ≥ 50 mm at selected stations in the Upper Vistula Basin in the period 1951–2015

negative trends were found at every station (Table 7). Seasonal precipitation totals experienced more uniform changes. Spatially coherent increase was identified in spring and autumn precipitation. However, significant trends were only detected in Rzeszów (+7.5 mm per 10 years in spring).

In the Upper Vistula Basin, precipitation ≥ 30 mm occurred, on average, several times a year with maximum at Kasprowy Wierch (9 days) and minimum in Sandomierz (1.3 days). The record-high frequencies of days with heavy precipitation fell in different years, depending on the station. However, they accumulated during previously recognized wet periods, namely in 1960s and in particular years of the 1995–2015 period. Tarnów was the only station with statistically significant increase in the frequency of days with precipitation ≥ 30 mm. This tendency was particularly visible between 1995 and 2010 (Fig. 14).

Days with precipitation ≥ 50 mm are very rare in the Upper Vistula Basin (Fig. 15). Even at the most rainy station, Kasprowy Wierch, they occurred on average only 3 times a year. At other stations the frequency of such days varied between 1.2 days (Zakopane) and 0.2 days (Sandomierz, Kielce). In the long-term course, daily precipitation ≥ 50 mm showed tendency to cluster in the second half of the 1960s or between 1965 and 1975 and in the last three pentads of the research period.

We performed detailed analysis of precipitation characteristics using daily data taken from 18 meteorological stations (Table 1; Fig. 1) located within the Tara Mountains and Podhale Basin. The meteorological series that we used end in different periods depending on station (Table 1). We investigated maximum 10-day precipitation total and maximum monthly precipitation total by calculating average

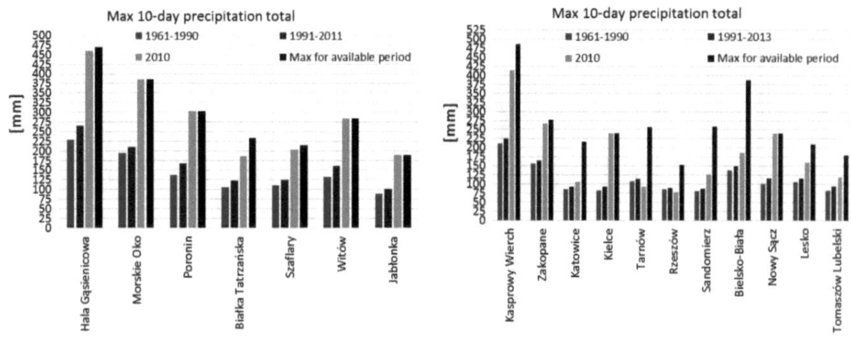

Fig. 16 Comparison of mean 10-day maximum precipitation total for period 1961–1990 with the mean value for the period 1991–2011 (*left*) or 1991–2013 (*right*); the value for 2010 year and the maximum value for all research period

indices for the WMO climatological standard period 1961–1990 and for the 1991–2011 (or 1991–2013 depending on tha data availability). These values were compared with the values of indices in the flood year of 2010, as well as with maximum value for the studied period.

Trend detections (Radziejewski and Kundzewicz 2004) for maximum 10-day precipitation totals showed a statistically significant increase at $\alpha < 0.05$ for Tomaszów Lubelski and Lesko. We also found weak ($\alpha < 0.25$) upward trends at four further stations including Hala Gąsienicowa, Poronin, Rzeszów and Tarnów. The comparison of average maximum 10-day precipitation totals for 1961–1990 with 1991–2013 (or 1991–2011) showed that in case of every station the index for the latest period was higher than that for the base period 1961–1990 (Fig. 16). The biggest difference was noted in Poronin (29 mm; 21.5 % of the average 1961–1990) while at other eight stations the increase reached about 10 % with minimum in Rzeszów: 4.2 %. At six stations the highest maximum 10-day precipitation total, just like annual totals, occurred in 2010 while at other five stations the index for 2010 was rated among the highest values.

Maximum monthly precipitation totals significantly increased at the level of 0.05 at stations: Lesko, Tarnów and Tomaszów Lubelski. Similar trends but significant at $\alpha < 0.1$ were also found at Hala Gąsienicowa and Rzeszów. On the other hand at three stations (Białka Tatrzańska, Szaflary and Witów) increases, however very weak, were also identified. At every station, average monthly precipitation for 1991–2013 (or 2011) was higher than that for the WMO base period 1961–1990. The biggest difference was recognised in Poronin (42 mm; 21.8 %). For other nine stations the differences were usually higher than 10 % but the smallest one—3.3 %—was noted at Kasprowy Wierch.

Figure 17 presents mean maximum monthly precipitation total for the normal period 1961–1990 and for the latest period 1991–2013. Also, the values of monthly precipitation total for the flood year 2001 and the maximum value from the entire

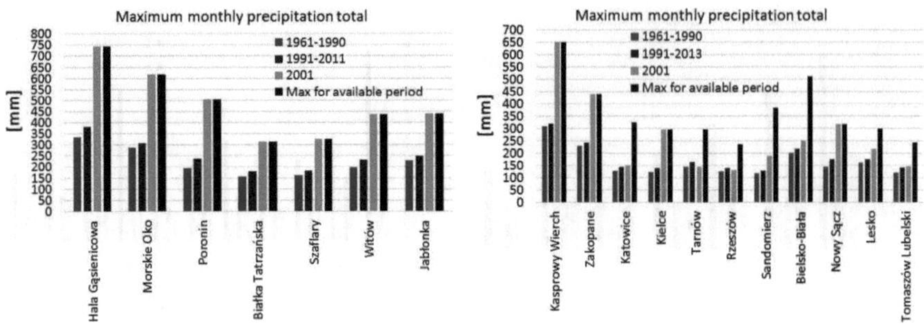

Fig. 17 Comparison between mean monthly maximum precipitation total for the period 1961–1990 and the mean value from period 1991–2011 (*left*) or 1991–2013 (*right*); the value for flood 2001 year and the maximum value for all research period

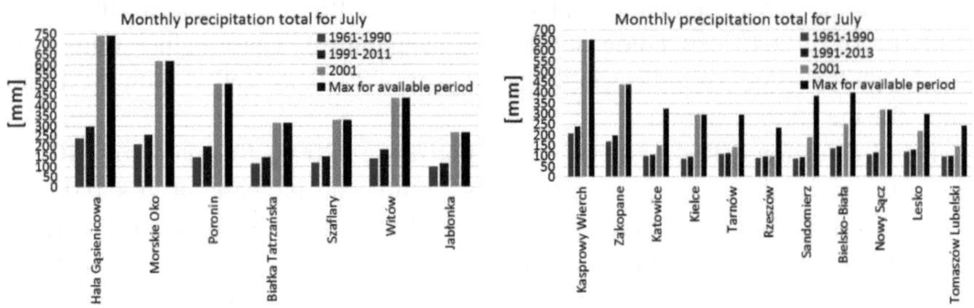

Fig. 18 Comparison between mean monthly precipitation total for July in the period 1961–1990 with the mean value from period 1991–2011 (*left*) or 1991–2013 (*right*); the value for July during flood 2001 year and the maximum value for all research period

period are shown. At every station located on the northern foothills of the Tatra Mountains, maximum monthly precipitation total was registered in the flood year 2001. However, we found no significant changes in monthly precipitation in July, when the totals are the highest of all monthly values. The mean July precipitation totals from the period 1991–2013 (or 2011) were higher than these from the WMO normal period. This increase was fairly evident at 10 stations, where the difference in precipitation totals was higher than 20 %. The highest increase was calculated for Poronin (54 mm; 37.9 %); for another six stations precipitation was recently higher by 10 % than in previous period. The lowest increase was detected in Tarnów (2.6 %).

Figure 18 depicts average monthly precipitation totals for July and extreme values for 2001 and the maximum for the entire research period. Beside July, May is the second month of large flood occurrence in the studied period. May precipitation has slightly increased in Tomaszów Lubelski and at stations: Lesko, Nowy Sącz, Rzeszów, Poronin and Zakopane. The flood during May and June 2010 was one of the most destructive floods in Poland. Precipitation in May 2010 was the

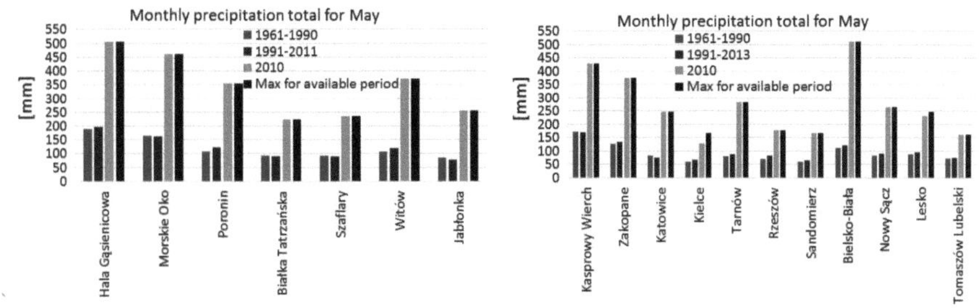

Fig. 19 Comparison between mean monthly precipitation total during May for period 1961–1990 with the mean value from period 1991–2011 (*left*) or 1991–2013 (*right*); the value for May during the flood 2010 year and the maximum value for all research period

highest of the whole available period at 16 stations. Furthermore, average May precipitation for the period 1991–2013 (or 2011) was higher than for the WMO base period at 12 stations, with the highest increase in Rzeszów: 15.9 % i.e. 11.2 mm (Fig. 19).

5 Links Between Temperature and Precipitation

Although there are several non-climatic factors influencing flood hazard, climate change is considered to be one of the principal drivers of the flood frequency. Climate models indicate that severe precipitation and rain-caused floods are likely to become more frequent due to climate warming at various spatial scales (from global via continental, national to regional and local) (IPCC 2013). However relations between precipitation and air temperature are not straightforward. Undebatable increase in air temperature is not currently accompanied by significant increase in precipitation. This is supported by correlation analysis that we performed between precipitation characteristics and summer air temperature (Fig. 20). The only significant ($\alpha < 0.01$) inverse correlation ($r = -0.33$) was found between temperature mentioned and the number of days with precipitation ≥ 30 mm. Decrease in precipitation characteristic in the periods of relatively high temperature in summer was also possible to notice in the long-term course of temperature and precipitation indices discussed previously in this chapter. Decrease in precipitation as a reaction on temperature increase is also proved by differences of temperature and precipitation indices calculated between the 1951–1987 and 1988–2015 periods presented in Table 8. Significant increase in air temperature and insignificant decrease in summer precipitation at majority of stations was found in summer. Trends in the frequency of days with precipitation ≥ 30 mm were insignificant and had different dirrections.

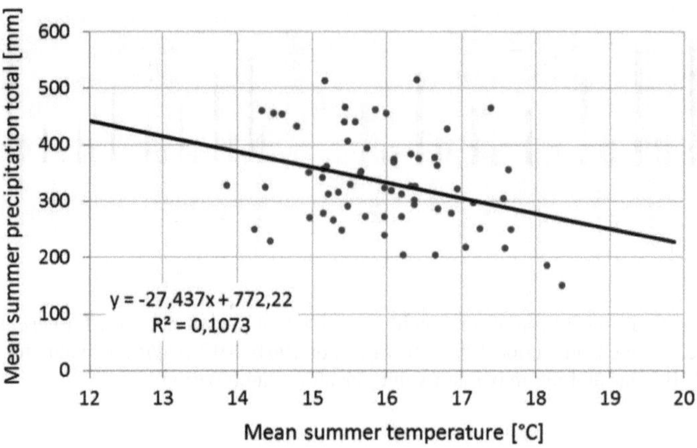

Fig. 20 Regression between mean summer precipitation totals and mean summer air temperature in the Upper Vistula Basin in the period 1951–2015. Mean is the arithmetical average from 10 stations

Figure 21 presents mean precipitation at nine stations, divided into daily precipitation intensity ranges (0–10, 20–20, … 40–50 mm, and above 50 mm) for two periods: 1961–1990 (WMO climatological standard normal) and more recent, warmer interval 1991–2013. Mean precipitation for the warmer interval 1991–2013 was slightly lower in the range of 40–50 mm, but much higher in the range above 50 mm. This illustrates the increase in precipitation intensity.

Figure 22 presents the mean annual number of days with precipitation in particular ranges for two periods, 1961–1990 and 1991–2013. Only for precipitation range 30–40 mm mean number of days is lower for warmer period 1991–2013. In all other cases mean annual number of days with precipitation was higher than in the base WMO period.

Figure 23 illustrates the time series of mean temperature and precipitation total in the warm half-year, from April to September. A broken-line trend can be spotted, consisting of two linear trends for sub-intervals, 1955–1984 and 1985–2013. In the former sub-interval, slight decrease in temperature and precipitation can be spotted, while in the latter sub-interval both variables are on the rise. However, looking at individual warm half-years, one can see that both time series are in counter-phase, i.e. a local minimum of one coincides with a local maximum of the other. Wet warm half-years 2001 and 2010 coincided with local temperature minima, while a dry half-year 2012 coincided with local temperature maximum.

Table 8 Comparison of selected climatic elements during the cool period (A 1951–1987) and warm period (B 1988–2015) identified on the basis of long-term course of air temperature in the Upper Vistula Basin

Element	Period	12 650 Kasprowy Wierch	12 625 Zakopane	12 600 Bielsko-Biała	12 690 Lesko	12 566 Kraków Balice	12 575 Tarnów	12 580 Rzeszów	12 585 Sandomierz	12 560 Katowice	12 570 Kielce	Average from 10 stations
Mean annual T (°C)	A	−0.8	5.1	7.8	7.3	7.7	8.2	7.6	7.7	7.8	7.2	6.6
	B	0.0	6.1	8.8	8.0	8.8	9.2	8.7	8.7	8.9	8.0	7.5
	B−A	+0.8***	+1.0***	+1.0***	+0.7***	+1.0***	+1.0***	+1.1***	+1.0***	+1.0***	+0.9***	+1.0***
Mean summer T (°C)	A	6.5	13.6	16.3	16.0	17.0	17.3	17.1	17.4	16.8	16.7	15.5
	B	7.7	14.9	17.6	17.0	18.3	18.6	18.3	18.4	18.0	17.6	16.6
	B−A	+1.2***	+1.2***	+1.3***	+1.0***	+1.3***	+1.2***	+1.2***	+1.0***	+1.2***	+0.9***	+1.2***
Mean annual P (mm)	A	1757.7	1122.2	996.9	796.6	669.3	702.1	610.2	578.1	715.1	629.0	857.7
	B	1740.9	1137.1	974.0	826.1	670.6	724.1	660.2	551.2	714.2	631.9	863.0
	B−A	−16.8	+14.9	−22.9	+29.5	+1.3	+21.9	+50.1	−27.0	−0.9	+2.9	+5.3
Mean summer P (mm)	A	622.7	501.9	413.4	323.5	269.4	291.7	241.5	236.7	277.2	238.4	341.6
	B	608.8	463.1	369.5	310.1	250.3	283.1	245.0	215.8	254.0	238.8	323.8
	B−A	−13.9	−38.7	−43.9	−13.4	−19.1	−8.6	+3.4	−21.0	−23.2	+0.4	−17.8
NoD $P \geq 30$ mm	A	9.2	5.3	4.1	2.3	1.5	1.7	1.3	1.5	1.5	1.4	3.0
	B	8.6	5.4	3.6	2.8	1.8	2.4	1.5	1.1	1.5	1.4	3.0
	B−A	−0.6	0.0	−0.5	+0.5	+0.3	+0.7	+0.2	−0.4	0.0	0.0	0.0

T—air temperature, *P*—precipitation, *NoD*—number of days, *A* 1951–1987, *B* 1988–2015, (statistical difference between mean air temperature tested with t-test and between mean precipitation with Mann-Whitney test), statistical significance: $^{*}\alpha \leq 0.05$, $^{**}\alpha \leq 0.01$, $^{***}\alpha \leq 0.001$

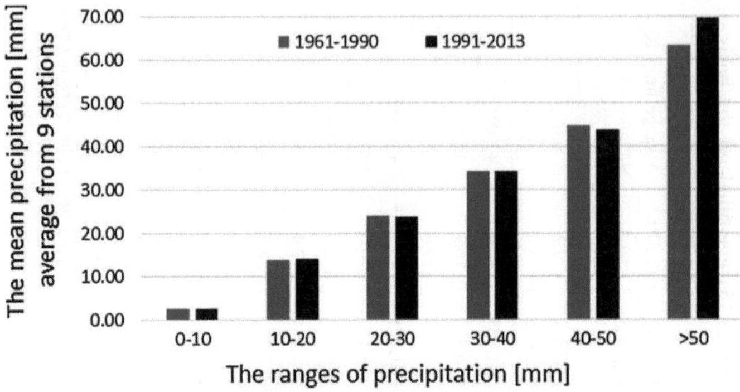

Fig. 21 Mean precipitation at nine stations, divided in daily precipitation intensity ranges (0–10, 20–20, … 40–50 mm, and above 50 mm) for two periods: 1961–1990 (WMO climatological standard normal) and more recent, warmer interval 1991–2013

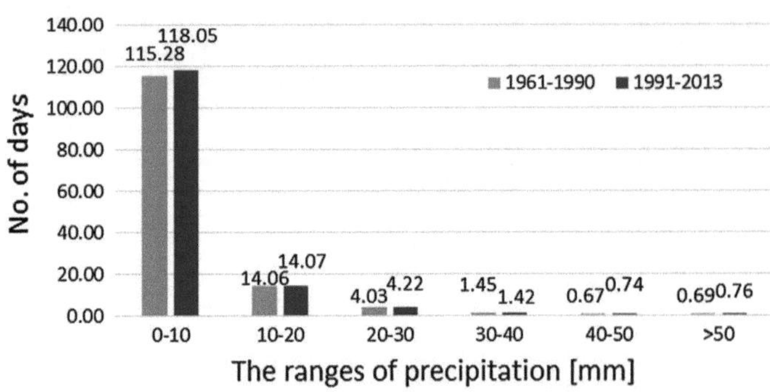

Fig. 22 Mean annual number of days with precipitation (mean from nine stations) for two periods: 1961–1990 and 1991–2013

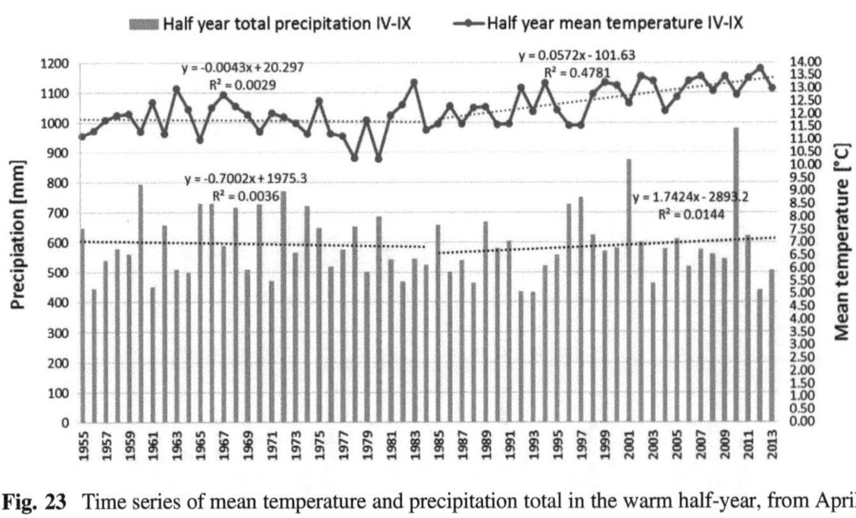

Fig. 23 Time series of mean temperature and precipitation total in the warm half-year, from April to September

6 Conclusions

The Upper Vistula Basin with the Tatra Mountains located in the south-eastern part of Poland are very specific in the scale of the country due to its natural environment which results from variegated altitude. In addition the Tatra Mountains play crucial role in flood precipitation generation being an orographic barrier for air masses flowing into the region from the north-eastern sector. The climatic condition and its changes in the Tatra Mountains can lead to changes in flood condition thus. Using data from 18 meteorological stations covering the period 1951–2015 this chapter discusses changes in air temperature and precipitation characteristics within the Upper Vistula Basin.

Air temperature in the Upper Vistula Basin, likewise on global scale, was significantly rising during the research period. Distinct warming on annual scale started in the first half of the eighties. Different pattern of long-term variability were found for particular seasons. Spring and summer air temperature has been significantly increasing, particularly since the beginning of the eighties, with the rate of 0.21 to 0.40 K per 10 years depending on season and station. Winter air temperature trends were much weaker, however significant at some stations while no significant changes were found in autumn. Upward trends in air temperature within the Upper Vistula Basin were also reflected in the frequency of thermally characteristic days which were significantly changing during the research period. Strong downward trends were found in the frequency of winter days (Tavg $\leq$ 0 °C) while trends in warm temperature characteristics (hot and very hot days) were positive.

Although there was significant warming in the flood seasons, no significant trends were found in the precipitation indices in the period 1951–2015 in spite of single stations. Moreover these insignificant trends were of various directions depending on station and season. March, September and August were the only months with coherent trends within the entire research area. On seasonal scale quite coherent increases were identified in spring and autumn. On the other hand trend analysis performed for shorter periods (to 2011 or 2013) revealed more coherent increases, but also insignificant. Additionally, the highest values of many precipitation indices occurred in flood years of 2001 and 2010. Important is that in the long-term course of daily precipitation $\geq$50 mm there was a tendency to their clustering in the second half of the sixties or between 1965 and 1975 and in the last three pentads of the research period.

Trend analysis for 18 stations revealed precipitation increases, but in most cases they were not statistically significant due to precipitation strong irregular natural variability. Every increase in precipitation, especially an increase in long lasting precipitation, i.e. in 10-day precipitation total and monthly precipitation total for the stations located in the area of the Upper Vistula basin can contribute in the future to more severe flash floods and landslides in small catchments of the mountainous tributaries of the Vistula (such as the River Dunajec) and it can generate flood events on lower area of Poland.

Significant increase in air temperature is not currently accompanied by significant increase in precipitation thus changes in precipitation are not directly related to changes in air temperature and they possibly vary in time. Relations between precipitation and air temperature are not straightforward. The results indicate that strong increase in air temperature is rather accompanied by decrease in precipitation frequency and amount. However, this applies only to overall precipitation totals and not to extreme events which are random and can occur unexpectedly. Flood precipitation can also occur during drier periods, as in the last decade of the research period.

Acknowledgments This study was carried out within the FLORIST project (Flood risk on the northern foothills of the Tatra Mountains), supported by a grant from the Swiss Government through the Swiss Contribution to the enlarged European Union (PSPB No. 153/2010). Temperature and precipitation data were made available by Institute of Meteorology and Water Management, National Research Institute (IMGW, PIB). Part of data for the period 2000–2015 was taken from synoptic data-base OGIMET.

References

Bielec-Bąkowska Z, Łupikasza E (2009) Frosty, freezing and severe freezing days and their synoptic implications in Małopolska, Southern Poland, 1951–2000. Bull Geogr Phys Geogr Ser 1:39–62

Bielec-Bąkowska Z, Piotrowicz K (2013) Temperatury ekstremalne w Polsce w latach 1951–2006 (Extreme temperatures in Poland 1951–2006). Prace Geograficzne, Kraków 132:59–98 (in Polish, summary in English)

Cebulak E (1992) Maksymalne opady dobowe w dorzeczu górnej Wisły (Maximum daily precipitation totals in the Upper Vistula Basin). Zeszyty Naukowe UJ, Prace Geograficzne 90:79–96

Cebulak E (1997) Variability of precipitation in selected regions of the Carpathians in the years 1951–1995. Geographia Pol 70:65–76

Cebulak E, Limanówka D (2007) Dni z ekstremalnymi temperaturami powietrza w Polsce (Days with extreme air temperatures in Poland). In: Piotrowicz K, Twardosz R (eds) Wahania klimatu w różnych skalach przestrzennych w Polsce. Instytut Geografii i Gospodarki Przestrzennej UJ, Kraków, pp 185–194

Chorynski A, Pinskwar I, Kron W, Brakenridge R, Kundzewicz ZW (2012) Catalogue of large floods in Europe in the 20th century. Chapter 3. In: Kundzewicz ZW (ed) Changes in flood risk in Europe. Special Publication No. 10. IAHS Press, Wallingford, Oxfordshire, pp 27–54

Degirmendžić J, Kożuchowski K, Żmudzka E (2004) Changes of air temperature and precipitation in Poland in the period 1951–2000 and their relationship to atmospheric circulation. Int J Climatol 24(3):291–310. doi:10.1002/joc.1010

Fortuniak K, Kożuchowski K, Żmudzka E (2001) Trendy i okresowość zmian temperatury powietrza w Polsce w drugiej połowie XX wieku (Trend and periodicity of changes in air temperature in Poland in the second half of 20th century). Przegląd Geofizyczny 46(4):283–303 (in Polish)

Hess M (1965) Piętra klimatyczne w polskich Karpatach Zachodnich (Vertical climatic zones in the polish Western Carpathians). Zeszyty Naukowe UJ, Prace Geograficzne, Kraków, vol 11, 267 pp

IPCC (2013) Climate change 2013: the physical science basis. In: Stocker TF, Qin D, Plattner G-K, Tignor M, Allen SK, Boschung J, Nauels A, Xia Y, Bex V, Midgley PM (eds) The contribution of working Group I to the Fifth Assessment Report of the Intergovernmental Panel on climate change. Cambridge University Press, Cambridge and New York, 1535 pp

Kożuchowski K, Żmudzka E (2001) Ocieplenie w Polsce: skala i rozkład sezonowy zmian temperatury powietrza w drugie połowie XX wieku (Warming in Poland: scale and seasonal distribution in changes of air temperature in the second half of 20th century). Przegląd Geofizyczny 46(1–2):81–90 (in Polish)

Kundzewicz ZW, Dobrowolski A, Lorenc H, Niedźwiedź T, Pińskwar I, Kowalczak P (2012) Floods in Poland. Chapter 17. In: Kundzewicz ZW (ed) Changes in flood risk in Europe. Special Publication No. 10. IAHS Press, Wallingford, Oxfordshire, pp 319–334

Limanówka D (1997) Variability of selected thermal characteristics of the air in the years 1951–1995. Geographia Pol 70:53–64

Łupikasza E (2010) Spatial and temporal variability of extreme precipitation in Poland in the period 1951–2006. Int J Climatol 30(7):991–1007

Niedźwiedź T (2003) Extreme precipitation events on the northern side of the Tatra Mountains. Geographia Pol 76(2):15–23

Niedźwiedź T, Obrębska-Starklowa B (1991) Klimat (Climate). Chapter 1.5. In: Dynowska I, Maciejewski M (eds) Dorzecze górnej Wisły. Część I (Upper Vistula Basin. Part I). PWN, Warszawa-Kraków, pp 68–84

Niedźwiedź T, Twardosz R, Walanus A (2009) Long-term variability of precipitation series in east central Europe in relation to circulation patterns. Theoret Appl Climatol 98(3–4):337–350. doi:10.1007/s00704-009-0122-0

Niedźwiedź T, Łupikasza E, Pińskwar I, Kundzewicz ZW, Stoffel M, Małarzewski Ł (2015) Variability of high rainfalls and related synoptic situations causing heavy floods at the northern foothills of the Tatra Mountains. Theor Appl Climatol 119(1–2):273–284. doi 10.1007/s00704-014-1108-0 (Open Access)

NOAA (2016) NOAA National Centers for Environmental Information, state of the climate: global analysis for annual 2015, published online January 2016. http://www.ncdc.noaa.gov/sotc/global/201513. Retrieved on 19 Feb 2016

Radziejewski M, Kundzewicz ZW (2004) Development, use and application of the HYDROSPECT data analysis system for the detection of changes in hydrological time-series for use in WCP-Water and National Hydrological Services, WCASP-65, Hydrospect, Version 2.0. User's manual. WMO, Geneva

Ruiz-Villanueva V, Stoffel M, Wyżga B, Kundzewicz ZW, Czajka B, Niedźwiedź T (2016) Decadal variability of floods in the northern foreland of the Tatra Mountains. Reg Environ Change 16(3):603–615. doi:10.1007/s10113-014-0694-9

Trepińska J (1969) Ogólna charakterystyka opadów w Krakowie 1916–1965 (General characteristics of precipitation in Cracow 1916–1965). Folia Geographica, series Geographica Physica 3:117–138

Trepińska J (1983) Climatic fluctuations in Cracow City, 1826–1975. Geographia Pol 45:59–70

Trepińska J (ed) (1997) Wahania klimatu w Krakowie (1792–1995) [Climate fluctuations in Kraków (1792–1995)]. Instytut Geografii UJ, Kraków, 204 pp (in Polish)

Trepińska J, Ustrnul Z, Kowanetz L (1997) Variability of air temperature in Central Europe in the years 1792–1995. Geographia Pol 70:43–52

Twardosz R (1999) Warunki pluwialne w Krakowie w latach 1792–1998 (Pluvial conditions in Cracow in the years 1792–1998). Czasopismo Geograficzne 70(2):221–232 (in Polish, summary in English)

Ustrnul Z, Czekierda D (2009) Atlas of extreme meteorological phenomena and synoptic situations in Poland. Instytut Meteorologii i Gospodarki Wodnej, Warszawa, p 182

Wibig J, Głowicki B (2002) Trends of minimum and maximum temperature in Poland. Clim Res 20:123–133

Change in Atmospheric Circulation Patterns

Tadeusz Niedźwiedź and Ewa Łupikasza

Abstract This chapter examines both variability and trends in atmospheric circulation favouring the occurrence of flood precipitation defined as daily totals ≥30, ≥50 and ≥100 mm in warm half-year (May–Oct) and in summer (JJA). We used a catalogue of circulation types created for the Upper Vistula Basin, and related circulation indices (zonal circulation index, meridional circulation index, cyclonicity index and NAO) covering the 1874–2015 period. Climatology of atmospheric circulation over the Upper Vistula Basin is discussed as a basis for further investigations. In order to select circulation types and indices impacting both the occurrence and long-term variability of flood precipitation, we calculated the frequency and conditional probability of high precipitation (≥50 mm) in circulation types and correlation between selected circulation characteristics and high precipitation frequency (≥30 mm). Trends in the frequency of circulation types and indices favouring the occurrence of high precipitation were calculated to assess current and possible future flood conditions. In summer and warm half-year the Upper Vistula Basin was usually under an influence of anticyclonic wedge (Ka circulation type) and cyclonic trough (Bc circulation type). Circulation types with the air flow from the west (Wa and Wc) were the most frequent of all advectional types. The occurrence and long-term variability in flood precipitation over the Upper Vistula Basin were strongly linked to the frequency of air advection from the north and north-east under an influence of low pressure system (Nc and NEc circulation types) and to Wi—zonal circulation index at both stations in summer (JJA) and to Ci—cyclonicity index at Kasprowy Wierch station in warm half-year

T. Niedźwiedź (✉) · E. Łupikasza
Faculty of Earth Sciences, University of Silesia in Katowice,
ul. Będzińska 60, 41-200 Sosnowiec, Poland
e-mail: tadeusz.niedzwiedz@us.edu.pl

E. Łupikasza
e-mail: ewa.lupikasza@us.edu.pl

Z.W. Kundzewicz et al. (eds.), *Flood Risk in the Upper Vistula Basin*,
GeoPlanet: Earth and Planetary Sciences, DOI 10.1007/978-3-319-41923-7_9

(May–Oct). Trends in majority of circulation characteristics favouring the occurrence of high precipitation and impacting its long-term variability were not statistically significant with an exception of the frequency of Nc type and Ci index in the warm half-year and Wi index in summer. Significant increase in the number of days with Nc circulation type and the cyclonic situations (Ci index) in warm half-year and intensification of air advection from the north-east in summer may lead to increase in the frequency of flood conditions in these seasons. Regardless insignificant trends, the variability in the NEc frequency considerably determines the occurrence and long-term variability of high precipitation thus was recognized as indicator of flood conditions. There was intensification of the air advection from the north (Nc) in the warm half-year (May–Oct) in the 50-year period between 1930 and 1980. High frequency of NEc type was found in seventies and eighties which coincided with the high frequency of floods in those decades in Poland and at the turn of the first and second decades of the 21st century. Low frequencies of these circulation types during 1982–1995 were in phase with relatively dry conditions in Southern Poland without the floods within the Upper Vistula Basin. In the further more wet years the great floods happened in July 1997 and May 2010.

Keywords Flood precipitation · Circulation patterns · Circulation indices · NAO · Circulation trends · The Upper Vistula Basin

1 Introduction

Atmospheric circulation defined as three dimensional motion of the air, is counted amongst crucial climatic drivers (Barry and Carleton 2001). Air masses flowing from various directions differ in physical properties thus determining the type of weather occurring over particular area and accompanying meteorological phenomena. It has been proved that the occurrence of heavy precipitation is strongly related to both direction of air advection and type of baric centre (Niedźwiedź 1981, 2003a). These relations vary regionally (Łupikasza 2013). Assessment of trends in flood precipitation is usually impossible due to its rare occurrence. Therefore knowledge about the changes in circulation patterns favouring flood precipitation events can give a hint on possible future trends in the events. This chapter discusses change in circulation patterns favouring flood precipitation occurrence. The climatology of atmospheric circulation serving as a basis for further sections is presented in Sect. 2. The linkage between flood precipitation occurrence and atmospheric circulation and its influence on long-term variability of high precipitation events are discussed in Sect. 3. Section 4 concerns changes in circulation types and indices that are favourable for flood precipitation occurrence thus indicating possible future changes in flood conditions. The chapter uses 142-year (1874–2015) long series of circulation types and indices and 65-year long (1951–2015) series of precipitation data which are discussed in subsequent sections.

2 Climatology of Circulation Types and Indices

Processes associated with many environmental problems, including water quality and quantity, are influenced by atmospheric circulation (Yarnal 1993). The impact of atmospheric circulation on precipitation in Poland so far has been presented by Twardosz and Niedźwiedź (2001), Degirmendžić et al. (2004), Niedźwiedź et al. (2009). Atmospheric circulation, as it is shown in the next section, also triggers flood precipitation which was recognized to occur mainly in warm part of the year with maximum in summer (Kundzewicz et al. 2012). This section discusses the climatology of atmospheric circulation in the periods of flood precipitation occurrence, namely in warm-half year (May–October) and in summer (JJA). We used a catalogue of circulation types created for the Upper Vistula Basin (Niedźwiedź 1981, 2016) and three regional circulation indices (Niedźwiedź 1978, 2000): zonal westerly circulation index (Wi), meridional southerly circulation index (Si) and cyclonicity index (Ci). The catalogue covers the period from September 1873 to December 2015 and consists of 21 circulation types denoted by direction of air advection (e.g. N—northern, SE—south-eastern, etc.) and the type of pressure system (a—anticyclonic, c—cyclonic). Sixteen of the 21 types are advectional types whereas the further four are non-advectional types (Ca—central anticyclonic, centre over or very near the Tatra Mountains, Ka—anticylonic wedge, ridge, Cc—centre of cyclone above or very near the Southern Poland, Bc—cyclonic trough). Days with fuzzy or complicated sea level pressure (SLP) pattern or with col were marked by x. The classification of advectional types is methodologically similar to the well-known Lamb (1972) classification for the British Isles with the only difference that transitional types were not included into Niedźwiedź's classification.

The method by Murray and Lewis (1966), originally used for the British Isles, was modified and adopted to calculate circulation indices for the Southern Poland. Regional circulation indices were calculated on the base of the circulation types frequencies (Niedźwiedź 2000). In order to do so points were assigned to each circulation type depending on the direction of air advection as follows:

- zonal westerly circulation index (Wi): +2 for W, +1 for NW and SW, −2 for E and −1 for NE and SE,
- meridional southern circulation index (Si): +2 for S, +1 for SW and SE, for N −2 and −1 for NW and NE,
- cyclonicity index (Ci): +2 for Cc and Bc, +1 for other cyclonic types (Nc, NEc, … etc.), −2 for Ca and Ka, −1 for other anti-cyclonic types (Na, NEa, … etc.).

Monthly values of every index were calculated as a total of the assigned points. The total was then divided by doubled number of days in a month (season or year) thus being expressed as percentage. The indices can theoretically vary from +100 to −100 % meaning:

- Wi index: westerly (Wi = +100 %) or easterly (Wi: −100 %) advection throughout an entire month,

- Si index: southerly (Si = +100 %) or northerly (Si = −100 %) advection throughout an entire month,
- Ci index: cyclonic (Ci = +100 %) or anticyclonic (Ci = −100 %) circulation types throughout an entire month.

Frequencies of particular circulation types in summer (JJA) and warm-half year (May–Oct) over the Upper Vistula Basin are shown in Fig. 1. The seasonal patterns are similar with clear maximum of anticyclonic wedge lying over the area during 16 % (JJA) or 14 % (May–Oct) of days. The second most frequent type—Bc—occurring with the frequency of 13 % (JJA) or 11 % (May–Oct) of days per season, is known to bring precipitation to the whole Poland (Morawska-Horawska 1971; Niedźwiedź 2003a, b). Among the advectional types, these with the air advection from the west dominated in case of both anti-cyclonic and cyclonic situations which results from location of the area within the moderate climatic zone of the western winds. The types constituted 9–10 % of both summer and warm half-year days. Over the Upper Vistula Basin the air quite often flowed from the northwest (5–7 % depending on baric system and season) (Fig. 1). The frequencies of other circulation types varied from 1 % (Cc type) to 5 % (Na, NEa and Ea types). The Nc and NEc types, which flood precipitation in the Upper Vistula Basin is usually associated to, were not very common (2–3 %), however, the maximum of their occurrence fell at the turn of the spring and summer (April, May, June, July) which is a flood prone period. The frequencies of circulation types in the 1951–2015 period did not differ much from these described above for the 1873–2015 period. The differences in the frequencies of majority of types did not exceed 1 % despite of Wa, Ca and Ka types. In the 1951–2015 the types Wa and Ka occurred less frequently while the type Ca was noted more frequently than in the 1873–2015 period.

Annual course of the circulation indices presented in Fig. 2 again indicates the domination of air advection from the west particularly from June to March with maximum in December (Wi: +26 %). This was only May (Wi: −4) when eastern airflow was more frequent than western one. The frequency of anticyclonic types was clearly higher than cyclonic types from August (Ci: −12 %) to October (Ci: −16 %) and in January (Ci: −15 %) with highest intensity in September (Ci: −17 %). April

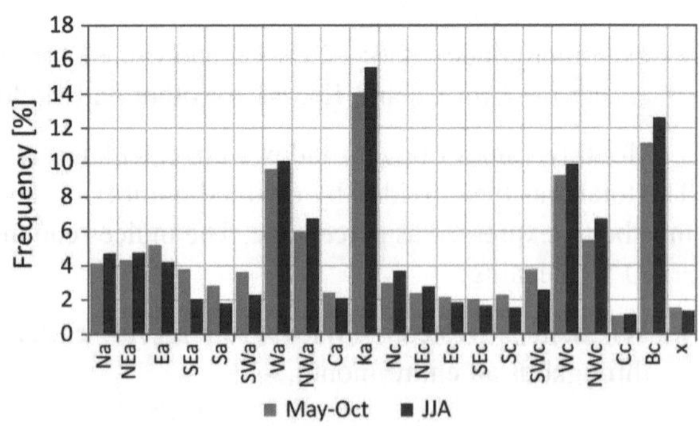

Fig. 1 Frequency of circulation types in summer (JJA) and warm half-year (May–Oct) over the Tatra Mountains in the 1873–2015 period

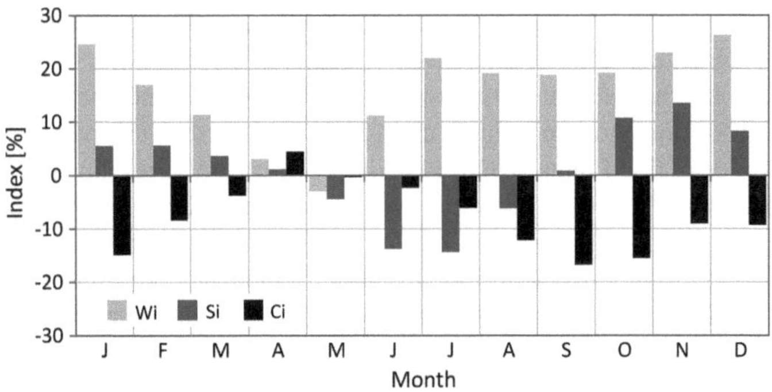

Fig. 2 Annual course of circulation indices over the Tatra Mountains (1873–2015). *Wi*—zonal westerly circulation index, *Si*—meridional southerly circulation index, *Ci*—cyclonicity index

was the only month with positive Ci index (+4 %) indicating, in this case slight prevalence of cyclonic types (Fig. 2). The intensity and direction of meridional advection also changed throughout the year. The Si index was positive from September to April, but noticeable prevalence of southerly advection was found in October (Si: +11 %), November (Si: +14 %) and December (Si: +8 %). Northerly advection prevailed from May to August and was the most intense in July (Si: −7 %). The atmospheric circulation was very variable in April and May when all circulation indices were closest to 0 ranging between +4 and −4 % (Fig. 2).

3 Relation Between Flood Precipitation and Circulation Patterns

Synoptic conditions favouring flood precipitation occurrence in Poland were mentioned for the first time by Kaczorowska (1933) and Milata (1955). They found that majority of extremely high, flood precipitation were noted during air advection from the northern sector under an influence of low pressure system or at days with centre of low (Cc type) or the through (Bc type) located over Tatras (Morawska-Horawska 1971; Cebulak 1992; Cebulak and Niedźwiedź 2000; Niedźwiedź 1999, 2003a, b; Mudelsee et al. 2004; Ustrnul and Czekierda 2001, 2009; Kundzewicz et al. 2012; Niedźwiedź et al. 2015; Ruiz-Villanueva et al. 2016). These circulation types were usually related to low pressure systems moving from the Adriatic Sea along the Vb van Bebber (1891) path. After reaching Hungary or Ukraine, the systems were usually blocked by high pressure system with its centre located over the Eastern Europe. This finally led to orographically forced convection of moist air along the northern slopes of the Tatra Mountains and resulting long-lasting rains (Niedźwiedź 2003a, b). Such a situation (Nc) happened in 30 June 1973 and caused record-breaking daily rainfall (300 mm) noted at Hala Gąsienicowa station (Cebulak 1992; Cebulak and Niedźwiedź 2000) and between

16 and 18 July 1934 when 3-day rainfall total reached 422 mm at the same station and led to damaging flood (Niedźwiedź 2003a, b).

We examined the relation between high precipitation, defined here as daily totals ≥50 and ≥100 mm, and atmospheric circulation in order to select patterns favouring the occurrence of flood rains. We used daily precipitation series covering the 1951–2015 period from Zakopane (844 m above sea level) and Kasprowy Wierch (1991 m above sea level) stations and the catalogue of circulation types and circulation indices (Niedźwiedź 2016) described in Sect. 2. Moreover we used North Atlantic Oscillation Index (NAO) which is proved to be an important teleconnection pattern triggering precipitation over the vast areas of northern hemisphere including Europe (Hurrel and van Loon 1997; Hurrel 2001). Precipitation data were made available by Institute of Meteorology and Water Management, National Research Institute (IMGW, PIB). Some data for the 2000–2015 period comes from synoptic data-base OGIMET (Valor and López 2016) and were used after detailed quality control. Average (1951–2015) annual number of days with precipitation ≥50 mm was very low reaching 1 day at Zakopane station (81 days in the whole period) and 3 days at Kasprowy Wierch station (192 days in 65 years). In the Tatra Mountains majority of days with high precipitation (≥50 mm) occurred from May to October (Zakopane: 95 % of days, Kasprowy Wierch: 89 % of days) with maximum in summer (Zakopane: 75 % of days, Kasprowy Wierch: 69 % of days). Therefore we performed the analysis for warm half-year (May–October) and summer (JJA).

Frequency and conditional probability of high precipitation occurrence (≥50 mm) in particular circulation types showed in Fig. 3, allowed selecting

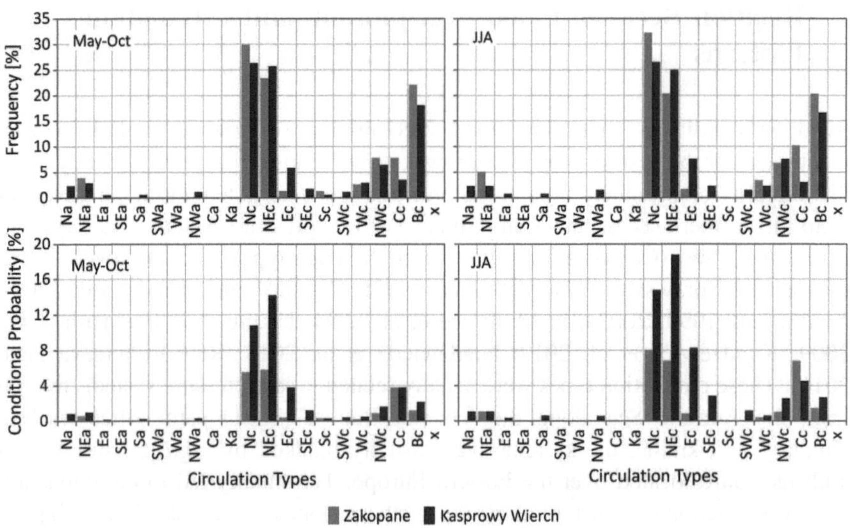

Fig. 3 Frequency (*upper panel*) and conditional probability (*lower panel*) of daily precipitation ≥50 mm in circulation types in warm half-year (May–Oct) and in summer (JJA) in the Tatra Mountains (1951–2015)

flood-generating patterns. Conditional probability is a method used to assess relationships between variables while taking into account the frequency of these variables. The statistic provides information about the degree to which each of the synoptic situations considered favours flood precipitation occurrence, and thus has a prognostic value. The probability of the occurrence of Event E1 (precipitation ≥50 mm) on condition of the occurrence of Event E2 (a specific circulation type) is expressed as (Wilks 2006):

$$\Pr\left\{\frac{E_1}{E_2}\right\} = \frac{\Pr\{E_1 \cap E_2\}}{\Pr\{E_2\}} * 100\,[\%] \quad (1)$$

The conditional probability of high precipitation in the circulation types is low, with the highest values not exceeding 8 % in Zakopane and 18 % at Kasprowy Wierch station, which is due to the relatively low frequency of precipitation ≥50 mm and to high frequency of circulation types, which occur every day. Despite this, the results obtained provide valuable information, which can be used in modelling precipitation at various spatial and temporal scales.

At both stations majority of high precipitation occurred at days with air advection from the north (Nc) and north-east (NEc) under an influence of low pressure system. The frequency of high precipitation in these types varied between 25 and 27 % at Kasprowy Wierch and 20–32 % in Zakopane depending on season and circulation type. Noticeable fraction of high precipitation was noted during cyclonic trough (Bc circulation type) located over the Upper Vistula Basin—20–22 % in Zakopane and 18–17 % at Kasprowy Wierch. Frequency of high precipitation in other cyclonic types did not exceed 8 % (May–Oct) and 10 % (JJA) in Zakopane and 6 % (May–Oct) and 8 % (JJA) at Kasprowy Wierch.

Conditional probability of high rainfall at days with Nc and NEc types was much higher than in other types, particularly at Kasprowy Wierch station. Unlike in case of frequency, the probability of high precipitation occurrence in type Cc was higher than in type Bc (Fig. 3). We assume that particular circulation type favours flood events occurrence if conditional probability of high rainfalls in that type exceeds by at least 50 % the probability of precipitation occurrence regardless of the circulation conditions. At both stations and seasons four circulation types met this condition, thus being recognized as favouring for flood precipitation occurrence—Nc, NEc, Cc and Bc (Fig. 3). At Kasprowy Wierch station Ec circulation type also favoured high precipitation events. Majority of flood precipitation (≥100 mm) that has occurred at Kasprowy Wierch station (10 of 15 events) and Zakopane (all events) were induced by Nc and NEc types (Tables 1 and 2).

Łupikasza and Niedźwiedź (2015) recognized that relations between high precipitation occurrence and air advection from the north (Nc type) and the north-eastern (NEc type) directions under an influence of low is stable in long-term course in both summer and warm-half year particularly at Kasprowy Wierch station. In Zakopane located in the centre of the city at the bottom of Tatras the relations discussed are more complicated and less stable in time but of a similar nature. Typical synoptic situations generating heavy precipitation within the Upper

Table 1 Synoptic conditions (*P*—precipitation, *CT*—circulation type, *AmT*—air mass type, *F*—atmospheric front) at days with high precipitation ($P \geq 100$ mm) at Kasprowy Wierch station in the period 1951–2015

Date	*P* (mm)	CT	AmT	F
30 June 1973	232.0	Nc	mP	Cold
8 July 1997	166.1	Nc	moP	Occlusion
18 July 1962	147.4	NEc	moP	Occlusion
14 July 1983	146.4	Nc	mP	Cold
15 May 2014	130.1	NEc	moP	No front
16 Aug 1988	118.7	Bc	Diff	Cold
10 June 2002	109.4	Ec	moP	Occlusion
23 July 2008	109.1	NEc	moP	No front
26 May 2015	106.9	Nc	moP	Various
23 July 1980	106.3	Nc	moP	No front
18 Juyl 1968	105.4	Cc	moP	No front
22 Aug 2009	103.8	Bc	Diff	Stationary
13 July 1960	103.6	NEc	moP	Various
21 June 1955	103.3	Bc	mP	Cold
6 Sept 2001	100.5	NEc	mP	No front

mP polar maritime air, *moP* polar maritime old (transformed) air, *Diff* different air masses during a day

Table 2 Synoptic conditions (*P*—precipitation, *CT*—circulation type, *AmT*—air mass type, *F*—atmospheric front) at days with high precipitation ($P \geq 100$ mm) at Zakopane station in the 1951–2015 period

Date	*P* (mm)	CT	AmT	F
30 June 1973	138.7	Nc	mP	Cold
18 July 1970	138.7	Nc	mP	No front
13 July 1960	116.5	NEc	moP	Various
23 July 2008	113.5	NEc	moP	No front
14 July 1983	109.2	Nc	mP	Cold
29 June 1958	109.1	Nc	mP	Occlusion
15 May 2014	104.2	NEc	moP	No front
8 July 1997	104.0	Nc	moP	Occlusion
18 July 1962	103.8	NEc	moP	Occlusion

mP polar maritime air, *moP* polar maritime old (transformed)

Vistula Basin (mainly in Tatras) are presented in Fig. 4. On 18 July 1970 stationary cyclonic system laying over Ukraine caused intense flow of polar maritime old and moist air mass from the north (Nc situation). The air mass was orographically lifted along the northern slopes of the Carpathian Mountains which produced prolonged three days intense rainfall. On that day rainfall amount reached 275 mm at Leskowiec station (870 m) in Skawa catchment and 234 mm at Stańcowa station in Orawa catchment (Niedźwiedź 2003a, b) (Fig. 4). Similar situation but with NE air flow happened on 16 July 1934 when the highest rainfall amounting to 285 mm was noted in Witów (Black Dunajec catchment). Slightly lower total of 255 mm was measured at Hala Gąsienicowa station (1520 m) located on the northern slopes of the Tatra Mountains. The highest-ever total for the whole Upper Vistula Basin was reached during one day duration rainfall on 30 June 1973 with Nc circulation

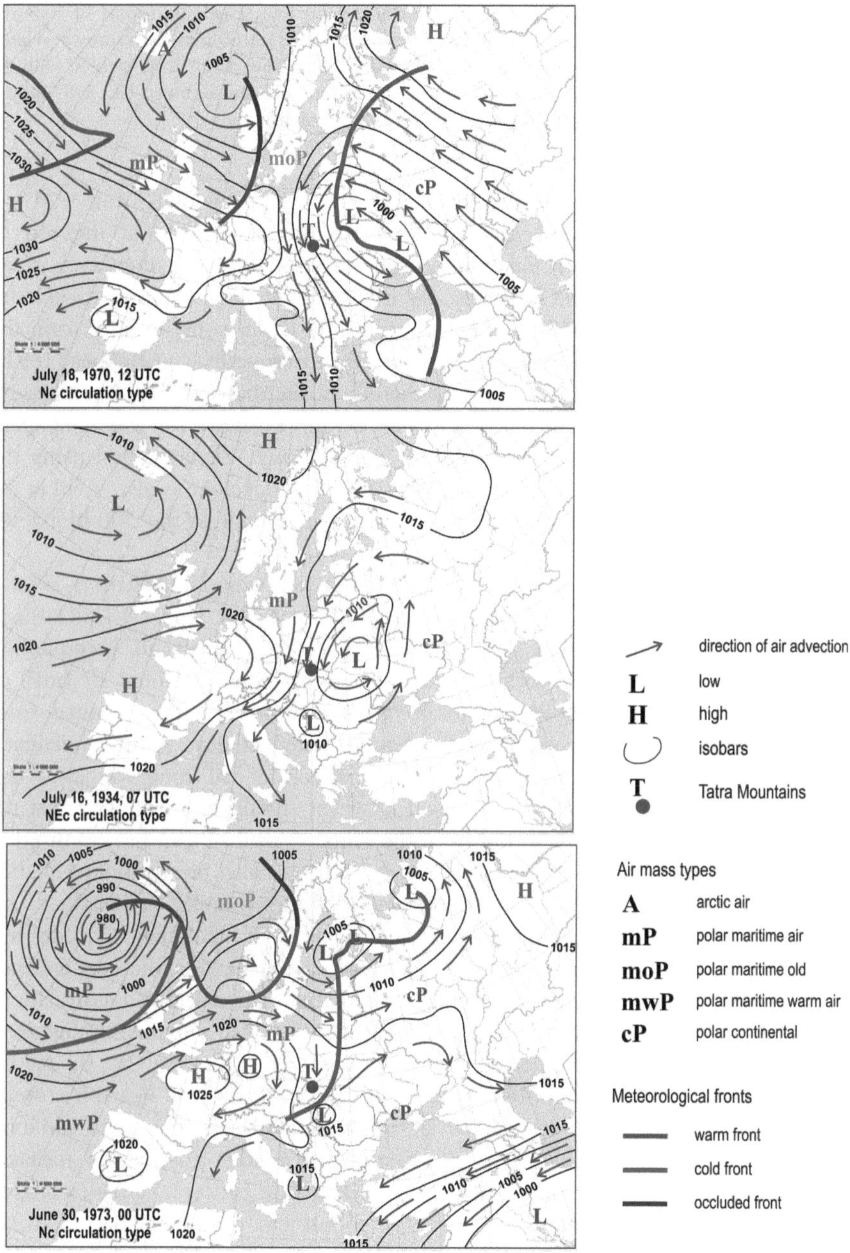

Fig. 4 Synoptic maps for selected days with precipitation ≥100 mm in the Upper Vistula Basin (based on published maps of Polish Meteorological Service: PIM, PIHM and IMGW PIB)

Fig. 5 Composite maps of averaged SLP (hPa) and geopotential heights (gpm) of 850 hPa ▶ (TB 850) and 500 hPa (TB 500) from 15 days with precipitation ≥100 mm at Kasprowy Wierch stations (see Table 1 for details), *Red arrows*—direction of air flow. Maps were created with the software and database of the NOAA/ESRL Physical Science Division, Boulder, CO, USA http://www.esrl.noaa.gov/psd/data/composites/day/ (color figure online)

type and cold front passing the area when 300 mm * 24 h^{-1} and 232 mm * 24 h^{-1} were, respectively, measured at Hala Gąsienicowa station and at the top of Kasprowy Wierch (1991 m). This rainfall was interrupted by the anticyclonic ridge coming from the France and Germany. As it has already been mentioned cyclones causing such extreme rainfall in the Upper Vistula Basin come usually form the Adriatic Sea along the Vb trajectory designated in Europe by van Bebber (1891). This is clearly seen in Fig. 5 presenting the composite maps of sea level pressure and baric topography of the levels of 850 and 500 hPa created as an average pattern from 15 days with precipitation ≥100 mm at Kasprowy Wierch station using the NCEP/NCAR reanalysis data (Kalnay et al. 1996). The cyclone is also vivid at the 850 hPa level while it takes the form of cyclonic trough at the 500 hPa level (Fig. 5).

Despite air advection and pressure system type, synoptic situation is also characterized by air mass and meteorological fronts. In the Upper Vistula Basin about 80 % of precipitation ≥50 mm (Table 3) and almost all precipitation ≥100 mm (Tables 1 and 2) formed within the moist polar maritime air (mP) or polar maritime old air (moP). About 40 % of high precipitation resulted from orographically forced convection since they occurred at days without meteorological fronts. That is occluded front out of all meteorological fronts which favours the most high precipitation occurrence. It however should be mentioned that the flood precipitation events (≥100 mm) in the Upper Vistula Basin has occurred at various meteorological fronts (Tables 1 and 2). It can also be mentioned that in Europe frontal precipitation clearly dominates in winter (Łupikasza 2013).

It is not the rule that circulation types favouring high precipitation occurrence also govern its long-term variability. Therefore, to check temporal dependence between flood rainfalls and atmospheric circulation we calculated correlations coefficients (Pearson, Spearman, Kendall) between selected circulation characteristics (number of days with selected circulation types: Nc, NEc, N+NEc, Bc and Cc +Bc, indices: Wi, Si, Ci and NAO) and precipitation. Parametric Pearson correlation coefficient measures the degree of linear relation between two variables whereas nonparametric Spearman and Kendall correlation coefficients measure nonlinear relations. Statistical significance of correlation between variables proved by three different coefficients enhance the robustness of the relation between extreme precipitation occurrence and atmospheric circulation. We used the number of days with precipitation ≥30 mm which is more frequent than precipitation ≥50 mm thus deliver more robust results. The correlation coefficients included in Table 3 indicated significant impact of Nc and NEc types on the long-term variability in high precipitation frequency while the role of Bc and Cc types was insignificant. This means that increase in the frequency of Nc and NEc types may

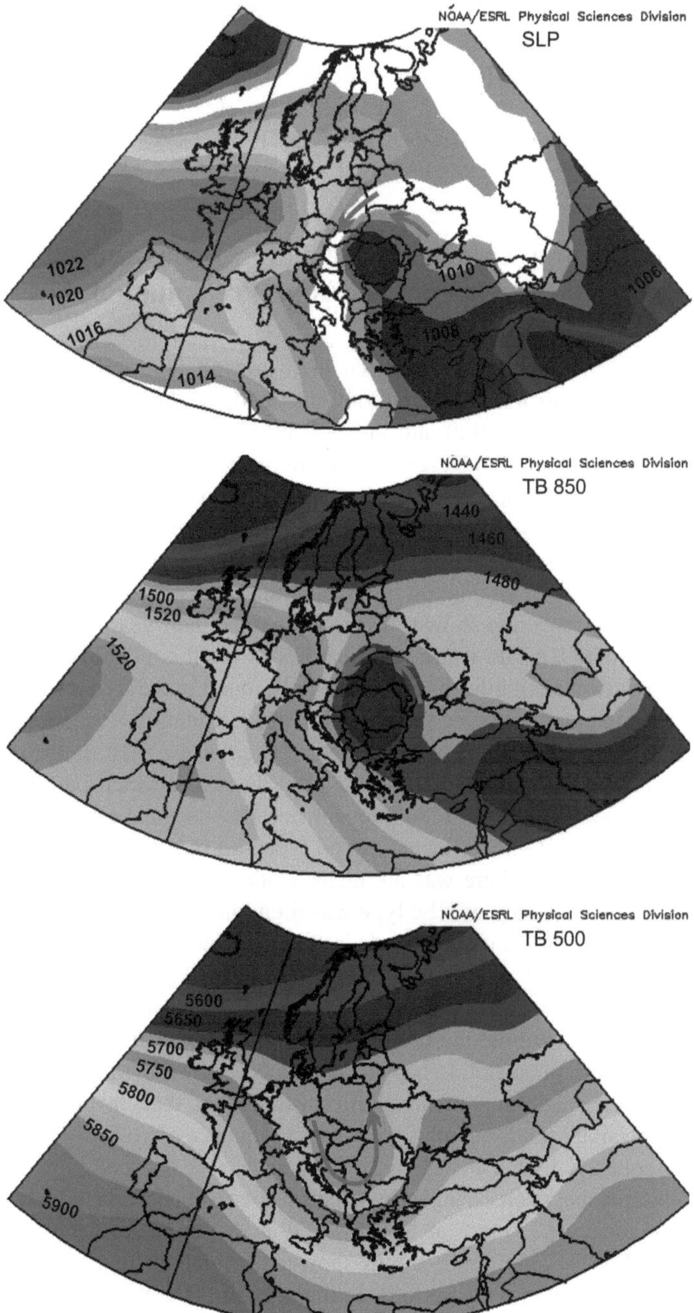
NOAA/ESRL Physical Sciences Division
SLP
1022
1020
1016
1014
1010
1008
1006
NOAA/ESRL Physical Sciences Division
TB 850
1440
1460
1480
1500
1520
1520
NOAA/ESRL Physical Sciences Division
TB 500
5600
5650
5700
5750
5800
5850
5900

result in more frequent flood conditions. In summer (JJA) high precipitation was also negatively correlated with W index revealing that intensification of eastern air flow may also increase the frequency of flood conditions. Increase in the frequency of cyclonic conditions impacted the frequency of high precipitation only at Kasprowy Wierch station in summer half-year (Table 3).

4 Variability and Trends in Circulation Types and Indices Impacting the Frequency of Flood Precipitation

Statistically significant relations between high precipitation and atmospheric circulation allowed us to make assumption that trends in the frequency of selected circulation types (Nc and NEc) and indices (Wi in JJA and Ci in May–Oct) will cause similar trends in flood precipitation frequency. Statistical significance of trends in circulation and precipitation characteristics was tested with Mann-Kendall method taking $\alpha = 0.05$ as a statistical significance level while trend slope was calculated with least squares method.

Long-term courses of circulation types and indices significantly influencing high precipitation occurrence are shown in Fig. 6 for warm half-year (May–Oct) and in Fig. 7 for summer (JJA). In warm half-year trends in all circulation characteristics were positive however significant ones were only found in the frequency of Nc type and Ci index indicating possible increase in high precipitation frequency. Since the beginning of the research period to the end of the thirties most of the circulation characteristics were lower than their long-term averages, which particularly concerns Nc type and Ci index. In subsequent decades the long-term course of circulation characteristics varied depending on type and index. In the 50-year period between 1930 and 1980 there was an intensification of the air advection from the north (Nc). High frequency of NEc type was found in seventies and eighties which coincided with high frequency of floods in those decades in Poland and at the turn of the first and second decades of the 21st century. Low frequencies of these circulation types during 1982–1995 was in phase with relatively dry conditions in the Southern Poland without the floods within the Upper Vistula Basin. In the next more wet years the great floods happened in July 1997 and May 2010. Cyclonic activity since the beginning of the thirties was higher than in previous decades (Fig. 6). In the sixties and seventies there was a sequence of years with the Ci index values consequently higher than long-term average.

The similarity between changes in selected circulation characteristics and precipitation frequency (≥30 mm at Kasprowy Wierch station) in warm half-year is evidenced in Fig. 8. This figure shows short-term trends in parameters discussed calculated for 30-year moving periods. In majority of 30-year periods there was an agreement between direction of trends in high precipitation frequency and circulation characteristics particularly for NEc type which was recognized as the most impactful indicator for high precipitation trends. Trends in the frequency of NEc

Table 3 Correlation coefficients between number of days with selected circulation characteristics (circulation types and circulation indices) and number of days with precipitation ≥30 mm

Period	R	Circulation types and circulation indices								
		Nc	NEc	N+NEc	Bc	Cc+Bc	NAO	Wi	Si	Ci
Zakopane										
May–Oct	P	**0.455*****	**0.416*****	**0.549*****	−0.064	0.029	−0.055	−0.244	−0.221	0.188
	S	**0.367****	**0.391****	**0.471*****	−0.099	0.011	−0.076	−0.227	−0.234	0.152
	K	**0.271****	**0.291****	**0.354*****	−0.072	−0.001	−0.060	−0.167	−0.169	0.115
JJA	P	**0.360****	**0.295***	**0.416*****	0.036	0.148	0.076	**−0.267***	−0.208	0.158
	S	**0.317***	**0.319***	**0.399****	0.019	0.170	0.115	**−0.257***	−0.236	0.095
	K	**0.250****	**0.243***	**0.305****	0.014	0.126	0.082	**−0.186***	−0.163	0.076
Kasprowy Wierch										
May–Oct	P	**0.329****	**0.573*****	**0.554*****	0.113	0.240	−0.064	−0.164	−0.222	**0.320***
	S	**0.288***	**0.530*****	**0.482*****	0.078	0.185	−0.110	−0.142	**−0.260***	**0.284***
	K	**0.216***	**0.398*****	**0.360*****	0.055	0.146	−0.074	−0.100	**−0.179***	**0.214***
JJA	P	**0.356****	**0.482*****	**0.525*****	0.073	0.223	−0.017	**−0.280***	−0.203	0.217
	S	**0.384****	**0.480*****	**0.526*****	0.048	0.219	−0.070	**−0.264***	−0.232	0.191
	K	**0.288****	**0.369*****	**0.397*****	0.031	0.155	−0.048	**−0.183***	−0.163	0.135

R correlation coefficients: *P* Pearson, *S* Spearman, *K* Kendall, *Nc*, *N+NEc* circulation types, *NAO* North Atlantic Oscillation index, *Wi* zonal (westerly) circulation index, *Si* meridional (southerly) circulation index, *Ci* cyclonicity index

Correlation coefficients significant at the level of ≤ 0.05 are bolded

Statistical significance levels: *$\alpha \le 0.05$, **$\alpha \le 0.01$, ***$\alpha \le 0.001$

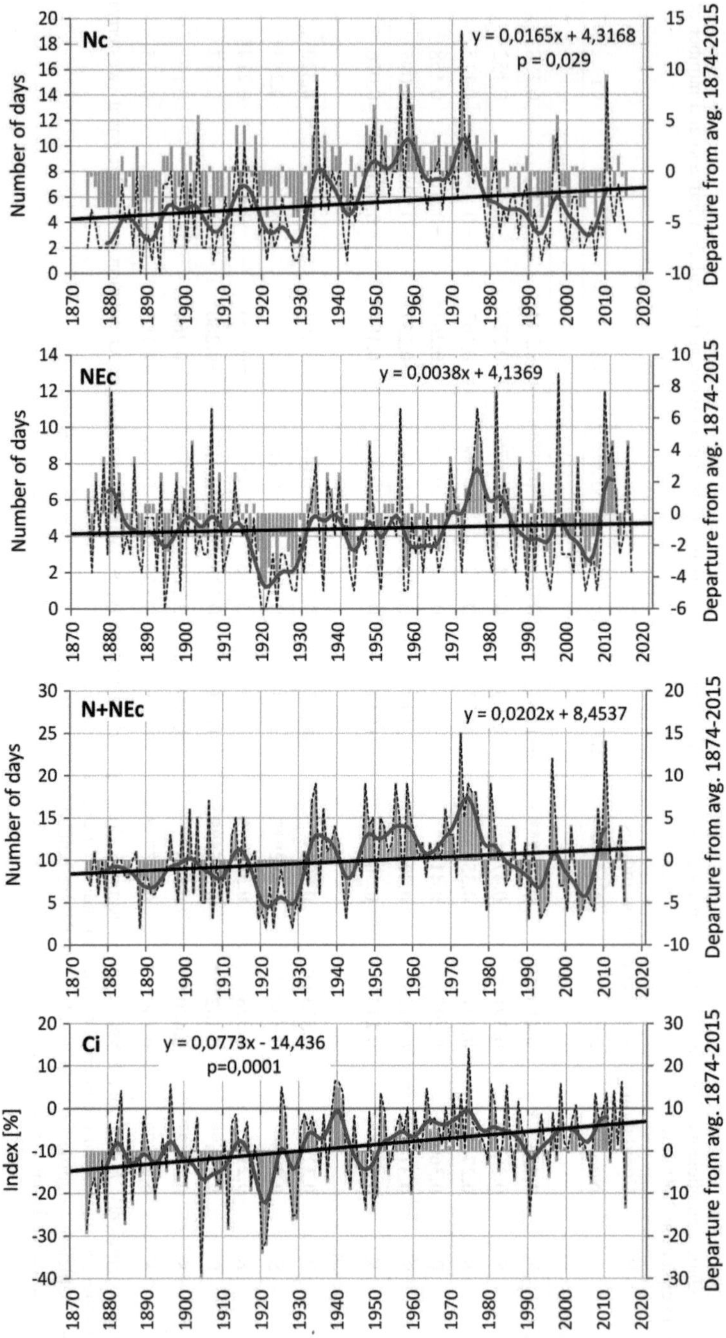
Nc
y = 0,0165x + 4,3168
p = 0,029
Number of days
Departure from avg. 1874-2015
NEc
y = 0,0038x + 4,1369
Number of days
Departure from avg. 1874-2015
N+NEc
y = 0,0202x + 8,4537
Number of days
Departure from avg. 1874-2015
Ci
y = 0,0773x - 14,436
p=0,0001
Index [%]
Departure from avg. 1874-2015
1870 1880 1890 1900 1910 1920 1930 1940 1950 1960 1970 1980 1990 2000 2010 2020

◄ **Fig. 6** Long-term course of the number of days with selected circulation types (Nc, NEc, N+NEc) and Ci index (*left axis*: *dashed dark grey line*—annual values, *solid thick grey line*—11-y Gauss filter, *straight black line*—linear trend) favouring the occurrence of high precipitation in Tatra Mountains and their departures from average 1874–2015 (*right axis*: *light grey bars*) in the warm half-year (May–Oct)

type can partly explain trends in high precipitation frequency. Regardless of insignificant trends the relations between these variables were clear.

Long-term variability in circulation types frequencies in summer resembled those in warm half-year but these growing trends in neither of types were significant. On the contrary, zonal westerly circulation index which impacts the frequency of high precipitation in summer was significantly dropping throughout the entire research period at both stations (Fig. 7). Such a downward tendency indicated intensification of air advection from the eastern sector and possible increase in flood precipitation (negative correlation with high precipitation). The eastern air flow was particularly intense (the lowest index values) in the sixties, seventies and in the first half of the eighties when the occurrence of flood events was enhanced while in the dry period of c.a. 1982–1996 Wi index was relatively high.

5 Conclusions

We investigated changes in atmospheric condition favouring the occurrence of flood precipitation in order to assess current and possible future changes in flood condition in the Upper Vistula Basin. We used the catalogue of circulation types created for the Upper Vistula Basin covering 143 years long period and related regional circulation indices as well as NAO index.

In summer and warm half-year the Upper Vistula Basin was usually under an influence of anticyclonic wedge (Ka circulation type) and cyclonic trough (Bc circulation type). The most frequent of all advectional types were those with air flow from the western direction which is a dominant wind within the moderate climate zone.

The occurrence of flood precipitation over the Upper Vistula Basin is strongly linked to the frequency of air advection from the north and north-east under an influence of low pressure system (Nc and NEc circulation types). The circulation types Nc and NEc which were recognized as favouring the most flood precipitation are not frequent within the Upper Vistula region; nevertheless the maximum of their occurrence falls in the flood prone period which is the turn of spring and summer. The linkage between high precipitation occurrence and Cc, Bc and Ec types (last type only at Kasprowy Wierch station) is also noticeable though weaker than in case of previously mentioned types. Unlike Cc, Bc and Ec types, the frequencies of Nc and NEc types also significantly trigger the long-term variability in flood precipitation frequency. Most flood precipitation events were associated with polar maritime or polar maritime old air masses and were a result of orographically forced

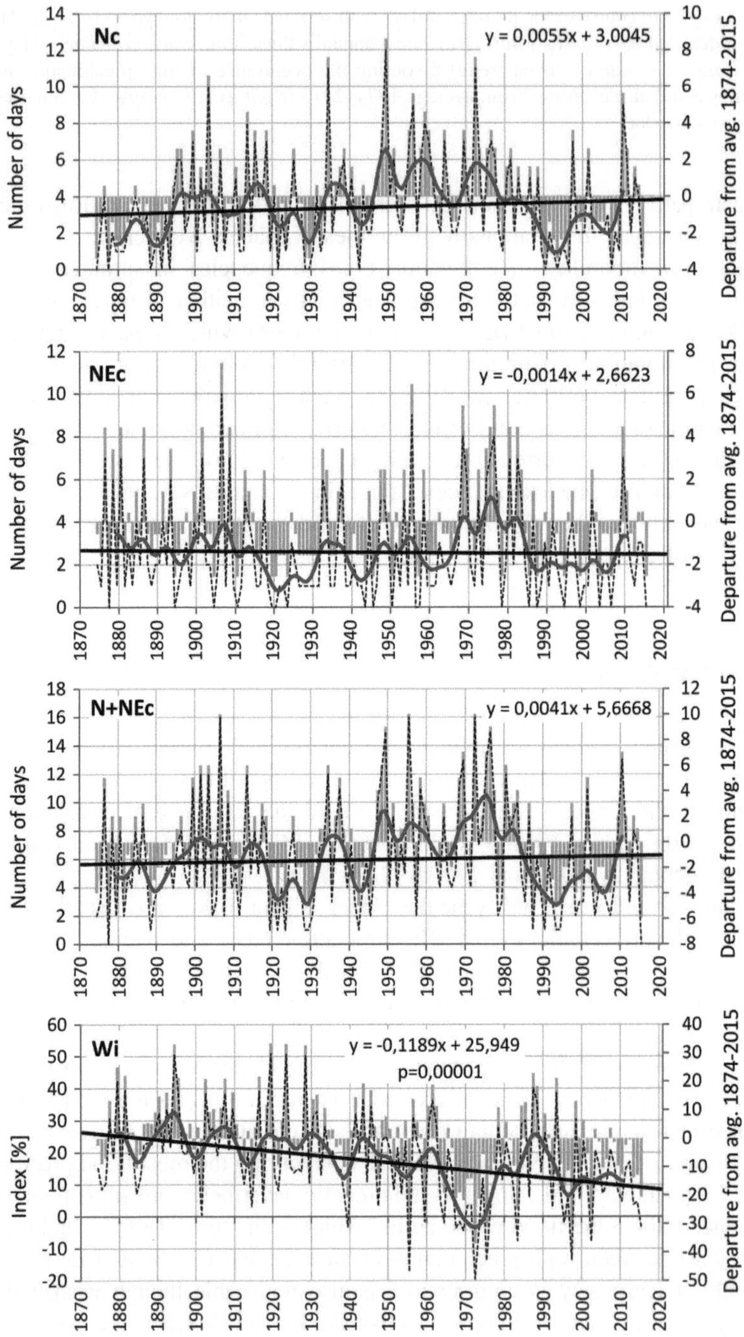

Fig. 7 Long-term course of the number of days with selected circulation types (Nc, NEc, N+NEc) and Wi index (*left axis*: *dashed dark grey line*—annual values, *solid thick grey line*—11-y Gauss filter, *straight black line*—linear trend) favouring the occurrence of high precipitation in Tatra Mountains and their departures from average 1874–2015 (*right axis*: *light grey bars*) in summer (JJA)

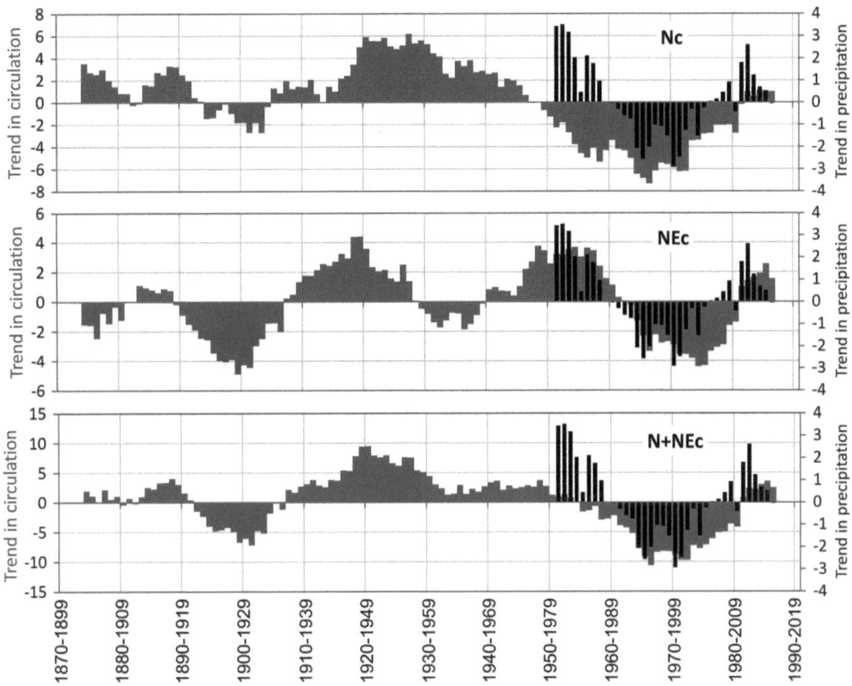

Fig. 8 Short-term trends in selected circulation types (Nc, NEc, N+NEc) favouring the occurrence of high precipitation in Tatra Mountains (*grey bars*) and in the number of days with precipitation ≥30 mm (*black bars*) calculated for 30-year moving periods between 1874 and 2015 (circulation types frequency) and between 1951 and 2015 (number of days with precipitation) in warm half-year

convection or were related to occluded front. The correlation between long-term courses of circulation indices and high precipitation was much lower compared to circulation types Nc and NEc. Significant connection was only found between long-term variability in high precipitation frequency and Wi index in summer (JJA) at both stations and Ci index at Kasprowy Wierch station in warm half-year (May–Oct).

Trends in majority of circulation characteristics favouring the occurrence of high precipitation and impacting its long-term variability were not statistically significant with an exception of the frequency of Nc type and Ci index in May–Oct and Wi index in summer. Significant increase in both variables (Nc frequency and Ci index) in warm half-year and intensification of air advection from the north-east (decreasing trend in Wi) in summer may lead to increase in the frequency of flood conditions in these seasons.

The results show that the occurrence and long-term variability of flood precipitation in the Tatra Mountains is most strongly linked to the frequency of air advection from the north-eastern sector under an influence of low pressure system which did not undergo significant changes. Regardless insignificant trends the variability in the frequency of NEc circulation type meaningfully determines the occurrence and variability in high precipitation occurrence thus was recognized as indicator of flood conditions. Such atmospheric conditions (NEc circulation type), usually induced by stationary low pressure systems located over Ukraine and originated from the Adriatic Sea, in combination with orography have caused the most damaging floods in the Upper Vistula Basin. Despite the atmospheric circulation the orography plays crucial role in generating flood precipitation not only in the Tatra Mountains but also in the other high mountains of the Europe (Sénési et al. 1996; Rotunno and Ferretti 2001; Pradier et al. 2004; Seibert et al. 2007).

Acknowledgments This work was supported by the project FLORIST (Flood risk on the northern foothills of the Tatra Mountains; PSPB no. 153/2010). Precipitation data were made available by Institute of Meteorology and Water Management, National Research Institute (IMGW, PIB). Part of data for the period 2000–2015 was taken from synoptic data-base OGIMET.

References

Barry GB, Carleton M (2001) Synoptic and dynamic climatology. Routledge, London

Cebulak E (1992) Wpływ sytuacji synoptycznej na maksymalne opady dobowe w dorzeczu górnej Wisły (Influence of synoptic situation on the maximum daily precipitation in the Upper Vistula Basin). Folia Geogr, Series Geogr Phys 23:81–95 (in Polish)

Cebulak E, Niedźwiedź T (2000) Zagrożenie powodziowe dorzecza górnej Wisły przez wysokie opady atmosferyczne (Flood hazard in the upper Vistula river basin through the high precipitation). Monografie Komitetu Gospodarki Wodnej PAN 17:55–70 (in Polish, summary in English)

Degirmendžić J, Kożuchowski K, Żmudzka E (2004) Changes of air temperature and precipitation in Poland in the period 1951–2000 and their relationship to atmospheric circulation. Int J Climatol 24(3):291–310. doi:10.1002/joc.1010

Hurrel JW (2001) The North Atlantic oscillation. Science 291:603–605

Hurrel JW, van Loon H (1997) Decadal variations in climate associated with the North Atlantic oscillation. Clim Change 36:301–326

Kaczorowska Z (1933) Przyczyny meteorologiczne letnich wezbrań Wisły (Causes of the summer Vistula floods). Prace PIM 2:1–54 (in Polish)

Kalnay E, Kanamitsu M, Kistler R, Collins W, Deaven D, Gandin L, Iredell M, Saha S, White G, Woolen J, Zhu Y, Chelliah M, Ebisuzaki W, Higgins W, Janowiak J, Mo KC, Ropelewski C, Wang J, Leetmaa A, Reynolds R, Jenne R, Joseph D (1996) The NCEP/NCAR 40-years reanalysis project. Bull Am Meteorol Soc 77:437–471. Data available at http://www.esrl.noaa.gov/psd/data/reanalysis/reanalysis.shtml

Kundzewicz ZW, Dobrowolski A, Lorenc H, Niedźwiedź T, Pińskwar I, Kowalczak P (2012) Floods in Poland. Chapter 17. In: Kundzewicz ZW (ed) Changes in flood risk in Europe. Special Publication No. 10. IAHS Press, Wallingford, Oxfordshire, pp 319–334

Lamb HH (1972) British Isles weather types and a register of daily sequence of circulation patterns, 1861–1971. HMSO, London, Geophysical Memoir, vol 116, 85 pp

Łupikasza E (2013) Typy genetyczne opadów ekstremalnych w Europie oraz ich uwarunkowania synoptyczne (1950–2008) [Genetic types of extreme precipitation in Europe and their synoptic determinants (1950–2008)]. Pracownia Komputerowa Jacka Skalmierskiego, Katowice, 378 pp (in Polish, summary in English)

Łupikasza E, Niedźwiedź T (2015) Wieloletnia zmienność typów cyrkulacji sprzyjających występowaniu wysokich opadów atmosferycznych w Tatrach (1873–2014) [Long-term

variability in circulation types favouring the occurrence of high precipitation in the Tatra Mountains (1873–2014)]. In: Lorenc H, Ustrnul Z (eds) Klimat a Społeczeństwo i Gospodarka. Polskie Towarzystwo Geofizyczne Oddział Warszawski, Instytut Meteorologii i Gospodarki Wodnej Państwowy Instytut Badawczy, Warszawa, pp 93–111 (in Polish, summary in English)
Milata W (1955) Synoptyka wielkich opadów atmosferycznych w Karpatach (Synoptics of the great atmospheric precipitation in Carpathians). Przegląd Meteorologiczny i Hydrologiczny 8 (3–4):210–217 (in Polish)
Morawska-Horawska M (1971) Meteorologiczne przyczyny powodzi w Polsce Południowej w lipcu 1970 r. (Meteorological causes of the July 1970 flood in southern Poland). Przegląd Geofizyczny 16(24)4:299–315 (in Polish)
Mudelsee M, Börngen M, Tetzlaf G, Grünewald U (2004) Extreme floods in Central Europe over the past 500 years: role of cyclone pathway "Zugstrasse Vb". J Geophys Res 109(D23101): 1–21
Murray R, Lewis RP (1966) Some aspects of the synoptic climatology of the British Isles as measured by simple indices. Meteorol Mag 95:193–203
Niedźwiedź T (1978) The changes of the atmospheric circulation indices over the Polish West Carpathians during last 27 years. Veröffentlichungen der Schweizerischen Meteorologischen Zentralanstalt, Zürich 40:19–22
Niedźwiedź T (1981) Sytuacje synoptyczne i ich wpływ na zróżnicowanie przestrzenne wybranych elementów klimatu w dorzeczu górnej Wisły (Synoptic situations and its influence on the spatial differentiation of selected climatic elements in the Upper Vistula Basin). Rozprawy Habilitacyjne UJ, vol 58, 165 pp (in Polish)
Niedźwiedź T (1999) Rainfall characteristics in southern Poland during the severe flooding event of July 1997. Studia Geomorphologica Carpatho-Balcanica, Kraków 33:5–25
Niedźwiedź T (2000) Variability of the atmospheric circulation above the Central Europe in the light of selected indices. In: Obrębska-Starkel B (ed) Reconstructions of climate and its modelling. Institute of Geography of the Jagiellonian University, Cracow, Prace Geograficzne, vol 107, pp 379–389
Niedźwiedź T (2003a) The extreme precipitation in Central Europe and its synoptic background. Polish National Committee of International Geosphere—Biosphere—Global Change Programme Polish Academy of Science, Warszawa. Papers on Global Change IGBP 10:15–29
Niedźwiedź T (2003b) Extreme precipitation events on the northern side of the Tatra Mountains. Geographia Pol 76(2):13–21
Niedźwiedź T (2016) Catalogue of synoptic situations in the upper Vistula river basin (1873.09-2015.12). Computer file available at: Department of Climatology, Faculty of Earth Sciences, University of Silesia, Bedzinska 60, 41-200 Sosnowiec, Poland; tadeusz.niedzwiedz@us.edu.pl; available also on line in http://klimat.wnoz.us.edu.pl
Niedźwiedź T, Twardosz R, Walanus A (2009) Long-term variability of precipitation series in east central Europe in relation to circulation patterns. Theor Appl Climatol 98(3–4):337–350. doi:10.1007/s00704-009-0122-0
Niedźwiedź T, Łupikasza E, Pińskwar I, Kundzewicz ZW, Stoffel M, Małarzewski Ł (2015) Variability of high rainfalls and related synoptic situations causing heavy floods at the northern foothills of the Tatra Mountains. Theor Appl Climatol 119(1–2):273–284. (Open Access). doi:10.1007/s00704-014-1108-0
Pradier S, Chong M, Roux F (2004) Characteristics of some frontal stratiform precipitation events south of the Alpine chain during MAP. Meteorol Atmos Phys 87:197–218
Rotunno R, Ferretti R (2001) Mechanisms of intense Alpine rainfall. J Atmos Sci 58:1732–1749
Ruiz-Villanueva V, Stoffel M, Wyżga B, Kundzewicz ZW, Czajka B, Niedźwiedź T (2016) Decadal variability of floods in the northern foreland of the Tatra Mountains. Reg Environ Change 16(3):603–615. doi:10.1007/s10113-014-0694-9
Seibert P, Frank A, Formayer H (2007) Synoptic and regional patterns of heavy precipitation in Austria. Theor Appl Climatol 87(1):139–153
Sénési SP, Bougeault P, Cheze JL, Cosentino P, Thepenier R (1996) The Vaison-La-Romaine flash flood: mesoscale analysis and predictability issues. Weather Forecast 11:417–442

Twardosz R, Niedźwiedź T (2001) Influence of synoptic situations on the precipitation in Kraków (Poland). Int J Climatol 21(4):467–481

Ustrnul Z, Czekierda D (2001) Circulation background of the atmospheric precipitation in Central Europe (based on the Polish example). Meteorol Z 10(2):103–111

Ustrnul Z, Czekierda D (2009) Atlas of extreme meteorological phenomena and synoptic situations in Poland. Instytut Meteorologii i Gospodarki Wodnej, Warszawa, p 182

Valor GB, López JMG (2016) OGIMET—Professional information about meteorological conditions in the world (SYNOP messages available on-line on the web site: http://www.ogimet.com). Last access 17 Jan 2016

van Bebber WJ (1891) Die Zugstrassen der barometrischen Minima. Meteorol Z 8:361–366

Wilks DS (2006) Statistical methods in the atmospheric sciences. Academic Press, Cambridge, 627 pp

Yarnal B (1993) Synoptic climatology in environmental analysis: a primer. Belhaven Press, London, 195 pp

Climate Reconstruction from Tree-Rings in the Tatra Mountains

Ryszard J. Kaczka, Barbara Spyt, Karolina Janecka, Tadeusz Niedźwiedź and Zdzisław Bednarz

Abstract This chapter examines the long-term variability of summer (June–July) air temperature and summer humidity (precipitation and Standardised Precipitation Evapotranspiration Index, SPEI) in the region of the Tatra Mountains, which represents natural climate conditions, free of strong anthropogenic influences. The reconstruction of temperature is available for the period since the beginning of the 17th century and reconstruction of humidity related parameters since the beginning of the 18th century by means of the methods based on the tree-ring chronologies. The main proxies utilized for temperature reconstruction were tree-ring widths of Norway spruce (*Picea abies* (L.) H. Karst) and Stone pine (*Pinus cembra* L.) growing in the timberline ecotone. The precipitation and SPEI were reconstructed based on Scots pine (*Pinus sylvestris* L.) tree-ring widths of trees growing at ~1000 m a.s.l. The reconstruction of summer temperature from tree-rings pointed to a relatively cold interval as a part of the Little Ice Age (from the mid 16th to late 19th centuries). In the 20th and at the beginning of the 21st centuries, general increase of air temperature was observed. However, in this recent warm period and during earlier main climatic periods, temperature conditions were not uniform. Analysing series of summer temperature (the 17th–21st centuries) several shorter warm and cool fluctuations were observed. The reconstructed humidity variables exhibited less variability. This is the first attempt of precipitation reconstruction in mountains regions based on the tree-ring chronologies. But the correlation between flood events and humid periods is poor due to the predominant character of the flood caused by short term intensive precipitation of short duration.

R.J. Kaczka (✉) · B. Spyt · K. Janecka · T. Niedźwiedź
Faculty of Earth Sciences, University of Silesia, Bedzinska 60, 41-200 Sosnowiec, Poland
e-mail: ryszard.kaczka@us.edu.pl

Z. Bednarz
Department of Forest Biodiversity, Institute of Forest Ecology and Silviculture, University of Agriculture, Al. 29 Listopada 46, 31-425 Kraków, Poland

Z.W. Kundzewicz et al. (eds.), *Flood Risk in the Upper Vistula Basin*, GeoPlanet: Earth and Planetary Sciences, DOI 10.1007/978-3-319-41923-7_10

Keywords Dendroclimatology · Tree-ring width · Temperature · Precipitation · SPEI · Climate reconstruction · Tatra Mountains

1 Introduction

Different annually resolved tree-ring proxies such as tree-ring width (TRW), maximum wood density, and stable isotopes are widely used to reconstruct past variations in local- to hemispheric-scale temperature means (e.g. Briffa et al. 2001; Jones and Mann 2004; Wilson et al. 2016).

Such natural proxy evidence generally derives from sites located in the high northern latitudes or high-elevations (Fritts 1976; Schweingruber 1996). For Europe, multi-centennial to millennial tree-ring records (Büntgen et al. 2005, 2006, 2013) and network analyses (Frank and Esper 2005a, b) predominantly derive from the Alps, whereas only little evidence exists for the Carpathian arc (Kaczka and Büntgen 2006; Büntgen et al. 2007, 2013; Ponocná et al. 2016) and the lower part of the Upper Vistula Basin (Szychowska-Krąpiec 2010). These examples show clearly that knowledge about regions and processes can be improved much more by studying the local network than singular sites. Even though few TRW based studies introducing the growth/climate response analyses were conducted at local scale (Feliksik 1972, 1992; Bednarz 1976, 1983, 1984, 1996, 2015; Bednarz et al. 1998–1999; Szychowska-Krąpiec 1998; Kaczka 2004; Niedźwiedź 2004; Bednarz and Niedźwiedź 2006; Kaczka and Büntgen 2006; Savva et al. 2006; Büntgen et al. 2007; Popa and Kern 2009; Kaczka and Czajka 2014; Kaczka et al. 2015), studies considering the entire mountain system and climate reconstructions are broadly missing. The initial analyses of growth/climate response for a wider area (Kaczka and Büntgen 2006; Sidor et al. 2015) as a function of geographical settings provide the evidence that multi-site approach allows building the reliable tree ring based reconstructions of summer temperature (Büntgen et al. 2007).

The potential of TRW as precipitation proxy was also probed in the Carpathians (Büntgen et al. 2012; Kaczka et al. 2012; Brzęk et al. 2014) but such reconstructions are still missing. The detailed information about the short- and long-term trends of humidity-related variables would play an important role in understanding the temporal dynamics of natural hazards since summer precipitation has caused most of the floods in the Tatra Mountains region (Kundzewicz et al. 2014; Niedźwiedź et al. 2015). The flood risk at the northern foothills of the Tatra Mountains is considered as one of the main natural hazards. The regular climatic and hydrological records are scattered and lack of data poses the challenge to build applicable models explaining the local and regional flood processes (Parajka et al. 2010; Marchi et al. 2010; Ballesteros-Cánovas et al. 2015a, b; Ruiz-Villanueva et al. 2016). That gap can be bridged using historical records (Figs. 1 and 2) and different proxies including tree rings.

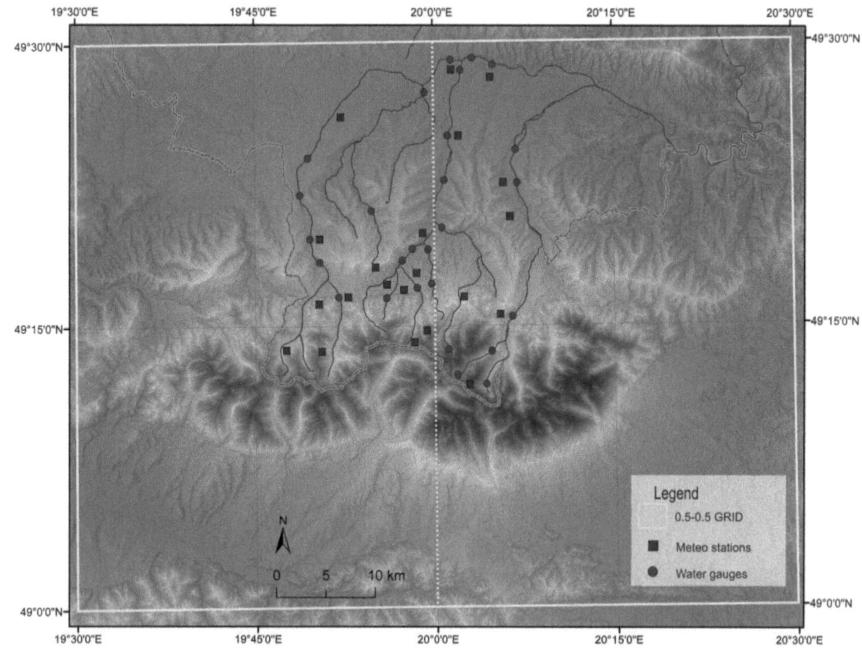

Fig. 1 Location of meteorological stations and flow gauges on the northern slopes and at the foothills of the Tatra Mountains (compare with Ballesteros-Cánovas et al. 2015). The spatial extent of the gridded data employed for dendroclimatic analyses and reconstructions (CRU TS3.23 (Harris et al. 2014))

2 Geographical Setting

The northern part of the studied region can be described as the Tatra Mountains and the foothills traditionally known as Podhale (Kondracki 1994). The Tatras are the highest mountain system (with Gerlach being the highest peak, 2655 m a.s.l., located in Slovakia) in the Carpathian arc. The Tatra Mountains are bio-geographically a part of the Western Carpathians (Warszyńska 1995).

Widespread forest stands that reach from the montane to sub-alpine zones are typical landscape elements with thermally induced treeline between 1500 and 1700 m a.s.l. along the north-south gradient. The dominant coniferous species of the sub-alpine zone is Norway spruce (*Picea abies* (L.) H. Karst: PCAB), occasionally Stone pine (*Pinus cembra* L.: PICE). The high-located forest of the Tatras has been under long-term human impact since at least the 15th century. The human activity was mainly related to logging and shepherding, affecting the structure of the mountain forest and lowering treeline of sub-alpine zone. Long-lasting selective logging resulted in decreasing the numbers of the Stone pine stands. The Scots pine (*Pinus sylvestris* L.: PISY) sites are constituted by small isolated groups of trees

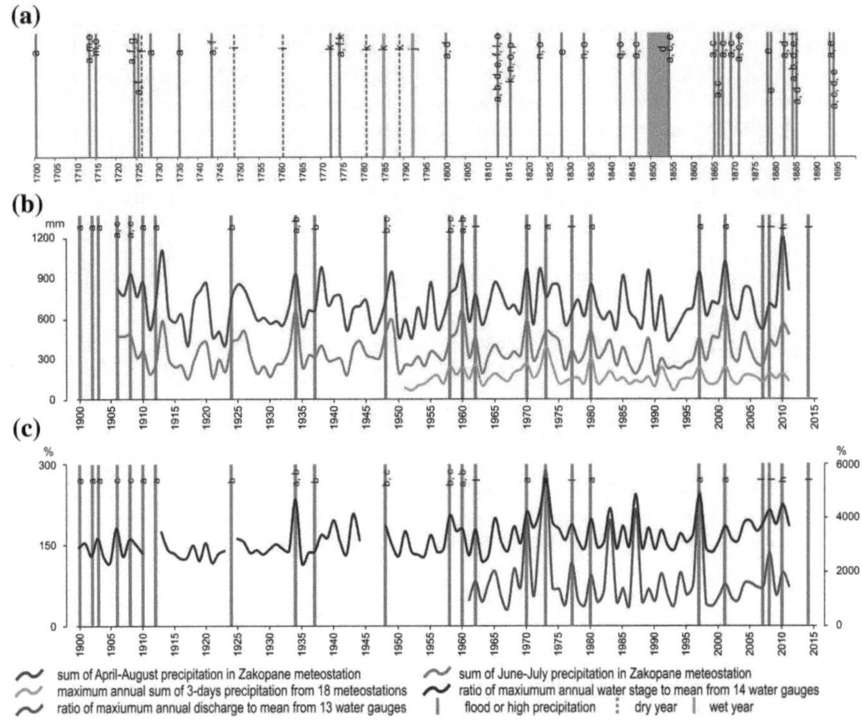

Fig. 2 The comparison of existing precipitation, water stage and discharge data (based on the network shown in Fig. 1) and historical records (*a*—Kotarba (2004), *b*—Wierchy (1923–2015), *c*—Bielański (1984), *d*—Stolarczyk (1915), *e*—Pamiętnik Towarzystwa Tatrzańskiego (1876–1920), *f*—Siemionow (1992), *g*—Generisch (1807), *h*—Woźniak (2013), *i*—Niedźwiedź (2004), *j*—Gustawicz (1883), *k*—Szewczuk (1939), *l*—Rajwa (2014), *m*—Komoniecki (1704), *n*—Adamczyk (1991a), *o*—Bednarz (2015), *p*—Adamczyk (1991b), *q*—Więckowski (1834))

located at the altitude of appr. 1000 m a.s.l. and related to combination of carbonate substrate soil and southern exposure (*Erico-Pinion*, Perzanowska 2010).

The Tatra Mountains represent natural climate conditions free of strong anthropogenic influences. The winter conditions are affected by polar-continental air masses arriving from the east and northeast while in the other seasons weather is predominated by oceanic polar-maritime air masses from the west. Foothills in the Western Carpathians are characterized by mean July temperature of ~19 °C compared to ~22 °C in the southern part, and 4.1 °C compared to 5.4 °C respectively for the highest locations (~2500 m a.s.l.). According to altitude, the mean annual temperature drops from about 6 °C at the level 600–650 m a.s.l. (Carpathian foothills, Orawa-Nowy Targ Basin) to −4 °C on the highest peak of the Tatra Mountains (Hess 1974; Niedźwiedź 1992). After Hess (1974), six vertical climatic belts of 2 °C width can be distinguished in the Western Carpathians and

Tatra Mountains. They are strictly connected with types of vegetation zones. For dendroclimatic investigations, the forest zone below the treeline (1550 m a.s.l.; mean annual temperature 2 °C, in July 10 °C) is the most important. A cool climatic belt (annual temperature between 2 and 4 °C) is located below this line and above the level of ~1150 m a.s.l. It is covered by the coniferous forest with *Picea abies* (L.) H. Karst. A moderately cool belt (from 4 to 6 °C) with mixed forest is located at the level 650–1150 m a.s.l.

The annual precipitation is lower than in the other mountains of Central Europe, such as the Alps and Czech Massive, due to the effect of the distance of around 1000 km to the Atlantic Ocean and the influence of those mountains on the mild air, thus increasing the continental character of upper-lands. The highest rates of annual precipitation (>2000 mm) are reported from the summits of the Tatra Mts and their northern slopes. In the Podhale region, the annual precipitation varies from below 800 mm in Orawa-Nowy Targ Basin to ~1100–1200 mm in Zakopane (Niedźwiedź 1992).

3 Materials and Methods

3.1 Tree-Ring Data

The tree-ring network spans across a northern part of the Tatra Mountains of ~55 km covering the Western, High and Belianske Tatras (Fig. 3). The network was established based on 10 PCAB and 5 PICE sites from high-elevation, near local timberline and six sites of PISY located in montane forest (Fig. 4). The data sets include tree-ring width series that were originally developed and downloaded from the International Tree-Ring Data Bank (ITRDB). The qualities of TRW

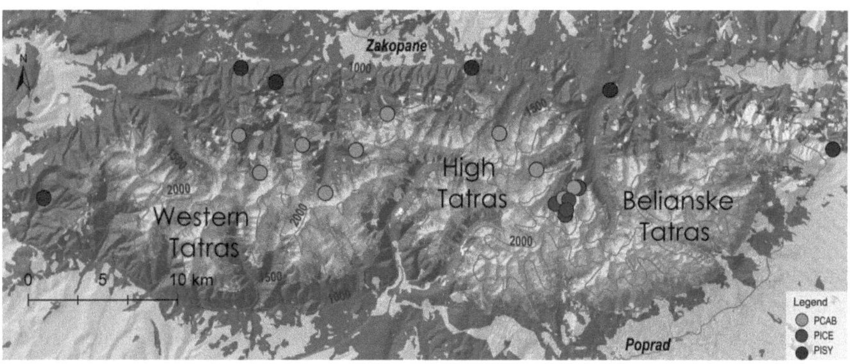

Fig. 3 Location of tree-ring sampling sites in the Tatra Mountains: *PCAB*—Norway spruce (*Picea abies* (L.) H. Karst), *PICE*—Stone pine (*Pinus cembra* L.), *PISY*—Scots pine (*Pinus sylvestris* L.)

Fig. 4 Three types of the sampled forest: **a** Subalpine spruce forest, *Plagiothecio-Piceetum tatricum*, **b** Stone pine stands, *Pino cembrae-Piceetum*, **c** semi-dry Scots pine forest, *Erico-Pinion*

measurements were visually checked using CDendro program (Larsson 2003) and screened for possible mistakes and missing rings using statistical analyses of COFECHA program (Holmes 1983). To remove non-climatic, age-related growth trends from raw measurement series (Fritts 1976), spline-detrending standardization was applied to individual series using ARSTAN software (Cook 1985). Indices were calculated as ratios from 200-year cubic smoothing splines (Cook and Peters 1981). Signal strength of the chronologies was assessed using a 'moving window' approach of the inter-series correlation (*Rbar*) (Fritts 1976), and the expressed population signal (*EPS*; Wigley et al. 1984). For chronology development, series were averaged using the bi-weight robust mean (Cook 1985), and then truncated at a minimum sample size of five series and commonly accepted threshold of EPS = 0.85 (Wigley et al. 1984) (Fig. 4).

3.2 Climatic Data and Growth/Climate Response

The records of monthly resolved gridded (0.5° × 0.5°) temperature, precipitation and Standardised Precipitation Evapotranspiration Index (SPEI, Vicente-Serrano et al. 2010; Beguería et al. 2014) means from dataset CRU TS3.23 (Harris et al. 2014) were used. Two sets of boxes (19.5–20.5°E, 49.0–49.5°N) of gridded data covering the region were established (Fig. 1). The growth/climate response and reconstructions were computed employing the temperature and precipitation anomalies calculated against the commonly used 1961–1990 climatological normal period (New et al. 2000).

Growth/climate response analyses using an 18-month window from May of the year prior to the tree growth, up to October of the growing season, and various seasonal means (April–September = A–S, May–September = M–S, June–September = J–S, April–August = A–A, May–August = M–A, June–August = J–A, June–July = J–J) emphasized the dendroclimatic 'response' (Fritts 1976) of each chronology. Monthly and seasonal growth response of TRW chronologies to temperature and precipitation was computed using Pearson's correlation coefficient.

3.3 The Climate Reconstruction

The direct use of regression model for climate reconstruction caused a variance reduction effect and loss of variance therefor the scaling technique was employed. The means and standard deviations of a proxy (TRW series) were brought to the same level as the corresponding values of climate records (temperature and precipitation) over an established common period (Esper et al. 2005). The overlap period was used as the calibration period spanning over 1902–1956 and was the same for all three reconstructions (PCAB, PICE and PISY). The model was checked against the climate data over the verification interval, 1957–2013. The adequacy of the calculations was tested using the reduction of error (RE) and coefficient of efficiency (CE) (Fritts 1976). The Durbin-Watson (WD) statistics were computed to detect the autocorrelation in the residuals (Durbin and Watson 1951).

4 The Results and Discussion

4.1 The Tree-Ring Width Chronologies

The three coniferous chronologies were developed covering the last three centuries. The PCAB chronology was established based on 575 TRW series from 10 sites located at or near the timberline in the Polish part of the Tatra Mountains (Fig. 5). The spruce chronology is the longest, spanning the interval 1637–2013, after truncation according to the series replication and EPS. The PICE chronology represents smaller and more concentrated sites (341 series from five sites) and covers a shorter period (1735–2013). Both species represent the subalpine environment but Stone pine grows often above the spruce-dominated timberline and often creates the treeline. Therefore PICE is more exposed to severe climate conditions (Carrer et al. 2007; Janecka and Kaczka 2015). Nevertheless, the growth of both conifers is similar and for common period of 1735–2013 standard chronologies correlate at 0.45 ($p = 0.01$). The PISY chronology was established utilizing 135 series derived from 6 sites of relatively low elevation and characterized by carbonated bedrock of cliff-like locations. The signal strength of all chronologies was assessed by Rbar and ranged from 0.28 to 0.60, whereas the EPS values varied amongst 0.84–0.99, indicating the internal consistency in the common variance of chronologies.

The inter-annual variability allows the identification of pointer years, caused by singular events, as summer cooling effects due to the radiative forcing of volcanic eruptions (e.g. 1816 and 1912/13) (Briffa et al. 2001, 2004). Two timberline chronologies showed this kind of short-term changes whereas PISY chronology did not demonstrate both cooling periods. The only one pronounced, well synchronised

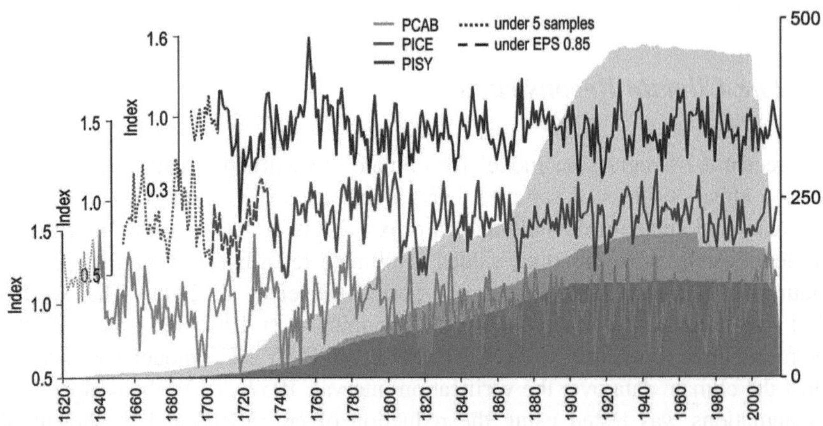

Fig. 5 Three (*PCAB, PICE, PISY*) standard chronologies and sample depth. The *solid line* represented the period exhibiting EPS value over threshold of 0.85 (Wigley et al. 1984) and/or trucked at <5 series

growth reduction for all three chronologies occurred in 1718, which could also be linked to volcanic activity (Crowley 2000; Robock 2000). The lowest growth rates are observed during the Dalton Minimum and coincide with an increased volcanic activity in the early 19th century and followed by improved radial growth momentarily interrupted by light decreases in the 1910–20s, and 70–80s. Interestingly, only PCAB chronology indicated the highest productivity within the last two decades. The previously constructed TRW chronologies emphasized the growth reduction in the 1970s and 1980s as a result of air pollution (Büntgen et al. 2007; Bednarz 2015; Treml et al. 2015). In this work, highly replicated chronologies do not exhibit this issue.

4.2 Growth/Climate Response

The response of tree-ring width differs between two subalpine and low elevation species. The growth of PCAB and PICE is driven by summer temperature whereas the growth of PISY mainly depends on humidity (precipitation and SPEI). The highest correlations between TRW and temperatures were found for June-July of the current year for PCAB chronology ($r = 0.69$, $p < 0.01$) (Fig. 6). The PICE standard chronology correlates with temperature of the same season (June–July) but with lower value ($r = 0.49$, $p < 0.01$). The growth of both species responds to several factors, but temperatures of June and July are the most important. The impact of temperature of longer seasons, such as April–September, May–October and April–October is also significant. The correlation with previous year autumn and current year late winter differentiates PCAB and PICE chronologies. Only Norway spruce chronology correlates positively with October in previous year ($p < 0.01$).

Similar results were reported in literature from several high elevation sites (Oberhuber 2004; Frank and Esper 2005b; Büntgen et al. 2007). The growth of the recent season could be linked with the amount of the reserves inherited from previous year warm autumn. The negative correlation with March temperature is individual feature of PICE chronology. The warm March decreases the possibility to produce wide ring in PICE stems ($r = 0.28$, $p < 0.01$). Vaganov et al. (1999) reported similar phenomena from the Ural Mountains and pointed out the danger posed by early onset of vegetation provoked by a warm end of winter and the beginning of spring. Precipitation has little influence on the growth of both subalpine species. The PCAB chronology correlates positively with January, March and January–March precipitation ($p < 0.01$) but the relationship is much lower than for summer temperature. The clear influence of June–July temperature signal allowed to select it as a reconstructed factor. The PISY chronology exhibits different relation with climate. The correlations with temperature are rather low while both humidity parameters (precipitation and SPEI) drive the growth of Scots pines. The PISY chronology exhibits statistically significant correlation values for precipitation and availability of water for the entire growing season, including spring

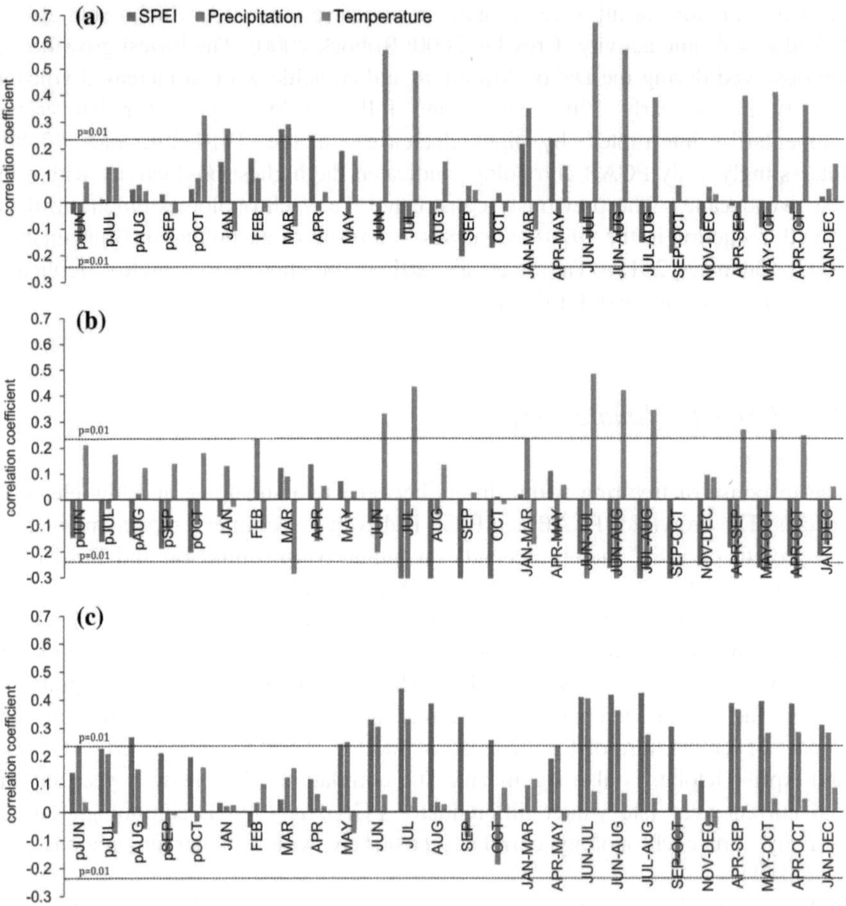

Fig. 6 The growth/climate response of three chronologies: **a** PCAB, **b** PICE, **c** PISY

and autumn and even annual means. The strongest relationship in the PISY chronology is present for summer precipitation, especially June–July (r = 0.41, $p < 0.01$) and July SPEI (r = 0.44, $p < 0.01$). The June–July precipitation constitutes an appropriate climatic factor for the TRW reconstruction.

One of the most important features of growth/climate response is the temporal stability of such correlation. The results of the 21-years running correlation between June–July temperature in case of PCAB and PICE and June–July precipitation reveal the changes over 1902–2013 period. The temperature signal registered in TRW of spruce is stable for most of the testing period (Fig. 7a). There are some fluctuations and except for the 1960, the values of Pearson correlation remain significant at $p = 0.05$. Although the stationary correlation between PICE chronology and June–July temperature is comparable to PCAB chronology, the

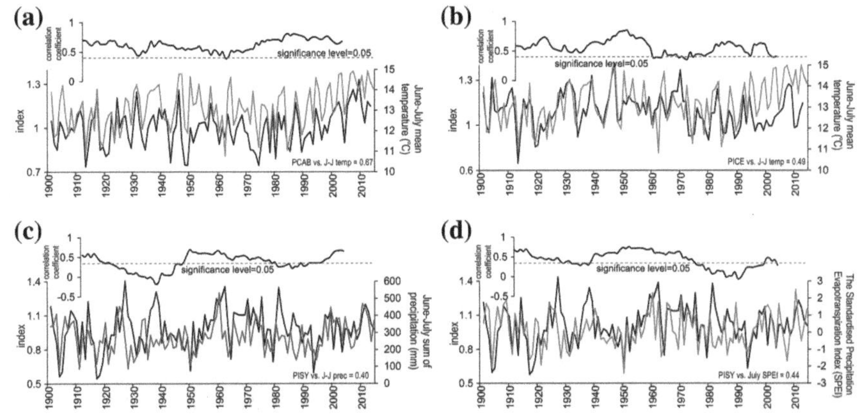

Fig. 7 Comparison of chronologies and selected climate factors over common period of 1901–2013: **a** PCAB and June–July temperature, **b** PICE and June–July temperature, **c** PISY and June–July precipitation, **d** PISY and June–July SPEI. The temporal stability of TRW/climate response was tested using 21-year running correlation

signal is much less stable over time. The results are not significant for longer period between 1960 and 1980 and also in the beginning of the 21st century (Fig. 7b). The lack of temporal stability of climate signal carried by TRW series suggests also poor performance of temperature reconstruction. The PISY chronology also exhibits low temporal stability of climatic signal for both precipitation and SPEI (Fig. 7c, d, respectively).

4.3 The Climate Reconstruction

The climate variables revealing the highest correlations with TRW chronologies were selected to develop climate reconstruction. In case of PCAB, June–July temperature was chosen based on the fact that this variable explains 69 % of variance and is stable over the tested period of ~100 years. Although slightly lower variance is explained (49 %) and temporal stability is weaker, the mean temperature of the same season was selected for reconstruction based on the PICE chronology. Since the PISY chronology reveals completely different relationship with climate, the July SPEI and June–July precipitation (explaining 44 and 41 % of variance, respectively) were applied for climate reconstruction. The temporal stability of both is not fully satisfactory but they are also somehow complementary. The period of lacking significant correlation for precipitation showed high values for SPEI. The calibration and verification statistics (Table 1) exhibit positive values for the RE and CE indicating the predictive skills of the model. The value of DW suggests low autocorrelation of residuals. The results of the same procedure performed for PISY chronology and humidity time series (July SPEI and June–July

Table 1 Statistics of the calibration (1902–1956) and verification (1957–2013) for PCAB and PICE June–July temperature, and PISY June–July precipitation reconstructions

Chronology	Period	Calibration		Verification		
		R^2	DW	R^2	RE	CE
PCAB	1902–1956	0.59	1.1	0.58	0.58	0.55
	1957–2013	0.75	1.8	0.58	0.37	0.33
PICE	1902–1956	0.57	1.3	0.56	0.52	0.45
	1957–2013	0.59	1.5	0.58	0.51	0.55
PISY	1902–1956	0.54	1.7	0.52	0.55	0.53
	1957–2013	0.51	1.6	0.54	0.57	0.49

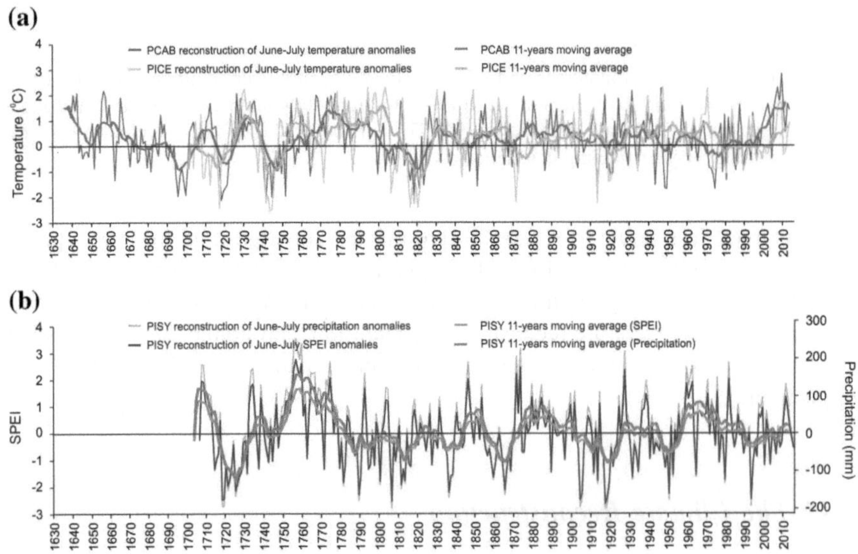

Fig. 8 Climate reconstruction of the Tatras Region, over the period 1705–2013 AD: **a** Reconstructed June–July mean temperature anomalies based on PCAB standard chronology and PICE standard chronology, **b** reconstructions based on PISY standard chronology June–July precipitation and July SPEI anomalies

precipitation) show higher agreement for SPEI than for precipitation but the calibration and verification statistics present the predictive skills of both models (Table 1).

Reconstructions of summer temperature, based on the Norway spruce and Stone pine, reveal high similarity (long-term trends) and also several differences (in short-term trends and performance in the most recent interval) (Fig. 8). The PCAB reconstruction is longer and covers the cold period, ~1640–1700, induced by the Late Maunder Minimum (Eddy et al. 1976; Niedźwiedź 2010) followed by a short interval of warmer summers in the three first decades of 1700s. During the cold phase 1643–1651, two major flood occurred (1650 and 1651) (Siemionow 1992).

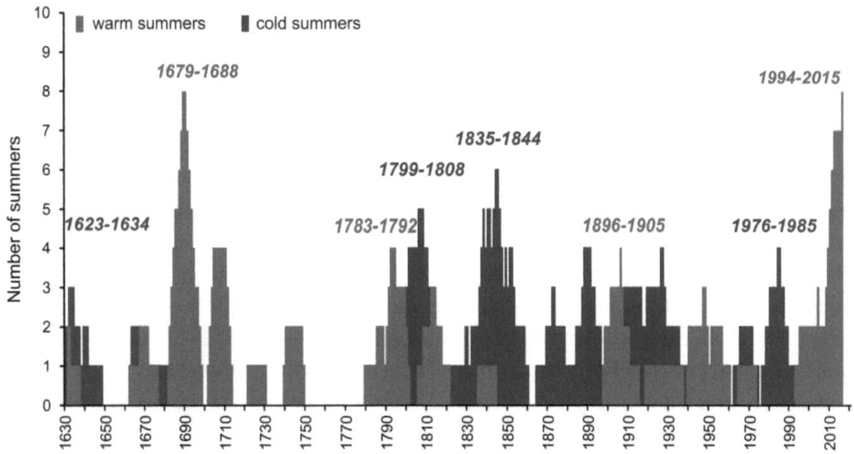

Fig. 9 Variability of the number of extreme warm and cold summers (June–August) in the Tatra Mountains (Hala Gąsienicowa 1520 m a.s.l.) during the consecutive 10 year periods

This was succeeded by warmer period of the 1660s, which was also noted in historical record (Bokwa et al. 2001) followed by the next cooling and warming (1670s and 1680s, respectively) (Fig. 9). The next cold period lasted also one decade but the temperature decrease was more severe, by almost 2 °C comparing with 1 °C for earlier variations. The year 1695 was very cold, due the eruption of Hekla, Serua and Aboino volcanoes (Grove 1988). This agrees with reports from other locations in Europe.

The next similar cooling happened in the 1710s, with the minimum in 1718 when the temperature dropped by 2 °C and reached the minimum of the reconstructed period. That cooling interval, with duration reaching almost a decade, could be associated with a series of volcanic activities in 1715–1718. The historical records inform about heavy snowfalls in July and August 1724 and again August 1725 (Siemionow 1992). The following years were warmer but can be characterized by oscillations of temperature (from +2 °C in 1726 to 0.0 °C in 1729). The common period of both temperature reconstructions started from 1737. The 1740 marked the beginning of the next cool phase, which has lasted till 1753. The volcano activity increased in 1739 and 1740 and could have resulted in decrease of temperature in the following years. The 1750s and 1770s were the next period of temperature changes around 0.7 °C with exception of 1771. The PICE reconstruction registered pronounced cooling whereas PCAB showed this year as 0.3 °C above the average. The remaining part of the 18th century was relatively warm. The eruption of Laki volcano (1783) could be linked with short but severe temperature drop (up to 1.0 °C in 1886) captured by the PICE chronology.

The next significant cooling at the beginning of 1800s was related to the Dalton solar minimum (Eddy et al. 1976) and major volcanic activity (Crowley 2000; Robock 2000). The Tambora eruption on April 1815 (VEI = 7; Newhall and Self 1982) played the most crucial role in inducing the coldest period with most severe and longest in duration temperature drop. Depending on the reconstruction, the temperature dropped by 2.4 °C (PICE) or 1.9 °C (PCAB) in comparison with the average. The Stone pine chronology registered the minimum temperature in 1816 while Norway spruce in 1818. The delay in the response of severe volcanic eruption-related cooling was discussed by Esper et al. (2015) and D'Arrigo et al. (2013). The cooling was probably less pronounced in Central Europe than in Western Europe but the term "year without summer" (1816–1818) undoubtedly applies also to the Tatra Region (Grove 1988; Harrington 1992; Bednarz and Trepińska 1992). The following decades showed sine-like cycle of dropping and rising temperature but with slightly positive long-term trend. The noticeable cooler periods occurred in 1843–44, 1851–54 and 1864–1871 mainly registered by PICE chronology. They were also mentioned by the historical sources reported the crop yield failures (Stolarczyk 1915). The cool summers mentioned above belong to the second phase of millennial climate variability named the Little Ice Age (1575–1850), cf Grove (1988). In reconstructed temperature series this cooling is prolonged to 1870 (Fig. 8a). The similar picture is observed in the tree-ring chronology derived from Norway spruce growing on Babia Góra Mountain (Bednarz 2015). After temperature reconstruction for Hala Gąsienicowa (1520 m a.s.l.) near the treeline in the Tatra Mountains, the end of the Little Ice Age can be placed around 1895 (Niedźwiedź 2004).

The beginning of the 1900s marked the end of the Little Ice Age in the Tatras Region, which lasted longer than in the other parts of Europe (Niedźwiedź 2004). Well documented cooling caused by the Katmai eruption in 1912 interrupted the warm period. The reaction to this event was delayed by one year and the summer of 1913 was colder by more than 2 °C in the Tatras. The summer of 1913 is the coldest one also in the other temperature reconstructions (Bednarz 1976, 1983, 1984, 2015; Bednarz et al. 1998–1998; Niedźwiedź 2004). This drop of temperature was registered by both species but PICE reacted stronger also creating strong signal in the form of pale rings (Janecka and Kaczka 2015).

The following decades again showed since-like cycle of dropping and rising temperature but with higher amplitude which could be attributed to the increase of climate continentality (Niedźwiedź 2004). The 1960s and 1970s were the most recent colder phases but less pronounced in comparison to the Little Ice Age period. The PICE chronology revealed problems to perform well during this period. The recent warming has been present in the Tatras since the end of 1980s, which is documented by instrumental data (Żmudzka 2009, 2010) and behaviour of the PCAB reconstruction. Interestingly, the TRW of PICE failed to capture this change. Stone pine often grows above the timberline in harsher conditions where locally and globally induced changes of temperature influence the growth stronger.

In the instrumental data on the treeline on Hala Gąsienicowa (1520 m a.s.l.) the last cool summer (8.3 °C) was indicated in 1984. After 1987 11-year average temperature increased from the level of 9.9 to 11.4 °C during the last 11 years (2005–2015). The summer of 2015, with temperature 12.9 °C was the warmest in the whole temperature series (where the long-term average is 10.0 °C). In Fig. 9, the variability of numbers of extreme warm (>11 °C) and cool (<9 °C) summer seasons (above and below the 10 and 90 ‰) on the treeline has been presented in consecutive decades since the beginning of the 18th century (Niedźwiedź 2004, data updated to 2015). In the last decade (2005–2015), eight summers were extremely warm. The four warmest summers are also noted in the decades 1896–1905 and 1783–1792. The six coolest summers are located in the decade 1835–1844 and five in the years 1799–1808 during the final phase of the Little Ice Age. The last decade with four coolest summers spans over the years 1976–1985.

The summer precipitation and Standardized Precipitation Evapotranspiration Index, SPEI, were reconstructed based on the PISY chronology. The changes of humidity-related climate variables are less pronounced over the reconstructed period of 1710–2013 (Fig. 8b). The cold period of the Late Maunder Solar Minimum (Eddy et al. 1976) is also the time of increase of summer humidity with high SPEI index (3.02) in 1707. In the Western Carpathians, after historical chronicles, three consecutive summers 1713–1715 were rainy and cold (Bednarz 2015). Relatively short, but deep dry period 1717–1726 preceded the long wetter summer conditions in 1733–1779 with the highest SPEI index (5.54) in 1756. The most humid conditions (SPEI > 2.00) in 1751–1775 coincide with the large increase of temperature after the cool period 1740–1752 (compare Fig. 8a). The mega drought in Western Europe, reconstructed by Cook et al. (2015), is also visible by decreasing the humidity after ~1770 (Fig. 8b) to the SPEI below −1.00 in 1789. The local but more severe drought happened in amount of summer precipitation and SPEI in 1806 (−1.34) and 1836 (−1.01). Extreme cooling during the Dalton Minimum of solar activity at the beginning of the 19th century (Fig. 8a and Bednarz 2015) is not visible in the SPEI variability. But in the historical sources, damaging floods are indicated in Nowy Targ in 1813, 1816, 1823 and 1834 (Bednarz 2015). Significant increases of SPEI index (3.08 in 1846) are strongly marked during 1840–1853 in the final cool phase of the Little Ice Age. This wet period was also reported in historical documents and tree-ring reconstructions from Moravia, Czech Republic (Büntgen et al. 2011). The last 30 years of the 19th century (1871–1900) can be distinguished as a very humid interval, with maximum SPEI in 1871 (3.26) and 1873 (3.48). In the 20th century, three wet periods (1924–1942, 1954–1987 and 2006–2010) and three drier intervals concentrated around the years 1917, 1950 and 1993 were observed.

The correlation between flood events and reconstructed or registered humid periods (Fig. 10) are difficult due to the predominant character of the flood caused by short-term intensive precipitation. Lack of the synchronicity between the

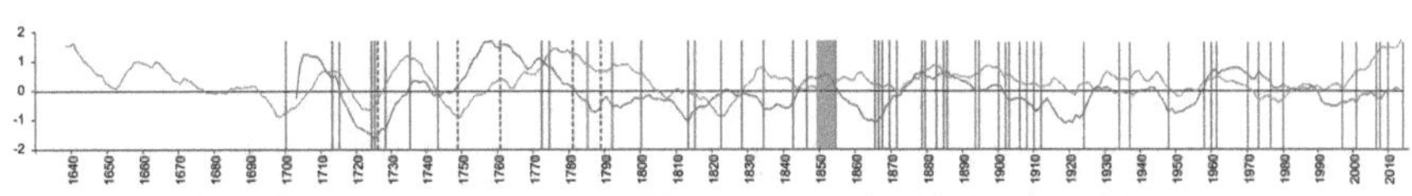

Fig. 10 Compression between the historical and instrumental flood records and reconstruction of summer temperature and precipitation

long-term and even medium-term changes of humidity and flood occurrences poses an important challenge for predicting the occurrence of floods and managing of this natural risk. The comparison of instrumental data enriched by historical records of floods with climate reconstructions demonstrates that even during drier period major flood can happen (as in intervals: 1721–1730 or 1865–1872).

5 Conclusion

The reconstruction of climate factors different than temperature in mountain regions of temperate climate is challenging but allows us to pose more complete image of paleoclimate. The coupling of regional reconstruction of humidity and temperature for the Tatra Mountains spanning an interval in excess of 300 years proved to be a promising approach to reconstruction of climate data. The temperature reconstruction based on two subalpine species demonstrates benefits of combining the records of Norway spruce and Stone pine proxies. The former better register long-term changes and latter are more susceptible to short-term cooling, clearly indicating events of more extreme nature.

The compilation of all performed climate reconstructions for the last three centuries allows us to conclude:

- The 17th–18th period of the Little Ice Age was characterized by complex changes of climate. The cold phases were followed by warm periods and single warm and cold years were frequent. The beginning of the 19th century was associated with a more pronounced cooling caused by the Dalton Solar Minimum and volcanic activity (Tambora eruption in 1815). The second part of the 19th century was also relatively cold and the end of the Little Ice Age in the Tatras Region could be placed at the beginning of the 20th century.
- Considering the entire 20th century, the recent warming is less pronounced. The significant increase of summer temperature has started since the 1980s and the 1990s and was registered by Norway spruce TRW whereas PICE failed to capture it.
- After 1985, cool summers (below 90 ‰) did not occur and in the decade 2006–2015 the number of the warmest summers (above 10 ‰) increased to eight.
- The temperature reconstructed from PCAB and PICE TRW reveals largely individual character of the Tatras in comparison with the Alps and Europe. This problem was also reported by Büntgen et al. (2007) and D'Arrigo et al. (2008).
- The short-term global events are better synchronised. The short (one or few years) decreases of summer temperature, mainly caused by volcanic activity are clearly present in the climate of the studied region. The 1718 cooling is the first and severe event, followed by 1816–20, 1912–13 and 1974–1980 periods.
- The correlation between flood events and reconstructed humid periods is poor due to the predominant character of the flood caused by short and intensive precipitation.

- Lack of the synchronicity between both long- and short-term changes of humidity and flood occurrences poses an important challenge for direct implementation to predict the floods and manage that natural risk. The comparison of instrumental data enriched by historical records of floods with climate reconstructions proves that even during drier period major floods can occur.

Acknowledgments This contribution has been realized in the framework of the FLORIST (Flood risk on the northern foothills of the Tatra Mountains) project, PSPB no. 153/2010, through a grant from Switzerland through the Swiss Contribution to the enlarged European Union. Temperature data were made available by the Institute of Meteorology and Water Management, National Research Institute (IMGW, PIB). Part of data for the period 2000–2015 was taken from synoptic data-base OGIMET.

References

Adamczyk M (1991a) Dzieje miasta Nowego Targu. Podhalańskie Towarzystwo Przyjaciół Nauk w Nowym Targu, Nowy Targ, pp 495

Adamczyk M (1991b) Zakopane-wieś [in:] Dutkowa R. (ed.) Zakopane czterysta lat dziejów. Tom I

Ballesteros-Cánovas JA, Stoffel M, Spyt B, Janecka K. Kaczka RJ, Lempa M (2015a) Paleoflood discharge reconstruction in Tatra Mountain streams. Geomorphology. doi:10.1016/j.geomorph.2015.12.004

Ballesteros-Cánovas JA, Czajka B, Janecka K, Lempa M, Kaczka RJ, Stoffel M (2015b) Flash floods in the Tatra Mountain streams: frequency and triggers. Sci Total Environ 511:639–648

Bednarz Z (1976) Wpływ klimatu na zmienność szerokości słojów rocznych limby (*Pinus cembra* L.) w Tatrach (The effect of climate on the variability in the width of growth tree-rings in stone pine (*Pinus cembra* L.) in the Tatra Mountains). Acta Agraria et Silvestria, Series Silvestris 16:17–34 (in Polish, summary in English)

Bednarz Z (1983) Dendroclimatological investigations in the Tatra Mountains. Zeszyty Naukowe UJ, Prace Geograficzne 57:127–131

Bednarz Z (1984) The comparison of dendroclimatological reconstructions of summer temperatures from the Alps and Tatra Mountains from 1741–1965. Dendrochronologia 2:63–72

Bednarz Z (1996) June–July temperature variation for the Babia Góra National Park, Southern Poland, for the period 1650–1910. Zeszyty Naukowe Uniwersytetu Jagiellońskiego, Prace Geograficzne 102:523–529

Bednarz Z (2015) Climate reconstruction. Plate II.7. In: Dąbrowska K, Guzik M (eds) Atlas of the Tatra Mountains. Abiotic nature. Tatra National Park, Zakopane

Bednarz Z, Niedźwiedź T (2006) Dendrochronologia świerka (*Picea Abies* (L.) Karst) z Parku Narodowego Wysokie Taury (Austria) (Dendrochronology of Norway spruce (*Picea Abies* (L.) Karst) from the High Tauern National Park (Austria)). In: Trepińska J, Olecki Z (eds) Klimatyczne Aspekty Środowiska Geograficznego. Instytut Geografii i Gospodarki Przestrzennej Uniwersytetu Jagiellońskiego, Kraków, pp 231–246 (in Polish, summary in English)

Bednarz Z, Trepińska J (1992) Climatic conditions of 1815 and 1816 from tree-ring analysis in the Tatra Mountains. In: Harington CR (ed) The year without a summer? World climate in 1816. National Museum of Science, Ottawa, Canada, pp 418–421

Bednarz Z, Jaroszewicz B, Ptak J, Szwagrzyk J (1998–1999) Dendrochronology of Norway spruce (*Picea abies* (L.) Karst) in the Babia Góra National Park, Poland. Dendrochronologia 16–17: 45–55

Beguería S, Vicente-Serrano SM, Reig F, Latorre B (2014) Standardized precipitation evapotranspiration index (SPEI) revisited: parameter fitting, evapotranspiration models, tools, datasets and drought monitoring. Int J Climatol 34(10):3001–3023

Bielański AK (1984) Materiały do historii powodzi w dorzeczu górnej Wisły. Politechnika Krakowska, Kraków: pp 32–119

Bokwa A, Limanówka D, Wibig J (2001) Pre-instrumental weather observations in Poland in the 16th and 17th centuries. In: Jones PD, Ogilvie AEJ, Davies TD, Briffa KR (eds) History and Climate: Memories of the Future. Springer US, pp 9–27

Briffa KR, Osborn TJ, Schweingruber FH, Harris IC, Jones PD, Shiyatov SG, Vaganov EA (2001) Low-frequency temperature variations from a northern tree ring density network. J Geophys Res Atmos 106(D3):2929–2941

Briffa KR, Osborn TJ, Schweingruber FH (2004) Large-scale temperature inferences from tree rings: a review. Global Planet Change 40(1):11–26

Brzęk L, Kaczka RJ, Czajka B (2014) Zróżnicowanie sygnału klimatycznego w przyrostach sosny zwyczajnej Pinus sylvestris L. z lasów klasy Erico-Pinetea w Tatrach. Studia i Materiały Centrum Edukacji Przyrodniczo-Leśnej 3(16):221–229

Büntgen U, Esper J, Frank DC, Nicolussi K, Schmidhalter M (2005) A 1052-year tree-ring proxy for Alpine summer temperatures. Clim Dyn 25(2–3):141–153

Büntgen U, Frank DC, Nievergelt D, Esper J (2006) Summer temperature variation in the European Alps, AD 755–2004. J Clim 19(21):5606–5623

Büntgen U, Frank DC, Kaczka RJ, Verstege A, Zwijacz-Kozica T, Esper J (2007) Growth responses to climate in a multi-species tree-ring network in the Western Carpathian Tatra Mountains, Poland and Slovakia. Tree Physiol 27:689–702

Büntgen U, Brázdil R, Dobrovolný P, Trnka M, Kyncl T (2011) Five centuries of Southern Moravian drought variations revealed from living and historic tree rings. Theor Appl Climatol 105(1–2):167–180

Büntgen U, Kaczka RJ, Trnka M, Rigling A (2012) Ensemble estimates reveal a complex hydroclimatic sensitivity of pine growth at Carpathian cliff sites. Agric For Meteorol 160: 100–109

Büntgen U, Kyncl T, Ginzler Ch, Jacks DS, Esper J, Tegel W, Heussner K-U, Kyncl J (2013) Filling the Eastern European gap in millennium-long temperature reconstruction. PNAS 110 (5):1773–1778

Carrer M, Nola P, Eduard JL, Motta R, Urbinati C (2007) Regional variability of climate-growth relationships in *Pinus cembra* high elevation forests in the Alps. J Ecol 95(5):1072–1083

Cook ER (1985) A time series analysis approach to tree-ring standardization. PhD thesis, University of Arizona, Tucson, p 171

Cook ER, Peters K (1981) The smoothing spline: a new approach to standardizing forest interior tree-ring width series for dendroclimatic studies. Tree Ring Bull

Cook ER, Seager R, Kushnir Y, Briffa KR, Büntgen, U, Frank D, Baillie M (2015) Old World megadroughts and pluvials during the Common Era. Sci Adv 1(10):e1500561

Crowley TJ (2000) Causes of climate change over the past 1000 years. Science 289(5477): 270–277

D'Arrigo R, Wilson R, Liepert B, Cherubini P (2008) On the 'divergence problem' in Northern forests: a review of the tree-ring evidence and possible causes. Global Planet Change 60 (3):289–305

D'Arrigo R, Wilson R, Anchukaitis K (2013) Volcanic cooling signal in tree-ring temperature records for the past millennium. J Geophys Res Atmos 118(16):9000–9010. doi:10.1002/jgrd.50692

Durbin J, Watson GS (1951) Testing for serial correlation in least squares regression. II. Biometrika 38(1/2):159–177

Eddy JA, Gilman PA, Trotter DE (1976) Solar rotation during the Maunder Minimum. Sol Phys 46(1):3–14

Esper J, Frank DC, Wilson RJ, Briffa KR (2005) Effect of scaling and regression on reconstructed temperature amplitude for the past millennium. Geophys Res Lett 32(7)

Esper J, Schneider L, Smerdon JE, Schöne BR, Büntgen U (2015) Signals and memory in tree-ring width and density data. Dendrochronologia 35:62–70

Feliksik E (1972) Studia dendroklimatologiczne nad świerkiem (*Picea excelsa* L.), Cz. 1 i 2 (Dendroclimatic studies of spruce (Picea excelsa L.), part 1 and 2. Acta Agraria et Silvestria, Series Silvestria 12:39–83

Feliksik E (1992) Wplyw warunkow klimatycznych na wielkosc przyrostow radialnych modrzewia europejskiego [Larix decidua Mill.] wystepujacego w Karpatach. Sylwan 136 (05):61–67

Frank DC, Esper J (2005a) Characterization and climate response patterns of a high elevation, multi species tree-ring network for the European Alps. Dendrochronologia 22:107–121

Frank D, Esper J (2005b) Temperature reconstructions and comparisons with instrumental data from a tree-ring network for the European Alps. Int J Climatol 25(11):1437–1454

Frank D, Esper J, Zorita E, Wilson R (2010) A noodle, hockey stick, and spaghetti plate: a perspective on high resolution paleoclimatology. Wiley Interdiscip Rev Clim Change 1 (4):507–516. doi:10.1002/wcc.53

Fritts C (1976) Tree-ring and climate. Academic Press, London, New York, San Francisco, p 567

Generisch C (1807) Reisein die Karpathen mit vorzüglicher Rücksicht auf das Tatra Gebirge, Neue Beiträge zur Topographie und Statistik des Königreichs Ungarn, Wien-Triest

Grove JM (1988) Little ice age. Chapman and Hall, New York, p 498

Gustawicz B (1883) Pomiary tatrzańskie, Pamiętnik Towarzystwa Tatrzańskiego. Kraków

Harington CR (1992) The year without a summer? World climate in 1816. National Museum of Science, Ottawa, Canada

Harris IPDJ, Jones PD, Osborn TJ, Lister DH (2014) Updated high-resolution grids of monthly climatic observations—the CRU TS3. 10 Dataset. Int J Climatol 34(3):623–642

Hess M (1974) Piętra klimatyczne w Tatrach (Vertical climatic zones in the Tatra Mountains). Czasopismo Geograficzne 45(1):131–146 (in Polish, summary in English)

Holmes RL (1983) Computer-assisted quality control in tree-ring dating and measurement. Tree-ring Bull 43(1):69–78

Janecka K, Kaczka RJ (2015) Tree rings of *Pinus cembra* L. in the Tatra Mts as a proxy of significant volcanic eruptions in the last 280 years. TRACE—Tree rings in archaeology, climatology and ecology 13:49–55 (GFZ Potsdam, Scientific Technical Report STR15/06, Potsdam)

Jones PD, Mann ME (2004) Climate over past millennia. Rev Geophys 42(2)

Kaczka RJ (2004) Dendrochronologiczny zapis zmian klimatu Tatr od schyłku małej epoki lodowej na przykładzie Doliny Gąsienicowej (Dendrochronological records of climate fluctuation during last 400 years in the Tatra Mountains). In: Kotarba A (ed) Rola małej epoki lodowej w przekształcaniu środowiska przyrodniczego Tatr. IGiPZ PAN Warszawa, Prace Geograficzne 197:89–113 (in Polish, summary in English)

Kaczka RJ, Büntgen U (2006) Spatial autocorrelation and growth/climate response of a high elevation spruce network along the Carpathian arc. In: Haneca K, Verheyden A, Beeckman H, Gartner H, Helle G, Schleser G (eds) Tree rings in archaeology, climatology and ecology, vol 5. Association for Tree Ring Research, Potsdam, pp 103–112

Kaczka RJ, Czajka B (2014) Intensywność odbicia światła niebieskiego jako nowy nośnik informacji w badaniach dendrochronologicznych. Studia i Materiały Centrum Edukacji Przyrodniczo-Leśnej w Rogowie 16(40):274–282

Kaczka RJ, Brzęk Ł, Czajka B (2012) Wpływ czynników klimatycznych na wzrost sosny pospolitej w Tatrach Reglowych. Studia i materiały Centrum Edukacji Przyrodniczo-Leśnej, Rogów 1(30):84–90

Kaczka RJ, Czajka B, Łajczak A (2015) The tree-ring growth responses to climate in the timberline ecotone of Babia Góra Mountain. Geographia Polonica 88(2):163–176

Komoniecki A (1704) Chronografia albo Dziejopis Żywiecki. Żywiec (ed.) 1987

Kondracki J (1994) Geografia Polski: mezoregiony fizyczno-geograficzne. Wydawnictwo Naukowe PWN

Kotarba A (2004) Zdarzenia geomorfologiczne w Tatrach Wysokich podczas małej epoki lodowej. (Geomorphic events in the High Tatra Mountains during the Little Ice Age). In: Kotarba A (ed) Rola małej epoki lodowej w przekształcaniu środowiska przyrodniczego Tatr (Effect of the Little Ice Age on transformation of natural environment of the Tatra Mountains). PAN IGiPZ, Kraków, pp 9–55 (in polish with english summary)

Kundzewicz Z, Stoffel M, Kaczka R, Wyżga B, Niedźwiedź T, Pińskwar I, Ruiz-Villanueva V, Łupikasza E, Czajka B, Ballesteros-Canovas J, Małarzewski Ł, Choryński A, Janecka A, Mikuś P (2014) Floods at the northern foothills of the Tatra Mountains—A Polish-Swiss research project. Acta Geophys 62(3):620–641

Larsson L-A (2003) CDendro: Cybis Dendro dating program. www.cybis.se

Marchi L, Borga M, Preciso E, Gaume E (2010) Characterisation of selected extreme flash floods in Europe and implications for flood risk management. J Hydrol 394(1):118–133

New M, Hulme M, Jones PD (2000) Representing twentieth century space-time climate variability: Part II: Development of 1901–96 monthly grids of terrestrial surface climate. J Climate 13:2217–2238

Newhall CG, Self S (1982) The volcanic explosivity index (VEI): an estimate of explosive magnitude for historical volcanism. J Geophys Res 87(C2):1231–1238

Niedźwiedź T (1992) Climate of the Tatra Mountains. Mt Res Dev 12(2):131–146

Niedźwiedź T (2004) Rekonstrukcja warunków termicznych lata w Tatrach od 1550 roku (Reconstruction of summer temperature in the Tatra Mountains since 1550). In: Kotarba A (ed) Rola małej epoki lodowej w przekształcaniu środowiska przyrodniczego Tatr. IGiPZ PAN Warszawa, Prace Geograficzne 197:57–88 (in Polish, summary in English)

Niedźwiedź T (2010) Summer temperatures in the Tatra Mountains during the Maunder Minimum (1645–1715). In: Przybylak R, Majorowicz J, Brázdil R, Kejna M (eds) The Polish Climate in the European Context: An Historical Overview (Chapter 19). Springer Science + Business Media BV, pp 397–406. doi:10.1007/978-90-481-3167-9_19

Niedźwiedź T, Łupikasza E, Pińskwar I, Kundzewicz ZW, Stoffel M, Małarzewski Ł (2015) Variability of high rainfalls and related synoptic situations causing heavy floods at the northern foothills of the Tatra Mountains. Theor Appl Climatol 119(1–2):273–284

Oberhuber W (2004) Influence of climate on radial growth of *Pinus cembra* within the alpine timberline ecotone. Tree Physiol 24(3):291–301

Pamiętnik Towarzystwa Tatrzańskiego (1876–1920). Rocznik Towarzystwa Tatrzańskiego

Parajka J, Kohnová S, Bálint G, Barbuc M, Borga M, Claps P, Cheval S, Dumitrescu A, Gaume E, Hlavčová K, Merz R, Pfaundler M, Stancalie G, Szolgay J, Bloschl G (2010) Seasonal characteristics of flood regimes across the Alpine-Carpathian range. J Hydrol 394(1):78–89

Perzanowska J. 2010. Górskie reliktowe laski sosnowe Erico-Pinion. W: Mróz W. (red.). Monitoring siedlisk przyrodniczych. Przewodnik metodyczny, GIOŚ, Warszawa, cz. I: 32–58

Ponocná T, Spyt B, Kaczka RJ, Büntgen U, Treml V (2016) Growth trends and climate responses of Norway spruce along elevational gradients in East-Central Europe. Trees—structure and function (in print)

Popa I, Kern Z (2009) Long-term summer temperature reconstruction inferred from tree-ring records from the Eastern Carpathians. Clim Dyn 32(7–8):1107–1117

Rajwa A (2014) Powodzie w Tatrach i na Podtatrzu. Tatry TPN, 3(49):56–61

Robock A (2000) Volcanic eruptions and climate. Rev Geophys 38(2):191–219

Ruiz-Villanueva V, Stoffel M, Wyżga B, Kundzewicz ZW, Czajka B, Niedźwiedź T (2016) Decadal variability of floods in the northern foreland of the Tatra Mountains. Reg Environ Change 16:603–615

Savva Y, Oleksyn J, Reich PB, Tjoelker MG, Vaganov EA, Modrzyński J (2006) Interannual growth response of Norway spruce to climate along an altitudinal gradient in the Tatra Mountains, Poland. Trees—structure and function 20(6):735–746

Schweingruber FH (1996) Tree rings and environment: dendroecology. Paul Haupt AG Bern

Sidor CG, Popa I, Vlad R, Cherubini P (2015) Different tree-ring responses of Norway spruce to air temperature across an altitudinal gradient in the Eastern Carpathians (Romania). Trees 29 (4):985–997

Siemionow A (1992) To i owo o Tatrach, Tom 1, Ciekawostki fizjograficzne z dziedziny ekstremaliów morfologicznych i geologicznych, zjawisk meteorologicznych, optycznych, flory i fauny Tatr. Biblioteka Tatrzańsko-Zakopiańska z kozicą. pp 160

Stolarczyk J (1915) Kronika Parafii Zakopiańskiej (1848–1890), wydał i wstępem opatrzył Adam Wrzosek. Nakładem Wydawcy–Druk WL Anczyca i Sp., Kraków

Szychowska-Krąpiec E (1998) Spruce Chronology from Mt Pilsko Area (Żywiec Beskid Range) 1641–1995 AD. Bull Pol Acad Sci Earth Sci 46(2):75–86

Szewczuk J (1939) Kronika klęsk elementarnych w Galicji w latach 1772–1848. Bujak F (ed.) Badania z Dziejów Społecznych i Gospodarczych 35. Sprawozdania Towarzystwa Naukowego we Lwowie 17:235–237

Szychowska-Krąpiec E (2010) Long-term chronologies of pine (*Pinus sylvestris* L.) and fir (*Abbies alba* MILL.) from the Małopolska Region and their palaeoclimatic interpretation. Folia Quaternaria 79:1–124 (Polish academy of Art and Sciences, Commission for Quaternary Palaeography, Cracow)

Treml V, Ponocná T, King GM, Büntgen U (2015) A new tree-ring-based summer temperature reconstruction over the last three centuries for east-central Europe. Int J Climatol 35:3160–3171

Vaganov EA, Hughes MK, Kirdyanov AV, Schweingruber FH, Silkin PP (1999) Influence of snowfall and melt timing on tree growth in subarctic Eurasia. Nature 400(6740):149–151

Vicente-Serrano SM, Beguería S, López-Moreno JI (2010) A multiscalar drought index sensitive to global warming: the standardized precipitation evapotranspiration index. J Clim 23(7): 1696–1718

Warszyńska J (ed) (1995) Karpaty Polskie: przyroda, człowiek i jego działalność. Uniwersytet Jagielloński

Wierchy (1923–2015). Rocznik Polskiego Towarzystwa Tatrzańskiego-Polskiego Towarzystwa Turystyczno-Krajoznawczego

Więckowski XM (1834) Kronika Parafii Poronińskiej (1845–1945). Łasut J (ed.) Poronin 1997

Wigley TM, Briffa KR, Jones PD (1984) On the average value of correlated time series, with applications in dendroclimatology and hydrometeorology. J Climate Appl Meteorol 23(2): 201–213

Wilson R, Anchukaitis K, Briffa KR, Büntgen U, Cook E, D'Arrigo R, Hegerl G (2016) Last millennium northern hemisphere summer temperatures from tree rings: Part I: The long term context. Q Sci Rev 134:1–18

Woźniak A (2013) Opady w 2010 roku w Karpatach Polskich na tle wielolecia 1881–2010. Prace Geograficzne, 133:35–48

Żmudzka E (2009) Changes of thermal conditions in the Polish Tatra Mountains. Land Anal 10:140–146

Żmudzka E (2010) Changes in thermal conditions in the high mountain areas and contemporary warming in the central Europe. Miscellanea Geogr 14:59–70

Variability of Flood Frequency and Magnitude During the Late 20th and Early 21st Centuries in the Northern Foreland of the Tatra Mountains

Virginia Ruiz-Villanueva, Bartłomiej Wyżga, Zbigniew W. Kundzewicz, Tadeusz Niedźwiedź, Ewa Łupikasza and Markus Stoffel

Abstract Much of the flood risk in Poland is related to the Upper Vistula River Basin, and its right-bank tributaries on the northern foreland of the Tatra Mountains significantly contribute to the total flood damage. Therefore, the question whether the magnitude and frequency of floods in this region have changed in the past

V. Ruiz-Villanueva (✉) · M. Stoffel
Dendrolab.ch, Institute of Geological Sciences, University of Bern, Baltzerstrasse 1+3, 3012 Bern, Switzerland
e-mail: virginia.ruiz@dendrolab.ch

M. Stoffel
e-mail: markus.stoffel@dendrolab.ch

V. Ruiz-Villanueva · M. Stoffel
Climatic Change and Climate Impacts, Institute for Environmental Sciences, University of Geneva, Boulevard Carl Vogt 66, 1205 Geneva, Switzerland

B. Wyżga
Institute of Nature Conservation, Polish Academy of Sciences, al. Mickiewicza 33, 31-120 Kraków, Poland
e-mail: wyzga@iop.krakow.pl

B. Wyżga · T. Niedźwiedź · E. Lupikasza
Faculty of Earth Sciences, University of Silesia, ul. Będzińska 60, 41-200 Sosnowiec, Poland
e-mail: tadeusz.niedzwiedz@us.edu.pl

E. Łupikasza
e-mail: ewa.lupikasza@us.edu.pl

Z.W. Kundzewicz
Institute for Agricultural and Forest Environment, Polish Academy of Sciences, ul. Bukowska 19, 60-809 Poznań, Poland
e-mail: kundzewicz@yahoo.com

M. Stoffel
Department of Earth Sciences, University of Geneva, rue des Maraîchers 13, 1205 Geneva, Switzerland

Z.W. Kundzewicz et al. (eds.), *Flood Risk in the Upper Vistula Basin*, GeoPlanet: Earth and Planetary Sciences, DOI 10.1007/978-3-319-41923-7_11

decades is of high interest. This chapter focused on the inter-decadal variability of magnitude, frequency, and seasonality of floods since the mid-20th century using a multi-temporal approach in which trends are fitted to several combinations of start and end years in a record. The correlation between flood parameters and different large-scale climate indices for the Northern Hemisphere was calculated, as well as trends of intense precipitation indices, number of days with snow cover, cyclonic circulation types, air temperature and moisture conditions were calculated to explain the detected trends. Other potential external drivers, such as catchment and channel changes were also considered. Results show that floods in the area exhibit considerable inter-decadal variability, however, certain patterns are apparent. Less frequent floods, although perhaps more extreme, are now likely to occur, with a shift in the seasonality, decreasing flood magnitudes in winter and increasing during autumn and spring.

Keywords Hydrological extremes · Flood hazard · Climate change · Carpathians · Poland

1 Introduction

The seasonality of streamflow in mountainous basins has been found to be extremely sensitive to global warming (Diaz et al. 2003; Barnett et al. 2005; Bates et al. 2008; Marty 2008; Allamano et al. 2009) and the concern about the increase of flood risk in these areas is rapidly raising in the scientific literature (Olsen et al. 1998; Palmer and Räisänen 2002). The question whether the magnitude and frequency of floods have changed due to climate change or other drivers of change is of high interest (Merz et al. 2012). The costs of extreme weather events have exhibited a rapid upward trend in recent decades, at every spatial scale (Barredo 2007, 2009: IPCC 2009; Kron 2012). Therefore, the search for trends (or lack thereof) in flood data series has become of scientific interest and practical importance in the last years as it is essential for planning flood-protection systems, where system design has been traditionally based on the assumption of stationarity of river flow (Milly et al. 2008). Within the framework of the World Climate Programme-Water and other, national and international projects, important activities have been undertaken in the analysis of long time-series of hydrological observations to detect signals of change (Kundzewicz and Robson 2000, 2004; Robson 2002; Kundzewicz et al. 2004, 2005; Lindström and Bergström 2004; Radziejewski and Kundzewicz 2004; Svensson et al. 2004, 2005; Birsan et al. 2005; Petrow and Merz 2009). Several studies on flood trends in Europe can be found, and they usually cover large regions or entire countries (Robson et al. 1998; Helms et al. 2002; Lammersen et al. 2002; Robson 2002; Lindström and Bergström 2004; Pfister et al. 2004a, b; Mudelsee et al. 2006; Pinter et al. 2006; and national and regional chapters in the monograph edited by Kundzewicz 2012). Although there is as yet no proof that the extreme flood events of recent years are a direct consequence of climate change, they may give an indication of what can be expected: the frequency and intensity of floods in large parts of Europe are projected to increase, even if there is a considerable uncertainty in projections (cf. discussion in Kundzewicz et al. 2016).

However, there are other important factors that may alter flood variability independently from streamflow trends and even in the absence of climate change (Slater et al. 2015). Flooding is a complex phenomenon integrating the influence of atmospheric variables over a watershed (Kundzewicz and Schellnhuber 2004). Therefore, land-use changes (e.g., reforestation and urbanization) may induce changes of terrestrial (hydrological and ecological) systems and control the rainfall–runoff relations, hence impacting on floods. In addition, changes in the capacity of river channels to convey flood flows may also change flood frequency (Wyżga 1997; Slater et al. 2015; Wyżga et al. 2016).

There are still some issues related to the data and to the methodology of change and trend detection (Kundzewicz and Robson 2004). A precise understanding of the hydroclimatic characteristics of mountain regions is complicated by a lack of observational data at the spatial and temporal resolution adequate for hydroclimatic research in regions of complex topography. In the particular case of streamflow studies, in order to detect a weak, if any, climate change component in the process of river flow, it is necessary to eliminate other influences and use data from pristine (baseline) river basins. Baseline conditions are rare, and human influence is often strong even in mountainous regions where the mountain tourism economy, largely linked to winter sports, has soared over the last 50 years.

The region of the foreland of the Tatras has been typified by high population density, with human activity in the northern foothills initiated relatively early (during the 18th century a considerable agricultural and pastoral activity was developed in the region). Over the 20th century, the use of the land changed, as the percentage of arable land decreased significantly and forest cover increased (Wyżga et al. 2012). In 1955 the establishment of the Tatra National Park restricted developments and the state of Tatra forests improved considerably (Jahn 1979). However, other types of human impacts appeared in the foreland of the high-mountain massif: urbanization, channelization of streams and rivers, in-channel gravel mining and channel incision (Korpak 2007; Zawiejska and Wyżga 2010).

During the 20th century, 41 significant floods took place in the Carpathian part of the Vistula River Basin (Cebulak and Niedźwiedź 2000). Nationwide, in the last two decades two extreme flood events (1997 and 2010) caused damages reaching or exceeding the level of 1 % of natural Polish GDP and dozens of people lost their lives (Kundzewicz et al. 2012, 2014). These events also affected the right-bank tributaries of the Upper Vistula located in the study area.

This chapter aims to analyse the flood variability in the Tatra Mountains foreland on decadal timescales and to examine, via a multi-temporal trend analysis, the extent to which observed trends in two fixed periods are influenced. The potential drivers (i.e. trend attribution) of the identified trends in floods are investigated by correlation analyses with large-scale climate indices, by identifying trends in precipitation, snow cover duration, and cyclonic circulation types, and by analysing changes in basin attributes.

2 Study Site and Data Acquisition

The study sites are located in the northern foreland of the Polish Tatra Mountains. The highest peak Rysy, located SE of Zakopane, reaches 2499 m a.s.l. The Tatras originate from the Alpine orogeny. Their highest parts are predominantly underlain by granite and metamorphic rocks, while in the lower parts carbonate rocks dominate. Glacial transformation in the Pleistocene and periglacial processes in the Holocene produced a system of alpine cliffs and talus slopes. They indeed resemble the Alps landscape-wise, representing the only form of alpine landscape in the entire arc of the Carpathians. In the high, crystalline part of the Tatras, thick and highly porous debris cones predominate on slopes and glacial till in the valley floors, both enabling effective infiltration of water. On the northern slopes of the Tatras, limestone and dolomite bedrock favours deep, karstic water circulation. In both areas, lithological conditions slow down runoff and result in a considerable proportion (40–70 %) of underground supply of streams draining the mountains (Łajczak 1988). The foothills of the Polish Tatra Mountains are underlain primarily by flysch. This geological setting, together with considerable historical deforestation of the area results in greater flashiness of runoff (Wyżga et al. 2012).

The upper tree line (1550 m a.s.l.) is consistent with the annual isotherm of 2 °C. The subalpine belt is covered with mountain pine (*Pinus mugo*) and encompasses elevations ranging from 1550 to 1850 m a.s.l., where mean annual temperature drops to 0 °C. A belt of alpine meadows extends from 1850 to 2200 m a.s.l. Above 2200 m a.s.l. (seminival belt), bare rock and lichens predominate, and snow precipitation is more frequent than rainfall, with snow cover duration of approximately 230 days per year.

Precipitation recorded on the northern slopes of the Tatra Mountains is the highest in Poland. For the period 1951–2012, mean annual precipitation at Kasprowy Wierch (1991 m a.s.l.) was 1765 mm, but the record-high annual maximum precipitation at this station was 2599 mm in 2001 (with monthly maximum of 651 mm in July 2001). Even higher annual precipitation values were recorded in 2001 at other high-elevation locations of the Tatra Mountains, namely 2628 mm at Hala Gąsienicowa and 2770 mm at Dolina Pięciu Stawów. The flood-triggering rainfall types in this mountainous area are: (i) 2–5 days-long rainfall with intensity higher than 50–100 mm per day (maximum 300 mm day^{-1}), usually connected with circulation types Nc, NEc, NWc, Cc and Bc, often with cyclones moving by track Vb of van Bebber classification from the Adriatic Sea through Hungary to Ukraine or eastern Poland; and (ii) heavy downpours connected with local thunderstorms, with intensity up to 100–200 mm h^{-1}, causing local flash floods.

Streamflow data series of the length of 40–60 years from 14 water-gauge stations in the northern foreland of the Tatra Mountains were analysed (Fig. 1). Two different groups of stations were distinguished: (i) a first group formed by headwater basins with variable but generally lower degree of human impact on the environment, drainage areas ranging from 34 to 210 km^2 and station altitudes between 396 and 967 m (indicated in bold in Table 1), and (ii) a second group composed of the larger basins with drainage areas from 430 to 4300 km^2 and station altitudes ranging from 277 to 581 m, with higher degree of anthropogenic disturbance. It should be emphasized that the headwater basins exhibit a wide spectrum of catchment aspect, which makes this group of basins an indicator of regional environmental changes, including climatic ones, rather than of a change in the predominant pattern of atmospheric circulation over the area. In addition, this first group shows short records of streamflows, while the second group, representing the larger basins, has longer records (going back to 1951).

While most stations have the catchments located entirely or mostly on the Polish territory, the River Poprad joining the Dunajec upstream of Nowy Sącz (station 14 in Table 1) drains mainly the territory of northern Slovakia. Some environmental changes that occurred in that area in the past few decades differed from those recorded in the Polish part of the Dunajec drainage basin. However, it has not been possible for us to acquire the complete image of environmental factors that might have been responsible for the changes in floods on the Poprad. Nevertheless, it is interesting to verify whether flood trends identified for the Nowy Sącz station are consistent with those that typify the upstream stations characterizing the runoff from the Polish part of the Dunajec basin. Table 1 summarizes the most important characteristics of the studied drainage basins derived from topographic maps and data of the Polish Hydrometeorological Service (IMGW-PIB).

The main criteria for station selection (Pfaundler 2001; Birsan et al. 2005; Stahl et al. 2010) are: (a) availability of data, (b) at least 30 years of continuous and complete observations (most of stations have 45 years of records), and (c) spatial independence between station records. Spatial independence for stations located along the same river was ensured by always choosing the upstream station. The downstream station was additionally selected only in the case of a substantial increase in drainage area between the stations or if a tributary exists between them. These conditions provide a good compromise between the assumed independence of station records and a relatively high number of stations spread fairly evenly throughout the northern foreland of the Polish Tatra Mountains.

Time series of annual and seasonal maxima, peak over threshold magnitude and peak over threshold frequency (Petrow and Merz 2009) were analysed. The analysis of floods may differentiate the flood generation mechanisms (e.g. snowmelt vs rainfall), without treating all floods as one category. Therefore, four climatological seasons may be defined and analysed separately: Winter (DJF), Spring (MAM), Summer (JJA) and Autumn (SON) and the seasonal maximum discharge was

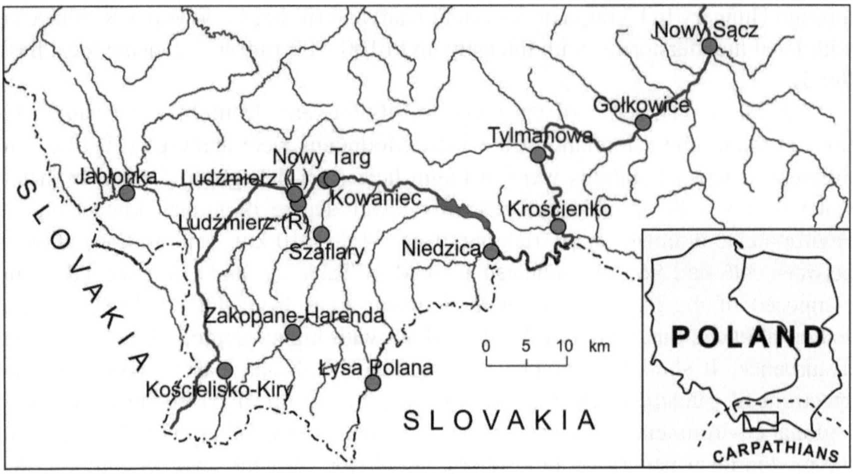

Fig. 1 Location of the studied river catchments in the northern foreland of the Polish Tatra Mountains

analysed for each (SMW, SMSp, SMSu, and SMA for winter, spring, summer and autumn, respectively). These parameters are summarized in Table 2.

3 Flood Trend Detection and Attribution

3.1 Flood Trend Detection

For all available time series, trends can be detected and classified according to their statistical significance as: (i) tendency, a statistically still unproven development, (ii) trend, a statistically proven development (at least 80 % significance), or (iii) strong trend, a statistically well-founded development (at least 95 % significance).

First of all, an exploratory data analysis (EDA) was carried out. This study involved mainly plotting graphs, and allowed to appreciate some features in data, as well as to assess the first hypotheses to be confirmed by the statistical analysis. In addition, the linear regression gradient plot in the EDA allowed testing of potential trends of the mean.

Trend analysis in this study is conducted by the nonparametric Mann–Kendall (MK) test. The application of the MK test to hydrological series has been discussed in detail by Kundzewicz and Robson (2004) and is summarised as follows:

Table 1 Physiographic characteristics of the 14 studied basins.

	Station (river)	Highest elevation (m a.s.l.)	Lowest elevation (m a.s.l.)	Relief (m)	Mean elevation (m a.s.l.)	Catchment area (km^2)	Catchment aspect	Mean catchment slope	Time period analyzed
1	**Kościelisko-Kiry** ***Kościeliski Stream***	**2158**	**922**	**1236**	**1452**	**34.5**	**N**	**0.133**	**1951–2011**
2	**Zakopane-Harenda** ***Cicha Woda***	**2096**	**765**	**1331**	**1320**	**58.4**	**N**	**0.126**	**1961–2011**
3	**Łysa Polana** ***Białka***	**2628**	**967**	**1661**	**1660**	**63.1**	**N**	**0.151**	**1961–2011**
4	**Ludźmierz (R)** ***Rogoźnik Stream***	**1124**	**596**	**528**	**832**	**125**	**N**	**0.029**	**1971–2011**
5	**Szaflary** ***Biały Dunajec***	**2301**	**638**	**1663**	**1295**	**210**	**N**	**0.072**	**1961–2011**
6	**Niedzica** ***Niedziczanka***	**1259**	**497**	**762**	**857**	**136.4**	**NE**	**0.049**	**1971–2011**
7	**Jabłonka** ***Czarna Orawa***	**1368**	**610**	**758**	**938**	**136**	**S**	**0.041**	**1961–2011**
8	**Ludźmierz (L)** ***Lepietnica***	**1312**	**598**	**714**	**908**	**50.4**	**SW**	**0.056**	**1971–2011**
9	**Tylmanowa** ***Ochotnica Stream***	**1288**	**396**	**892**	**791**	**107.6**	**E**	**0.050**	**1971–2011**
10	Nowy Targ *Czarny Dunajec*	2176	581	1595	1207	432	NE	0.043	1951–2011
11	Kowaniec *Dunajec*	2301	577	1724	1245	681	NE	0.046	1951–2011
12	Krościenko *Dunajec*	2628	416	2212	1199	1580	NE	0.040	1951–2011
13	Gołkowice *Dunajec*	2628	314	2314	1089	2047	NE	0.033	1951–2011
14	Nowy Sącz *Dunajec*	2655	277	2378	1100	4341	NE	0.031	1951–2011

Relief is calculated as the difference between the highest and the lowest elevation
Aspect is the averaged aspect of the basin (*N* North; *NE* North-East; *S* South; *SW* South-West). The smallest basins (group 1) are shown on bold

Table 2 Flood variables analysed in this study

	Abbreviation	Explanation
Flood variable	AMAX	Annual maximum discharge
	POTF	Peak over threshold (third quartile) frequency
	POTM	Peak over threshold (third quartile) magnitude
	SMW	Seasonal maximum discharge for winter
	SMSp	Seasonal maximum discharge for spring
	SMSu	Seasonal maximum discharge for summer
	SMA	Seasonal maximum discharge for autumn

Considering a sample (x_1, …, x_n) with size n, the MK statistic, S, is given by:

$$S = \sum_{i=1}^{n-1} \sum_{j=i+1}^{n} \operatorname{sign}(x_j - x_i)$$

Under the null hypothesis that there is no trend within the time series:

$$\mathrm{E(S)} = 0,$$

$$\mathrm{Var(S)} = \mathrm{n(n-1)(2n+5)/18}.$$

The test statistic is the standardised value calculated as

$$Z = \begin{cases} \frac{S-1}{\sqrt{\mathrm{Var}(S)}} & \text{if } S > 0 \\ 0 & \text{if } S = 0 \\ 0 & \text{if } S < 0 \end{cases}$$

The MK test has been widely used for the trend analysis of climatologic and hydrologic time series because: (i) it is a non-parametric test and does not require the data to be normally distributed, and (ii) the test has low sensitivity to abrupt breaks due to inhomogeneous time series. According to this test, the null hypothesis H0 assumes that there is no trend (the data is independent and randomly ordered) and this is tested against the alternative hypothesis H1 assuming that there is a trend. It has been demonstrated that the presence of serial correlation may lead to an erroneous rejection of the null hypothesis (Type I error; e.g. Yue and Pilon 2003). Here, detrending was accomplished by using the so called Zhang's method, described in Wang and Swail (2001). It is known that the prewhitenning reduces the rejection rate when there is no trend. However, it has been found that this may cause the deflation of an existing trend (Yue and Wang 2002), so that a trend that is not strong enough may escape detection. To check for the influence of prewhitening on the results, both original data as well as prewhitened data were analysed, and only the results obtained with the prewhitening approach were taken into consideration.

The MK Z statistic and the significance (i.e. *p*-value) were calculated from the available time series for each flood indicator, applying the MK test to two fixed time periods (1951–2011 and 1971–2011) and to moving windows (with a minimum window length of 30 years) following the approach proposed by Hannaford et al. (2013) and Ruiz-Villanueva et al. (2016).

3.2 *Flood Trend Attribution*

The goal of trend studies should not be just the detection of these changes in recorded time series and the discussion of possible causes, but they should be a tool for testing hypothesis about the influences of these drivers on floods (Merz et al. 2012). The detected trends were explained by the correlation with changes in another variable (meteorological variables, such as intense precipitation indices, maximum daily precipitation, number of days with snow cover and cyclonic circulation types, and external drivers). In the large-scale climate variability, different climate indices were analysed: the North Atlantic Oscillation (NAO), the East Atlantic (EA), the East Atlantic/Western Russia (EA/WRUS) and the Scandinavia

Table 3 Large-scale climate indices, meteorological indices and terrestrial factors analysed in this study

	Abbreviation	Explanation
Large-scale climate indices	NAO	North Atlantic Oscillation
	EA	East Atlantic
	WP	West Pacific
	EP-NP	East Pacific-North Pacific
	EA/WRUS	East Atlantic/Western Russia
	PNA	Pacific/North American
	SCA	Scandinavia pattern
	TNH	Tropical/Northern Hemisphere
	POL	Polar/Eurasia pattern
	PT	Pacific transition
Meteorological indices	MDP	Maximum daily precipitation
	Snow	Number of days with snow cover
	CCT	Cyclonic circulation types
External drivers	Urbanization	
	Deforestation/afforestation	
	Agricultural practices	
	River channelization	
	River incision	
	Construction of dikes or reservoirs	

pattern (SCAND). For this latter group of large-scale climate indices, we used correlation with the flood parameters.

In addition, changes in catchment and river characteristics were investigated, such as urbanization, deforestation and reforestation, alteration in agricultural management practices and construction of dikes or reservoirs. Table 3 summarizes all the factors analysed in this study.

4 Results and Discussion

4.1 Annual Maximum Discharge

The MK test applied at each site for the two fixed periods showed a majority of non-significant trends, although some stations showed significant trends (28 % of the stations for the period 1971–2011, and 33 % for 1951–2011, p-value < 0.2), most of them being positive trends. For the most recent period, other 14.3 % of the stations showed weak increasing tendencies (with 0.2 > p-value < 0.5 in the MK test) and other 21.4 % weak decreasing tendencies, while for the last 50 years, 22.2 % of the stations showed weak increasing tendencies. For the longest period, 50 % of the sites exhibited weak positive tendencies. The significant positive trends were found both in the first and the second group of stations, those installed in small headwater basins in the upper part of the mountainous area, but also in the larger basins. Figure 2 shows the exploratory analysis applied to the standardized annual maxima for all the studied stations.

As Fig. 3 shows, different periods are analysed depending on availability of data. For example, the analysis applied to the Czarny Dunajec River at the Nowy Targ station (starting in 1951) resulted in a graph with numerous combinations of 30-year time windows (large number of pixels), while the result for the Lepietnica at Ludźmierz (data starting in 1970) shows a graph with less number of pixels (less time window combinations).

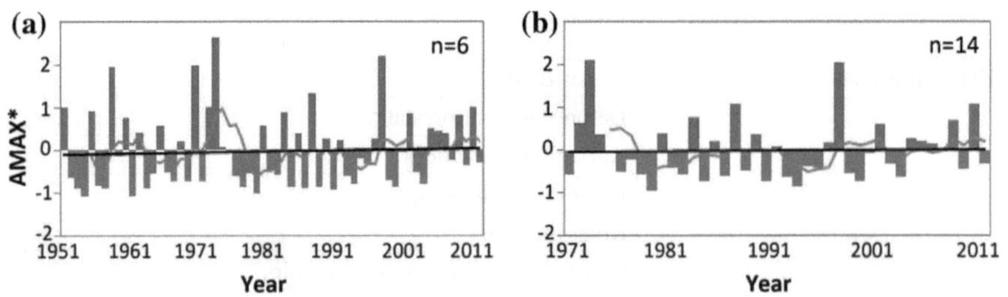

Fig. 2 Standardised annual maximum discharge (AMAX*) averaged over the studied stations for the periods 1951–2011 (**a**) and 1971–2011 (**b**). *Black line* is the linear trend, *grey curve* is the 5-year moving average, and n means the number of river catchments

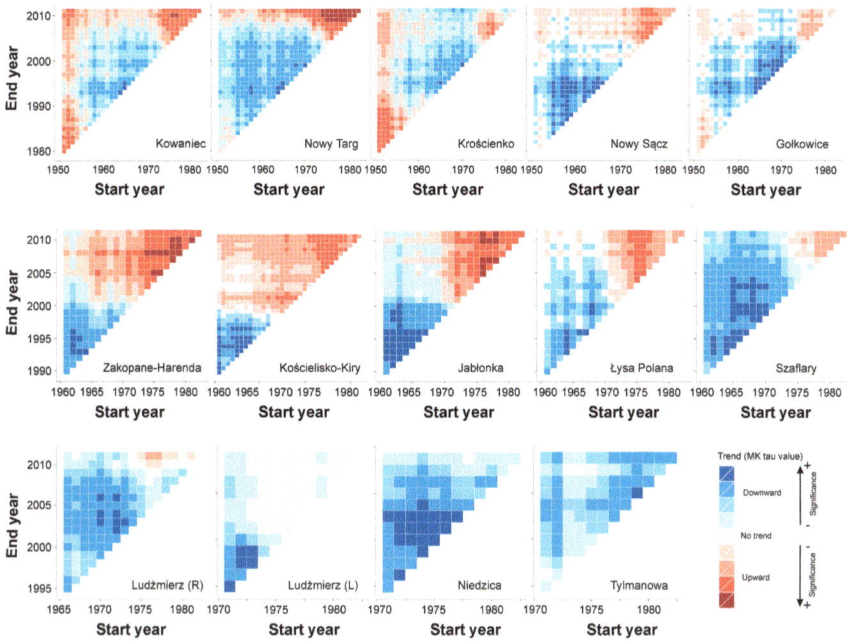

Fig. 3 Multi-temporal trend analysis for the annual maximum discharge (AMAX) for river catchments (for details on station names refer to Table 1). *Blue* and *red* cells correspond to negative and positive tau values, respectively (the darker the colour the more significant the trend)

Some conclusions can be extracted from this multi-temporal analysis. First of all, blue colours are more frequent than red colours, meaning that negative trends are more frequent than positive, especially for the most recent period (as shown on the Rogoźnik Stream at Ludźmierz and on the Lepietnica at Ludźmierz, the Niedziczanka at Niedzica and the Ochotnica Stream at Tylmanowa—Fig. 3). Second, positive trends (red colours) appear more significantly in few stations, but only in two cases the positive trend is stable over the longest time window (the Dunajec at the Kowaniec and Krościenko stations). In other cases, blue and red colours are divided along the time window, which represents a change in the general tendency, as shown in the Cicha Woda at Zakopane, the Kościeliski Stream at Kościelisko and the Czarna Orawa at Jabłonka.

4.2 *Flood Frequency and Seasonality*

The recorded data revealed an annual mean of 3 events with discharges higher than the third quartile (POTF) for the entire region. Some years (i.e., 1955, 1962, 1965, 1974, 1978, 1980, 1985, 1989, 2006, and 2010) were particularly rich in flood

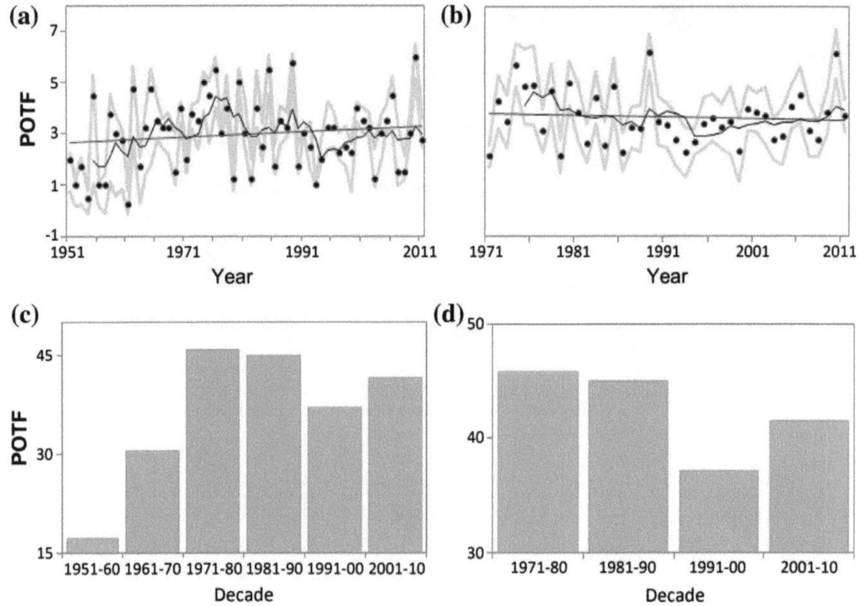

Fig. 4 Average peak over threshold (3rd quartile) frequency (POTF) plots for the periods: 1951–2011 (**a**), 1971–2011 (**b**). The plots show the mean value (*black circles*), standard deviation (*grey borders*), 5-year moving average (*black curve*) and linear regression (*black line*). Decadal averaged POTF for the periods: 1951–2011 (**c**), 1971–2011 (**d**)

events, with more than 4 events on average, and even 6 events in 2010. In contrast, in some other years (i.e., 1952, 1954, 1956, 1957, 1961, 1979, and 1993) none or just one POTF event was recorded. Interestingly, the decades when the most extreme events were recorded are not always those richest in flood events, as shown during the 1990s.

The averaged POTF for the entire studied region reveal an increasing trend (up to 35 %) of the number of events for the longest time period 1951–2011, but this tendency seems to decrease for the recent past (Fig. 4).

At-site results of the MK test for the long-term period (1951–2011) showed positive trends of POTF for 50 % of the sites, while for the most recent fixed period (1971–2011) 21 % of the sites revealed significant positive trends and 28 % negative trends (mainly present in the data from the larger basins). One example of these negative trends is illustrated by the Czarny Dunajec River (at Nowy Targ) where a reduction in the frequency and magnitude of flood flows was defined for the fourth quarter of the 20th century (1974–2000) as presented by Zawiejska and Wyżga (2010).

The seasonal discharge for the two fixed time periods illustrates importance of time window in the analysis of flood trends. As Fig. 5 shows, the EDA shows that winter discharge does not present any trend in the longest period (1951–2011), while it shows a decrease in the most recent period (1971–2011).

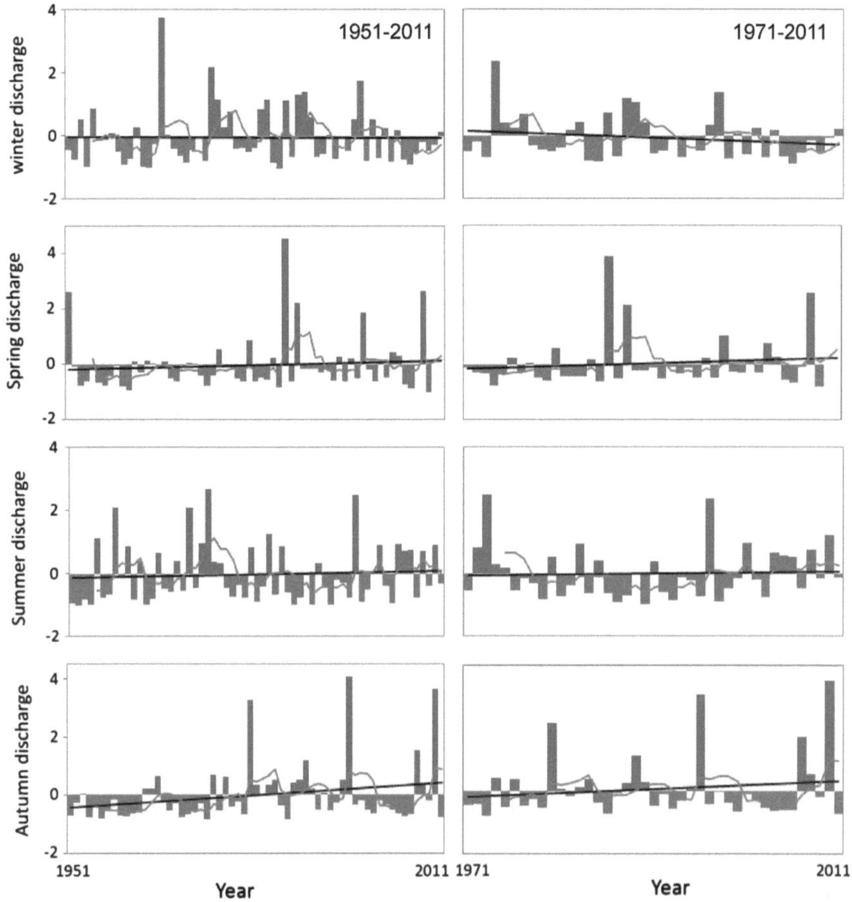

Fig. 5 Standardised seasonal maximum discharge averaged over the studied river catchments for the periods 1951–2011 and 1971–2011. *Black line* is the linear trend and *grey curve* is the 5-year moving average

The multi-temporal analysis allowed deciphering the variability (Fig. 6). Similar patterns in autumn and winter, and different in summer and spring may be observed. Darker colours show statistically significant trends, and they are more frequent in autumn and less frequent in spring. In spring, only two river catchments showed significant positive trends in the most recent period, and none for the longest period; otherwise weak, insignificant, trends were found. These stations, where strong trends were identified, belong to group 1 (small headwater basins). For autumn (SMA) significant positive trends were identified for 21 % of the river catchments in the period 1971–2011 and 50 % for the longest period, whereas in only 14 % downward trends were detected for the recent past. Although strong positive discharge trends are visible for the longest period of the autumn season, this changes to

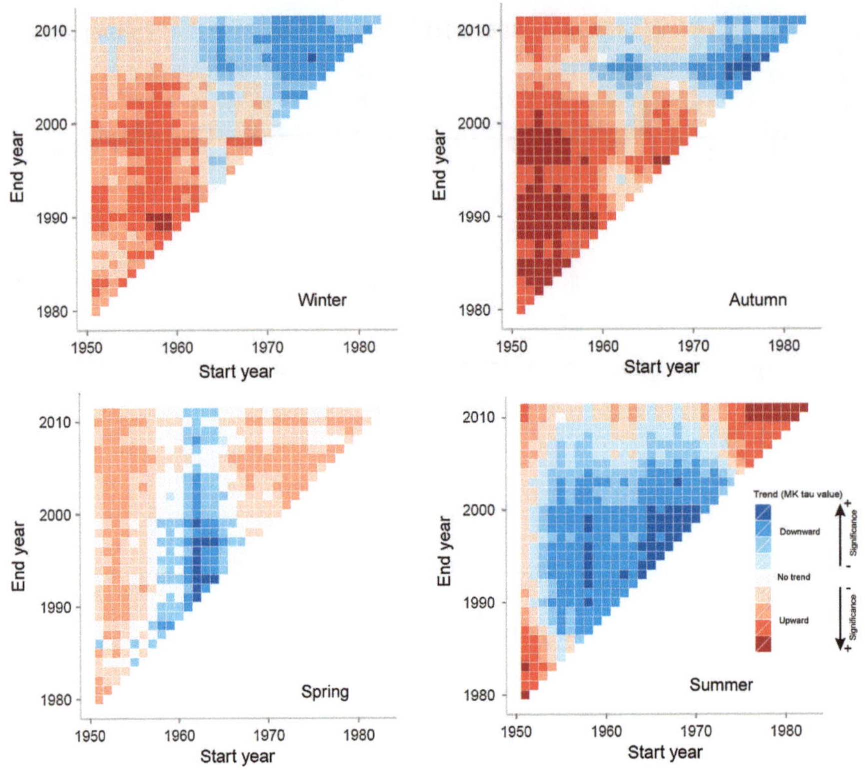

Fig. 6 Multi-temporal trend analysis for the maximum discharge in particular seasons. Legend explained in Fig. 3

a negative trend in the recent last years (Fig. 6). This pattern is even stronger in winter, when 35.7 % of the stations showed significant negative trends for the period 1971–2011, whereas only one significant positive trend and two weak negative tendencies were found for the years 1951–2011. The significant negative trends were identified in both the large and small basins. For the summer season, only 14 % of the stations showed significant positive trends for the last 40 years. The same was previously observed for the Elbe and Oder Rivers, where a significant decrease in winter floods (snowmelt related) was detected, while no significant change in summer floods was identified (Mudelsee et al. 2003). And changes of high river discharge in 20 river catchments in the Dunajec River basin revealed positive but not statistically significant trends of the majority of the runoff time series studied (1971–1990, 1984–2003, 1951–2003). The analysis was based on the normalized daily discharge data from 20 catchments with areas ranging from 58.4 to 5316 km^2. The high river discharge was defined as discharge greater than the fifth percentile of daily river discharges in a year or in a half-year seasons in particular years (Kasina et al. 2006–2007).

4.3 Summary of Flood Trends

The presented results showed that there has been no ubiquitous increase in the flood frequency and/or magnitude over the second half of the 20th century for the entire region analysed. However, significant trends, both positive and negative, were detected for a considerable proportion of the studied stations (Ruiz-Villanueva et al. 2016). In a regional perspective, decreasing trends of annual maxima (especially for the headwater small basins) and winter seasonal maxima were slightly dominant in the northern foreland of the Tatra Mountains during the last 60 years, while increasing frequency of floods was observed. It is also evident that changes in floods were not uniformly distributed between sites and between seasons. It is partially consistent with earlier studies conducted at various scales. Kundzewicz et al. (2004, 2005) examined a set of 195 long-time series of annual values of maximum daily discharge worldwide, therein 70 series from European rivers, of which only 20 showed statistically significant changes. Therefore, the potential response of river runoff to climate change in Poland is still uncertain (Pociask-Karteczka 2011). However, changes in the hydrological regime, as expressed by the increasing contribution of mean and high rates of discharge and a decreasing contribution of low rates of discharge have been observed, particularly in the second part of the 20th century (Pociask-Karteczka et al. 2010; Pociask-Karteczka 2014).

There have been many other national studies analysing long time-series of high river flow records in Europe, aimed to detect signals of change. There are 14 national and regional chapters in the book edited by Kundzewicz (2012), examining the changes in flood hazard in European countries and regions, but, typically, no ubiquitous, geographically organized, uniform increasing patterns of climate-driven changes in the flood magnitude/frequency could be detected. However, less frequent is the multi-temporal analysis which shows the inter-decadal variability and highlights considerable variability of trends in time (Petrow et al. 2009; Schmocker-Fackel and Naef 2010). According to Hannaford et al. (2013), fixed periods are essential for trend analysis, but a question could arise as to how representative is the found trend? With the approach used in this study, the decadal variability and determining some representative changes over long timescales were investigated.

4.4 Potential Drivers of Flood Trends

4.4.1 Meteorological Drivers

According to previous studies, after a dry period between 1951 and 1957 a frequency of high daily precipitation totals increased in the years 1958–1980 (Niedźwiedź et al. 2015). Analysis of the number of days with precipitation sum of ≥ 50 mm in Poland in the years 1971–2000 leads to the conclusion that the number of such days is growing at a rate of 2 days per decade, especially in the southern and central parts and locally in the north of the country. These trends are persistent in the first decade of the 21st century. Moreover, the number of days with precipitation sum of ≥30 mm is also increasing, while the number of days with precipitation sum of ≥20 mm is increasing almost all over the country at a rate of 4 days per decade (Lorenc et al. 2012). This increase was associated with an occurrence of the highest flood discharges in the region. The last decade may also be considered wet. In turn, decreases in maximum daily rainfall in the years 1981–1996 resulted in a significant lowering of flood discharges of the studied rivers. This would be in agreement with our findings regarding the frequency of flood events above the threshold of the third quartile, although they do not entirely explain discovered trends.

Statistically significant trends were not found for the maximum 24 h precipitation (Łupikasza 2010), although some short-term fluctuations might be evident with clustering of extreme events (Starkel 2001). These findings mean that annual high precipitation is not the only factor responsible for the flood variability in the studied region. The occurrence of floods also depends on seasonal precipitation totals. Recent data indicates a clear decrease in the summer precipitation (JJA) since 2011 with the summer of 2015 being the driest of the last 65 years. In summer 2015 a seasonal total precipitation reached only 182 mm (34 % of the 1951–2015 total) in Zakopane while 281 mm (46 % of the average) at the Kasprowy Wierch station. On the other hand, extremely high precipitation generating flood can also occur during dry periods.

The number of days with snow cover and maximum snow depth might be important to the streamflow contribution, especially in spring when snow-melt floods are expected. A previous study revealed a statistically significant decreasing trends in the number of days with snow cover for the period 1961–1990 (Falarz 2002), while at the decadal scale the duration of snow cover was more stable particularly in the years 1954–1960, 1970–1980 and 1990–1998 (Falarz 2013). The decrease in the number of days with snow cover could mean that snow melting takes place earlier in the year, which corresponds with a slight increase in flood discharge in spring and a decrease in winter, as it has been observed.

There is no significant trend in the summer discharge, however, the frequency of floods in summer is usually linked to three cyclonic circulation types (Niedźwiedź 2003a, b): north cyclonic (Nc), north-east cyclonic (NEc) and cyclonic trough (Bc), and the frequency of these types of air circulation increased by about 20 % over the second half of the 20th century (Niedźwiedź et al. 2015). There is a significant decrease ($p < 0.001$) in meridional westerly circulation index (Wi) which is inversely correlated with the frequency of the summer flood precipitation over the Tatra Mountains. Decreasing trends in this index mean the intensification of air advection from the eastern sector including the air advection from NE direction. A significant increase in the frequency of cyclonic situations is noticeable in the warm half-year (May–October). These trends may indicate a possible increase in the occurrence of floods. However, flood precipitation occurrence is determined, among others, by synoptic conditions in particular years and may not follow the general trends. For example, extremely dry 2015 year was accompanied by an exceptionally low value of cyclonicity index (Niedzwiedz and Lupikasza 2016).

Temperature is a factor indirectly influencing streamflow as it is responsible for the evaporation and moisture. The study carried out by Przybylak et al. (2007) described an increase of the maximum temperature in terms of average annual values and the number of days exceeding commonly used threshold values, i.e., 25 °C (hot days), 30 °C (very hot days) and 35 °C (extremely hot days). According to this work, this trend is particularly strong from the beginning of the 1990s, while extremely hot days occurred very rarely prior to 1985. Summer 2015 brought exceptionally high frequency of hot days in the southern Poland that reached 51 days in Katowice and 34 days in Zakopane. In that year, the number of very hot summer days exceeded 28 days in Katowice. Przybylak et al. (2007) also analysed the moisture index (defined as the difference between precipitation and potential evapotranspiration) and they reported statistically significant increasing trends of extremely dry and very dry days, with an increase of 1–2 days before the 1990s to greater than 5 days afterwards. On the other hand, they observed statistically significant decreasing trends of the extremely wet days and very wet days, decreasing from 3 to 6 days in the period 1951–1980 to 1–3 days for the last 25 years. These findings could explain the trends discovered in the annual maximum discharge data sets.

4.4.2 Large-Scale Climate Drivers

Ten indices characterizing the large-scale climate variability patterns for the studied region (Table 3) were correlated with the flood parameters (for details, see Ruiz-Villanueva et al. 2016). Four out of the 10 indices were selected: the East Atlantic (EA), the East Atlantic/Western Russia (EA/WRUS), the North Atlantic Oscillation (NAO) and the Scandinavia pattern (SCA). The criterion for selection was the strength of the links between these indices and the flood variables found in our study or indicated in literature. Table 4 shows Pearson correlation coefficient (R) computed between selected climatic indices and flood-related variables for the two studied periods.

The EA pattern shows a general upward tendency for the period 1951–2011 towards the positive phase that is associated with above-average precipitation over northern Europe. A slight positive correlation of EA with flood magnitudes (AMAX) was found.

In contrast, the EA/WRUS pattern shows a linear negative trend for the last 60 years, with a tendency of its negative phase in the annual scale (related to above-average precipitation), and in all seasons except winter when the trend is towards its positive phase (reflecting below-average precipitation over Europe). In effect, in the annual scale, a positive correlation with flood magnitudes (AMAX) in the region has been revealed, while the negative trend of winter floods could be linked to the winter trend of this climate index. Similarly to EA/WURS, the SCA pattern also shows a negative linear trend, possibly related to above-average

Table 4 Pearson correlation R computed between selected climatic indices and analysed hydrological variables for the two studied periods

Index		Correlation (Pearson's R, Kendall's tau significance)			
		1951–2011		1971–2011	
Annual	Tendency	AMAX	POTF	AMAX	POTF
NAO	Up	**−0.19**	**−0.18**	**−0.35**	**−0.32**
EA	Up	+0.11	No cor.	+0.025	No cor.
EA/WRUS	Down	+0.13	+0.04	+0.051	**+0.24**
SCA	Down	−0.045	+0.11	−0.075	**+0.20**
Summer		SMSu		SMSu	
NAO	Down	−0.018		−0.04	
EA	Up	+0.07		+0.092	
EA/WRUS	Down	**+0.25**		+0.13	
SCA	Down	**+0.33**		**+0.36**	
Spring		SMSp		SMSp	
NAO	Up	+0.01		+0.014	
EA	Up	+0.037		+0.071	
EA/WRUS	No trend	−0.058		−0.033	
SCA	Down	−0.069		−0.05	
Autumn		SMA		SMA	
NAO	Down	**−0.22**		**−0.25**	
EA	Up	+0.085		+0.021	
EA/WRUS	Down	−0.16		**−0.27**	
SCA	Down	+0.031		+0.053	
Winter		SMW		SMW	
NAO	Up	+0.11		**+0.16**	
EA	Up	**+0.22**		**+0.19**	
EA/WRUS	Up	+0.14		+0.12	
SCA	Down	−0.11		−0.09	

Tendency represents the linear trend observed in the EDA. Statistically significant correlations (based on Kendall tau test p-value < 0.05) are shown in bold. Notation explained in Table 3

precipitation across central and southern Europe. However, no correlation with examined flood variables for the studied region was found, although a link between extreme discharges of some Carpathian rivers and the SCA index has been noted in other studies (Pociask-Karteczka et al. 2003; Pociask-Karteczka 2007).

The NAO pattern identified as one of the essential patterns of climate variability in the Northern Hemisphere, exhibited an increasing tendency (to the positive phase) at the annual scale for a longer period (1951–2011) that seems to slow down over the last decades. Strong positive phases of the NAO tend to be associated with below-average precipitation over southern and central Europe. Both examined flood characteristics (AMAX and POTF) are significantly correlated with negative NAO phases. This would explain the increase of both flood variables over the longer period (1951–2011) and the decrease of flood frequency in the shorter period (1971–2011). At the seasonal scale, the increase in flood magnitude in autumn is also related to more persistent negative NAO phase, whereas the decrease in the magnitude of winter floods is related to an upward trend identified in the seasonal NAO index (associated with less snow cover).

A lag between the response of river discharge and the behaviour of the NAO index can be apparent at a monthly resolution (cf. Trigo 2011). This aspect was already investigated by relating spring maxima to winter NAO index to study snowmelt-induced floods in Poland (Kaczmarek 2003). For the Dunajec catchment, Pociask-Karteczka and Nieckarz (2010) found various relationships between 10-day high flows in spring and summer and the NAO index in preceding winter. In order to analyse this possible lag, we plotted AMAX together with the values of seasonal NAO index and analysed possible correlations for the longer period. We found a negative correlation between AMAX and NAOW (NAO winter) and positive, statistically significant, correlation with NAOA (NAO autumn). Therefore, extremely low NAOW value corresponds with the river flow increase. This is in agreement with the findings of Pociask-Karteczka (2007) who observed that considerable flooding of the Vistula and Odra occurs after winters for which the values of the NAO index are extremely low.

It may be concluded that the interpretation of links between large-scale climate anomalies and flood indices in the studied region remains a challenging research area that requires further studies since flood management practices are likely to benefit from improved medium-term (e.g. seasonal) weather forecasts that depend on forecasting of these indices, and particularly of the NAO index (Salgueiro et al. 2013). Nevertheless, it must be remembered that heavy, flood-inducing precipitation in the Upper Vistula Basin depends more on local air circulation than on macroscale circulation. Despite atmospheric circulation, orography is of crucial significance in this complex terrain.

4.4.3 Other Potential Drivers

The results have indicated the relationships between flood trends and changes in some atmospheric variables, but the latter only partly explain the variability of streamflow. Therefore, non-climatic changes in catchment and river parameters must also be taken into account.

During the past decades some environmental changes occurred in the studied catchments that influence the conditions of flood runoff (Table 5). Land use and land cover changed and forestry has been changing in the northern foreland of the Tatra Mountains, mainly due to transformations of land use systems after 1990 (e.g. land reforms, policy changes, socioeconomic transformation, accession to the European Union, international agreements) and land use legacies from Austro-Hungarian and Socialist times. Main processes include farmland abandonment and forest expansion and agricultural parcelization. A contribution of arable land decreased considerably in favour of forested areas (Wyżga et al. 2012), with the land use changes slowing down runoff and reducing peak flows. In the former Czechoslovakia where a large part of the Poprad catchment is located, collectivization of farms shortly after the World War II resulted in the formation of large plots of cultivated land that accelerated runoff processes, especially as a consequence of rapid development of gullies (Stankoviansky and Barka 2007). Development of the town of Zakopane increased the share of paved surfaces in the catchment of Cicha Woda, hence accelerating the runoff from the area. Since 2004 a few events of widespread windthrow of mature spruce forest have occurred in the Slovak and Polish parts of the Tatra Mountains, followed by bark beetle infestation (Kopecká and Nováček 2009). As the forest damage occurred in the area with high intensities and totals of precipitation, it must have increased and accelerated runoff from the deforested hillslopes, especially during a few years after the events.

Another group of impacts on the conditions of flood runoff encompasses direct human interventions in the studied river channels (Table 5). Channelization works on the Dunajec, carried out in the 1950s–1970s, resulted in the narrowing and considerable simplification of its channel (Zawiejska and Wyżga 2010). Since the

Table 5 Summary of main regional environmental changes since the mid-20th century and their effects on flood runoff

Regional environmental change	Time of change	Effects on floods
Reduction in the proportion of arable land coupled with increase in forest cover	Second half of the 20th century	Change in albedo, increased evapotranspiration, slowed down runoff from hillslopes
Farm collectivization (Slovak part of the Poprad catchment)	The early 1950s	Enhanced slope gullying due to the elimination of agricultural terraces, accelerated runoff from hillslopes
Urbanization	Second half of the 20th century	Significant effect on the Cicha Woda catchment where the town of Zakopane is located
Widespread forest windthrows followed by bark beetle infestation in the Tatras	2004 and the following years	Change in albedo, decreased evapotranspiration, accelerated water runoff from hillslopes
Large-scale, in-channel gravel mining	The 1950s–1960s	Channel incision leading to the concentration of flood runoff in river channels and the reduction of water storage in floodplain areas
River training works (simplification of channel pattern, channel narrowing)	The 1950s–1970s in the Dunajec, since the 1960s in headwater watercourses	Shortening of the travel time of flood waves, channel incision, both leading to increased peak flows in the downstream reaches
Construction of Czorsztyn Reservoir	1997	Reduction of peak flows of flood waves

1960s streams and rivers in the headwater catchments have also been subjected to channelization works that substantially (up to a fifth of the original value) narrowed the watercourses and replaced their former multi-thread channels by single, artificial ones along a majority of their courses (Korpak 2007; Zawiejska and Wyżga 2010; Wyżga et al. 2012).

During the 1950s–1960s, large-scale gravel mining was conducted in rivers, especially in the Czarny Dunajec. In the following decades, it was replaced by widely distributed, illegal extraction of cobbles from the river channels (Zawiejska and Wyżga 2010; Zawiejska et al. 2015). A considerable shrinkage of sediment available for fluvial transport, resulting from the gravel mining, together with an increase in transport capacity of the watercourses caused by their channelization (Wyżga 2001) induced rapid channel incision (Wyżga 2008; Wyżga et al. 2016). To date, up to 3.5 m of bed degradation have occurred in the rivers of the study area (Zawiejska and Wyżga 2010). The effect of channel incision on the resultant increase in channel capacity and reduction in the frequency of floodplain inundation was especially large in upper river courses where the initial capacity of their channels was small (Wyżga et al. 2016). Flood magnitudes in the foothill and foreland reaches of Polish Carpathian rivers were demonstrated to have increased as a result of upstream channel straightening and incision (Wyżga 1997, 2008).

The above discussion indicated complex temporal and spatial patterns of the impacts of non-climatic environmental changes on flood runoff from the catchments in the northern foreland of the Tatra Mountains. In the headwater catchments, runoff processes have been predominantly affected by the counteracting effects of forest cover increase that gradually progressed over the second half of the 20th century (but was interrupted by the windthrow events from the last 10 years), and those of channelization and channel incision. In the group of the five largest catchments, the scale of catchment reforestation diminished towards lower areas more suitable for agriculture, and thus its effect on runoff most likely lowered with increasing catchment size. In turn, the effects of channelization and channel incision on flood runoff should increase with the length of modified channels, hence being more pronounced in the stations located downstream. While the flood record in the largest catchments integrates the effects of all upstream-operating factors, analysis of the results for individual stations may reveal some specific drivers. For instance, a significant negative trend in flood frequency coupled with a significant positive trend in flood magnitude was found for the Dunajec River at Gołkowice. The station is located a few tens of kilometres downstream of the Czorsztyn Reservoir that started to operate in 1997. The reduction in the frequency of largest floods is a typical effect of the operation of dam reservoirs. At the same time, the increase in flood magnitude at some distance from the reservoir might be attributed to the loss of floodplain water storage along the river reach (Wyżga et al. 2016) where deep channel incision took place in the second half of the 20th century (Zawiejska and Wyżga 2010).

5 Concluding Remarks

The present chapter aims to analyse the inter-decadal variability of magnitude, frequency, and seasonality of floods in the Tatra Mountains foreland, on decadal timescales, since the mid-20th century. A multi-temporal approach was employed and trends were fitted to several combinations of start and end years in a record. This allowed identifying the extent to which observed trends in two fixed periods are influenced. The presented results showed that there has been no ubiquitous increase in the flood frequency and/or magnitude over the second half of the 20th century for the entire region analysed. However, even if a majority of non-significant trends were found, significant trends (both positive and negative) were also detected for a considerable number of the studied river catchments. Analysis was made for annual as well as seasonal values. It was found that changes in floods were not uniformly distributed between sites and between seasons (more details in Ruiz-Villanueva et al. 2016).

In order to explain the detected trends, the flood parameters of interest were correlated with 10 indices describing the large-scale climate variability for the Northern Hemisphere and compared with trends found in intense precipitation indices, number of days with snow cover, cyclonic circulation types, air temperature and moisture conditions. Links with four indices, i.e., NAO, EA, EA/WRUS, SCA, expressed by Pearson correlation, were interesting and promising. Other external, non-climatic, drivers, such as catchment and channel changes were also considered in the studied region. Results show that floods in the area exhibit considerable inter-decadal variability, however, certain patterns are apparent. Less frequent floods, although perhaps more extreme, are now likely to occur, with a shift in the seasonality, decreasing flood magnitudes in winter and increasing during autumn and spring.

According to Merz et al. (2012) it may be stated that different drivers, climatic and non-climatic, act in parallel and interact in a catchment, while changes in flood behaviour are the integral response of the catchment to these different drivers and to their interactions.

Acknowledgments This work was supported by the project FLORIST (Flood risk on the northern foothills of the Tatra Mountains; supported by a grant from the Swiss Government through the Swiss Contribution to the enlarged European Union (PSPB No. 153/2010). Authors thank Professor Joanna Pociask-Karteczka for the revision which improved the early version of this chapter.

References

Allamano P, Claps P, Laio F (2009) An analytical model of the effects of catchment elevation on the flood frequency distribution. Water Resour Res 45:W01402

Barnett T, Adam J, Lettenmaier D (2005) Potential impacts of a warming climate on water availability in snow-dominated regions. Nature 438:303–309

Barredo JI (2007) Major flood disasters in Europe: 1950-2005. Nat Hazards 42(1):125–148
Barredo JI (2009) Normalised flood losses in Europe: 1970–2006. Nat Hazards Earth Syst Sci 9:97–104
Bates B, Kundzewicz Z, Wu S, Palutikof J (eds) (2008) Observed and projected changes in climate as they relate to water. In: Climate change and water, IPCC Tech Pap VI, Intergovernmental Panel on Climate Change and Secretary, Geneva, Switzerland, pp 13–31
Birsan MV, Molnar P, Burlando P, Pfaundler M (2005) Streamflow trends in Switzerland. J Hydrol 314:312–329
Cebulak E, Niedźwiedź T (2000) Zagrożenie powodziowe dorzecza górnej Wisły przez wysokie opady atmosferyczne (The flood hazard in the upper Vistula basin by the high atmospheric precipitation), Monografie Komitetu Gospodarki Wodnej PAN 17, Oficyna Wydawnicza Politechniki Warszawskiej, Warszawa, pp 55–70 (in Polish)
Diaz H, Eischeid J, Duncan C, Bradley R (2003) Variability of freezing levels, melting season indicators, and snow cover for selected high-elevation and continental regions in the last 50 years. Clim Change 59:33–52
Falarz M (2002) Long-term variability in reconstructed and observed snow cover over the last 100 winter seasons in Cracow and Zakopane (southern Poland). Clim Res 19:247–256
Falarz M (2013) Seasonal stability of snow cover in Poland in relation to the atmospheric circulation. Theoret Appl Climatol 111:21–28
Hannaford J, Buys G, Stahl K, Tallaksen M (2013) The influence of decadal-scale variability on trends in long European streamflow records. Hydrol Earth Syst Sci 17:2717–2733
Helms M, Büchele B, Merkel U, Ihringer J (2002) Statistical analysis of the flood situations and assessment of the impact of diking measures along the Elbe (Labe) river. Journal of Hydrology 267:94–114
IPCC (2009) IPCC expert meeting on detection and attribution related to anthropogenic climate change. The World Meteorological Organization Geneva, Switzerland 14–16 Sept 2009
Jahn A (1979) On the Holocene and present-day morphogenetic processes in the Tatra Mountains. Stud Geomorph Carpatho-Balcan 13:111–129
Kaczmarek Z (2003) The impact of climate variability on flood risk in Poland. Risk Anal 23:559–566
Kasina M, Pociask-Karteczka J, Nieckarz Z (2006–2007) Tendencies of high flows in the Dunajec River basin in the second half of the XX c. Folia Geogr Series Geogr-Phys 37–38:3–33 (in Polish)
Kopecká M, Nováček J (2009) Forest fragmentation in the Tatra Region in the period 2000-2006. Land Anal 10:58–63
Korpak J (2007) The influence of river training on mountain channel changes (Polish Carpathian Mountains). Geomorphology 92:166–181
Kron W (2012) Changing flood risk – a reinsurer's viewpoint. In: Kundzewicz ZW (ed) Changes in flood risk in Europe. IAHS Press, Wallingford, pp 459–490
Kundzewicz ZW (ed) (2012) Changes in flood risk in Europe, Special Publication No. 10. IAHS Press, Wallingford, Oxfordshire, UK
Kundzewicz ZW, Robson A (eds) (2000) Detecting trend and other changes in hydrological data. World Climate Programme—Water, World Climate Programme Data and Monitoring, WCDMP-45, WMO/TD:1013, Geneva, May 2000
Kundzewicz ZW, Robson AJ (2004) Change detection in river flow records—a review of methodology. Hydrol Sci J 49:7–19
Kundzewicz ZW, Schellnhuber HJ (2004) Floods in the IPCC TAR perspective. Nat Hazards 31:111–128
Kundzewicz ZW, Graczyk D, Maurer T, Przymusińska I, Radziejewski M, Svensson C, Szwed M (2004) Detection of change in world-wide hydrological time-series of maximum annual flow WCASP-64. WMO, Geneva
Kundzewicz ZW, Graczyk D, Maurer T, Przymusińska I, Radziejewski M, Svensson C, Szwed M (2005) Trend detection in river flow time-series: 1 annual maximum flow. Hydrol Sci J 50:797–810

Kundzewicz ZW, Dobrowolski A, Lorenc H, Niedźwiedź T, Pińskwar I, Kowalczak P (2012) Floods in Poland. In: Kundzewicz ZW (ed) Changes in flood risk in Europe. Intern Assoc Hydrol Sci Spec Publ 10:319–334

Kundzewicz ZW, Stoffel M, Kaczka RJ, Wyżga B, Niedźwiedź T, Pińskwar I, Ruiz-Villanueva V, Łupikasza E, Czajka B, Ballesteros-Canovas JA, Małarzewski Ł, Choryński A, Janecka K, Mikuś P (2014) Floods at the northern foothills of the Tatra Mountains—a Polish-Swiss research project. Acta Geophys 62:620–641

Kundzewicz ZW, Krysanova V, Dankers R, Hirabayashi Y, Kanae S, Hattermann FF, Huang S, Milly PCD, Stoffel M, Driessen PPJ, Matczak P, Quevauviller P, Schellnhuber H-J (2016) Differences in projections of changes in flood hazard in Europe—their causes and consequences for decision making. Submitted to Hydrol Sci J (in review)

Łajczak A (1988) Precipitation and runoff in the Polish Tatra Mts in the light of many-year measurements. Czas Geogr 59:137–170 (in Polish)

Lammersen R, Engel H, van de Langemheen W, Buiteveld H (2002) Impact of river training and retention measures in flood peaks along the Rhine. J Hydrol 267:115–124

Lindström G, Bergström S (2004) Runoff trends in Sweden 1807-2002. Hydrol Sci J 49:69–83

Lorenc H, Cebulak E, Głowicki B, Kowalewski M (2012) The structure of heavy rainfall causing danger to society, the environment and the Polish economy. In: Lorenc H (ed) Natural disasters and the security of the country (in Polish)

Łupikasza E (2010) Spatial and temporal variability of extreme precipitation in Poland in the period 1951-2006. Intern J Climat 30:991–1007

Marty C (2008) Regime shift of snow days in Switzerland. Geophys Res Letters 35:L12501

Merz B, Kundzewicz ZW, Delgado J, Hundecha Y, Kreibich H (2012) Detection and attribution of changes in flood hazard and risk. In: Kundzewicz ZW (ed) Changes in flood risk in Europe, Special Publication No 10. IAHS Press, Wallingford, pp 435–458

Milly PCD, Betancourt J, Falkenmark M, Hirsch RM, Kundzewicz ZW, Lettenmaier DP, Stouffer RJ (2008) Stationarity is dead: whither water management? Science 319(5863):573–574

Mudelsee M, Börngen M, Tetzlaff G, Grünewald U (2003) No upward trends in the occurrence of extreme floods in central Europe. Nature 425(6954):166–169

Mudelsee M, Deutsch M, Börngen M, Tetzlaff G (2006) Trends in flood risk of the River Werra (Germany) over the past 500 years. Hydrol Sci J 51:818–833

Niedźwiedź T (2003a) Extreme precipitation events on the northern side of the Tatra Mountains. Geographia Polonica 76:13–21

Niedźwiedź T (2003b) The extreme precipitation in Central Europe and its synoptic background. Pap Global Change IGBP 10:15–29

Niedzwiedz T, Lupikasza E (2016) Change in atmospheric circulation patterns. In: Kundzewicz ZW, Stoffel M, Niedźwiedź T, Wyżga B (eds) Flood risk in the Upper Vistula Basin. Springer, Cham (this volume)

Niedźwiedź T, Łupikasza E, Pińskwar I, Kundzewicz ZW, Stoffel M, Małarzewski Ł (2015) Variability of high rainfalls and related synoptic situations causing heavy floods at the northern foothills of the Tatra Mountains. Theor Appl Climatol 119:273–284

Olsen R, Lambert J, Haimes Y (1998) Risk of extreme events under nonstationary conditions. Risk Anal 18:497–510

Palmer T, Räisänen J (2002) Quantifying the risk of extreme seasonal precipitation events in a changing climate. Nature 415:512–514

Petrow T, Merz B (2009) Trends in flood magnitude, frequency and seasonality in Germany in the period 1951–2002. J Hydrol 371:129–141

Petrow T, Zimmer J, Merz B (2009) Changes in the flood hazard in Germany through changing frequency and persistence of circulation patterns. Nat Hazards Earth Syst Sci 9:1409–1423

Pfaundler M (2001) Adapting, analysing and evaluating a flexible index flood regionalisation approach for heterogeneous geographical environments. Dissertation, Technische Wissenschaften ETH Zürich 14253

Pfister L, Kwadijk J, Musy A, Bronstert A. Hoffmann L (2004a) Climate change, land use change and runoff prediction in the Rhine–Meuse basins. River Res Appl 20:229–241

Pfister L, Drogue G, Idrissi AEl, Iffly JF, Poirier C, Hoffman L (2004b) Spatial variability of trends in the rainfall-runoff relationship: a mesoscale study in the Mosel Basin. Climatic Change 66:67–87

Pinter N, van der Ploeg RR, Schweigert P, Hoefer G (2006) Flood magnification on the River Rhine. Hydrol Process 20:147–164

Pociask-Karteczka J (2007) River hydrology and the North Atlantic Oscillation—a general review. Ambio 35:312–314

Pociask-Karteczka J (2011) River runoff response to climate changes in Poland (East-Central Europe). IAHS Publication 344, pp 182–187

Pociask-Karteczka J (2014) Alpine type of water circulation in the Tatra Mountains. In: Cieśliński R, Jereczek-Korzeniowska K (ed) Scientific research in water cycle in the 20th and 21st centuries, Gdańsk, pp 308–329 (in Polish)

Pociask-Karteczka J, Nieckarz Z (2010) Extreme flood events in the Dunajec River drainage basin (Carpathian Mts.). Folia Geogr Series Geogr-Phys 41:49–58

Pociask-Karteczka J, Nieckarz Z, Limanówka D (2003) Prediction of hydrological extremes by air circulation indices. IAHS Publication 280, pp 134–141

Pociask-Karteczka J, Baścik M, Nieckarz Z (2010) Spatial diversity and volatility runoff from the catchment area of the Tatra. In: Kotarba A (ed) The nature of the Tatra National Park and man. Science and the Management of the Tatras and Their Surroundings, Zakopane, pp 123–130 (in Polish)

Przybylak R, Vízi Z, Araźny A, Kejna M, Maszewski R, Uscka-Kowalkowska J (2007) Poland's climate extremes index, 1951–2005. Geographia Polonica 80:47–58

Radziejewski M, Kundzewicz ZW (2004) Detectability of changes in hydrological records. Hydrol Sci J 49:39–51

Robson AJ (2002) Evidence for trends in UK flooding. Phil Trans R Soc Lond A 360:1327–1343

Robson AJ, Jones TK, Reed DW, Bayliss AC (1998) A study of national trend and variation in UK floods. Int J Climatol 18:165–182

Ruiz-Villanueva V, Stoffel M, Wyżga B, Kundzewicz ZW, Czajka B, Niedźwiedź T (2016) Decadal variability of floods in the northern foreland of the Tatra Mountains. Reg Environ Change 16:603–615

Salgueiro MJM, Barriendos M, Garcia P, Benito G (2013) Flood magnitudes in the Tagus River (Iberian Peninsula) and its stochastic relationship with daily North Atlantic Oscillation since mid-19th century. J Hydrol 502:191–201

Schmocker-Fackel P, Naef F (2010) Changes in flood frequencies in Switzerland since 1500. Hydrol Earth Syst Sci 14:1581–1594

Slater LJ, Singer MB, Kirchner JW (2015) Hydrologic versus geomorphic drivers of trends in flood hazard, Geophys Res Lett 42:370–376

Stahl K, Hisdal H, Hannaford J, Tallaksen LM, van Lanen HA, Sauquet E, Demuth S, Fendekova M, Jódar J (2010) Hydrol Earth Syst Sci 14:2367–2382

Stankoviansky M, Barka I (2007) Geomorphic response to environmental changes in the Slovak Carpathians. Studia Geomorphologica Carpatho-Balcanica 41:5–28

Starkel L (2001) Extreme rainfall and river floods in Europe during the last millennium. Geographia Polonica 74(2):69–79

Svensson C, Kundzewicz ZW, Maurer T (2004) Trends in flood and low flow series (POT). WCASP-66. WMO Geneva

Svensson C, Kundzewicz ZW, Maurer T (2005) Trend detection in river flow series: 2. Flood and low-flow index series. Hydrol Sci J 50:811–824

Trigo RM (2011) The impacts of the NAO on hydrological resources of the Western Mediterranean. In: Vicente-Serrano SM, Trigo RM (eds) Hydrological, socioeconomic and ecological impacts of the North Atlantic Oscillation in the Mediterranean Region. Adv Global Change Res 46:41–56

Wang XL, Swail VR (2001) Changes of extreme wave heights in Northern Hemisphere oceans and related atmospheric circulation regimes. J Climate 14:2204–2221
Wyżga B (1997) Methods for studying the response of flood flows to channel change. J Hydrol 198:271–288
Wyżga B (2001) A geomorphologist's criticism of the engineering approach to channelization of gravel-bed rivers: Case study of the Raba River, Polish Carpathians. Environ Manage 28:341–358
Wyżga B (2008) A review on channel incision in the Polish Carpathian rivers. In: Habersack H, Piégay H, Rinaldi M (eds) Gravel-bed rivers VI: from process understanding to river restoration. Elsevier, Amsterdam, pp 525–555
Wyżga B, Zawiejska J, Radecki-Pawlik A, Hajdukiewicz H (2012) Environmental change, hydromorphological reference conditions and the restoration of Polish Carpathian rivers. Earth Surf Proc Landforms 37:1213–1226
Wyżga B, Zawiejska J, Radecki-Pawlik A (2016) Impact of channel incision on the hydraulics of flood flows: examples from Polish Carpathians rivers. Geomorphology 272C. doi:10.1016/j.geomorph.2015.05.017
Yue S, Pilon P (2003) Interaction between deterministic trend and autoregressive process. Water Resour Res 39:1077
Yue S, Wang CY (2002) Regional streamflow trend detection with consideration of both temporal and spatial correlation. Int J Climatol 22:933–946
Zawiejska J, Wyżga B (2010) Twentieth-century channel change on the Dunajec River, southern Poland: patterns, causes and controls. Geomorphology 117:234–246
Zawiejska J, Wyżga B, Radecki-Pawlik A (2015) Variation in surface bed material along a mountain river modified by gravel extraction and channelization, the Czarny Dunajec, Polish Carpathians. Geomorphology 231:353–366

Deciphering Flood Event Information from Tree-Ring Data in the Tatra Mountains: Implications for Hazard Assessment

Juan Ballesteros-Cánovas, Barbara Spyt, Karolina Janecka, Ryszard J. Kaczka and Markus Stoffel

Abstract Mountains and their foothills are areas where intense floods are characterized by high discharge and where they occur more often than in lowlands. Furthermore, due to the fact that they are mainly caused by short-lasting heavy rainfall events, they are difficult to predict. The dense network of meteorological stations and river gauges is crucial to understand hydrological processes and to forecast floods. Most mountain regions suffer from a scarcity of instrumental data that are long enough for scientific purposes. In such cases where historical data or instrumental records are lacking, floods in forested catchments can be analysed by using growth series from trees growing along stream channels. In the streams draining the northern slopes of the Tatra Mountains, it was possible to reconstruct the occurrence and magnitude of paleofloods using tree-ring data. More than 1100 increment cores were sampled from 218 *Picea abies* and *Abies alba* trees growing at 6 stream sectors and allowed determination of 480 growth disturbances and a definition of the magnitude of 47 flood events between A.D. 1866 and 2012. In this region, floods are triggered by intense and prolonged rainfall events in spring and

J. Ballesteros-Cánovas (✉) · M. Stoffel
Climatic Change and Climate Impacts, Institute for Environmental Sciences, University of Geneva, 1227 Carouge, Switzerland
e-mail: juan.ballesteros@dendrolab.ch

M. Stoffel
e-mail: markus.stoffel@dendrolab.ch

J. Ballesteros-Cánovas · M. Stoffel
Dendrolab.ch, Institute of Geological Sciences, University of Bern, 3012 Bern, Switzerland

B. Spyt · K. Janecka · R.J. Kaczka
Faculty of Earth Sciences, University of Silesia, 40007 Katowice, Poland
e-mail: barbaraspyt@wp.pl

R.J. Kaczka
e-mail: kaczka@wnoz.us.edu.pl; ryszard.kaczka@us.edu.pl

M. Stoffel
Department of Earth and Environmental Sciences, University of Geneva, 1205 Geneva, Switzerland

Z.W. Kundzewicz et al. (eds.), *Flood Risk in the Upper Vistula Basin*, GeoPlanet: Earth and Planetary Sciences, DOI 10.1007/978-3-319-41923-7_12

summer. These paleoflood records also allowed construction of a regional flood analysis and to reduce the uncertainties in the flood frequency assessment.

Keywords Dendrochronology · Proxy data · Flood information · Hazard assessment

1 Introduction

Floods are one of the most common natural processes worldwide. In mountain streams, the catchment predisposition and steep channel characteristics contribute to a quick hydrological response of catchment and very powerful channel responses. Floods in mountain streams, therefore, are characterized by recurrent, highly-turbulent and sediment-laden flows (Wohl 2000, 2006). These characteristics make mountain streams highly hazardous and flood events therein difficult to forecast (Borga et al. 2011, 2014). Floods in mountain environments have caused large amounts of losses and fatalities in the past, with dramatic and recurrent episodes in all mountain ranges worldwide (Weingartner et al. 2003; Kundzewicz et al. 2014a).

The scientific-technical analyses of floods in mountains require long-term records on both occurrence and flow magnitude (Brázdil et al. 2006). However, the widespread and characteristic scarcity of instrumental data and basic hydrological information in mountain terrains clearly hampers these analyses. Augmenting information about past flood events by using historical archives and indirect evidence in the paleorecord is therefore critical to assess flood hazard, evaluate discharge trends or decipher a link to climate change. Paleoflood records based on tree-ring analyses are, therefore, highly convenient in forested mountain terrains (Ballesteros-Cánovas et al. 2015a, b). The interaction between floods and vegetation growing along a mountain stream can leave datable evidence of past floods in tree rings, which allows tracking of past process activity with high spatial and temporal accuracy. Before the formal definition of paleohydrology (Baker 2008), Sigafoos (1964) already stated that the information contained in tree-ring records of riverine trees is unique and constitutes a valuable source of data for hydrological analysis (Ballesteros-Cánovas et al. 2014). Since then, several efforts have been done to improve the tree-ring analyses to determine past flood records (Ballesteros-Cánovas et al. 2015b for a recent review). All these experiences support the initial hypothesis launched by Sigafoos (1964) about the economic value of tree-rings records for paleoflood analyses, specifically in ungauged or poorly gauged sites. However, a pedagogic explanation about the utility of this kind of information is still needed within the hydrological community (Baker 2008).

Particularly in the Tatra Mountains, flood risk is an important issue and has been a known problem for long time. However, despite the fact that first gauging stations were installed in several streams and rivers in the Tatra Mountains foreland in the late 19th century, a lack of flood processes interpretation is still recognized because of the highly fragmented and incomplete gauging network in the region (Fig. 1).

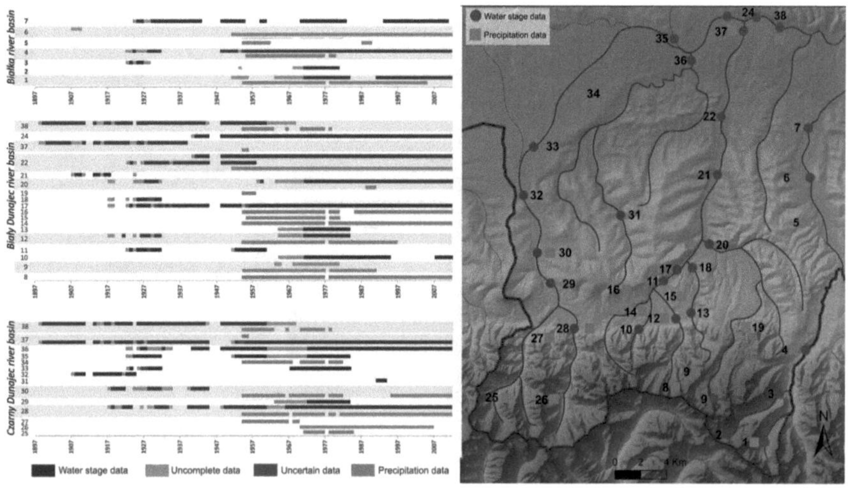

Fig. 1 Spatial and temporal distribution of meteorological and flow gauge dataset on the northern slope and foothills of the Polish Tatra Mountains. After Ballesteros-Cánovas et al. (2015a), published by permission of Elsevier

This problem is even larger as some crucial gauging stations have been relocated after the occurrence of intense flood events (e.g., in 1934), which destroyed several stations and prevented direct data comparison. Moreover, the political and administrative instability of the region—related to both World Wars—also resulted in several gauging network changes and significant losses of records (Szczepański et al. 1996).

Here we synthesize the major research outcomes obtained during the Florist project and report on the use of tree-ring series to decipher the flood history of the Tatra Mountains. We first present the foundation of paleoflood research based on tree rings, provide some fundamental bibliography and then summarize some of the key findings of Ballesteros-Cánovas et al. (2015a, 2016) dealing with the flood dating and peak discharge reconstruction in the region.

2 Fundamentals of Tree-Ring-Based Paleoflood Reconstructions

The methodological basis of paleoflood reconstructions based on tree rings is based on the "process–event–response" concept defined by Shroder (1978). This concept defines the *process* as any geomorphic agent, in this specific case floods, the *event* as a specific impact which the process can induce to a tree; and the *response* the tree's reaction to this specific disturbance. The main disturbances caused by floods include impact scars, tilted stems and abnormal stem morphologies, eroded roots,

Fig. 2 Examples of main dendrogeomorphic evidence related with floods in the Tatra Mountains: **a** scar on trees and **b** exposed roots

re-sprouting and anatomical abnormalities caused by prolonged inundation without needed to invoke mechanical damages (Stoffel and Bollschweiler 2009; Stoffel et al. 2010; Stoffel and Corona 2014). Hydrological processes also control riparian vegetation pattern; these have been used to understand the frequency and duration of flooding events (Kozlowski 2002; Kames et al. 2016).

At the level of forest stands, flood activity modulates the riparian forest. Extreme flood events may eliminate trees along the channel and flood plain. Survivor trees may benefit from less competition, more light, nutrients and water availability (Osterkamp et al. 2012; Stoffel and Wilford 2012). At the same time, new plants may germinate on bare surfaces. These vegetation changes condition the growth rate and the age structure of the forest, which has been used to date the minimum age of new landforms and its linkage with fluvial dynamic (Sigafoos 1964).

At the scale of individual trees, and in the case of Tatra Mountain streams, we most commonly observe mechanical damage in the form of stem wounds (Fig. 2a). Previous studies have reported that in this region, the transport of boulders and woody debris during intense flood events can impact trees, removing the bark and destroying the meristematic cambium tissue (Zielonka et al. 2008). Scars on trees can, however, also be generated by falling trees or impacts during logging, which is quite widespread in the region. After wounding, trees react upon this disturbance by compartmentalizing the wounded surface (Stoffel and Klinkmüller 2013; Ballesteros et al. 2014) so as to minimize rot and decay. Therefore, the partial removal of bark and cambium tissues during mechanical wounding will favor the formation of callus pads at the edge of open wounds and dissimilar growth disturbances according to tree species.

Conifers growing in the Tatras Mountains respond to wounding in a way which allows them to minimize the risk of cell embolism and cavitation and by forming smaller cells. In addition, *Abies*, *Larix* and *Picea* have been reported to form tangential rows of traumatic resin ducts (TRD) after mechanical disturbance (Stoffel

2008; Schneuwly et al. 2009a, b; Ballesteros et al. 2010a), as well as a significant reduction of both ring width and the size of the lumen of earlywood tracheid (Arbellay et al. 2012a, b; Ballesteros-Cánovas et al. 2010a). Specifically, TRDs normally form tangentially within the current growth ring and close to the wound, but can also be observed in subsequent rings (Bollschweiler et al. 2008; Stoffel and Hitz 2008). This anatomical feature represents a nonspecific defense response to injury that compartmentalizes the wood, thereby preventing the attack of fungi and insects, as well as the loss of cell moisture (Shigo 1984). The occurrence of TRD in the tree-ring records represents a valuable indicator of the existence of "hidden" scars, and can even provide information on the seasonality of events (Stoffel and Beniston 2006; Schneuwly-Bollschweiler and Stoffel 2012). In broadleaved trees, the main reaction of trees to wound damages is a significant decrease in mean vessel size within rings formed during flood conditions (Arbellay et al. 2012b; Ballesteros-Cánovas et al. 2010b).

Tilted trees are also a common feature in this region. The sudden, unidirectional pressure on the stem generated by geomorphic events (e.g., deposition of material by a debris flow) as well as the destabilization of the root-plate system by bankfull erosion cause that the stems of many trees close to the channel are rotated (Fig. 2b). Trees are responding to tilting with the formation of reaction wood (Stoffel and Bollschweiler 2008, 2009). This anatomical feature is recognized macroscopically through the occurrence of eccentric growth rings in the stem. The recognition of compression wood in the tree-ring records allows determination of the moment of tilting, and consequently dating the flood event. By using dendro-mechanical models, the magnitude of the tilting can also correlated with the flow energy and consequently provides information about the peak discharge at a given cross-section (Ballesteros-Cánovas et al. 2015c).

In the tree-ring records, growth suppression can be recognized in trees suffering from crown decapitation or stem burial (Procter et al. 2012; Kogelnig-Mayer et al. 2013). It has been reported that crown or branch losses reduce photosynthetic activity, which in turn affects the growth rate. Similarly, stem burial may prevent water supply to the roots which in turn will lead to a decrease in growth rates, although the opposite reaction has been reported as well in case when a thin layer of nutrient-rich sediments is deposited around the trunk (Kui and Stella 2016; Friedman et al. 2005).

Finally, the last botanical evidence of past floods in the Tatra Mountains is exposed roots, which are caused by bankfull failures (Osterkamp et al. 2012; Stoffel et al. 2012). Exposed roots present clear signal in the tree-ring records, reflecting sudden or continuous erosion process by noticeable changes in wood anatomy (Stoffel et al. 2013).

The large set of potential evidence coming from trees can be used in paleoflood reconstructions, but also points to differences in the intensities of different responses. Stoffel and Corona (2014) provide a detailed overview on criteria used to assign signal intensity based on the anatomical features for different species as well as on the existing indices and minimum sampling sizes. In the case of paleoflood studies, flood scars on trees are the most useful botanical indicators of paleofloods since they do not only provide information on past events with seasonal resolution, but also represent paleostage indicators (i.e. height of the scar; Baker 2008) in flow discharge estimations (Ballesteros-Cánovas et al. 2015a).

Based on literature review (n = 52 studies performed in North America and Europe; Ballesteros-Cánovas et al. 2015b), the number of sampled trees per study was 104 ± 122 trees, of which 60 % were broadleaved trees and 40 % conifer trees.

3 Paleoflood Reconstruction in Tatra Mountain Streams Using Tree Rings

3.1 Study Site

The Tatra Mountains (Tatras) are located in the Carpathian arc (max. elevation 2655 m a.s.l.) between Poland and Slovakia. The Tatras are a crystalline and metamorphic core mountain range covered by nappes of Mesozoic sedimentary rocks. During the Pleistocene, the Tatras underwent at least three glaciations, which strongly reshaped the region, left conspicuous relief forms and moraine deposits. This mountain terrain is highly susceptible to geomorphic processes such as debris flows, snow avalanches and floods, which have caused several disasters during the last century. The study was conducted in four streams, i.e. Rybi Potok (RP), Roztoka (RS),

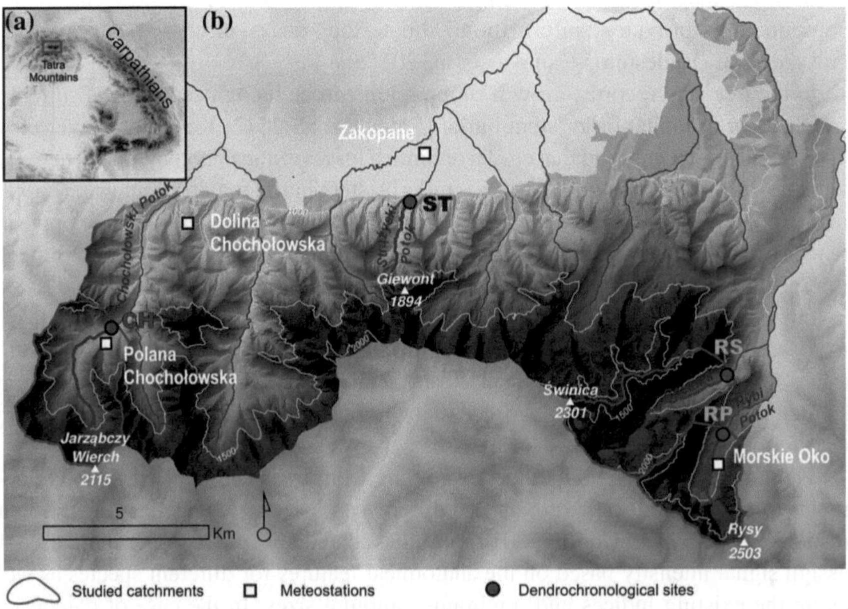

Fig. 3 **a** Location of the study site in the Carpathians. **b** Location of the investigated catchments together with meteorological stations

Strążyski (ST) and Chochołowski streams (CH), covering the western-central and eastern parts of the northern slopes of the Tatra Mountains (Fig. 3).

In this region, climate is influenced by regional air mass oscillations and local topography, with a predominance of polar marine (65 % of annual incidents) and polar continental air masses (25 %; Niedźwiedź et al. 2015). The Tatras also form a considerable barrier to air mass movements resulting in heavy rainfall events with 24-h sums of up to 300 mm (30 June 1973; Niedźwiedź 1992). Annual precipitation varies from 1100 mm at the foothills (Zakopane, 844 m a.s.l.) to 1660 mm at timberline (Hala Gąsienicowa, 1550 m a.s.l.) and 1721 mm on the summits (Lomnicky stit, 2635 m a.s.l.). The most effective precipitation events that result in flash floods are largely concentrated in the summer months as shown in Fig. 4.

Vegetation at the study sites is mostly formed by subalpine forests of Norway spruce (*Picea abies* (L.) Karst.) at higher altitudes and a mixed forest with large proportions of *P. abies*, and Silver fir (*Abies alba* Mill.) at lower altitudes. In general, stream channels are formed in gravel and loamy moraine deposits, which cover granitic and pegmatitic bedrock (Bac-Moszaszwili et al. 1979).

The land use history of the Tatras starts with Medieval ore mining, through pasturing to intense logging in the 18th and 19th centuries. The local forests have been used intensively for grazing, with peaks in grazing pressure during the 19th and mid-20th centuries, changing the characteristics of soil, vegetation, and forests, and leading to an increase in flood risk. This region has been also traditionally affected by intensive logging associated with the steel industry in the second half of the 18th century. The Tatra National Park was enacted in 1954 but pasturing locally continued until 1978 and logging remains permitted, at least in some areas of the

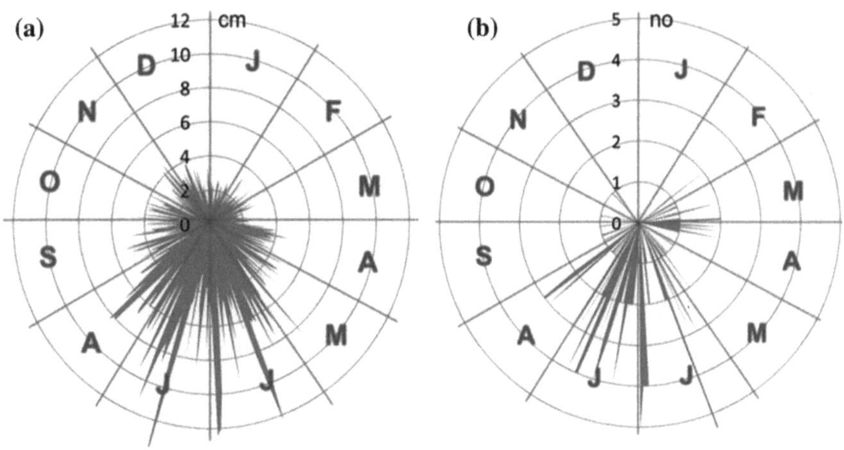

Fig. 4 Hydrometeorological situation at the northern foothills of the Tatra Mountains: **a** Average precipitation (cm) from 19 meteorological stations (28 years: 1954–1982); **b** Number of water flow larger than 1.5 times than averages as measured at Nowy Targ water gauge for the period 1898–1983 (Ballesteros-Cánovas et al. 2015a). After Ballesteros-Cánovas et al. (2015a), published by permission of Elsevier

park. The long history of timber exploitation—as fuel for the local industry and as construction material—has substantially changed the character of forests in the study region. The use of the valley floor and stream channels as transportation routes furthermore intensified this process.

3.2 Temporal Reconstruction of Floods in Tatra Mountain Streams

The reconstruction of past flood events was carried out in six representative stream sectors, previously selected due to the lack of signs of human activity or the influence by other geomorphic processes (e.g., landslides, snow avalanches). A detailed geomorphic survey and inspection of trees were performed to select trees presenting suitable past evidence of flood activity. As stated above, work focused on scars on trees due to (i) their suitability to date scars using TRDs; and (ii) the utility of scar height to estimate peak discharge. Trees growing in suitable positions (i.e. exposed to the flow) and/or with visible damage were sampled using increment borers and by following standard dendrogeomorphic sampling procedures (Stoffel and Corona 2014). Additional cores, wedges or even cross-sections were taken from dead trees. Undisturbed specimens of *P. abies* and *A. alba* were also sampled to build 5 reference chronologies. We also recorded specific information from each tree such as tree height, remarkable growth characteristics, diameter at breast height (DBH), photographs and tree location (GPS).

After field data acquisition, samples were mounted on woody supports, sanded and polished. We scanned all samples with a resolution of 3200 dpi for imagery analysis using CooRecorder and CDendro (Cybis Elektronik and Data AB; Larsson 2003a, b). The reference cores were cross-dated against site chronologies employing both visual and statistical techniques. Mistakes and growth discrepancies (i.e., missing, wedging, and false rings) were identified and corrected. Growth disturbances (GD) were analysed and dated under a stereomicroscope, with a special focus on injuries and related features such as TRDs, including their location in the tree ring, and callus tissues. Other GD such as abrupt growth suppression or release as well as the occurrence of compression wood were only used to support the interpretation of events.

The definition of flood events was based on the weighted index proposed by Kogelnig-Mayer et al. (2011), which considers the number of GD as well as their intensity for each year. The threshold used to distinguish flash flood signals from noise was set to $W_{it} \geq 0.5$ and $GD \geq 2$, as suggested in previous work for hydrogeomorphic processes (e.g., Schneuwly-Bollschweiler et al. 2013).

A total of 1111 increment cores were sampled from 218 *P. abies* and *A. alba* trees affected by past flood activity. They allowed identification of 480 GD and definition of 47 flood events between A.D. 1866 and 2012 (Fig. 5). Our results reveal that the catchment presenting the largest floods activity is RP with 23

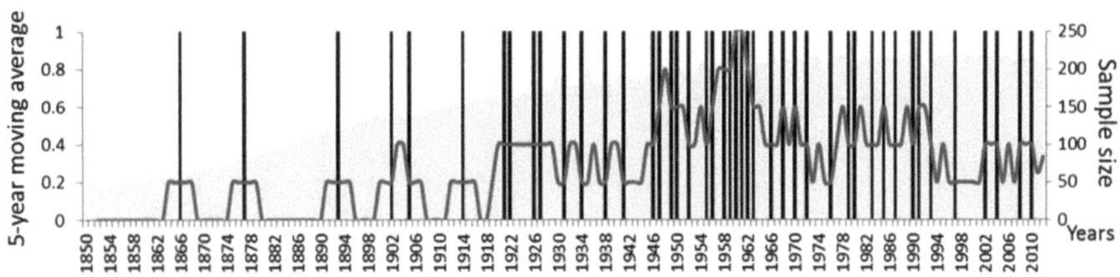

Fig. 5 Reconstructed chronology of major floods for the northern slopes of the Tatra Mountains and 5-year moving average of process activity highlighting periods of enhanced and limited flood activity. After Ballesteros-Cánovas et al. (2015a), published by permission of Elsevier

reconstructed events since A.D. 1866 (0.15 events year^{-1} with a sample size = 97), whereas the smallest activity was observed at DR with 4 reconstructed floods since A.D. 1926 (0.04 events year^{-1} with a sample size = 21). At DCH and ST, tree rings allowed the reconstruction of 19 (0.24 events year^{-1} with sample size = 73) and 9 (0.12 events year^{-1} with sample size = 22) since AD 1934 and 1938, respectively (Table 1).

Our results also suggest that a large majority of dated events (82 %) was limited to individual catchments, and no years can be found with reconstructed floods in all catchments. The flood with the largest spatial replication took place in 1970 and was documented in DCH, ST, and RP, whereas in ten other years with floods signs were found in only two catchments. The temporal occurrence of floods in the Tatras also allows distinction of at least three periods with enhanced flood activity. As shown in Fig. 5, greater activity occurred between 1946 and 1949 (5-year average: 0.8 event year^{-1}), 1955–1963 (5-year average: 0.7 event year^{-1}), and 1979–1987 (5-year average: 0.6 event year^{-1}).

3.3 *Scar-Based Peak Discharge Reconstruction in Tatra Mountain Streams*

After the analysis of flood occurrence in the Tatra Mountains (in terms of temporal frequency), we performed a reconstruction of flood magnitude based on a selected subset of trees presenting the most suitable evidence in the form of clear scars. The specific criteria to select the trees were (i) the presence of scars previously assigned to a specific flood event (see point 3.2); and (ii) scars facing the direction of flow and (iii) exhibiting a shape typical of flood impacts (see Ballesteros-Cánovas et al. 2010a, b for details). During fieldwork, each tree containing scars was accurately positioned with a GPS (Trimble GeoXT 6000, precision <1 m). We also measured scar height as well as the distance of trees with respect to the thalweg with a laser distance meter.

For each tree position, we computed peak discharge (q)–water stage (W) relationship W (q, i) by means of the two-dimensional hydrodynamic model

Table 1 Overview of sample size (n), growth disturbances (GD; n) and weighted index (W_{it}) for each year of the time series and for the four streams analysed

	RP			DR			DCH			ST		
Year	Sample size	N° GD	Wit	Sample size	N° GD	Wit	Sample size	N° GD	Wit	Sample size	N° GD	Wit
1866	20	2	0.66									
1877	31	4	1.39									
1893	55	2	0.94									
1903	70	3	1.21									
1905	72	2	0.52									
1914	82	3	0.68									
1921	88	4	0.66									
1922	88	6	1.00									
1926	*89*	*8*	*3.93*	*15*	*3*	*0.60*						
1927	89	7	1.96									
1931	*90*	*3*	*0.82*	*15*	*3*	*1.00*						
1934							50	2	0.54			
1938	*93*	*5*	*1.46*							*12*	*4*	*4.00*
1941							53	3	0.94			
1946							55	3	1.17			
1947										14	2	2.00
1949	94	4	0.96									
1950							58	2	0.63			
1952										15	2	1.87
1955							61	2	0.74			
1956	94	6	2.22									
1958	96	22	32.2									

(continued)

Table 1 (continued)

	RP			DR			DCH			ST		
Year	Sample size	N° GD	Wit	Sample size	N° GD	Wit	Sample size	N° GD	Wit	Sample size	N° GD	Wit
1959	*96*	*3*	*0.51*							*17*	2	*0.71*
1960							63	2	0.74			
1961	*96*	*5*	*0.84*				*63*	*5*	*1.19*			
1962							65	3	0.51			
1963							65	2	0.70			
1966	96	4	1.09									
1968							67	2	0.77			
1970	*96*	*4*	*0.55*				*68*	*6*	*2.86*	*19*	2	*1.05*
1972							70	9	8.26			
1976										21	3	2.14
1979	96	7	1.98									
1980							73	4	1.42			
1983							*73*	*3*	*0.56*	*21*	*3*	*2.43*
1985							73	3	1.19			
1987							73	4	1.36			
1990				21	3	3.00						
1991							73	2	0.93			
1993							73	2	0.56			
1997	*95*	*4*	*0.75*							22	*3*	*0.75*
1997												
2002	*93*	*6*	*1.43*				*73*	*6*	*1.20*			
2004				21	2	0.95						
2008	91	3	0.71									
2010										22	2	0.55

Years highlighted in italic refer to reconstructed major flood events (Ballesteros-Cánovas et al. 2015a)

IBER, which is a numerical model simulating turbulent-free, unsteady surface flows and solves depth-averaged, two-dimensional shallow water (2D Saint-Venant) equations using a finite volume method with a second-order roe scheme. In terms of digital terrain information we used LiDAR data with 1×1 m resolution as a basis for the mesh. Bed friction was evaluated using Manning's n roughness coefficient, which was initially assessed in the field by using homogenous roughness units (Chow 1959). In this study, we used Manning's n of 0.04 for the main channel and 0.1 for the overbank sections. To compute inlet water discharge (i.e. steady flow) into each study reach, we modelled successive inlet water discharges. At each tree, we used the best match between modelled water table and scar height to define the mean and standard deviation for the estimated average peak discharge of each flood event.

A total of 55 trees showed visible scars inflicted by sediment and/or wood transported during past flood events (namely 25 in DCH, 22 in RP, 6 in ST, and 3 in DR). Our results point out that the largest event took place in RP in 1903 with a reconstructed peak discharge of 115.9 ± 59.2 m^3 s^{-1} (Table 2). By contrast, the smallest discharge of 11.1 ± 4.9 m^3 s^{-1} was reconstructed for the 1934 event in catchment ST. This flood event was also dated in catchment DCH with reconstructed peak discharge of almost 39.7 ± 11.8 m^3 s^{-1}. On average, the range of discharges for reconstructed events was generally much higher in DCH and RP (92.8–39.7 and 115.9–28.6 m^3 s^{-1}, respectively) than in ST (45.8–11.1 m^3 s^{-1}) and DR (24.1 ± 7.6 m^3 s^{-1}). In terms of uncertainties related to the reconstructions, our methodology points to an average standard deviation of almost 41 %. The largest deviation (up to 80 %) was found for the 1997 event in ST, whereas the lowest deviation was observed at DCH and for an event in 1983 (~16 %). The detailed analysis of the impact of the hydrogeomorphic position of trees used in the reconstruction suggested that the trees showing lower deviation were located: (i) in sections with straight channel configurations or on the internal side of channel bends (10–20 %), (ii) far away from neighbouring trees (23 %), or (iii) next to the boundary of bankfull channel (~20 %). Trees showing the largest variability in scar heights were, by contrast, located on the external side of channel bends (up to 40 %), in areas with high tree density (41 %), as well as within the central channel or in overbank positions (up to 30 %).

4 Implications for Paleoflood Reconstructions in the Tatra Mountains

4.1 *Climate Triggers and Flood Variability*

The temporal flood reconstruction as well as the analyses of the location of TRDs within the tree rings have allowed an assessment of the seasonality of events, at least if they occurred during the growing seasons between late spring and early fall.

Table 2 Peak discharge reconstructions based on scar heights and water levels as modelled for each stream reach and each reconstructed (i.e. tree-ring data) event considered

Site	Year	Scar height (cm)	EPD (m^3/s)	RPD (m^3/s)	Site	Year	Scar height (cm)	EPD (m^3/s)	PD (m^3/s)
DCH	1934	70	48.1	39.7 ± 11.8	RP	1970	90	21.9	39.6 ± 9.5
DCH	1934	170	31.3		RP	1970	80	35.3	
DCH	1955	90	29.0	56.0 ± 23.9	RP	1903	80	74.0	115.9 ± 59.2
DCH	1955	130	45.3		RP	1903	190	157.8	
DCH	1955	90	83.7		RP	1949	140	71.8	59.4 ± 17.5
DCH	1955	90	66.1		RP	1949	40	47.0	
DCH	1963	110	69.8	92.8 ± 25.1	RP	1926	118	114.1	70.4 ± 34.6
DCH	1963	120	89.1		RP	1926	110	70.7	
DCH	1963	100	119.6		RP	1926	90	67.3	
DCH	1972	125	65.1	67.8 ± 28.9	RP	1926	50	29.7	
DCH	1972	115	83.5		RP	1958	90	52.3	86.2 ± 47.7
DCH	1972	130	58.5		RP	1958	100	47.8	
DCH	1972	50	29.2		RP	1958	90	82.1	
DCH	1972	100	39.4		RP	1958	70	162.8	
DCH	1972	155	86.1		RP	1958	120	59.5	
DCH	1972	60	113.2						
DCH	1983	120	52.3	46.7 ± 7.9	ST	1938	80	14.5	11.1 ± 4.9
DCH	1983	105	41.1		ST	1938	90	7.6	
DCH	1991	60	28	44.8 ± 23.7	ST	1947	80	14.5	20.7 ± 8.8
DCH	1991	70	61.5		ST	1947	50	53.4	
DCH	2002	30	31.2	40.4 ± 12.8	ST	1952	80	25.4	20.8 ± 6.5
DCH	2002	50	49.4		ST	1952	80	16.2	
DR	1990	85	44.8	24.1 ± 7.6					
DR	1990	40	19.1						
DR	1990	45	8.4						

EPD estimated peak discharge at each tree location; *RPD* reconstructed peak discharge for specific event (Ballesteros-Cánovas et al. 2016)

Based on a comparison with records from the closest rain gauge stations, we described the potential hydrometeorological triggers of floods by analysing the maximum 1-, 3- and 5-day April–October rainfall events. We found that for 1-day events, rainfall totals ranged from 28.8 to 168.4 mm (mean 80.2 ± 34.3 mm), whereas the 3- and 5-day precipitation sums ranged between 50.3 and 247.7 mm (mean 130.4 ± 51.3 mm) and 65.2 and 258.1 mm (mean 149.4 ± 53.5 mm), respectively (Table 2). The statistical test used (nonparametric Friedman test) confirmed the existence of non-significant differences of rainfall threshold values between the catchments. The most intense event was recorded in DCH with a daily precipitation total of 333.5 mm (24 July 1980). The most replicated event (ST, RP, DCH) occurred in 1970 and has not only been confirmed by flood records in Zakopane, but also corresponds to the second largest recorded rainfall in the Tatra Mountains on 20 July 1970, when the 1-, 3-, and 5-day accumulated rainfall totals were 150.1, 206.4, 243.2 mm, respectively (Table 3).

These results confirm the observation by Niedźwiedź et al. (2015) who suggest that rainfall above the threshold of 50 mm day^{-1} can cause floods in the Tatras (in our study we provide an average value of 80.2 mm, but with a lower minimum at 28.8 mm). The seasonality and evidence of intense daily rainfalls during May and October are in agreement with local observations and previous studies (Kotarba 2004; Niedźwiedź et al. 2015). Rain-on-snow and sudden snowmelt due to warmer springs may trigger floods as well, as stated by Zielonka et al. (2008). However, our results suggest that the main weather mechanism leading to floods is related to prolonged rainfall induced by North cyclonic circulation (Nc) (Niedźwiedź 1992; Niedźwiedź et al. 2015). This situation implies generalized and widespread occurrence of heavy rainfalls, explaining the situations with floods such as the well-replicated event of 1970, for which we observe the highest daily rainfall sums in all meteorological stations. At the same time, we also observe a large number of asynchronous flood events between catchments. While differences in catchment characteristics may play a role in flood generation, we also think that local effects, such as the presence of rainfall shadow areas (i.e. orographic effect; Borga et al. 2014) may play an important role. This hypothesis is supported by the coefficient of variation on rainfall measurement observed in rain gauge stations along the Tatra Mountains. For instance, the year 1972 was characterized by prolonged rainfall, affecting mostly the Central and Western parts of the Tatra Mountains with maximum 3-day precipitation recorded at the Hala Gąsienicowa station (218.1 mm) and reconstructed floods in catchments DCH and ST (CV:038). This rainfall shadow effect has been described for other mountain areas as well (Buytaert et al. 2006). Our results, therefore, imply that significant space-time variability exists in catchment response.

Reconstructed peak discharges confirm dissimilar catchment responses as well. The specific discharge analyses of both gauged and reconstructed flow records clearly support the observed variability of catchment responses. On the other side, we also observe that reconstructed floods from the first half of the 20th century were generally larger than those measured by the gauging stations during the second half of the century, especially in catchments ST and DCH. These observations can be explained by the evolution of land uses in the Tatra Mountains. Therefore, the strong human impact from the eighteenth to the first half of the twentieth century

Table 3 Precipitation records (1-, 3- and 5-day totals) associated with dated flash flood events

Site	Year	Likely event date	Precipitation station	1-day (mm)	3-day (mm)	5-day (mm)
DCH	1934	19.07	Historic event (no data)	N/A	N/A	N/A
DCH	1955	5.08	Dolina Chochołowska	66.4	85.2	112.2
RP	1956*	19.06	Morskie Oko	37.4	50.3	65.2
RP	1958	29.06	Morskie Oko	168.4	238.2	248.4
RP	1959	30.06	Morskie Oko	83.4	148.9	156.2
ST	1959*	1.06	Dolina Chochołowska	93.4	133.0	134.1
DCH	1960	13.07	Dolina Chochołowska	110.6	113.6	153.0
DCH	1961	30.07	Dolina Chochołowska	45.7	115.5	115.5
DCH	1962	18.07	Dolina Chochołowska	97.4	128.2	132.0
DCH	1963	5.10	Dolina Chochołowska	53.4	89.8	96.6
RP	1966	25.06	Morskie Oko	47.3	52.0	91.4
DCH	1968	18.07	Dolina Chochołowska	74.9	90.5	99.7
DCH	1970	18.07	Historic event (no data)	N/A	N/A	N/A
ST	1970	18.07	Zakopane	138.7	192.1	218.8
RP	1970	18.07	Morskie Oko	150.1	206.4	243.2
DCH	1972*	21.08	Polana Chochołowska	92.1	187.6	214.5
ST	1976	17.09	Zakopane	41.3	72.3	85.6
RP	1979*	27.06	Morskie Oko	28.8	86.3	88.5
DCH	1980	24.07	Polana Chochołowska	84.5	247.7	258.1
ST	1983	14.07	Zakopane	109.2	179.3	190.2
DCH	1983	14.07	Polana Chochołowska	84.7	118.9	134.7
DR	1990*	25.05	Morskie Oko	59.0	73.0	87.4
ST	1997	08.07	Zakopane	104.0	166.0	205.3
RP	1997	08.07	Morskie Oko	84.0	134.1	153.0

(continued)

Table 3 (continued)

Site	Year	Likely event date	Precipitation station	1-day (mm)	3-day (mm)	5-day (mm)
RP	2002	14.08	Morskie Oko	58.0	89.8	107.2
DR	2004	28.07	Morskie Oko	72.2	150.1	168.2
RP	2008	23.07	Zakopane	113.5	155.0	185.2
ST	2010	27.07	Zakopane	68.1	119.1	151.6

* represent years with a CV larger than 0.35, which indicate a larger variability in the measured precipitation over the northern slopes of the Tatra Mountains (Ballesteros-Cánovas et al. 2015a)

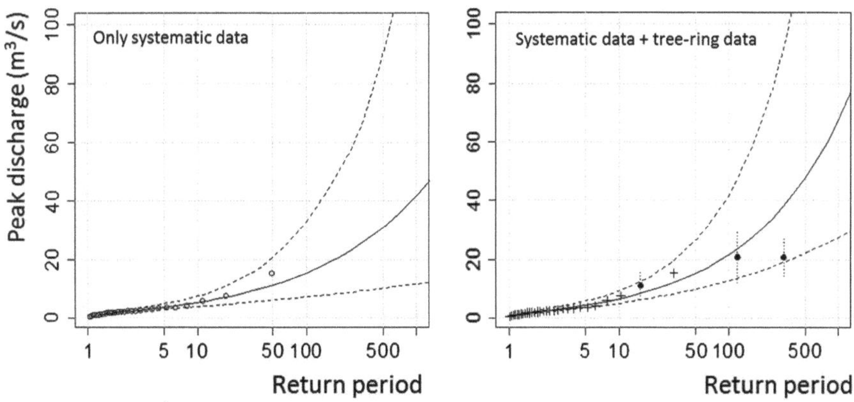

Fig. 6 Flood frequency distribution based on systematic flow gauge series (1) and the series consisting of systematic flow data and the reconstructed paleodischarges (2) After Ballesteros-Cánovas et al. (2016), published by permission of Elsevier

resulted in increased soil erosion and surface runoff, thereby favouring the formation of braided channels. Our results may also support the idea that, in general, average peak discharges could have been reduced due to the impact of intense forest recovery on runoff in the study region during the second half of the twentieth century (Wyżga et al. 2012).

4.2 Implications for Flood Hazard Assessments

The paleoflood records provided in this study have implications in terms of flood hazard definition. Despite the fact that existing historical archives describe the occurrence of intense floods triggered by summer precipitation since at least the seventeenth century (Krzemień 1991; Starkel 1996; Kotarba 2004; Gorczyca et al. 2013), the existing gauging network in the Tatra Mountains and their foreland is highly fragmented and incomplete, which therefore renders proper hazard definition and makes long-term flood studies based on systematic records very difficult. Our results provide two main outcomes which are considered very useful for an improved understanding of flood hazards in the region.

First, we provide unique and quantitative rainfall thresholds associated with long-term flood occurrences. These rainfall thresholds should be used as a baseline for the determination of probabilities of threshold exceedances under future climate conditions. By using regional climate model (RCMs) output and downscaling procedures, the expected future frequency of intense precipitation events could be compared with our data so as to provide data on changes in the duration, severity, and frequency of intense precipitation events leading to the triggering of floods in the region over the next few decades. This is specifically relevant as evidence exists on an increase in the activity, at least in the long term, of cyclonic circulation types which are responsible for floods in the region (Niedźwiedź et al. 2015).

Secondly, the inclusion of non-systematic paleohydrological data, derived from tree-ring analysis, has an important impact on the results of the flood frequency analysis. Through the use of Bayesian Markov Monte Carlo Chain algorithms (Gaume et al. 2010; Viglione et al. 2013), we combined the reconstructed flood magnitude with the existing flow gauge series to determine changes in the flood frequency quartiles. Our results highlight that the largest changes in flood frequency (up to 25.5 % for T = 100 years) have occurred at Strążyska (ST) (Fig. 6). This implies that the flood hazard at this site would have been underestimated had the assessment of flood frequency distributions been based solely on systematic data. This finding is of high importance as the Strążyski Stream (ST) drains the catchment upslope of the main population center of the Polish Tatras, Zakopane. Consequently, our results should be taken into account for the design of hydraulic infrastructure and the definition of hazard planning. Our results also help to substantially reduce uncertainties in the estimation of the 100-year flood. Under a stochastic flood risk assessment approach, this uncertainty could be included in future risk assessments in the area, as it has been done in other regions in the past (Apel et al. 2004; Ballesteros et al. 2013).

Finally, and similarly to what is done on the southern side of the Tatra Mountains (Slovakian sector; Gáal et al. 2010), our paleoflood records could be included to deliver regional flood analysis, given that both datasets pass the homogeneity test based on the Hosking and Wallis (1997) algorithm. If so, a combined analysis on either side of the main divide could lead to a completely new setting for flood hazard assessments at the regional scale, which could then have clear and important implications for risk-based land management.

5 Conclusions

The lack of data on past floods and their impacts on the understanding of flood processes and their impacts in the Tatra Mountains and their foreland has been one of the key drivers of the FLORIST project (Kundzewicz et al. 2014b). Tree-rings studies, as the ones presented here, have been applied to Tatra Mountain streams to overcome some of the past drawbacks related with this lack of data. Despite some difficulties to deal with and to combine non-systematic (paleoflood) and systematic (gauged) records, the application of tree-ring records in Tatra Mountain streams has undoubtedly contributed to an improved understanding of past process activity, hydrometeorological triggers of floods; flood variability during the last century, and have facilitated redefinition of flood hazards in the region.

Acknowledgments This contribution has been realized in the framework of the FLORIST (Flood risk on the northern foothills of the Tatra Mountains) project, PSPB no. 153/2010, through a grant from Switzerland through the Swiss Contribution to the enlarged European Union.

References

Arbellay E, Corona C, Stoffel M, Fonti P, Decaulne A (2012a) Defining an adequate sample of earlywood vessels for retrospective injury detection in diffuse-porous species. PLoS ONE 7: e38824

Arbellay E, Fonti P, Stoffel M (2012b) Duration and extension of anatomical changes in wood structure after cambial injury. J Exp Bot 63:3271–3277

Apel H, Thieken AH, Merz B, Blöschl G (2004) Flood risk assessment and associated uncertainty. Nat Hazard Earth Syst 4(2):295–308

Bac-Moszaszwili M, Burchardt J, Głazek J, Iwanow A, Jaroszewski W, Kotański Z, Lefeld J, Mastella L, Ozimkowski W, Roniewicz P, Skupiński A, Westwalewicz-Mogilska E (1979) Mapa Geologiczna Tatr Polskich (Geological Map of the Polish Tatra), 1: 30,0000. Wydawnictwa Geologiczne, Warszawa, Poland

Baker VR (2008) Paleoflood hydrology: origin, progress, prospects. Geomorphology 101(1):1–13

Ballesteros-Canovas JA, Stoffel M, Bodoque del Pozo JM, Bollschweiler M, Hitz OM, Diez-Herrero A (2010a) Changes in wood anatomy in tree rings of Pinus pinaster Ait. following wounding by flash floods. Tree Ring Res 66:93–103

Ballesteros-Canovas JA, Stoffel M, Bollschweiler M, Bodoque del Pozo JM, Díez-Herrero A (2010b) Flash-flood impacts cause changes in wood anatomy of Alnus glutinosa, Fraxinus angustifolia and Quercus pyrenaica. Tree Physiol 30:773–781

Ballesteros-Cánovas JA, Sanchez-Silva M, Bodoque JM, Díez-Herrero A (2013) An example of integrated approach to flood risk management: the case of Navaluenga (Central Spain). Water Resour Manage 27(8):3051–3069

Ballesteros-Canovas JA, Rodriguez-Morata C, Garófano-Gómez V, Rubiales JM, Sánchez-Salguero R, Stoffel M (2014) Unravelling past flash flood activity in a forested mountain catchment of the Spanish Central System. J Hydrol. doi:10.1016/j.jhydrol.2014.11.027

Ballesteros-Cánovas JA, Czajka B, Janecka K, Lempa M, Kaczka RJ, Stoffel M (2015a) Flash floods in the Tatra Mountain streams: frequency and triggers. Sci Total Environ 511:639–648

Ballesteros-Cánovas JA, Stoffel M, St. George S, Hirschboeck K (2015b) A review of flood records from tree rings. Prog Phys Geog 1–23. doi:10.1177/0309133315608758

Ballesteros-Cánovas JA, Márquez-Peñaranda JF, Sánchez-Silva M, Díez-Herrero A, Ruiz-Villanueva V, Bodoque JM, Eguibar MA, Stoffel M (2015c) Can tree tilting be used for paleoflood discharge estimations? J Hydrol 529:480–489

Ballesteros-Cánovas JA, Stoffel M, Spyt B, Janecka K, Kaczka RJ, Lempa M (2016) Paleoflood discharge reconstruction in Tatra Mountain streams. Geomorphology 272C. doi:10.1016/j.geomorph.2015.12.004

Bollschweiler M, Stoffel M, Schneuwly DM, Bourqui K (2008) Traumatic resin ducts in Larix decidua stems impacted by debris flows. Tree Physiol 28:255–263

Borga M, Anagnostou EN, Blöschl G, Creutin JD (2011) Flash flood forecasting, warning and risk management: the HYDRATE project. Environ Sci Policy 14(7):834–844

Borga M, Stoffel M, Marchi L, Marra F, Jakob M (2014) Hydrogeomorphic response to extreme rainfall in headwater systems: flash floods and debris flows. J Hydrol 518:194–205

Brázdil R, Kundzewicz ZW, Benito G (2006) Historical hydrology for studying flood risk in Europe. Hydrol Sci J 51(5):739–764

Buytaert W, Celleri R, Willems P, Bièvre BD, Wyseure G (2006) Spatial and temporal rainfall variability in mountainous areas: a case study from the south Ecuadorian Andes. J Hydrol 329 (3):413–421

Chow V (1959) Open channel hydraulics. McGraw-Hill, New York

Friedman JM, Vincent KR, Shafroth PB (2005) Dating floodplain sediments using tree-ring response to burial. Earth Surf Proc Land 30:1077–1091

Gaál L, Szolgay J, Kohnová S, Hlavčová K, Viglione A (2010) Inclusion of historical information in flood frequency analysis using a Bayesian MCMC technique: a case study for the power dam Orlík, Czech Republic. Contrib Geophys Geodesy 40(2):121–147

Gaume E, Gaál L, Viglione A, Szolgay J, Kohnová S, Blöschl G (2010) Bayesian MCMC approach to regional flood frequency analyses involving extraordinary flood events at ungauged sites. J Hydrol 394(1):101–117

Gorczyca E, Krzemień K, Wrońska-Wałach D, Sobucki M (2013) Channel changes due to extreme rainfalls in the polish carpathians. In: Loczy D (ed) Geomorphological impacts of extreme weather. Springer, Netherlands, pp 23–35

Hosking JRM, Wallis JR (1997) Flood frequency analysis: an approach based on L-moments

Kames S, Tardif JC, Bergeron Y (2016) Continuous earlywood vessels chronologies in floodplain ring-porous species can improve dendrohydrological reconstructions of spring high flows and flood levels. J Hydrol 534:377–389

Kogelnig-Mayer B, Stoffel M, Bollschweiler M, Hübl J, Rudolf-Miklau F (2011) Possibilities and limitations of dendrogeomorphic time-series reconstructions on sites influenced by debris flows and frequent snow avalanche activity. Arctic Antarct Alp Res 43:649–658

Kogelnig B, Stoffel M, Schneuwly-Bollschweiler M (2013) Four-dimensional growth response of mature Larix decidua to stem burial under natural conditions. Trees 27:1217–1223

Kotarba A (2004) Zdarzenia geomorfologiczne w Tatrach Wysokich podczas małej epoki lodowej. (Geomorphic events in the High Tatra Mountains during the Little Ice Age) In: Kotarba A (ed) Rola małej epoki lodowej w przekształcaniu środowiska przyrodniczego Tatr (Effect of the Little Ice Age on transformation of natural environment of the Tatra Mountains). PAN IGiPZ, Kraków, pp 9–55 (in polish with english summary)

Kozlowski TT (2002) Physiological-ecological impacts of flooding on riparian forest ecosystems. Wetlands 22(3):550–561

Krzemień K (1991) Dynamika wysokogórskiego system fluwialnego na przykładzie Tatr Zachodnich. (Dynamics of the high-mountain fluvial system with the Western Tatra Mts. as example).Rozprawy Habilitacyjne (Profesure desideratum). Nr 215. Kraków. UJ. ISSN (in polish with english summary)

Kui L, Stella JC (2016) Fluvial sediment burial increases mortality of young riparian trees but induces compensatory growth response in survivors. Forest Ecol Manage 366:32–40

Kundzewicz ZW, Kanae S, Seneviratne SI, Handmer J, Nicholls N, Peduzzi P, Mechler R, Bouwe LM, Arnell N, Mach K, Muir-Wood R, Brakenridge GR, Kron W, Benito G, Honda Y, Takahashi K, Sherstyukov B (2014a) Flood risk and climate change: global and regional perspectives. Hydrol Sci J 59(1):1–28

Kundzewicz Z, Stoffel M, Kaczka R, Wyżga B, Niedźwiedź T, Pińskwar I, Ruiz-Villanueva V, Łupikasza E, Czajka B, Ballesteros-Canovas J, Małarzewski Ł, Choryński A, Janecka A, Mikuś P (2014b) Floods at the northern foothills of the Tatra Mountains—a polish-swiss research project. Acta Geophys 62(3):620–641

Larsson L-A (2003a) CooRecorder: image co-ordinate recording program. www.cybis.se

Larsson L-A (2003b) CDendro: Cybis Dendro dating program. www.cybis.se

Niedźwiedź T (1992) Climate of the Tatra Mountains. Mt Res Dev 12(2):131–146

Niedźwiedź T, Łupikasza E, Pińskwar I, Kundzewicz ZW, Stoffel M, Małarzewski Ł (2015) Variability of high rainfalls and related synoptic situations causing heavy floods at the northern foothills of the Tatra Mountains. Theor Appl Climatol 119(1–2):273–284

Osterkamp WR, Hupp CR, Stoffel M (2012) The interactions between vegetation and erosion: new directions for research at the interface of ecology and geomorphology. Earth Surf Proc Land 37:23–36

Procter E, Stoffel M, Bollschweiler M, Neumann M (2012) Exploring debris-flow history and process dynamics using an integrative approach on a dolomitic cone in western Austria. Earth Surf Proc Land 37:913–922

Schneuwly DM, Stoffel M, Dorren LKA, Berger F (2009a) Three-dimensional analysis of the anatomical growth response of European conifers to mechanical disturbance. Tree Physiol 29:1247–1257

Schneuwly DM, Stoffel M, Bollschweiler M (2009b) Formation and spread of callus tissue and tangential rows of resin ducts in Larix decidua and Picea abies following rockfall impacts. Tree Physiol 29:281–289
Schneuwly-Bollschweiler M, Corona C, Stoffel M (2013) How to improve dating quality and reduce noise in tree-ring based debris-flow reconstructions. Quat Geochronol 18:110–118
Shigo AL (1984) Compartmentalization: a conceptual framework for understanding how trees grow and defend themselves. Annu Rev Phytopathol 22(1):189–214
Shroder JF (1978) Dendrogeomorphological analysis of mass movement on Table Cliffs Plateau. Utah Quat Res 9(2):168–185
Sigafoos RS (1964) Botanical evidence of floods and flood-plain deposition. US Geol Surv Prof Pap 485A:35
Starkel L (1996) Geomorphic role of extreme rainfalls in the Polish Carpathians. Studia Geomorphologica Carpatho-Balcanica 30:21–39
Stoffel M (2008) Dating past geomorphic processes with tangential rows of traumatic resin ducts. Dendrochronologia 26(1):53–60
Stoffel M, Beniston M (2006) On the incidence of debris flows from the early Little Ice Age to a future greenhouse climate: a case study from the Swiss Alps. Geophys Res Lett 33:L16404
Stoffel M, Bollschweiler M (2008) Tree-ring analysis in natural hazards research—an overview. Nat Hazards Earth Syst Sci 8:187–202
Stoffel M, Hitz OM (2008) Snow avalanche and rockfall impacts leave different anatomical signatures in tree rings of Larix decidua. Tree Physiol 28(11):1713–1720
Stoffel M, Bollschweiler M (2009) What tree rings can tell about earth-surface processes: teaching the principles of dendrogeomorphology. Geogr Compass 3(3):1013–1037
Stoffel M, Wilford DJ (2012) Hydrogeomorphic processes and vegetation: disturbance, process histories, dependencies and interactions. Earth Surf Proc Land 37:9–22
Stoffel M, Klinkmüller M (2013) 3D analysis of anatomical reactions in conifers after mechanical wounding: first qualitative insights from X-ray computed tomography. Trees 27:1805–1811
Stoffel M, Corona C (2014) Dendroecological dating of geomorphic disturbance in trees. Tree-Ring Res 70:3–20
Stoffel M, Bollschweiler M, Butler DR, Luckman BH (eds) (2010) Tree rings and natural hazards: a state-of-art (vol 41). Springer Science & Business Media, Berlin
Stoffel M, Casteller A, Luckman BH, Villalba R (2012) Spatiotemporal analysis of channel wall erosion in ephemeral torrents using tree roots—an example from the Patagonian Andes. Geology 40(3):247–250
Stoffel M, Corona C, Ballesteros Canovas JA, Bodoque del Pozo JM (2013) Dating and quantification of erosion processes based on exposed roots. Earth Sci Rev 123:18–34
Szczepański W, Czarnecka H, Mierzwiński A (1996) Atlas posterunków wodowskazowych dla potrzeb państwowego monitoringu środowiska: posterunki wodowskazowe IMGW
Viglione A, Hosking JR, Laio F, Miller A, Gaume E, Payrastre O, Salinas JL, N'guyen CC, Halbert K (2013) Package 'nsRFA'. Non-supervised Regional Frequency Analysis. CRAN Repository. http://r.adu.org.za/web/packages/nsRFA/nsRFA.pdf
Weingartner R, Barben M, Spreafico M (2003) Floods in mountain areas—an overview based on examples from Switzerland. J Hydrol 282(1):10–24
Wohl E (2000) Mountain rivers. No. 14. Amer. Geophysical Union
Wohl E (2006) Human impacts to mountain streams. Geomorphology 79(3):217–248
Wyżga B, Zawiejska J, Radecki-Pawlik A, Hajdukiewicz H (2012) Environmental change, hydromorphological reference conditions and the restoration of Polish Carpathian rivers. Earth Surf Proc Land 37(11):1213–1226
Zielonka T, Holeksa J, Ciapała S (2008) A reconstruction of flood events using scarred trees in the Tatra Mountains, Poland. Dendrochronologia 26(3):173–183

History of Floods on the Upper Vistula

Elżbieta Nachlik and Zbigniew W. Kundzewicz

Abstract The history of floods on the Upper River Vistula in Poland is presented. Information on floods in the pre-instrumental period is derived from documentary sources. It is only since the re-establishment of Polish independence in 1945 that instrumental data make it possible to analyse floods on relatively homogeneous, uniform and gap-free observational material. In order to extend the flood-related records, recourse must be made to historical hydrology. Sources of flood-related information are examined and the particularities of the situation, driven by the history of Poland, are explained. Interpretation is offered of changes in flooding, caused by land use, river training and climate change. Changing characteristics of floods of different generation mechanisms (convective or advective rainfall, snowmelt, ice-jam) are discussed.

Keywords Floods · Historical hydrology · Poland · Pre-instrumental records · River Vistula

1 Introduction

Poland, a country covered by two large river basins, the Vistula and the Odra, has frequently suffered from the destructive abundance of water. Floods have occurred many times, from the earliest days of documented history until the present, leading

E. Nachlik (✉)
Institute of Water Engineering and Water Management, Technical University of Cracow, Cracow, Poland
e-mail: elzbieta.nachlik@iigw.pl

Z.W. Kundzewicz
Institute for Agricultural and Forest Environment, Polish Academy of Sciences, Bukowska 19, 60-809 Poznań, Poland
e-mail: kundzewicz@yahoo.com

Z.W. Kundzewicz
Potsdam Institute for Climate Impact Research, Telegrafenberg, 14412 Potsdam, Germany

Z.W. Kundzewicz et al. (eds.), *Flood Risk in the Upper Vistula Basin*,
GeoPlanet: Earth and Planetary Sciences, DOI 10.1007/978-3-319-41923-7_13

to fatalities and high material losses. Floods still remain the main natural disaster in Poland, raising considerable concern. The material losses in recent severe floods were very high, up to the level of US$2.3–3.5 billion of flood damage, nationwide, in the particularly dramatic event in the summer of 1997 (Kundzewicz et al. 1999). In the present paper, an attempt is undertaken to synthesize the available information on the history of floods on the Upper Vistula and its tributaries since the Middle Ages to the present.

Since 1997, there were two more big disastrous floods, in 2001 and 2010, in the Upper Vistula Basin. Direct material losses in 2010 flood in the Upper Vistula basin reached a value of nearly US$1.6 billion, which exceeded the total value of material losses in this region during the floods in 1997 and 2001 (cf. Bojarski and Nachlik 2014).

Beyond direct and indirect material damage and fatalities, large floods lead to serious social damage—ill health, stress and social disruption.

2 General Information on the Vistula and Its Floods

The River Vistula (*Wisła* in Polish), flowing entirely on the territory of Poland is the longest river draining to the Baltic Sea and, at 1024 km, the longest Polish river. The area of its drainage basin is 193,866 km^2 (of which 168,775 km^2 is in Poland) (cf. Czarnecka 2005).

The River Vistula flows northwards from its source on the Barania Góra in the Beskidy Mountains in the Carpathian range to its mouth in the Baltic Sea. The highland-dominated Upper Vistula basin extends from the source downstream to the San confluence. Throughout the course of the Vistula, there have been natural channel changes superimposed on man-made river regulations and transformations of the landscape (land-use and land-cover changes) in the river basin.

There are several important towns located on the Upper Vistula and its tributaries, hence, there has been a large damage potential related to urban flooding. In the light of recent floods, the flood protection system for urban areas is found to be in need of re-assessment, overhaul and strengthening.

Floods on the Upper Vistula and its tributaries can be caused by a number of generating mechanisms, determined by the time of flood occurrence, location and territorial range. Three principal classes of flood generation mechanisms can be distinguished, namely: intense and/or long-lasting rain, snowmelt, and ice-related phenomena. The sub-division of rain-induced floods into those generated by convective rains, occurring locally on small areas, typically in the summer, particularly from late June to the end of July, and advective precipitation, which can cover the whole river basin of the Upper Vistula is important. Snowmelt-induced floods result from rapid melting of the snow pack, sometimes amplified by rainfall. Such floods, which mostly occur in March and April, but can occur during the whole winter, can have a very extensive territorial range. Ice-jam mechanisms, related to the freeze-up and break-up periods, have frequently caused winter and spring floods. Floods caused by advective and frontal precipitation covering large areas are typical in most of the Upper Vistula basin.

On the Upper Vistula, flood levels can be up to 5–6 m above the low-water level. There are two seasons when floods on the Vistula are unlikely; they have been very rare in the second half of April and during the whole of May and the first half of June (except for May and June 2010) and also in October–November. The flood defences in the entire Vistula basin include embankments of approx. 4700 km in length, protecting approx. 530,000 ha. The total length of embankments on the Upper Vistula and its tributaries is 2200 km, protecting an area of 240,000 ha. There are several storage reservoirs in the Upper Vistula Basin, playing a role in flood protection on the mountain tributaries of the Vistula, including Porąbka and Tresna on the Soła, Czorsztyn and Rożnów on the Dunajec and, the largest, Solina (with capacity of 460×10^6 m^3) on the San. There is also a storage reservoir on the River Vistula itself—Goczałkowice on the Mała Wisła (Small Vistula). Smaller reservoirs in the Upper Vistula basin have also a significant impact on reducing the size of floods. They are localized so that their flood reserve capacity have the effect of lowering the culmination of a wave on the River Vistula. The total volume of flood reserve of these reservoirs reaches 240–380 million cubic meter. However, this is not sufficient, as compared to the needs (cf. The Upper Vistula River Basin Flood Protection Program 2011).

3 Sources of Information on Upper Vistula Floods

The River Vistula is deeply rooted in the Polish tradition, being a subject of particular reverence to the Polish nation. It played a considerable role in the turbulent history of Poland. Over many centuries, Poland was a large and flourishing kingdom. In its less fortunate period (1772–1795), the country as well as the Upper Vistula Basin, was partitioned between three powerful neighbour countries and incorporated into Austria, Prussia (later Germany) and Russia. From 1795 to 1918, when Poland did not exist as an independent country, the Vistula flowing through the Austrian, the Russian and the Prussian (later German) partition, was, in fact, an international water course. In 1918, Poland regained its independence. During the Second World War (1939–1945), when Poland lost its independence, the Vistula Basin was under both German and Soviet occupation. The history of the country has had a bearing on information about historical river floods. The three countries which partitioned Poland and the Vistula Basin in the late 18th to early 20th centuries did not have common measurement units, e.g. in the Russian Empire, unique units of length were used, as well as the Julian calendar. These cross-boundary effects have to be taken into account in hydrological analyses, as observations on the River Vistula and its catchment were carried out by the hydrological services of three different countries. Due to the complex history of the Polish state, and destruction during both World Wars, only post—World War II flood records since the 1950s can be interpreted reliably as an uninterrupted time series (cf. Cyberski et al. 2006).

This adds to the problem of short instrumental records in general. Indeed, extending the instrumental records of river stage with the help of other data is necessary. Two classes of approaches can be used: proxy data (indices linked indirectly to water flow/stage including tree scars caused by ice-jam floods or by wooded debris carried by water, and palaeohydrology), or documentary information (written documents and flood boards).

Documentary records are indeed very important for improving understanding, and extending the instrumental observation records. However, it is difficult to harmonize flood-related information originating from different sources bearing different credibility and accuracy. Having recognized the caveats, one can state that there exists a pool of information on the history of the Vistula River floods. Since the early Middle Ages until the late 16th century, flood data on the Vistula were available only as narrative, hand-written documentary records, mostly in Latin. Those descriptions included only the most dramatic natural disasters and were often based on casual sources. A common entry was a qualitative impact-related statement that a flood in a given year caused "huge damage and destruction", "flooding fields, poor harvest"; sometimes further characterization of the event was given. Typically, statements on flood water stage and the extent of inundated areas cannot be found. However, sometimes water levels can be estimated from statements referring to existing topographic points or structures, such as information on damaged churches and destroyed bridges. Discussion of methodology of historical hydrology with respect to floods can be found in Brazdil et al. (2006).

There is much documentary information describing floods on the Vistula in the region of Cracow. Primary sources of data in this period are old chronicles, archives, annals and diaries by Długosz, Bielski, Kromer and others. Studies in this area, carried out at the Jan Kazimierz University in Lwów before the World War II, led to pioneering publications (cf. review in Cyberski et al. 2006). A wealth of historical flood-related information can be found in Mikulski (1955) and Girguś and Strupczewski (1965). However, the historical material typically concerns only the high-impact events recorded in places with high damage potential (towns, ports and river crossings). That is, only patchy information is available, while, in contrast, ice jams were likely to occur over long sections of rivers.

Furthermore, the homogeneity of the information is distorted by the many natural and anthropogenic changes in the river regime, land use and river regulation. Later, in the 16th and 17th centuries, with the wide dissemination of printed material, the number of commentaries about floods increased, written by direct observers in numerous diaries and notes. Signs of high waters survived on walls of churches and secular buildings. The oldest authentic flood board, built in the year of occurrence of the flood it marks, and dating back to 1671, is located on the monastery of St Agnes Church in Cracow. Some flood boards were erected years after the flood events they commemorate (cf. Fig. 1).

Many flood marks have deteriorated, and many of them have disappeared. A landmark in the collection of flood related information was one of the largest

Fig. 1 Flood board on the cloister of the Norbertanki Convent, in Cracow, putting the then recent (1813) flood in perspective, by comparing it with the 1593 and 1697 floods. The board bears an inscription in old Polish: "*For information, in the year 1593 in the month of July and in the year 1697 in the month of August the water flooding was by 20 in.* [508 mm] *greater than the inundation which occurred on 26 August 1813, and is commemorated by the monument below*"

floods on the Vistula, in August 1813, after which more systematic observations of water levels were started in the Austrian partition of the Polish lands. How observations on the River Vistula carried out by the hydrological services of the three countries sharing Poland should be related to the hydrographic state of rivers and hydrotechnical constructions is difficult to verify at present. In the 19th century, intense regulation of the River Vistula started and subsequently influenced the time series of floods. Since the end of the 19th century, photographic and cartographic documentation has become an indispensable source of information about flood magnitude and damage.

Soon after Poland regained independence, in 1919, the Institute of Hydrology and Meteorology (State Hydrometeorological Service) was founded, which gave rise, in 1945, to the Polish Hydrometeorological Institute, that is now the Institute of Meteorology and Water Management—State Research Institute (IMGW-PIB). This institution has subsequently been responsible for measurements of precipitation and river stage.

The accuracy, homogeneity and uniformity of the data have only been sufficient for objective analysis of floods since the 1950s. Estimates of earlier floods have to be treated with caution, due to non-uniformity and non-homogeneity of the available source material and the small number of observations on floods. Source information on older floods is typically restricted to stage, without any attempt to compute the corresponding river flow (due to the lack of past rating curves). For many historical floods, however, information is only qualitative, so that even the water level is not known. Even if the number of floods, e.g. mentioned on record, is typically used as indicator, it is not a representative, robust and reliable aggregate characteristic, due to inherent limitations. For instance, in the case of ice-related flooding, the number of registered floods depends on the location of the ice-jam flood with regard to settlements, because events far away from larger settlements may have remained unnoticed.

4 Floods in the Upper Vistula River Basin Until 1972

The relatively high runoff from the Upper Vistula basin is a result of the hypsometric, geological and climatic conditions. This is an area of mainly mountainous and piedmont terrains with a dense river network. The 16th century map in Fig. 2 illustrates the features of flood generation. The map clearly shows the topographic

Fig. 2 A 16th century map illustrating development of a part of the Upper Vistula basin (sub-basins of large right-hand Carpathian tributaries, Soła and Skawa). Source: *Atlas of Abraham Ortelius, Theatrum Orbis Terrarum*, Antwerp, 1575

characteristics: mountains and highlands dominating in the south, forests in the north, and there were many settlements all around. Mountains with high precipitation, steep slopes, thin permeable soil layers, low storage, high runoff coefficients, and also numerous settlements with low infiltration capacity and water storage, have played an important role in flood generation. The network of right-bank Carpathian tributaries of the Upper Vistula has been a source of threat for the whole basin. It is estimated that the River Soła makes, on average, the largest contribution to the generation of floods on the Upper Vistula (in approx. 50 % of all flood events), followed by the Dunajec (40 %), the Skawa (30 %) and the Raba (25 %). The topography, geology and precipitation conditions in drainage basins render these rivers the main sources of the largest floods in the Upper Vistula Basin. However, this does not mean that other tributaries can be neglected. Information on past floods focused on events affecting large settlements located along the main course of the River Vistula, usually Cracow and adjacent towns. Systematic documentation for Carpathian tributaries of the River Vistula is limited and dates back to the end of 19th century. Documents referring to earlier events are rather sporadic.

There have been many anthropogenic factors that have influenced flood risk in the Upper Vistula basin throughout the centuries. Up to the end of the 10th century, accelerated human settlement fostered movement of solid material from the slopes towards the river valleys. This process of transport and accumulation triggered changes in river valleys and beds and increase in the range of inundation during intense summer rainfall (Starkel 2000). From the 11th to the 18th centuries, increases in population led to massive deforestation. From the 14th to the 16th centuries, the cultivated areas increased from 35 to 60 % of the basin area (Gieysztor 1982). Population growth prompted anthropogenic interference with the hydrological cycle. For example, a large flood in 1270 in Cracow triggered construction of the first river regulation structures on the Vistula (Trafas 1992). In the western part of the basin, in Upper Silesia, expansion of coal mining since the 14th century affected the hydrological regime, via changes in landscape—mass forest clearance, construction of coal and metal ore flushing machines and construction of drainage drifts in mines (Czaja 1994). Many settlements were established, as can be seen in the 16th century map covering the Soła and Skawa watersheds (Fig. 2). However, settlements were typically located in such a way that the river's right to flood the valley was respected. In the 19th century, the population grew further, and this resulted in the increase of the vulnerability to flooding. Later on, river regulation and the transformation of the Upper Vistula system (bypasses shortening the river course) were undertaken and further intense urbanization took place, which increased the velocity of water movement. Water engineering responded to the needs of both the use of the river for transportation of products, and flood protection of the areas in the vicinity of the river, which were reclaimed for urbanization. There were no regulations to guide site planning for river valleys (flood risk areas). The most serious changes in the river hydraulics of the Vistula included the shortening of the river course by 40 % and the narrowing of the natural valley by river embankments. In the Carpathian part of the Upper Vistula, training walls were built for river and mountain streams. They served and still serve as protection against damage from river bed washout and erosion of river banks during floods.

The consistent development of flood protection systems dates back to the end of the World War I. Earlier works were performed irregularly since 1830 and included cleaning of the Vistula riverbed, elimination of meanders and river branches and translocation of some tributaries, for example, the River Rudawa. Based on records and data about floods since 1813, at the beginning of the 1920s, a plan of works was developed, also utilizing previous projects, such as the concept of the Vistula bypass in Cracow, dating back to 1900. After the large flood of 1934, design works and construction accelerated (Kędzior 1934). These included extensions of retaining walls and construction of river embankments, storage reservoirs and dams and river training in mountain areas. The World War II and occupation of Poland stopped these construction activities, which were continued from the 1950s onwards.

The chronology of floods goes back to the early days of the Polish statehood. Girguś and Strupczewski (1965) mention, among others, the following large events: 1118, with inundations caused by long-lasting rain in spring and summer; 1221, with multiple inundations between Easter and autumn; 1253, with period of long-lasting and intense rain between Easter (20 April) and St Jacob's Day (25 July); 1270, with wet period and flooding on St Mary Magdalene's Day (22 July), with many fatalities; 1427, with rain from St Bartholomeo Day (24 August) until the onset of frost. Documents mention many historic floods, caused by intense and/or long-lasting, precipitation in summer, which occurred on the days of Roman Catholic saints such as St John (24 June), St Margaret (13 July), St Mary Magdalene (22 July), St Jacob (25 July), St Ann (26 July) and St Bartholomeo (24 August).

The chronology of great floods in the Upper Vistula basin from the 10th century until 1960, was prepared from documentary sources, and presented in the monograph of Bielański (1997). The chronology is not uniform, since the principal interest was in floods in last two centuries.

Punzet and Czulak (1988) examined volumes of flood waves of the greatest events in the Upper Vistula Basin from 1903 to 1972. Their conclusion was that the process of flood generation is very complex and depends on the intensity and duration of rainfall, as well as its spatial distribution and the direction of movement of the precipitation zone. It also depends on the degree of saturation of the basin in the interval before the flood as well as the configuration of flood waves on the Vistula and its major tributaries. The coincidence of major floods on the Upper Vistula and its Carpathian tributaries leads to disastrous floods. For, example, in 1934, the coincidence of flood waves on the Upper Vistula and Dunajec caused a large-scale inundation in the Vistula basin.

Punzet and Czulak (1988) distinguished five catastrophic floods on the Upper Vistula: July 1903 flood, during which whole districts of the City of Cracow were under water (in some streets, the water depth was up to 150 cm); violent June/July 1925 flood; July 1934 flood that broke levees in many places and created a lake of 10 km^2 area; July 1960 and July 1970 floods. Table 1 illustrates the estimates of the three largest flood wave volumes in 1903–1972 in five cross-sections of the Upper Vistula. The three largest flood waves in Table 1 represent basically three events:

Table 1 Three largest flood wave volumes, in million cubic meter, in 1903–1972 in five cross-sections of the Upper Vistula

Rank	1	2	3
Cracow-Tyniec	850 (1903)	660 (1925)	620 (1960)
Upstream of Dunajec confluence	1200 (1903)	1050 (1960)	1000 (1934)
Downstream of Dunajec confluence	1900 (1903)	1720 (1934)	1560 (1960)
Upstream of San confluence	2680 (1934)	2400 (1960)	2230 (1903)
Downstream of San confluence	3750 (1934)	2750 (1960)	2600 (1903)

Year of flood occurrence given in parantheses. Estimates after Punzet and Czulak (1988)

1903, 1934, and 1960 (in addition to 1925 flood ranked as second greatest in Tyniec). Table 1 clearly shows the significant contribution of the large tributaries: the Dunajec and the San to flood waves on the River Vistula proper in 1903, 1934 and 1960.

5 Floods in the Upper Vistula River Basin After 1972

After 1972, for several years there were no larger floods in the Upper Vistula Basin, except for the 1980 summer flood in the basin of the River San. There were larger floods elsewhere in Poland, such as the 1978/1979 snowmelt flood, 1980 summer flood caused by abundant and long-lasting precipitation and 1982 ice-jam flood. However, in last 20 years or so, there have been several large floods in the Upper Vistula River Basin, e.g. in 1997, 2001, and 2010.

At present, flood risk is high, resulting from population increase, subsequent urbanization and insufficient structural defences protecting the settlements.

During the turn of the 20th/21st centuries, rapid development of urbanization in Poland as well as in the Upper Vistula Basin took place, especially in larger cities and their surroundings. This process includes the development of road transport, where the density of networks has largely increased. In the Upper Vistula Basin, this process is particularly noticeable in Cracow and in its region, in nearby cities located on the River Vistula. This change largely influenced the 2010 flood and will continue to change the nature of the flood and their generating mechanisms. In this situation, quantitative detection of the causes, nature and consequences of floods is badly needed and is being developed. The biggest problem is the inability to effectively drain rain water from the area of the city during the development of the flood in the river, as happened in Cracow in 2010. Figure 3 shows the location of storm overflows for rainwater drained sewers in the city center. Their height is similar to the height of the boulevard along the Vistula, and the water stage of the 100-year flood exceeds the elevation of a few meters. During high water levels on the Vistula these transfers are closed so that the waters of the River Vistula do not penetrate into the city. The effect of this situation is often the accumulation of rainwater in the city (cf. Fig. 4).

Fig. 3 Cracow city center—location of storm overflows for sewer and storm sewage system. Source: *A Municipal Water Supply and Sewerage in Cracow*

Fig. 4 Cracow city center—Rondo Grunwaldzkie in the summer of 2010. Source: *Elzbieta Nachlik*

Fig. 5 Sandomierz region—the effect of levee failure in May and June of 2010. Source: *Koncepcja ochrony przed powodzią Wisły i jej dopływów w rejonie Sandomierza i Tarnobrzega – Projekt Studialny* (2009)

The problem of urban development in the area of the Upper Vistula Basin also embraces intense urbanization of riparian areas and allowing the development of vegetation in the area of the floodplain. The result is an increase of the water level corresponding to a flood wave on the River Vistula, longer persistence of high river stage conditions and thus increase of the possibility of levee failure. That happened in May 2010 in the Sandomierz region. The flooding river interrupted the embankment in the narrow part of the high-water channel of the River Vistula. Then, in June of 2010—during the second climax, weakened by a long lasting load, embankments situated below also failed (see Fig. 5). As a result, low-lying area of Sandomierz remained flooded for weeks. The volume of the flood wave is estimated at 200 thousand cubic meter (cf. Koncepcja ochrony przed powodzią Wisły i jej dopływów w rejonie Sandomierza i Tarnobrzega – Projekt Studialny 2009). Evaluation of the causes and extent of the threat, but also of the ways of restricting the flood is complex and requires multipath actions, Essential is creation of retention capacity lowering the size and the culmination of the flood wave.

6 Concluding Remarks

The history of floods of the Upper Vistula is an important research area, yet it is a difficult one to study. Several projects on historical hydrology have been undertaken in Poland and, as a result, much qualitative information on historical floods has become available. Over a long time in the history of Poland, observations of Vistula floods were carried out by the services of three countries that partitioned the Polish Kingdom.

Floods on the Upper Vistula are a complex phenomenon, caused by several generating mechanisms, with a relative frequency of occurrence specific for the different parts of the basin. Different flood types have been subject to various changes throughout centuries. Finally, using various types of data (instrumental records, documentary information, indirect and proxy data) of largely differing accuracy and reliability is not a trivial matter.

Floods on the Upper Vistula show a multi-factor dependence: juxtaposition of changes in climate, geomorphological evolution, land use (runoff generation), development of dams and embankments and their maintenance. There is a clear nonstationarity, but the course of the climate may have been overshadowed by immense land-use and land-cover changes: deforestation, urbanization and river regulation. However, there is also a Little-Ice-Age track in winter flood records. Winter floods seem to have exhibited a decreasing frequency of snowmelt and ice jam floods in the warming climate over much of the Vistula basin. Yet, from time to time a severe winter occurs, with cold spells and high snow cover. Shifted timing of high flows (snowmelt floods shifted towards winter) has been also observed.

Rating the development and related transformations and their impact on the origin and nature of the flood risk in the Upper Vistula Basin, tend to: (i) the search for technological solutions to ensure the safety of the existing infrastructure (see Fig. 6), (ii) seeking retention spaces that would significantly lower the flood water

Fig. 6 Destruction of the low boulevard of the River Vistula in the vicinity of the Dębnicki Bridge, in Cracow, resulting of the 2010 flood. *Self Source (Politechnika Krakowska)*

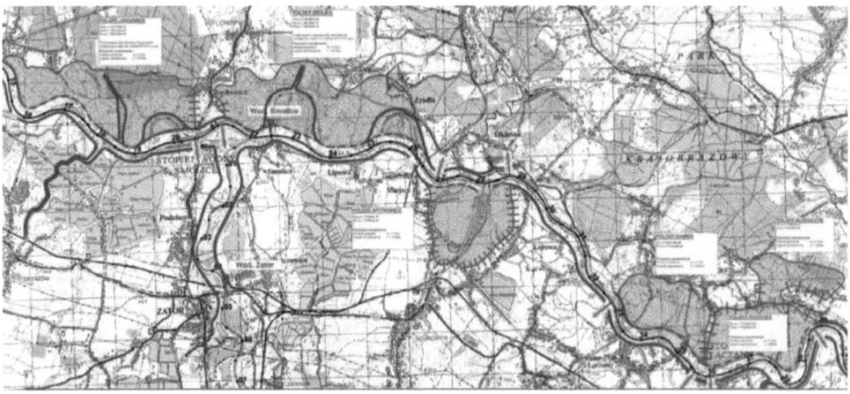

Fig. 7 The proposal of location of five flood polders in the valley of the River Vistula upstreams of Cracow for the reduction of flood peaks greater than 100-year flood: *Self Source (Politechnika Krakowska)*

level in urban areas (see Fig. 7) and (iii) the development of multi-scale action to improve the drainage, and conveyance of rainwater and flood water (cf. The Project Flood Risk Management Plan for the Upper Vistula Water Region 2015). This multi-stage and lengthy process that must be based on good and verified information and correct reference to the past.

Despite all the developments in different areas of technology, the flood risk continues to threaten riparian areas in the Upper Vistula basin. A look at the history of floods places the recent events in a broader perspective covering a very long period. The analysis of documentary information for the Vistula is a contribution to the growing pool of material on pre-instrumental floods in Central and Eastern Europe. It helps find area-covering events caused by the same weather patterns, and verify the dating.

References

Bielański AK (1997) Materiały do historii powodzi w dorzeczu górnej Wisły (Materials on the history of floods in the Upper Vistula Basin, in Polish). Monografia Politechniki Krakowskiej im. T. Kościuszki no 217, Kraków

Bojarski A, Nachlik E (2014) Upper Vistula Component description in conception note of the Odra-Vistula Flood Management Project, World Bank

Brázdil R, Kundzewicz ZW, Benito G (2006) Historical hydrology for studying flood risk in Europe. Hydrol Sci J 51(5):739–764

Cyberski J, Grześ M, Gutry-Korycka M, Nachlik M, Kundzewicz ZW (2006) History of floods on the River Vistula. Hydrol Sci J 51(5):799–817

Czaja S (1994) Przemiany stosunków wodnych w przestrzeni cywilizacyjnej Górnego Śląska od wieku XIV do wieku XVIII (Transformations of water conditions in the civilization space of Upper Silesia from the 14th to the 18th centuries, in Polish). *Rzeki – kultura, cywilizacja, historia,* 3

Czarnecka H (ed) (2005) Atlas podziału hydrograficznego Polski, cz. 2. Zestawienie zlewni, Instytut Meteorologii i Gospodarki Wodnej, Warszawa

Gieysztor A (1982) Wisła w średniowieczu (The Vistula in the Middle Ages, in Polish). In: Piskozub A (ed) *Wisła – Monografia rzeki* (Vistula—monograph of the river). Wydawnictwa Komunikacji i Łączności, Warszawa

Girguś S, Strupczewski W (1965) Wyjątki ze źródeł historycznych o nadzwyczajnych zjawiskach hydrologicznometeorologicznych na ziemiach polskich w wiekach od X do XVI (Excerpts from historical sources on unusual hydrological and meteorological phenomena on Polish territories from the 10th to the 16th centuries, in Polish), WKiŁ, Instrukcje i Podręczniki PIHM 87, Warszawa

Kędzior A (1934) W sprawie trwałego zabezpieczenia doliny Wisły i jej dopływów przed powodzią (On a durable flood protection of the Vistula valley and its tributaries, in Polish). Kraków

Koncepcja ochrony przed powodzią Wisły i jej dopływów w rejonie Sandomierza i Tarnobrzega – Projekt Studialny (2009) (The concept of protection against flood Vistula River and its tributaries in the Sandomierz and Tarnobrzeg region—study project, 2009), Contractors: Politechnika Krakowska, MGGP, Cermet-Bud, IMGW-PIB, Poland

Kundzewicz ZW, Szamałek K, Kowalczak P (1999) The great flood of 1997 in Poland. Hydrol Sci J 44(6):855–870

Mikulski Z (1955) Katastrofalne powodzie w Polsce (Catastrophic floods in Poland, in Polish). Czas. Geogr. 4:380–396

Projekt Planu Zarządzania Ryzykiem Powodziowym dla regionu wodnego Górnej Wisły (Project of Flood Risk Management Plan Upper Vistula water region) (2015), KZGW, Poland

Punzet J, Czulak J (1988) Bilanse odpływu wielkich wezbran Gornej Wisly w XX wieku. Wiadomości IMGW XI (XXXII) 3–4:3–16

Starkel L (2000) Heavy rains and floods in Europe during last millennium. Prace Geograficzne UJ 107:55–62 (Kraków)

Trafas K (1992) Zmiany biegu koryta Wisly pomiedzy ujsciem Przemszy a Sandomierzem (Changes of the course of the Vistula channel between the confluence of Przemsza and Sandomierz, in Polish). In: Wisla w dziejach i kulturze Polski. Zmiany biegu górnej Wisly i ich skutki. Wydawnictwa Uniwersytetu Warszawskiego

The Upper Vistula River basin flood protection program, by Nachlik E, et al (2011), Cracow

Palaeohydrology of the Upper Vistula River Basin

Leszek Starkel

Abstract This chapter briefly reviews the palaeogeomorphology and palaeohydrology of the Upper Vistula Basin. It starts from the history of the landscape of the Basin throughout the geological time scale. Next, changes in the hydrological regime since the Late Glacial to the present, are discussed. In particular, anthropogenic transformation of the hydrological regime of the Upper Vistula Basin in the last two millennia is examined. It can be stated that the floodiness (periods with more large floods) can be associated with both climatic and non-climatic variability and change.

Keywords Palaeogeomorphology · Paleohydrology · Landscape change · River Vistula · Poland

1 History of the Landscape of the Upper Vistula River Basin

Beginnings of the formation of the Upper Vistula River and its drainage basin are associated with the evolution of the Carpathian Mountains—the northernmost arc of the Alpine mountain system—and the formation of a fore deep in their northern foreland, where a Late Miocene sea poured. The first rivers dissecting the slopes of the mountains transported bulk alluvium towards the sea already in the Sarmatian and Pannonian periods. The coarse-grained Witów series is a remnant of this alluvium (Brud 2004). Along with the sea receding towards the east, a network of rivers began to form, that initially drained the mountain foreland towards the Black Sea, and in the western part (Silesia) towards the basin formed in the foreland of the Sudetes drained by the Odra river (Fig. 1a). In the Late Pliocene the meta-Carpathian rampart limiting the depression from the north was dissected by the Middle Vistula and the Wieprz rivers. Rivers draining Carpathian slopes from

L. Starkel (✉)
Department of Geoenvironmental Research, Institute of Geography and Spatial Organization, Polish Academy of Sciences, Św. Jana Str. 22, 31-018 Kraków, Poland

Z.W. Kundzewicz et al. (eds.), *Flood Risk in the Upper Vistula Basin*, GeoPlanet: Earth and Planetary Sciences, DOI 10.1007/978-3-319-41923-7_14

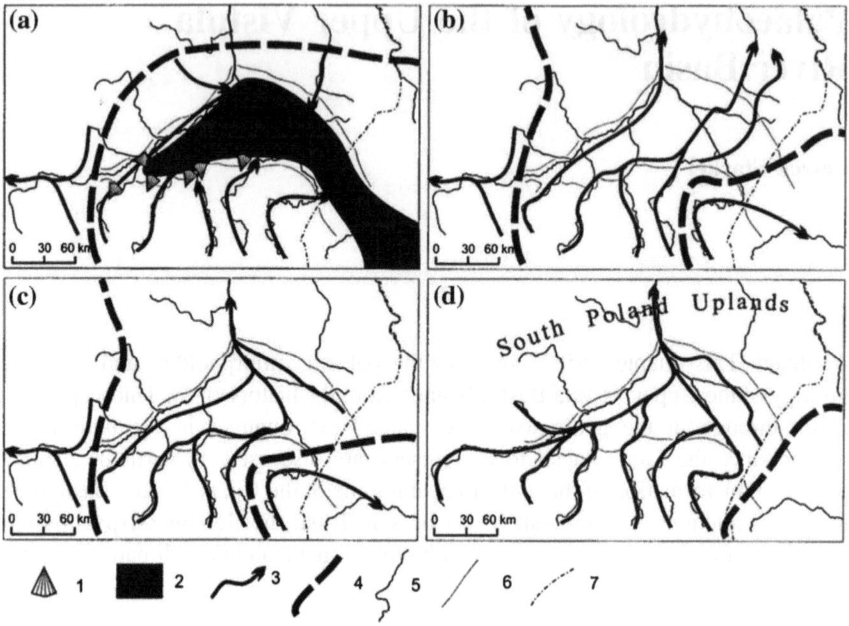

Fig. 1 Drainage pattern changes in the area of contemporary Upper Vistula Basin (compiled by Brud 2004). **a** Late Sarmatian, **b** Eopleistocene, **c** after the San Glaciation, **d** after the Oder Glaciation. *1* Fan-deltas formed into retreating sea, *2* marine basin during the Sarmatian-Pannonian, *3* palaeoflow directions, *4* watersheds divide, *5* contemporary drainage pattern, *6* boundaries of the Sandomierz Basin, *7* eastern state boundary

the Raba to the San were captured by a north-running river (Fig. 1b). This is indicated by pre-glacial gravels of Carpathian origin, that were found in the southern Masovia, north of the Holy Cross Mountains (Pożaryski et al. 1994). So, even before the invasions of Scandinavian ice sheets, the northern slopes of the Carpathians were drained towards the valleys of the middle Odra, middle Vistula and Dniester.

The ice sheet of San 2 glaciation (approx. 500 ka BP) entered not only into sub-Carpathian depressions, but also climbed onto the northern part of the Carpathian Foothills, in the western part leaning on the marginal zone of the Beskidy Mountains (Fig. 1c). During the ice-sheet recession, glacio-fluvial waters flowed out towards the Black Sea (Klimaszewski 1948). In the interglacial, drainage to the north, towards the emerging pre-Baltic basin, was renewed.

During the next, Odra glaciation, ice-sheet transgression in the western part had a similar range along the Upper Odra as during the previous glaciation, but to the east it did not crossed the meta-Carpathian rampart with its highest part—the Holy Cross Mountains. Waters from the glacier flowed out toward the east and permanently captured waters of the source reach of the Vistula as well as of the Soła and the Skawa rivers, directing them via the Cracow Gate to the Sandomierz Basin (Lewandowski

1996). Since then, the extent of the Upper Vistula basin, including also the southern slope of the upland belt, has not changed. During more recent glaciations and interglacials, the Vistula gradually deepened its valley and filled it with alluvium.

During cold periods, in the mountainous catchments, braided rivers carried out sediment delivered from deforested slopes, while the mountain glaciers developed only in the Tatras. In the interglacial periods, the deepening of valleys took place in the valley sections raised by tectonic movements. In the foreland basins, stages of sediment accumulation and of dissection of the covers occurred interchangeably (Klimaszewski 1948).

2 Changes in the Hydrological Regime Since the Late Glacial to the Present

Late Glacial period (13–10 ka BP ^{14}C, 15–11.5 ka BP) was characterized by a general warming with recurrent cold conditions, simultaneous disappearance of permafrost and the encroachment of forest communities. Change of the hydrological regime followed. Snowmelt floods and sediment delivery from slopes decreased in significance, whereas aquifers were reconstructed. Braided channels of multi-thread rivers were replaced by asymmetric channels with large meanders indicating substantial flows, known from the valleys of the Vistula and its tributaries, the San and the Wisłoka (Fig. 2). The radius of the meanders was 3–4 times greater than those formed in the Holocene and the width was 2–3 times greater, indicating the average discharge being 5–10 times greater and the bankfull discharge even more (Table 1). The age of decline in their activity, determined based on the dating of the bottom of organic sediments filling the cut-off bends, ranges from 12,500 to 9500 years BP ^{14}C (Starkel et al. 1991). Large paleomeanders of the Late Glacial age were not recorded in the valleys of the Soła and the Dunajec, as a result of steeper channel gradient and later development of forest cover in the higher parts of the Carpathians. These rivers transported coarse bedload and longer retained regime typical of braided river systems. Next fairly drastic reduction of flows followed with the development of forest cover, limited supply of sediment and diminishing role of snowmelt period. The occurrence of small paleomeanders is characteristic of different parts of the Holocene. The whole system of oxbow lakes was usually preserved at rapid channel avulsion (Fig. 3). Exceptionally, in the San valley, a small paleomeander remained, inserted in an older, large paleomeander (Fig. 2) dated to 8560 ± 100 BP ^{14}C, associated with the first flood phase of the Early Holocene (Szumański 1982).

During the Holocene, several humid phases were recorded, with frequent floods, higher sediment transport and channel avulsions. They were dated at several tens of sites (Fig. 4) for the following periods in radiocarbon years: 8.5–7.7, 6.5–6.0, 5.5–4.9, 4.4–4.1, 3.5–3.0, 2.7–2.4, 2.2–1.8 ka BP, as well as in the 5th–6th, 10th–11th, 14th–17th centuries (Starkel 1983; Kalicki 1991; Starkel et al. 1996). As demonstrated by the system of river channels in the Grobla forest (Fig. 3), such a phase started by

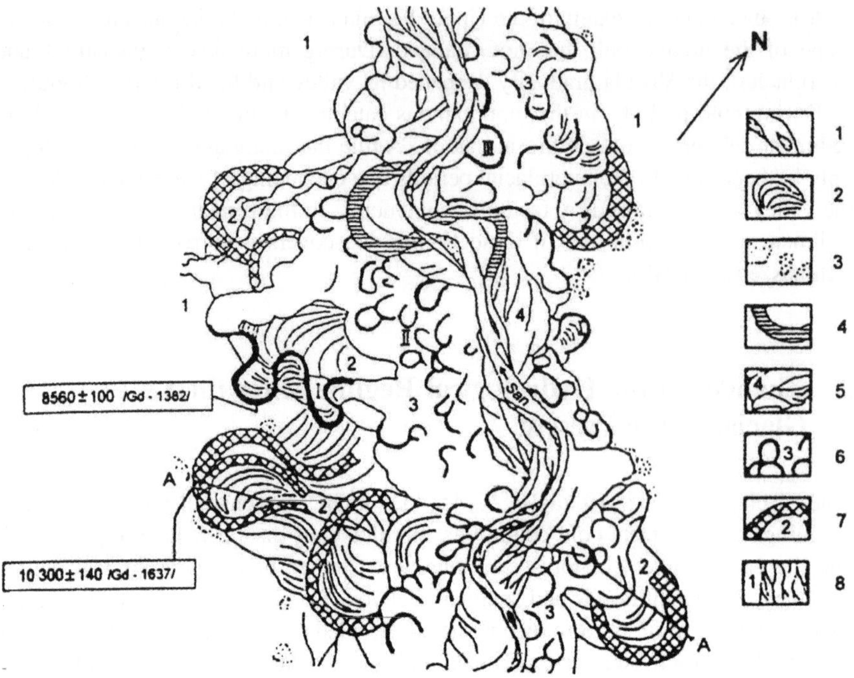

Fig. 2 Relief of the San valley floor in the Sandomierz Basin (after Szumański 1982). *1* San river channel with bars, *2* older meander bars, *3* dunes, *4* meanders cut after regulation in 1903, *5* floodplain with contours of braided channels (*4*), *6* Holocene terrace (*3*) with small palaeomeanders of three generations, *7* Late Glacial terrace (*2*) with large palaeomeanders, *8* Vistulian terrace (*1*) with contours of braided channels

Table 1 Parameters of selected paleomeanders and their reconstructed paleodischarges (after Starkel et al. 1996)

Valley locality *river*	Age (after ^{14}C or palynology)	Channel width *W* (m)	Meander radius *R* (m)	Q_{mean} (m^3/s)	Q_b (m^3/s)	References
Vistula Zabierzów Bocheński	early 19-th c.		580		1214	Trafas (1975)
Vistula Śmiłowice	>3090 ± 140	~90	~340	48	426	Kalicki unpublished
Vistula Grobla Forest (younger system)	>5090 ± 110	58 (45–70)	125–165 or sinuous	28.0	60–103	Starkel and Kalicki (1984), Gębica and Starkel (1987), Starkel et al. (1991)
Vistula Lasówka	>7980 ± 160	(30–50)	130–240	14.8	65–215	Kalicki (1991)
Vistula Borek	>9800 ± 80	(160–180)	670	176	1611	Kalicki unpublished

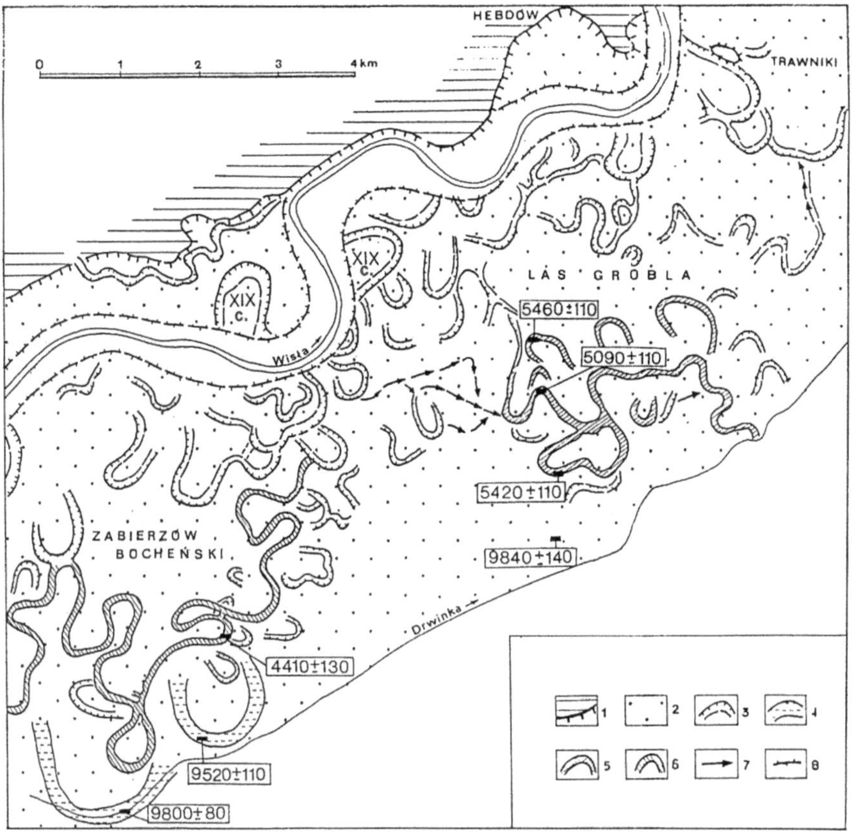

Fig. 3 Geomorphology of the Vistula valley floor near Zabierzów Bocheński and Grobla forest (after Starkel et al. 1991; Kalicki et al. 1996). *1* Loess terrace, *2* Holocene floodplain, *3* palaeochannels (various), *4* palaeomeanders from the Late Glacial—Holocene transition, *5* palaeochannels abandoned before 5 ka BP, *6* palaeochannels abandoned about 4.4 ka BP, *7* crevasse channels, *8* flood embankments

cutting-off bends of tortuous meandering system and a gradual widening and straightening of the channel, ending either by creating a braided system or, finally, a channel avulsion (Fig. 5).

Occurrence of sub-fossil black trunks of oaks in channel or channel-fill sediments, that were mainly delivered by bank erosion, is also an indicator of frequent flooding (Krąpiec 1996). The number of trunks from a given century, accumulated in the sediments exploited in a single gravel pit goes in several tens (Fig. 6). This happened in the second half of the 5th century, the first part of the 6th century and the end of the 11th century. Frequent flooding is also indicated by a number of abandoned channels of the same age (Fig. 7).

Examination of an alluvial fan of a small stream Maga, a tributary to the Wisłoka in Podgrodzie, allowed to find a series of flood deposits dated between 8390 ± 105

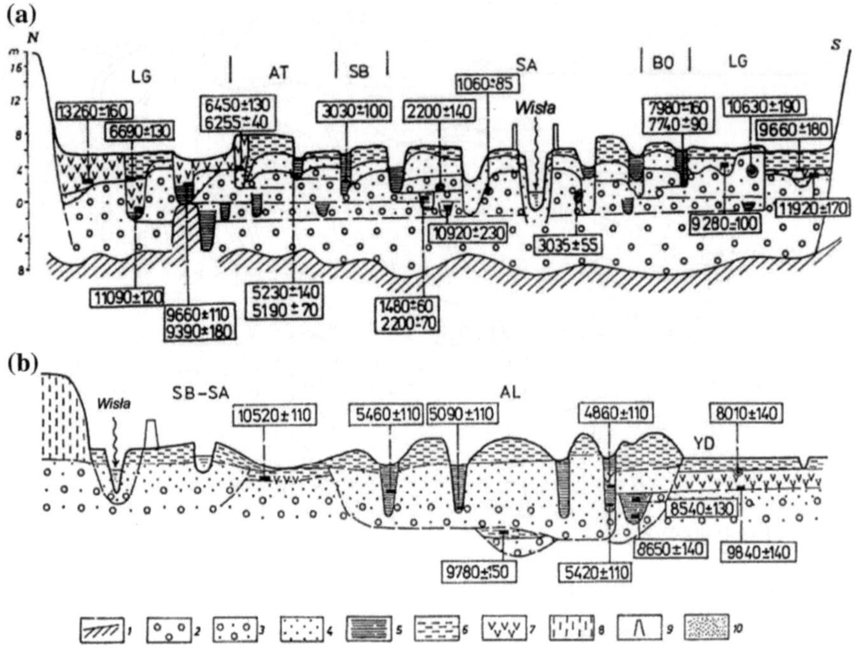

Fig. 4 **a** Synthetic cross-section of the River Vistula valley floor: (**a**) near Cracow (after Kalicki 1991), and **b** 20 km downstream of Cracow in the Grobla forest (after Starkel et al. 1991). *1* Miocene substratum, *2* gravels, *3* gravels and sands, *4* sands, *5* silts and other palaeochannel fills, *6* silty overbank deposits, *7* peat, *8* loess, *9* flood embankments; *LG* Late Glacial, *AL* Alleröd, *YD* Younger Dryas, *BO* Boreal, *AT* Atlantic, *SB* Subboreal, *SA* Subatlantic

and 7785 ± 145 BP ^{14}C containing about 95 flood events, grouped in 13 clusters, embracing several events, separated by layers of initial soil or peat indicating a break in sedimentation of mineral material lasting from a few to more than ten years (Niedziałkowska et al. 1977; Czyżowska 1997). It was a sedimentation not yet associated with human disturbance, older than the early Neolithic settlement. This site correlates with multiple profiles recording channel avulsions, mineral intercalations in peat (e.g. the Tarnawa peat bog in the Bieszczady Mountains; Ralska-Jasiewiczowa 1980), the formation of several major landslides in the Carpathians (Gil et al. 1974; Margielewski 2006), track of coarse debris flow in the lake sediments in the Tatra (Kotarba and Baumgart-Kotarba 1997), and finally raising the water level in Lake Gościąż known from its annual lamination of sediments, in the valley of the Lower Vistula (Ralska-Jasiewiczowa et al. 1998).

The greatest number of sites from particular wet phases with dozens of sediment dates and channel avulsions were found in the section of the expanding valley floor of the Vistula River between a gorge in the Cracow Gate and the mouth of the Raba (Kalicki 1991; Starkel et al. 1996; Gębica 1995; Figs. 3 and 4). In the material filling in one of paleochannels, records of two younger phases of flooding were identified (Kalicki et al. 1996).

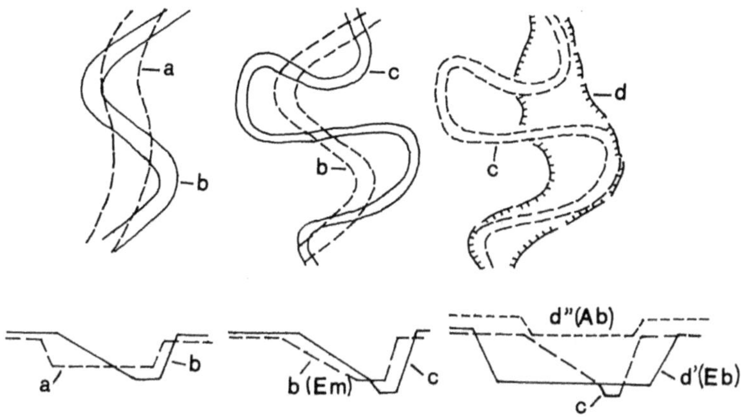

Fig. 5 Changes of channel pattern and cross-section during cyclic variations of hydrological regime (after Starkel 1983). *a* Straight channel of braided river, *b* sinuous channel, *c* deeper meandering channel modelled by erosion (Em), *d* straight braided channel formed during the phase with higher flood frequency transformed by incision (Eb) or by aggradation (Ab)

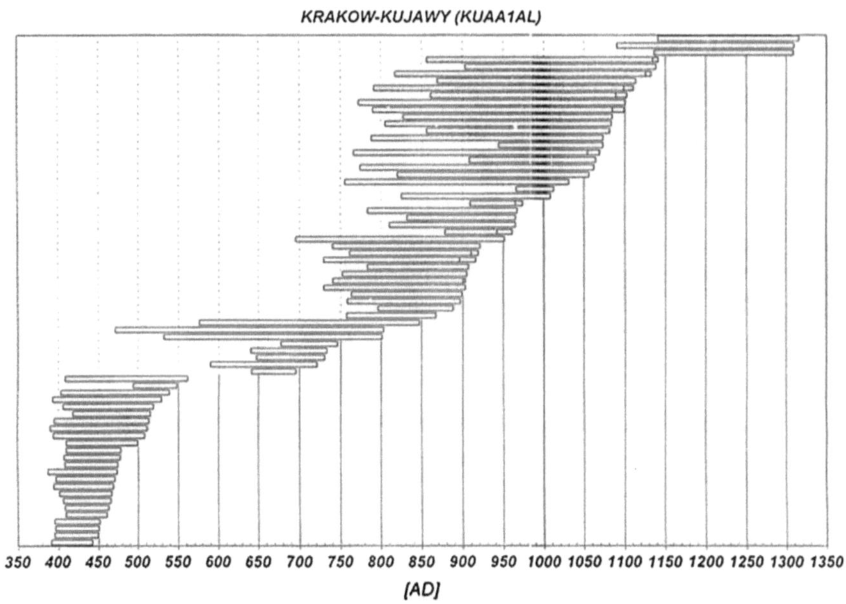

Fig. 6 Subfossil oak trunks from a gravel pit at Kujawy, indicating frequent flood years in the periods 450–550, 900–960 and 1050–1100 AD (after Krąpiec 1996)

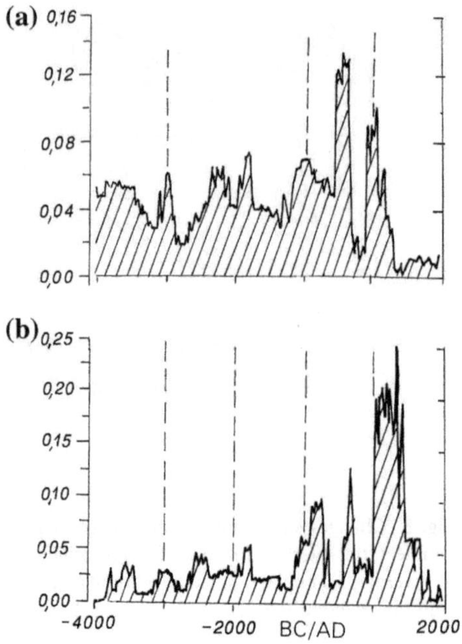

Fig. 7 Various frequency of dated palaeochannel fills (above **a**) and dated organic horizons covered by overbank deposits (below **b**). In first case **a** diagram—it indicates a humid phase in the 5th–6th centuries. AD, in the second case **b** diagram indicates heavy soil erosion in the 1st–3rd centuries and the 11th–14th centuries (based on Starkel et al. 2006)

Towards the east the outlets of Carpathian tributaries of the Vistula from the mountains are generally narrower, their valley sections are getting longer, younger alluvial sediments either overlie the older ones or are inserted in them. Increasing human interference since the Roman period makes it difficult to separate its impacts from purely climatic conditions.

Smaller left-hand tributaries of the Vistula, draining the upland belt largely underlain by carbonates and partly covered by loess, were characterized by smaller floods and gradual aggradation of floodplains forced by the aggradation of the Vistula fed with abundant material delivered by Carpathian tributaries.

3 The Role of Man in Transforming the Hydrological Regime in the Last Two Millennia

Human interference into the circulation of water and soil erosion in the Upper Vistula drainage basin began at the beginning of the Neolithic period. This is indicated by an Early Neolithic settlement on the edge of a loess terrace at Pleszów near Cracow over the old Vistula paleochannel abandoned prior to the year 6255 ± 40 BP. There are numerous pollens of wheat and barley in the dated level of the peat, as well as thin layers of silt washed away from loess soils (Wasylikowa et al. 1985). Layers of namints in the valleys of the Szreniawa and the Nidzica, densely populated by humans, also date back to the Neolithic (Śnieszko 1985).

The sharp increase in deforestation and cultivation observed in the Roman period is marked by an increase in the thickness of overbank facies sediments to 2–3 m and with increasing fraction from 6–10 to 4–8 phi, in numerous cases deposited on dated fossil soils (Figs. 3 and 4). Stumps of freshly cut oak or trunks processed by man are often buried there (Kalicki 1991). The collapse of agriculture in the 5th–7th centuries was synchronous with climate wettening and the collapse of the late Roman culture, and multiple channel avulsions recorded in the trunks found in oxbow-lake sediments (Fig. 7) (Krąpiec 1996).

Enhanced soil erosion occurred again in the 10th–11th centuries, resulting in the deposition of deluvial sediments, often encroaching on peaty soil, recorded in dozens of sites (especially in loess areas), (Starkel 1995).

Archaeological research in Cracow showed distinct settlement phases in the 10th and the 12th–13th centuries, interrupted by flood periods (Fig. 8). Deforestation of

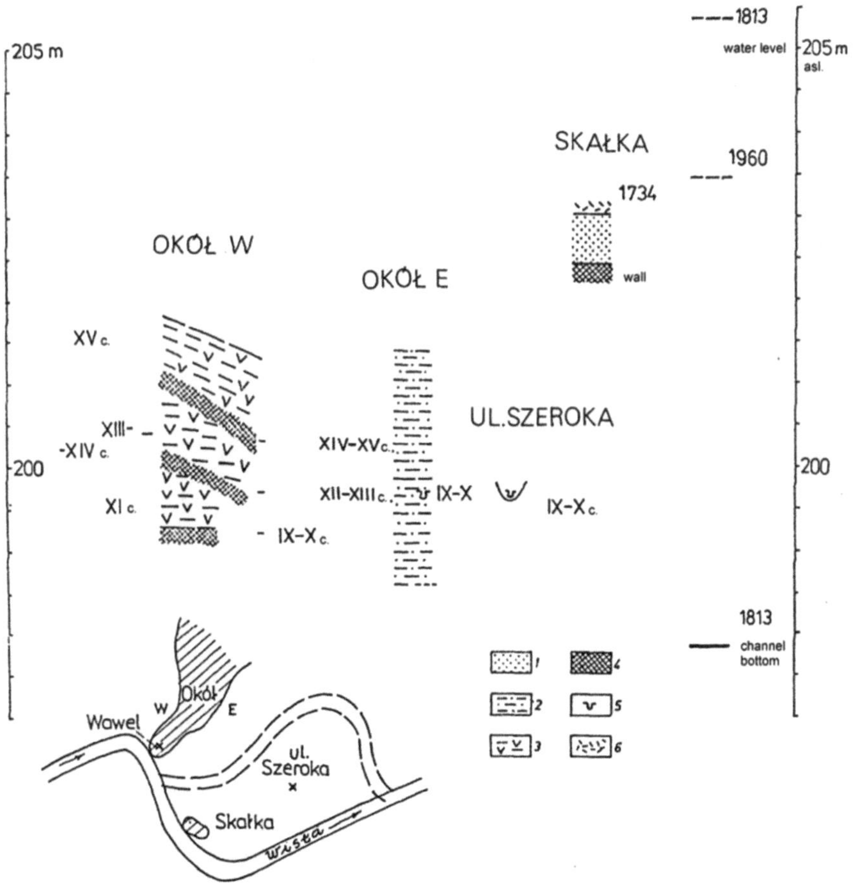

Fig. 8 Deposits of historical floods of Vistula in Cracow (based on the studies of Radwański studies, compiled by Starkel 2005). *1* Sand, *2* sandy silts, *3* silts with organic material, *4* cultural horizons, *5* anthropogenic hollows, *6* anthropogenic debris. On the right side of the figure, dated flood levels are indicated

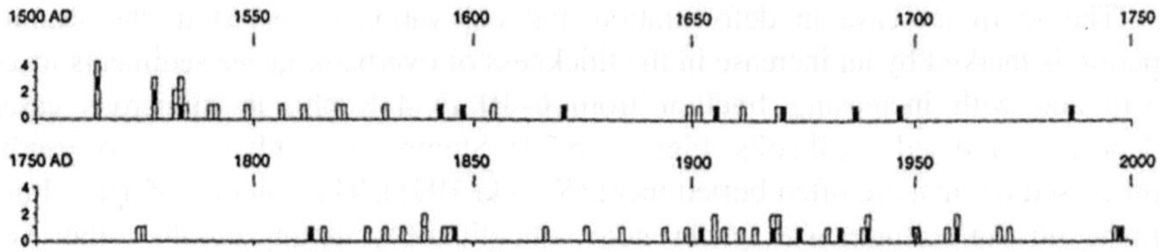

Fig. 9 Clusters of floods in the Vistula valley at the Carpathian foreland (after Starkel 1995); diagram based on historical records collected by Girgus and Strupczewski (1965) and Bielański (1984)

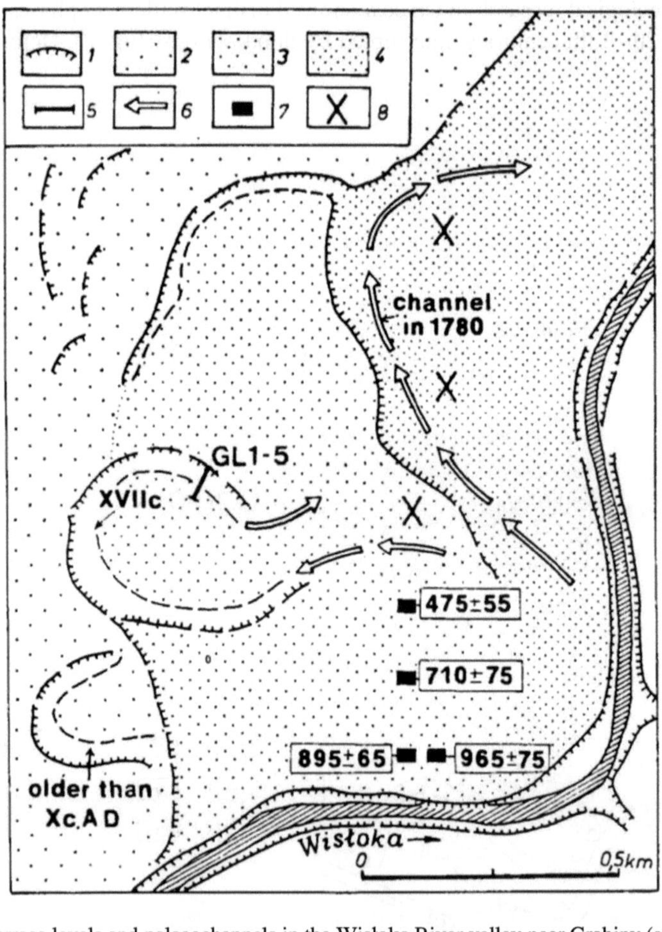

Fig. 10 Terrace levels and palaeochannels in the Wisłoka River valley near Grabiny (after Starkel 2001). *1* Edges of terraces, *2* terrace level 8–9 m, *3* terrace level 7 m, *4* terrace level 4 m, *5* corss-section, *6* course of palaeochannel, *7* dated buried trees

the Silesian Beskidy Mountains from the 10th century was especially clearly marked in the form of thick silty-sandy loams deposited on the tectonically lowered foreland (Niedziałkowska et al. 1985). In most mountainous areas, deforestation and the increase of water runoff accompanied the settlement of new areas until the 15th–17th centuries. In 1515, in Cracow, four floods were recorded; during three of them water burst into the Bernardine Church (Fig. 9). In the 17th and 18th centuries, when cultivation of potatoes and other root crops started, accumulation of overbank deposits increased (Klimek 1974), as did the size of meanders, and some meandering channels were transformed into braided ones (Fig. 10). Sometimes, meander parameters are close in size to the great paleomeanders of the Late Glacial. This is recorded by old topographical maps prepared since the late 18th century (Klimek and Trafas 1972), indicating multi-thread channels and the simultaneous tendency to straighten river course. Some sites record a rapid response of enhanced soil erosion to deforestation in the stage of settlement of the Bieszczady Mountains on the turn of the 15th and 16th centuries (Starkel 1995). A similar effect in the Bieszczady Mountains was due to a massive forest logging in the 18th–19th centuries—formation of alluvial cones at the mouths of small streams, synchronically with the widening of larger braided channels (Kukulak 2003). This period coincides in part with the Little Ice Age when higher frequency of floods was recorded, which undoubtedly contributed to the development of river braiding. A series of floods in the 19th century prompted authorities of Galicia to begin channel regulation works at the lower sections of the Upper Vistula valley in the foreland of the Carpathians (Fig. 11). Deepening straightened channels and the construction of flood levees near the channels did not sufficiently counteract the devastating floods, whereas flushing out of sediment down the river has led to the formation of a braided channel in the middle course of the River Vistula in the part of the Polish territory incorporated to the Russian Empire. Aggradation in a narrow inter-levee strip

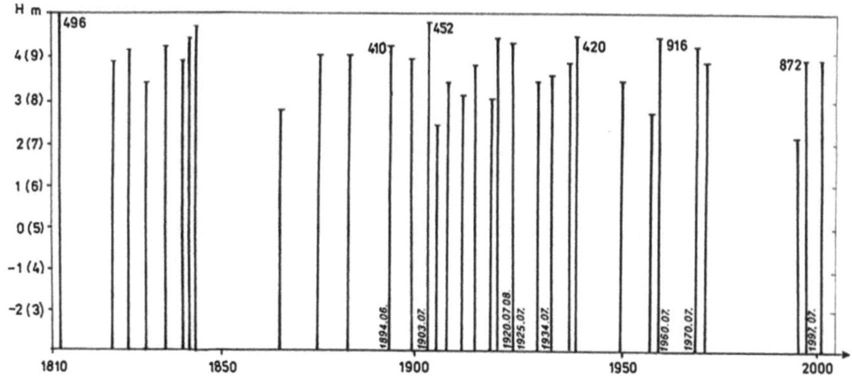

Fig. 11 Floods on the River Vistula in Cracow in the 19th and 20th centuries (after Bielański 1984). Note distinct clusterings of floods in the years 1825–1845 and 1903–1940 (reproduced from Starkel 1995)

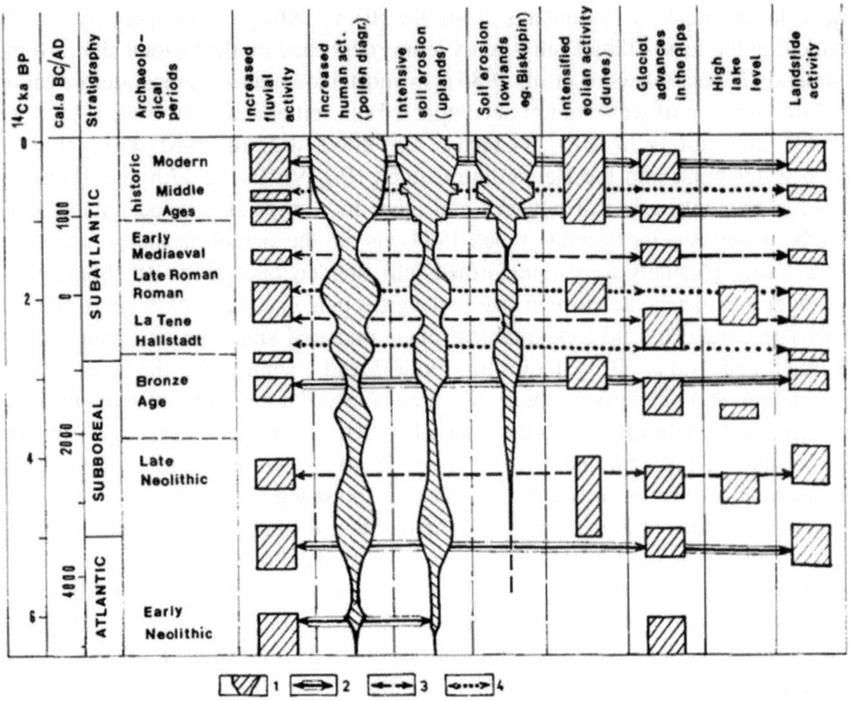

Fig. 12 Concordance and discordance of climatic and anthropogenic factors in the formation of phases of high-intensity geomorphic processes and extreme events on the territory of Poland (after Starkel 2005). *1* The width presents occurrence and intensity of events or factors, *2* concordance of climatic and anthropogenic factors in formation of active phase, *3* phase conditioned by more humid climate, *4* phase conditioned by human activity. Explanation of columns: *1* stratigraphy, *2* archeological periods, *3* fluvial activity, *4* human activity (after pollen diagrams), *5* soil erosion (uplands), *6* soil erosion (lowlands), *7* eolian activity, *8* advances of alpine glaciers, *9* high lake level, *10* landslide activity in the Carpathians

caused that, in turn, more and more land "taken out from the flood plain" is unexpectedly flooded. Only the construction of polders can counteract floodings.

Analysis of comparisons of climate change and periods of vigorous human activity in the last millennia in the Upper Vistula drainage basin allowed to distinguish three types of intense activity of rivers (Starkel 2005; Fig. 12):

(i) only climatically conditioned (greater frequency of floods);
(ii) conditioned anthropogenically (enhanced soil erosion and sediment transport);
(iii) coincidence of both factors (periods of most effective activity).

In the 20th century, in mountain valleys, one can observe the reverse trend to deepen the channels even in solid rock, which some experts try to explain (incorrectly) by general tectonic movements. In fact, the reasons are complex. Until recently, bed material was exploited directly from river channels. At the same time, a common phenomenon is recession of cultivation on hillslopes and restoration of forest communities, progressing at a different pace (Klimek 1983; Wyżga 2008). The effect of these various changes is deepening of river channels in the last century, sometimes by up to 2–3 m.

So, the roots of the Upper Vistula drainage basin can be regarded as historically connected with the Mediterranean domain of uplifted Alpide belt as well as shaped by the invasion of the Scandinavian ice sheets.

References

Bielański A (1984) Materiały do historii powodzi w dorzeczu Górnej Wisły. In: Fisher J (ed) Politechnika Krakowska, Kraków (in Polish)

Brud S (2004) Palaeogeography of the Western Sandomierz Basin in Late Neogene and early Quaternary times (Carpathian Foredeep, South Poland). Annales Soc Geologorum Poloniae 74:63–93

Czyżowska E (1997) Zapis zdarzeń powodziowych na pograniczu boreału i atlantyku w osadach stożka napływowego w Podgrodziu. Dokumentacja Geograficzna 5 (in Polish)

Gębica P (1995) Evolution of the Vistula valley and alluvial fans of the Raba and Uszwica rivers between Uście Solne and Szczurowa in the Vistulian and Holocene. In: Starkel L (ed) Evolution of the Vistula river valley during the last 15,000 years, part V. Geographical studies, IGiPZ PAN, Special Issue 8:31–50

Gębica P, Starkel L (1987) Evolution of the Vistula river valley at the northern margin of the Niepołomice Forest during the last 15,000 years. In: Starkel L (ed) Evolution of the Vistula river valley during the last 15,000 years, part II. Geographical studies, IGiPZ PAN, Special Issue 4:71–86

Gil E, Gilot E, Kotarba A, Starkel L, Szczepanek K (1974) An early Holocene landslide in the Niski Beskid and its significance for palaeogeographical reconstructions. Studia Geomorphologica Carpatho-Balcanica 8:69–83

Girguś R, Strupczewski W (1965) Wyjątki ze źródeł historycznych o nadzwyczajnych zjawiskach hydrologiczno-meteorologicznych na ziemiach polskich w wiekach od X do XVI. In: Rojecki A (ed) WKiŁ, PIHM, Warszawa p 209 (in Polish)

Kalicki T (1991) The evolution of Vistula river between Cracow and Niepołomice in late Vistulian and Holocene times. Geographical studies IGiPZ PAN, Special Issue 6:11–37

Kalicki T, Starkel L, Sala J, Soja R, Zernickaya VP (1996) Subboreal paleochannel system in the Vistula valley near Zabierzów Bocheński (Sandomierz Basin). In: Starkel L (ed) Evolution of the Vistula River valley turing the last 15,000 years, part VI. Geographical studies, IGiPZ PAN, Special Issue 9:129–158

Klimaszewski M (1948) Polskie Karpaty Zachodnie w okresie dyluwialnym. Prace Wrocławskiego Towarzystwa Naukowego B7, Wrocław (in Polish)

Klimek K (1974) The retreat of alluvial river banks in the Wisłoka valley. Geographia Polonica 28:59–75

Klimek K (1983) Erozja wgłębna dopływów Wisły na przedpolu Karpat. In: Kajak Z (ed) Ekologiczne podstawy zagospodarowania Wisły i jej dorzecza. PWN, Warszawa-Łódź (in Polish), pp 97–108

Klimek K, Trafas K (1972) Young Holocene changes in the course of the Dunajec river in the Beskid Sądecki Mts. (Western Carpathians). Studia Geomorphologica Carpatho-Balcanica 6:85–92

Kotarba A, Baumgart-Kotarba M (1997) Holocene debris-flow activity in the light of lacustrine sediment studies in the High Tatra Mountains, Poland. Paläoklimaforschung 19:147–158

Krąpiec M (1996) Dendrochronology of "black oaks" from river valleys in Southern Poland. In: Starkel L (ed) Hydrological changes of valley floor in the upper Vistula basin during late Vistulian and Holocene. Geographical studies, Special Issue 9:61–78

Kukulak J (2003) Impact of mediaeval agriculture on the alluvium in the San river headwaters (Polish Eastern Carpathians). Catena 51:255–266

Lewandowski J (1996) The main factors of Neogene and Quaternary morphogenetic evolution of the Silesian-Cracow region, S Poland. Acta Geographica Lodziensia 71:131–146 (in Polish, English summary)

Margielewski W (2006) Records of the Late glacial-Holocene palaeoenvironmental changes in landslide forms and deposits of the Beskid Makowski and Beskid Wyspowy Mts. area (Polish Outer Carpathians). Folia Quaternaria 76:149

Niedziałkowska E, Skubisz A, Starkel L (1977) Litology of the Eo- and Mezoholocene alluvia in Podgrodzie upon Wisłoka river. Studia Geomorphologica Carpatho-Balcanica 11:89–100

Niedziałkowska E, Gilot E, Pazdur M, Szczepanek K (1985) The evolution of the Upper Vistula valley in the region of Drogomyśl in the upper Vistulian and Holocene. Folia Quaternaria 56:101–132

Pożaryski W, Maruszczak H, Lindner L (1994) Chronostratygrafia osadów plejstoceńskich i rozwój doliny Wisły środkowej ze szczególnym uwzględnieniem przełomu przez wyżyny południowopolskie. Prace Państwowego Instytutu Geologicznego 147 (in Polish)

Ralska-Jasiewiczowa M (1980) Late Glacial and Holocene of the Bieszczady Mountains (Polish Eastern Carpathians). PWN, Warszawa

Ralska-Jasiewiczowa M, Goslar T, Madeyska T, Starkel L (eds) (1998) Lake Gościąż, Central Poland. A monographic study. W. Szafer Institute of Botany, Polish Academy of Sciences, Kraków

Śnieszko Z (1985) Paleogeografia holocenu w dolinie Sancygniówki. Acta Geographica Lodziensia 51 (in Polish)

Starkel L (1983) The reflection of hydrologic changes in the fluvial environment of the temperate zone during the last 15,000 years. In: Gregory KJ (ed) Background to palaeohydrology. Wiley, Chichester, pp 213–234

Starkel L (1995) Odbicie ekstremalnych wezbrań okresu historycznego w osadach rzecznych i stokowych w dorzeczu górnej Wisły. Acta Universitatis Nicolai Copernici, Geographia 27:13–20 (in Polish)

Starkel L (2001) Historia doliny Wisły (od ostatniego zlodowacenia do dziś). Monografie Geograficzne, IGiPZ PAN, 2 (in Polish)

Starkel L (2005) Anthropogenic soil erosion since the Neolithic time in Poland. Zeitschrift für Geomorphologie 139:189–201

Starkel L, Kalicki T (1984) Paleomeandry Wisły w uroczysku Grobla (Puszcza Niepołomicka). In: Holocen okolic Krakowa. Materiały Sympozjum, 18–20 June 1984, AGH, Kraków pp 58–63

Starkel L, Gębica P, Niedziałkowska E, Podgórska-Tkacz A (1991) Evolution of both the Vistula floodplain and lateglacial-early Holocene palaeochannel systems in the Grobla forest (Sandomierz Basin). In: Starkel L (ed) Evolution of the Vistula river valley during the last 15,000 years, part IV. Geographical studies, IGiPZ PAN, Special Issue 6:87–99

Starkel L, Kalicki T, Krąpiec M, Soja R, Gębica P, Czyżowska E (1996) Hydrological changes of Valley floors in Upper Vistula basin during the last 15,000 years. In: Starkel L (ed) Evolution of the Vistula river valley during the last 15,000 years, part VI. Geographical studies, IGiPZ PAN, Special Issue 9:7–128

Starkel L, Soja R, Michczyńska DJ (2006) Past hydrological events reflected in Holocene history of Polish rivers. Catena 66(1–2), 24–33

Szumański A (1982) The evolution of the Lower San River Valley during the late glacial and the Holocene. In: Starkel L (ed) Evolution of the Vistula river valley during the last 15000 years. Geographical studies, IGiPZ PAN, Special Issue 1:57–78

Trafas K (1975) Zmiany biegu koryta Wisły na wschód od Krakowa w świetle map archiwalnych i fotointerpretacji. Zeszyty Naukowe UJ, Prace Geogr 40 (in Polish)

Wasylikowa K, Starkel L, Niedziałkowska E, Skiba S, Stworzewicz E (1985) Environmental changes in the Vistula valley at Pleszów caused by Neolithic man. Przegląd Archeologiczny 33:19–55

Wyżga B (2008) A review on channel incision in the Polish Carpathian rivers during the 20-th century. In: Habersack H, Piegay H, Rinaldi M (eds) Gravel-bed rivers VI: from processes understanding to river restoration. Elsevier, Amsterdam, pp 525–555

Part IV
Projections and Adaptation

Projections of Precipitation in the Northern Foothills of the Tatra Mountains

Iwona Pinskwar, Adam Chorynski, Zbigniew W. Kundzewicz and Virginia Ruiz-Villanueva

Abstract Floods are usually related to extreme and/or long-lasting intense precipitation events. In warmer climate, future precipitation extremes could be higher than nowadays. Assessment of these future changes and climate adaptation to future flood risk are very important issues. In this study, four regional climate models and seven global climate models for two climate scenarios A1B and A2 were used to get better description of the range of changes in annual as well as extreme precipitation events. With help of the delta-change method, projections were made based on responses from regional and global climate models, for 11 precipitation stations in the Tatra Mountains in Poland, for which observation data for 1961–1990 were available. Analyses were made of various indices, such as annual totals, maximum 24 h, 5-day; 10-day, monthly maximum sum of precipitation and also numbers of days with intense precipitation equal or above the thresholds of 30 and 50 mm. It was found that all RCM and GCM models under examination project an increase in mean annual precipitation totals as well as in heavy precipitation in two future time windows considered (2061–2090 and 2080–2100).

Keywords Extreme precipitation · Climate models · Projections · Delta-change method

I. Pinskwar (✉) · A. Chorynski · Z.W. Kundzewicz
Institute for Agricultural and Forest Environment, Polish Academy of Sciences, Poznan, Poland
e-mail: iwona.pinskwar@isrl.poznan.pl

Z.W. Kundzewicz
Potsdam Institute for Climate Impact Research, Potsdam, Germany

V. Ruiz-Villanueva
Dendrolab.ch, Institute of Geological Sciences, University of Bern, Bern, Switzerland

Z.W. Kundzewicz et al. (eds.), *Flood Risk in the Upper Vistula Basin*,
GeoPlanet: Earth and Planetary Sciences, DOI 10.1007/978-3-319-41923-7_15

1 Introduction

The Intergovernmental Panel on Climate Change (IPCC) stated that there is high confidence that recent climate changes have had discernible impacts on physical and biological systems (IPCC 2007). Many General Circulation Models (GCMs) predict increases in frequency and magnitude of extreme climate events and variability of precipitation (IPCC 2007, 2012). This is likely to influence terrestrial water resources (Srikanthan and McMahon 2001; Xu and Singh 2004). In addition, the costs of extreme meteorological events have exhibited a rapid upward trend in recent decades and have large impacts on society (IPCC 2012). Floods are the most destructive natural hazard in Poland and most of the flood damage in the country occurs in the basins of two large rivers: the Upper Vistula and the Upper Odra. Abundant rainfall in the Upper Vistula Basin, especially in its mountainous, right-hand, tributaries, such as the rivers: Dunajec, Raba, Wisloka and Skawa, leads to violent and highly erosive flood events. Destructive floods occurred most frequently in July (in 1903, 1934, 1960, 1970, 1997, 2001), but also in other months between May and September (e.g. May and June 2010).

The Tatra Mountains (in Polish: Tatry), located in southern Poland and northern Slovakia, are the highest range of the massive of the Carpathian Mountains. The northern foothills of the Tatra Mountains belong to the drainage basin of the largest Polish river—the Vistula. Precipitation recorded on the northern slopes of the Tatra Mountains is the highest in Poland and largely contributes to flood risk generation over a larger area (Niedzwiedz et al. 2015). Hence, assessment of future changes in intense precipitation in the northern foothills of the Tatra Mountains is very important. Every increase in magnitude, duration, and frequency of extreme precipitation may lead to an increase in climate-related flood risk. The mean annual precipitation at Kasprowy Wierch (1991 m a.s.l.) for the period 1951–2013 was 1752 mm, nevertheless, the record-high annual maximum precipitation at this station was 2599 mm in 2001 (with monthly maximum of 651 mm in July 2001). Even higher annual precipitation values were recorded in 2001 at two other stations: 2628 mm at Hala Gasienicowa and 2770 mm at Dolina Pieciu Stawow (Kundzewicz et al. 2014). Furthermore in the next year with large flood, i.e. in 2010 the highest value of the annual sum of precipitation (in available periods) occurred at five lower located stations, i.e. in Zakopane: 1646 mm, Poronin: 1562 mm, Witow: 1518 mm; Szaflary: 1150 mm and Jablonka: 964 mm. For other four stations it was one of the highest values: the 2-nd highest record for Hala Gasienicowa: 2359 mm; the 3-rd record for Morskie Oko: 2256 mm (in 1974: 2290 mm; in 2001: 2271 mm); the 7-th record for Bialka Tatrzanska: 1015 mm (in 1970: 1128 mm; in 2001: 1062 mm) and the 10-th record for Kasprowy Wierch: 2117 mm (see Chapter "Observed Changes in Air Temperature and Precipitation and Relationship between them, in the Upper Vistula Basin").

Heavy precipitation plays a very important role as a flood generating factor. Therefore assessment of future changes of extreme precipitation events is necessary. This study presents projections of intensification of annual sum and precipitation extremes in the future, for 11 meteorological stations located in the area of the northern foothills of the Tatra Mountains. A range of indices of heavy precipitation are examined. The calculation is based on application of a delta-change method to results of simulation by global and regional climate models. Particular attention was paid to the assessment of future changes (2061–2090 for RCMs and 2080–2100 for GCMs) in precipitation during the warm part of the year (April–September), especially for summer months, when floods occurred most often.

2 Data and Study Site

This study makes use of four regional climate model (RCM) simulations from the ENSEMBLES project (Van der Linden and Mitchell 2009) with the resolution of 0.22° (25 km), for two time horizons, 1961–1990 and 2061–2090. The projections correspond to A1B SRES greenhouse gas emission scenario (Nakicenovic et al. 2000), available for the area of the study. C4IRCA3 (Community Climate Change Consortium for Ireland, Rossby Centre Atmospheric model ver. 3; Norrkoping, Sweden) is driving by global models (ECHAM5, HadCM3Q16), whose data are setting as boundary conditions for the regional model. The ETHZ CLM (Zurich, Switzerland) is driven by a global model HadCM3Q0; KNMIRACMO2 from the Royal National Meteorological Institute (de Bilt, the Netherlands) by ECHAM5, MIROC3.2-hires; and MPI-M-REMO from the Max Planck Institute (Hamburg, Germany) by ECHAM5. For the Hadley Centre Global Climate Model (HadCM3), multiple runs are represented by different climate sensitivities (Q0 = normal sensitivity, 3.50 K and Q16 = high sensitivity, 5.46 K).

To supplement results from this four-member RCM ensemble, seven additional GCM simulations (for A1B and A2 scenarios) were analysed, stemming from the KNMI Climate Explorer (http://climexp.knmi.nl). Projections of these models are available for the last two decades of the 21st century, i.e. 2081–2100. The global model BCCR BCM2 (Bergen Climate Model, BCM, version 2; Bjerknes Centre for Climate Research, BCCR, University of Bergen, Norway) has resolution 2.8° of latitude by 2.8° of longitude for time horizons: 1961–1990 and 2081–2099 (scenario A2). CCCMA CGCM3.1t63 is the third version of the Canadian Centre for Climate Modelling and Analysis (CCCma) Coupled Global Climate Model (CGCM3) and has resolution 2.8° by 2.8° (time horizons: 1961–1990 and 2081–2100; scenario A1B). GFDL's CM2 Global Coupled Climate Models were developed at NOAA's Geophysical Fluid Dynamics Laboratory (GFDL). Two versions of the coupled model called CM2.0 (scenarios: A1B and A2) and CM2.1 (scenario A2) have resolution of 2° latitude by 2.5° longitude and time horizons: 1961–1990 and 2081–2100. MIROC3.2 hires global model (Center for Climate System Research/National Institute for Environmental/Frontier Research Center for Global Change, Japan) has resolution 1.13° by 1.13° (time horizons: 1961–1990 and 2081–2100, scenario A1B). HadCM3 (abbreviation for Hadley Centre Coupled Model, version 3) is a coupled Atmosphere-Ocean General Circulation model (AOGCM) developed at the Hadley Centre in the United Kingdom. The resolution of this model is 2.5° by 3.75° (time horizons: 1961–1990 and 2080–2099, scenario A2).

Analysing simulations from an ensemble of various models allows to account for the uncertainty. The list of regional and Global Climate Models whose simulations were used in this study with references is given in Table 1.

Hence, the results from climate models are used to produce future projections of precipitation indices, by modification of observed precipitation data in the control period (1961–1990), from 11 meteorological stations in the Upper Vistula River Basin—situated on the northern foothills of the Tatra Mountains in basins of the rivers: Czarny Dunajec, Bialy Dunajec, and Bialka. For one station, Hala Ornak, data extend from 1970 to 1990; but even if the period is shorter, this station is very important from the point of view of the impact of precipitation on the formation of floods, so that it is retained. The Czarny Dunajec is considered as headwater of the Czarny Dunajec, one of the largest rivers of the Polish Carpathians. Czarny Dunajec rises at 1540 m a.s.l. in the Western Tatra massif as the Stream Chocholowski. The Stream Chocholowski collects water from Jarzabczy Wierch (2137 m a.s.l.), Wolowiec (2064 m. a.s.l.) and Rakon (1879 m. a.s.l.) (Czarnecka 1983). The Czarny Dunajec joins with the Bialy Dunajec in the town of Nowy Targ at the altitude of 578 m (Wyzga and Zawiejska 2010). The Malolacki stream (lower Cicha Woda and Zakopianka rivers) is headwater of the Bialy Dunajec. It rises at about 2000 m a.s.l. from the peaks of the Czerwone Wierchy massif. The Stream Biala Woda gives rise of the River Bialka whose source areas are situated in granite parts of the High Tatra massif (Czarnecka 1983). Jablonka station is located in Czarna Orawa Basin (The Danube Basin). Data from this station were used due to the limited availability of data in this area. Location of this station is close to the Dunajec River Basin and it enriches our knowledge of distribution of precipitation (Fig. 1; Table 2). According to Starkel (1991) this region is characterised by high values of excess water and accelerated outflow. In the north of this area, mountainous valleys are situated where a periodic retention takes place. The stations read: Kasprowy Wierch, Hala Gasienicowa, Morskie Oko, Hala Ornak (with shorter available period of records: 1970–1990), Kuznice, Zakopane, Witow, Poronin, Bialka Tatrzanska, Szaflary and Jablonka (in The Danube Basin). Locations of these stations are shown in Fig. 1.

3 Methods

3.1 Introduction to Downscaling

Global general circulation models (GCMs) are powerful tools for constructing future climate projections. However, such models currently operate at coarse spatial resolutions, hence are unable to resolve significant subgrid-scale features, including topography and land use. In brief, climate models are restricted in their usefulness

Table 1 RCM and GCM simulations used in this study

	Model output	Scenario	Projection	References
RCMs	C4IRCA3	A1B	2061–2090	Kjellström et al. (2005)
	ETHZ CLM	A1B	2061–2090	Böhm et al. (2006)
	KNMI RACMO2	A1B	2061–2090	Van Meijgaard et al. (2008)
	MPI-M-REMO	A1B	2061–2090	Jacob (2001), Jacob et al. (2001)
GCMs	BCCR BCM2	A2	2081–2099	Ottera et al. (2009)
	CCCMA CGCM3.1t63	A1B	2081–2100	http://www.ec.gc.ca/ccmac-cccma/default.asp?lang=En&n=1299529F-1
	GFDL-CM2.0	A1B	2081–2100	Delworth et al. (2006)
	GFDL-CM2.0	A2	2081–2100	
	GFDL-CM2.1	A2	2081–2100	
	MIROC3.2 hires	A1B	2081–2100	Hasumi and Emori (2004)
	UKMO HADCM3	A2	2080–2099	Collins et al. (2001)

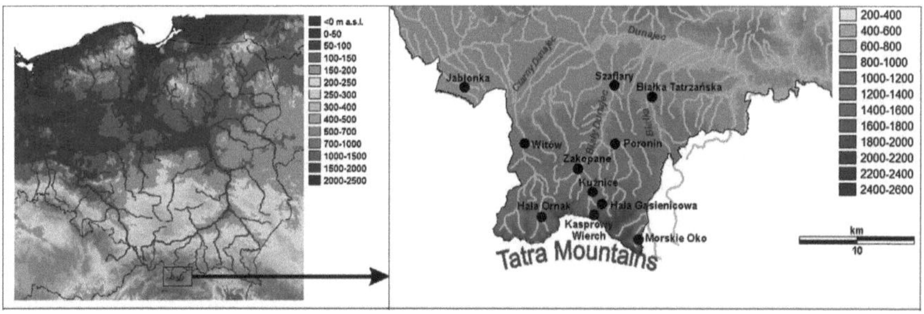

Fig. 1 Location of the studied area in Poland and location of the meteorological stations in the area of the northern foothills of the Tatra Mountains

for many subgrid-scale applications, such as hydrological modelling and impact assessment, essential for designing flood risk reduction and climate change adaptation policies (Wilby and Wigley 1997).

Hence, it is of utmost importance to solve the mismatch of scales, between low-resolution GCMs and high-resolution practical needs, by development of methodologies to infer small-scale (local or regional) information from large-scale information. All these methodologies are collectively called "downscaling techniques".

Increasing the resolution of climate models could improve the estimates of regional-scale phenomena but downscaling outputs from GCMs is the primary approach for addressing the inadequacies of large-scale models (Wilby and Wigley 1997). Climate variables simulated by individual GCMs often do not agree with observed time series. This poses a problem for using these simulations as input data

Table 2 Altitude of 11 meteorological stations and their location in the river basin

Station	Hs (m a.s.l.)	River basin		
		Name of the river	Area (km^2)	
Hala Ornak	1109	Czarny Dunajec	456.1	Dunajec Basin to the Bialka River tributary 1037.4 km^2
Witow	835	Czarny Dunajec		
Kasprowy Wierch	1991	Bialy Dunajec	224.0	
Hala Gasienicowa	1520	Bialy Dunajec		
Kuznice	1024	Bialy Dunajec		
Zakopane	857	Bialy Dunajec		
Poronin	773	Bialy Dunajec		
Szaflary	655	Bialy Dunajec		
Morskie Oko	1408	Bialka	229.9	
Bialka Tatrzanska	700	Bialka		
Jablonka	615	Czarna Orawa	135.4 to the Jablonka gauge	Danube

for hydrological impact studies. The correction procedures usually identify possible differences between observed and simulated climate variables (Hagemann et al. 2011), which provide the basis for correcting both control and scenario GCM runs with a transformation algorithm.

3.2 *Dynamical and Statistical Downscaling*

Many downscaling techniques have been proposed (Wilby and Wigley 1997), which can be grouped into two fundamental approaches: dynamic downscaling (DD) and statistical downscaling (SD) methods. Dynamic downscaling, nesting a fine-scale regional climate model (RCM) in a coarse scale model, produces spatially complete fields of climate variables, thus preserving spatial correlation as well as physically plausible relationships between variables. In the statistical downscaling, high-resolution (small-scale) predictands are obtained by identifying relationships in the observed climate between these predictands and large-scale predictors.

The relationships obtained in SD can be applied to GCM output for the future conditions, to produce projections.

Both approaches, dynamic downscaling and statistical downscaling have advantages and disadvantages, and both necessitate assumptions that indeed cannot be credibly verified for the climate change context (Wilby and Wigley 1997). The advantage of dynamic downscaling is simulating fine-scale physical processes, based on rigorous and universal laws. However, dynamic downscaling is very computationally intensive, restricting its use in impact studies and essentially impossible for multi-decade simulations with different global climate models and/or multiple greenhouse gas emission scenarios.

The main advantages of the statistical downscaling (Wilby et al. 1998) are:

(i) computational efficiency (empirical relationships between the coarse-resolution GCM and high-resolution local climate variables are used, hence calculations can be easily made for several variants: time horizons, climate models, and greenhouse gas emission scenarios);
(ii) specific information is provided for small areas or even point locations where observations are available.

The disadvantages of statistical downscaling, as compared to dynamic downscaling are:

(i) adequate historical observations of the studied variables (covering a sufficient time span and being of sufficient quality) are needed;
(ii) assumption of stationarity in the relationships between predictors and predictands is necessary, while it is well recognized that stationarity is dead (Milly et al. 2008, 2015).

The local detail available via statistical downscaling is assumed to be relevant also in the future and this assumption seems plausible. It is likely that in future as well as in the period for which observations are available, the system characteristics as well as variables of concern can be quite different for points which are located not long apart. In contrast, dynamic downscaling approaches typically provide spatial resolutions of the order of, say, 20 km, which can still be insufficient to resolve topography with enough detail and to show differences in the projected changes for points located close together.

Statistical downscaling methods cover such approaches as regression-type models (both linear and nonlinear), neural networks, weather generators for generating synthetic sequences of local variables, and techniques based on weather classification that draw on the attributes of models to simulate circulation patterns. The approach also includes analogue methods that seek similar states and perform a multiple regression analysis for the most analogous days or using the probability distribution of analogous days. Evaluation of SD is done most commonly through cross-validation with observational data.

3.3 Delta-Change Method

A delta-change method can be used to transform observations into a time series that is representative of future conditions consistent with the GCM climate change signal (Van Pelt et al. 2012). The delta method makes use of "change factors". The simplest form of the delta method (sometimes referred to as the "classical delta method") only considers changes in the mean that may vary seasonally or spatially. The assumption that one has to make is that changes at the (large) scale of the climate model (GCM) can be directly applied to the (local/point) scale of the time series. An "advanced delta method" (Van Pelt et al. 2012) does not only take changes in the mean into account but also the changes in the extremes. Then, rather than a proportional adjustment of observed values, a non-linear transformation is applied to the data.

We used the statistical delta-change approach, thus applying differences between the control and future RCM/GCM simulations to baseline observations by adding or scaling the mean precipitation change to each day. We obtain meteorological time series with the same resolution, characteristics and variance as in the past. The procedure can be described as follows cf. Van Pelt et al. (2012):

First, non-overlapping five-day precipitation sums are the starting point for the transformation. The observed five-day precipitation amounts P are transformed using:

- for the sum of precipitation below 90 quantile:

$$\mathrm{P}^* = \mathrm{a}\,\mathrm{P}^{\mathrm{b}} \tag{1}$$

- for the sum of precipitation above 90 quantile:

$$\mathrm{P}^* = \frac{\overline{E^F}}{\overline{E^C}}\left(P - P_{90}^O\right) + a\left(P_{90}^O\right)^b \tag{2}$$

where P is a historical five-day sum, P* the transformed (future) five-day sum, and a and b the transformation coefficients (parameters), P^{O} and P* represent the observed and transformed future precipitation data, C and F are representative for the control and future climates, and E^{C} and E^{F} represent the excess variable of the control and future period of the RCMs/GCMs. The amount E above the 90 % quantile is the excess:

$$\mathrm{E} = \mathrm{P} - \mathrm{P90} \tag{3}$$

To get the daily observation data the change factor R was applied for each 5-day sum period:

$$R = P^{*}/P \tag{4}$$

We performed this procedure for precipitation data recorded at 11 stations spatially distributed over the entire analysed region.

In this study, projections of annual and extreme precipitation in the northern foothills of the Tatra Mountains were calculated. To obtain insight into possible future changes, several precipitation indices were examined:

- annual sum of precipitation;
- 24 h maximum sum of precipitation;
- maximum sum of 5-day precipitation for warm season (April–September);
- maximum sum of 10-day precipitation for warm season (April–September);
- maximum monthly sum of precipitation for warm season (April–September);
- number of days with intense precipitation greater than or equal to 30 and 50 mm during warm season (April–September).

The 24 h maximum sum of precipitation and maximum sum of 5-day precipitation are especially important due to the possibility of causing violent floods and landslides in small catchments. The long-lasting and very intensive precipitation such as 10-day or monthly values can generate floods on larger area.

4 Results

Climatological models have greater difficulty in simulating precipitation as opposed to temperature because of complexity phenomenon involving the sub-grid scale features such as topography or land use that are often inadequate for assessing the correct location and intensity of precipitation. Especially this is noticeable for such complex and difficult area as northern foothills of the Tatra Mountains. Although there is some disagreement as to the range of changes, almost all the RCMs and GCMs model agree on the signs of change; i.e. increase in mean and extreme precipitation.

Figure 2 presents comparison of relative changes for three indices: (1) mean 10-day maximum precipitation; (2) mean monthly maximum precipitation; and (3) mean annual sum of precipitation (future versus control period). They were derived from raw RCM and GCM model output (the ratio between future and control period Pfmean/Pcmean) in comparison with relative change obtained from the transformation procedure (the ratio between transformed observation and observation P*mean/Pmean). The correction leads to an acceptable correspondence in the case of all indices (R^2 about 0.6).

Changes in mean values are often lower than changes in extreme events: for example the range of relative changes in mean annual sum of precipitation is between 1.2 and 1.4 while the range for relative changes for mean 10-day maximum precipitation and mean monthly maximum precipitation is up to 1.7 (Fig. 2).

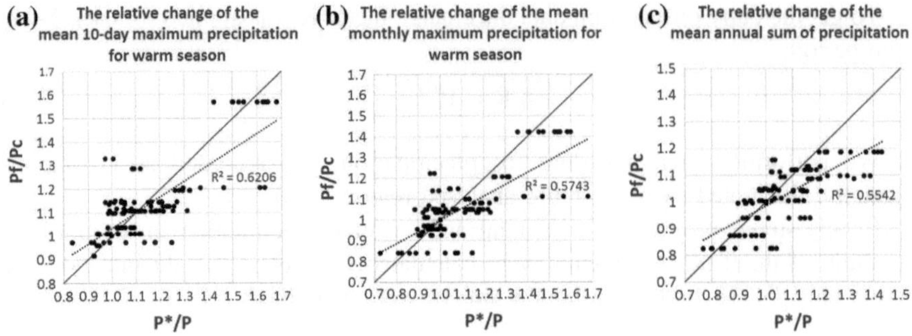

Fig. 2 The relative changes from RCMs and GCMs output versus the relative changes from transformed observations for 10 meteorological stations (without Hala Ornak) for three indices: **a** mean 10-day maximum precipitation for warm season (*left chart*); **b** mean monthly maximum precipitation for warm season (*middle chart*); and **c** mean annual sum of precipitation (*right chart*). Grey line represents the perfect correspondence (i.e. the 1:1 line)

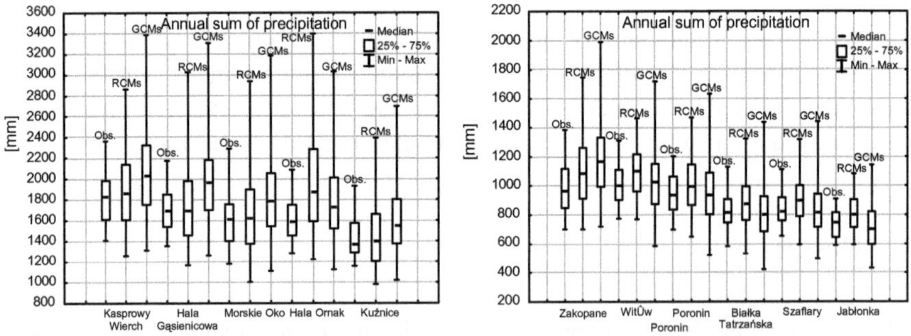

Fig. 3 Observed annual sum of precipitation for the control interval 1961–1990 and projected annual sum of precipitation for RCMs (2061–2090) and GCMs (2081–2100) for stations located at elevation above 1000 m a.s.l. (*left*) and stations situated below 1000 m a.s.l. (*right*)

In general, the analysed RCM and GCM models project an increase in the annual sum of precipitation for the studied area. The future median value may not change considerably for stations located above 1000 m a.s.l. for RCMs and is likely to be higher for these stations for GCMs. For stations at lower elevations, RCMs project an increase in the median value of the annual sum of precipitation while GCMs project a smaller increase or even decrease for stations: Bialka Tatrzanska, Szaflary and Jablonka. However, the future level of the 75 percentile and the maximum value of annual precipitation are projected to be higher than those observed in 1961–1990, in every case. Figure 3 shows box plots with observed annual sum of precipitation for the interval 1961–1990 and projected annual sum of precipitation for RCMs and GCMs for listed stations.

Figure 4 shows the return period plots (Gumbel distribution) of the maximum daily precipitation for the future projections obtained from the RCMs and GCMs used in this study for 11 meteorological stations and comparison with observed data. It is

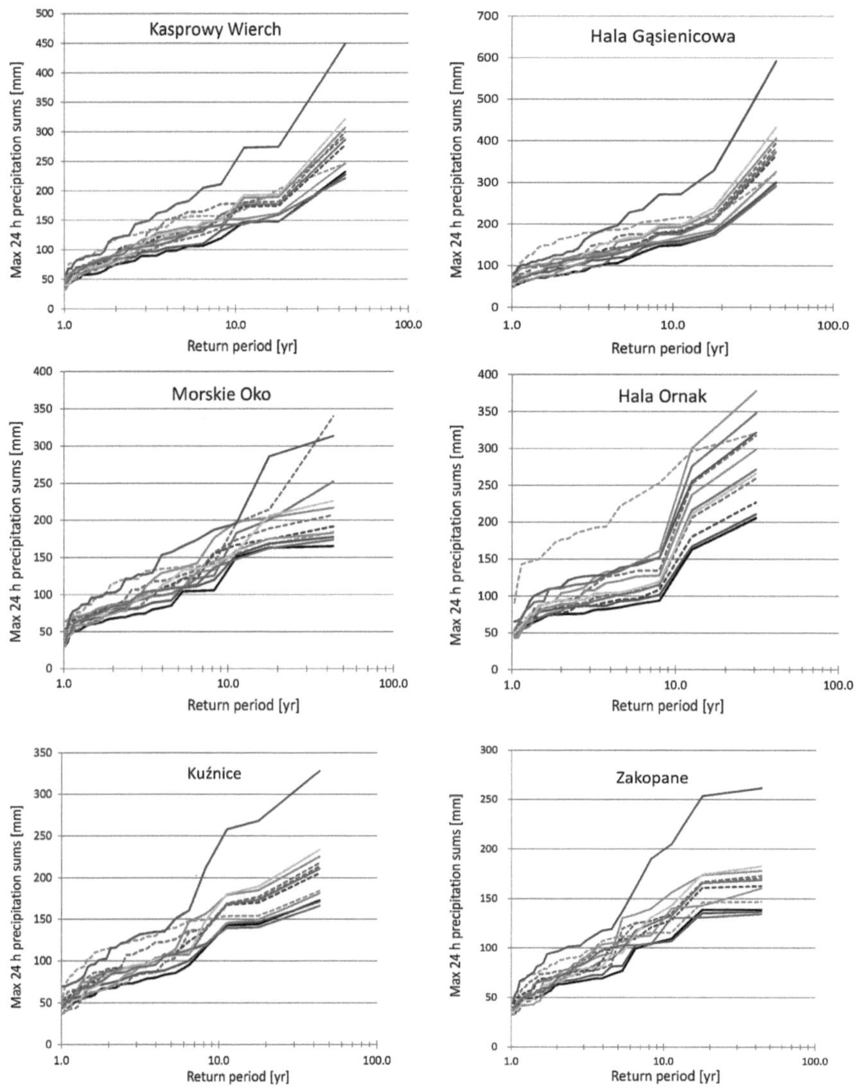

Fig. 4 The return period plots (Gumbel distribution) calculated for 30-year time series (21 for Hala Ornak) of the maximum daily precipitation for observed data (1961–1990) and the RCMs (2061–2090) and GCMs (2081–2100) for 11 meteorological stations situated on the northern foothills of the Tatra Mountains

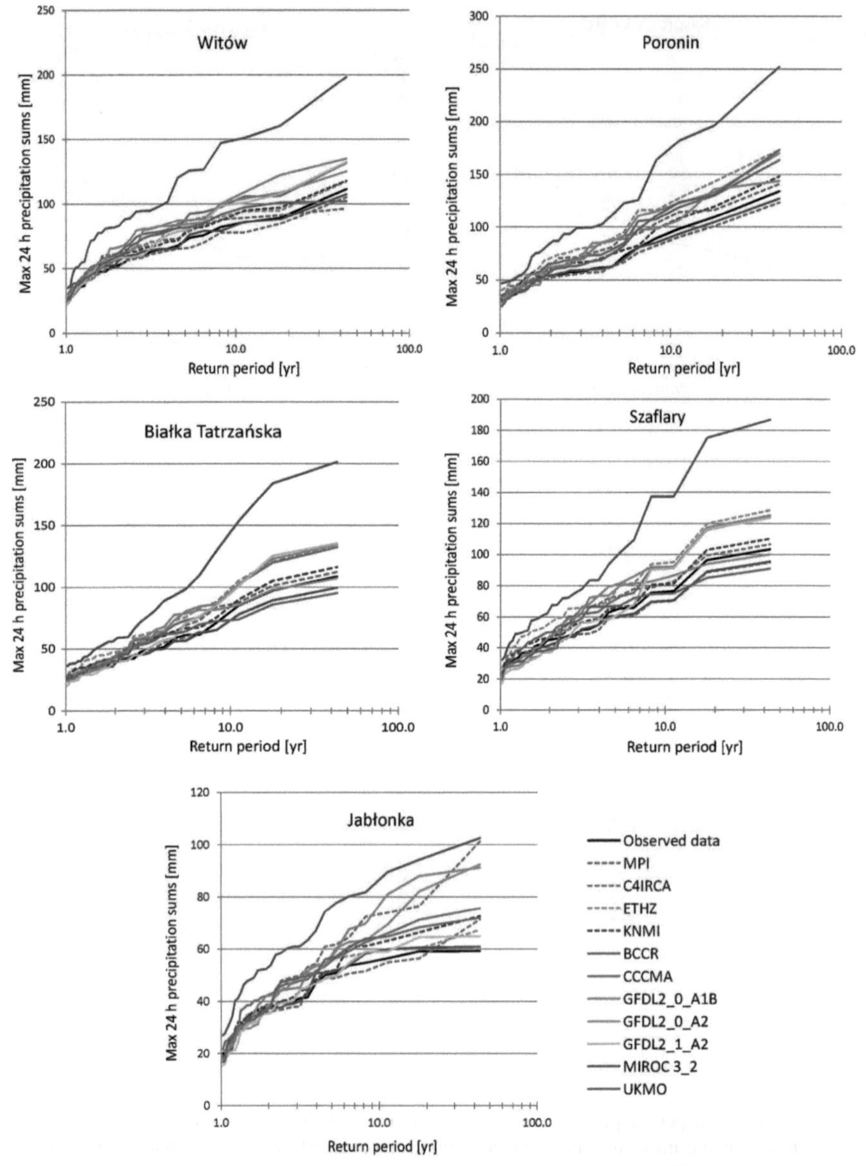

Fig. 4 (continued)

found that the range of projections of change in precipitation for the future across the four-member RCM ensemble is similar to the results from the GCM ensemble.

One of the highest values we can observe for the GCM model MIROC. For three stations, also the highest values are given by RCMs: MPI, ETHZ and GCMs: BCCR and GFDL2.0 (scenario A2). The plots show a similar change in higher values and a larger range between the values for the greater return periods. For some stations, two or three models (GCMs and RCMs) indicate a decrease in precipitation for the highest return period (for 10 stations the highest return period is 43.4 years, for Hala Ornak it is 30.6 years). But most of RCMs and GCMs models suggest an increase in the maximum 24 h precipitation.

Table 3 illustrates the maximum daily precipitation and 5-day sum of precipitation at each of the stations and the year of occurrence of this maximum precipitation for the available period of observations. Both daily precipitation and 5-day precipitation totals are of importance for flood hazard.

According to the observed data, most of 24-h maximum precipitation, as well as maximum 5-day precipitation appeared in examined control period: 1961–1990. Only in five cases, the maximum value of these indices occurred in 1997 (the year with large flood) and in 2007. The maximum values of 24-h precipitation for stations: Poronin in 1997 and for Jablonka in 2007 are at the level of future maximum values downscaled based on control period 1961–1990 (Fig. 4).

Figure 5 presents box plots with observed maximum 5-day precipitation for the reference period 1961–1990 and projected value of this index for RCMs and GCMs for particular stations. As previously, for maximum 5-day precipitation we can observe that the maximum value for Poronin in 1997 is nearly at the level of the maximum value projected by RCMs.

The future changes in maximum 5-day precipitation during warm season obtained by RCMs and GCMs are likely to be stronger. For all stations, increases of the median value, the 75 percentile and also the extreme values are projected. The future

Table 3 The maximum daily precipitation and the year of occurrence of this maximum precipitation, as well as the maximum 5-day precipitation for the available period of observations for 11 meteorological stations

Station	The available period	Max 24-h precipitation (mm)	Year of occurrence	Max 5-day precipitation (mm)	Year of occurrence
Kasprowy Wierch	1951–2013	232.0	1973	388.7	1980
Hala Gasienicowa	1951–2011	300.0	1973	425.2	1980
Morskie Oko	1954–2011	168.4	1958	296.3	1970
Hala Ornak	1970–2011	205.3	1973	343.6	1980
Kuznice	1954–1996	172.7	1973	263.5	1980
Zakopane	1951–2013	138.7	1970, 1973	235.9	1972
Witow	1954–2011	111.3	1970	200.4	1972
Poronin	1951–2011	156.1	1997	248.2	1997
Bialka Tatrzanska	1951–2011	112.0	1997	188.3	1997
Szaflary	1951–2011	103.4	1970	179.1	1997
Jablonka	1955–2011	70.1	2007	144.7	1972

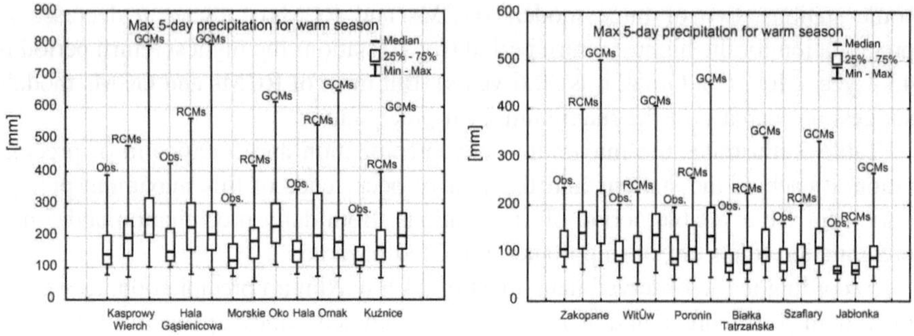

Fig. 5 The observed maximum 5-day precipitation for the reference interval 1961–1990 and projected maximum 5-day precipitation for RCMs (2061–2090) and GCMs (2081–2100) for stations located at elevation above 1000 m a.s.l. (*left*) and below 1000 m a.s.l. (*right*)

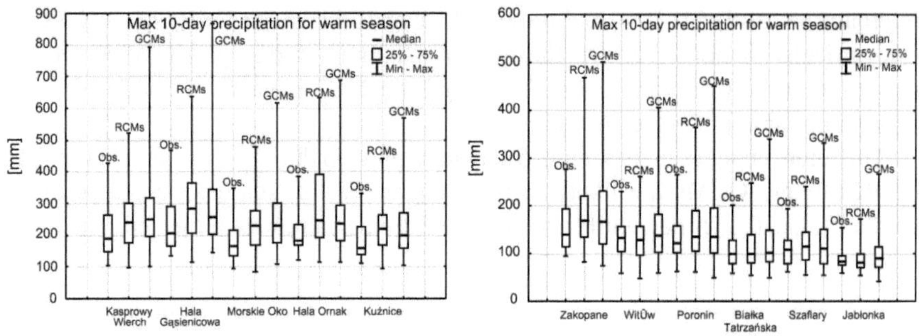

Fig. 6 Observed maximum 10-day precipitation for 1961–1990 and projected maximum 10-day precipitation for RCMs (2061–2090) and GCMs (2081–2100) at stations located at the elevation above 1000 m a.s.l. (*left*) and below 1000 m a.s.l. (*right*)

median of maximum 5-day precipitation for high stations (above 1000 m a.s.l.) and also for Zakopane could be on the level of 75 ‰ (for the control period 1961–1990).

A similar situation can be observed for maximum 10-day precipitation for the warm season. The changes of median for higher located stations are projected to be higher than observed in 1961–1990. Present 75 ‰ can become a median in the future for projections (with both RCMs and GCMs). In the case of lower placed meteorological stations, the median value of maximum 10-day precipitation could be on the similar level as the one observed in the 20th century. The future 75 ‰ and maximum value of this index are likely to be higher for all stations (Fig. 6).

The RCM and GCM simulations show increase in median value as well as in 75 percentile and maximum value for maximum monthly sum of precipitation in the warm season for higher located stations and Zakopane. For other studied stations, the increases of this index could be lower. Figure 7 presents observed ranges for maximum monthly sum of precipitation and projected changes according to RCMs and GCMs.

Similar behavior of projections of changes in the number of days with intense precipitation (greater than or equal to 30 and 50 mm) during the warm season can be observed (Fig. 8). For the stations above 1000 m a.s.l. for both thresholds (i.e.

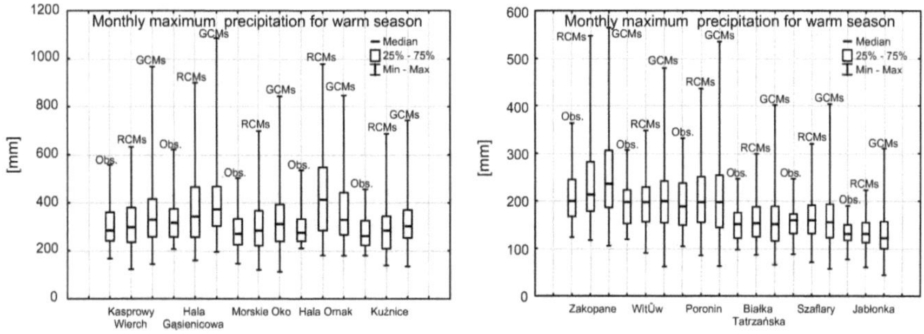

Fig. 7 Observed maximum monthly sum of precipitation for 1961–1990 and projected value for RCMs (2061–2090) and GCMs (2081–2100) at stations located at the elevation above 1000 m a.s. l. (*left*) and below 1000 m a.s.l. (*right*)

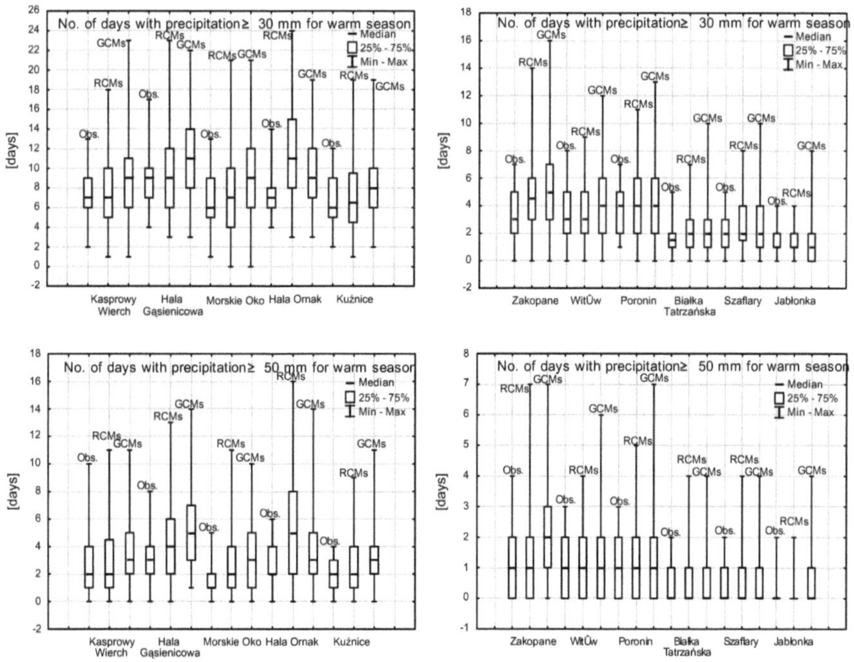

Fig. 8 Observed number of days with 24 h precipitation equal to or greater than the threshold of: 30 mm (*top*) and 50 mm (*bottom*) for the warm season, for 1961–1990 and projected value for RCMs (2061–2090) and GCMs (2081–2100) for stations located at the elevation above 1000 m a. s.l. (*left*) and below 1000 m a.s.l. (*right*)

30 and 50 mm), regional and global models show an increase in median value and also in higher values of these indices. The projected changes are lower for other stations. In the future, occurrence of days with intense precipitation (above 30 and 50 mm) may be very rare. Projections show increase rather in extreme value.

5 Discussion and Conclusions

This study allows to get insight into possible future changes in annual sum of precipitation as well as in intense and extreme precipitation in the northern foothills of Tatra Mountains. Use of four regional climate models and seven global climate models for two SRES climate scenarios A1B and A2 made it possible to obtain the likely range of changes in precipitation during last decades of the 21st century. In most cases, the range of projections of annual and extreme precipitation for the future across the results from the GCM ensemble is slightly higher than across the four-member RCM ensemble. Projections obtained from GCMs cover further period of the 21st century, i.e. 2081–2100 when changes in precipitation, in association with the greater increases of temperature simulations in the end of the 21st century could be more pronounced than for RCMs projections for earlier period 2061–2090.

The knowledge about changes in precipitation extremes is very important for climate adaptations to future flood risk (Zhang et al. 2013). Use of 11 climate model simulations covers a range of plausible changes in extremes.

Models in this study belong to generation of CGCMs which were submitted to the Coupled Model Intercomparison Project Phase 3 (CMIP3) in support of the Intergovernmental Panel on Climate Change (IPCC) Fourth Assessment Report (AR4) (Kharin et al. 2007). The recent generation of global climate models participating in the Coupled Model Intercomparison Project Phase 5 (CMIP5) are employing the new radiative forcing scenarios, called Representative Concentration Pathways (RCPs, Moss et al. 2010). However, apart from the different forcing scenarios in the CMIP5 protocol, the performance of the CMIP5 multi-model ensemble in simulating temperature and precipitation climate extremes is comparable to that in the CMIP3 ensemble (Kharin et al. 2013).

All RCM and GCM models under examination project an increase in mean annual precipitation totals as well as in heavy precipitation in future time horizons studied. However, the changes in mean precipitation were not as pronounced as for extreme indices, such as maximum 24 h, 5-day; 10-day, monthly maximum sum of precipitation and also number of days with intense precipitation (30 and 50 mm). The increasing change is projected especially for stations located above 1000 m a.s. l. For most stations located in the area of the northern foothills of the Tatra Mountains, the maximum observed daily precipitation occurs in summer, so that the increase in precipitation can result in more severe flash floods and landslides in small catchments and can generate flood events in basins of the mountainous, right-hand tributaries of the River Vistula (such as the River Dunajec).

Similarly changes can be found in the case of projected return period of late 20th century (1981–2000) 20-year return values of annual maximum 24-h precipitation in the study of Seneviratne et al. (2012). The 20-year return period for area of Central Europe for the reference interval is likely to become more frequent, e.g. a 1-in-10 to 1-in-15 year event by the end of 21st century (the median models value).

These findings are compatible with those in Tebaldi et al. (2006) where future changes produced by the IPCC-AR4 model ensemble in maximum 5-day precipitation, precipitation intensity measured as SDII (Simple Daily Intensity Index, defined as the annual total precipitation divided by the number of wet days) and fraction of total precipitation exceeding the 95th percentile are considered significant by a majority of models across the mid- to high latitudes of the northern hemisphere.

Every increase in extreme precipitation can contribute to more severe floods in the future. However, recent studies indicate that the projection of precipitation extremes is associated with uncertainties related to GCMs, RCMs and statistical downscaling methods as well as by natural variability of the climate (Seneviratne et al. 2012). Nevertheless, records of precipitation during the latest large floods in the studied region, i.e. in 2001 and 2010 also show that the observed values of maximum monthly sum of precipitation reached new heights.

Acknowledgments This study was carried out within the FLORIST project (Flood risk on the northern foothills of the Tatra Mountains), supported by a grant from the Swiss Government through the Swiss Contribution to the enlarged European Union (PSPB No. 153/2010).

References

Böhm U, Kücken M, Ahrens W, Block A, Hauffe D, Keuler K, Rockel B, Will A (2006) CLM—the climate version of LM: brief description and long term application. COSMO Newsl 6:225–235

Collins M, Tett SFB, Cooper C (2001) The internal climate variability of HadCM3, a version of the Hadley Centre coupled model without flux adjustments. Clim Dyn 17:61–81

Czarnecka H (ed) (1983) Atlas podzialu hydrograficznego Polski. Seria Atlasy i monografie IMGW, Warszawa (in Polish)

Delworth TL, Broccoli AJ, Rosati A, Stouffer RJ, Balaji V, Beesley JA, Cooke WF, Dixon KW, Dunne J, Dunne KA, Durachta JW, Findell KL, Ginoux P, Gnanadesikan A, Gordon CT, Griffies SM, Gudgel R, Harrison MJ, Held IM, Hemler RS, Horowitz LW, Klein SA, Knutson TR, Kushner PJ, Langenhorst AR, Lee H-Ch, Lin S-J, Lu J, Malyshev SL, Milly PCD, Ramaswamy V, Russell J, Schwarzkopf MD, Shevliakova E, Sirutis JJ, Spelman MJ, Stern WF, Winton M, Wittenberg AT, Wyman B, Zeng F, Zhang R (2006) GFDL's CM2 global coupled climate models part I: formulation and simulation characteristics. J Clim Spec Sect 19:643–674

Hagemann S, Chen C, Haerter JO, Heinke J, Gerten D, Piani C (2011) Impact of a statistical bias correction on the projected hydrological changes obtained from three GCMs and two hydrology models. J Hydrometeorol 12:556–578. doi:10.1175/2011jhm1336.1

Hasumi H, Emori S (eds) (2004) K-1 coupled GCM (MIROC) description. Center for Climate System Research (CCSR), University of Tokyo, National Institute for Environmental Studies

(NIES), Frontier Research Center for Global Change (FRCGC). http://ccsr.aori.u-tokyo.ac.jp/~hasumi/miroc_description.pdf

IPCC (2007) Climate change 2007: synthesis report. Contribution of Working Groups I, II and III to the Fourth Assessment Report of the Intergovernmental Panel on Climate Change. In: Pachauri RK, Reisinger A. (eds) IPCC, Geneva, p 104

IPCC (2012) A special report of working groups I and II of the intergovernmental panel on climate change. In: Field CB, Barros V, Stocker TF, Qin D, Dokken DJ, Ebi KL, Mastrandrea MD, Mach KJ, Plattner G-K, Allen SK, Tignor M, Midgley PM (eds) Managing the risks of extreme events and disasters to advance climate change adaptation. Cambridge University Press, Cambridge, New York, p 582

Jacob D (2001) A note to the simulation of the annual and inter-annual variability of the water budget over the Baltic Sea drainage basin. Meteorol Atmos Phys 77(1):61–73

Jacob D, Van den Hurk BJJM, Andræ U, Elgered G, Fortelius C, Graham LP, Jackson SD, Karstens U, Köpken Ch, Lindau R, Podzun R, Rockel B, Rubel F, Sass BH, Smith RNB, Yang X (2001) A comprehensive model inter-comparison study investigating the water budget during the BALTEX-PIDCAP period. Meteorol Atmos Phys 77(1):19–43

Kharin VV, Zwiers FW, Zhang X, Hegerl GC (2007) Changes in temperature and precipitation extremes in the IPCC ensemble of global coupled model simulations. J Clim 20:1419–1444

Kharin VV, Zwiers FW, Zhang X, Wehner M (2013) Changes in temperature and precipitation extremes in the CMIP5 ensemble. Clim Change 119:345–357

Kjellström E, Bärring L, Gollvik S, Hansson U et al (2005) A 140-year simulation of European climate with the new version of the Rossby Centre regional atmospheric climate model (RCA3). SMHI Rep Meteorol Climatol 108 (SMHI, Norrköping)

Kundzewicz ZW, Stoffel M, Kaczka RJ, Wyzga B, Niedzwiedz T, Pinskwar I, Ruiz-Villanueva V, Lupikasza E, Czajka B, Ballesteros-Canovas JA, Malarzewski L, Chorynski A, Janecka K, Mikuś P (2014) Floods at the Northern Foothills of the Tatra Mountains—a Polish-Swiss research project. Acta Geophys 62(3):620–641

Milly PCD, Betancourt J, Falkenmark M, Hirsch RM, Kundzewicz ZW, Lettenmaier DP, Stouffer RJ (2008) Stationarity is dead: whither water management? Science 319 (5863):573–574

Milly PCD, Betancourt J, Falkenmark M, Hirsch RM, Kundzewicz ZW, Lettenmaier DP, Stouffer RJ, Dettinger MD, Krysanova V (2015) On critiques of "Stationarity is dead: whither water management?". Water Resour Res 51(9):7785–7789

Moss RH, Edmonds JA, Hibbard KA, Manning MR, Rose SK, van Vuuren DP, Carter TR, Emori S, Kainuma M, Kram T, Meehl GA, Mitchell JFB, Nakicenovic N, Riahi K, Smith SJ, Stouffer RJ, Thomson AM, Weyant JP, Wilbanks TJ (2010) The next generation of scenarios for climate change research and assessment. Nature 463:747–756

Nakicenovic N, Alcamo J, Davis G, de Vries B, Fenhann J, Gaffin S, Gregory K, Grübler A, Jung TY, Kram T, La Rovere EL, Michaelis L, Mori S, Morita T, Pepper W, Pitcher H, Price L, Riahi K, Roehrl A, Rogner H-H, Sankovski A, Schlesinger M, Shukla P, Smith S, Swart R, van Rooijen S, Victor N, Dadi Z (2000) IPCC special report on emissions scenarios. Cambridge University Press, Cambridge, New York, p 599

Niedzwiedz T, Lupikasza E, Pinskwar I, Kundzewicz ZW, Stoffel M, Malarzewski L (2015) Variability of heavy rainfall events and related synoptic situations causing floods at the northern foothills of the Tatra Mountains. Theoret Appl Climatol 119:273–284

Ottera OH, Bentsen M, Bethke I, Kvamstø NG (2009) Simulated pre-industrial climate in Bergen Climate Model (version 2): model description and large-scale circulation features. Geosci Model Dev 2:197–212

Seneviratne SI, Nicholls N, Easterling D, Goodess CM, Kanae S, Kossin J, Luo Y, Marengo J, McInnes K, Rahimi M, Reichstein M, Sorteberg A, Vera C, Zhang X (2012) Changes in climate extremes and their impacts on the natural physical environment. In: Field CB, Barros V, Stocker TF, Qin D, Dokken DJ, Ebi KL, Mastrandrea MD, Mach KJ, Plattner G-K, Allen SK, Tignor M, Midgley PM (eds) Managing the risks of extreme events and disasters to advance climate change adaptation. A special report of working groups I and II of the

Intergovernmental Panel on Climate Change (IPCC). Cambridge University Press, Cambridge, New York, pp 109–123

Srikanthan R, McMahon TA (2001) Stochastic generation of annual, monthly and daily climate data: a review. Hydrol Earth Syst Sci 5(4):653–670

Starkel L (1991) Rzezba terenu. In: Dynowska I, Maciejewski M (eds) Dorzecze gornej Wisly. Cześć 1. Panstwowe Wydawnictwo Naukowe, Warszawa–Krakow, pp 42–54 (in Polish)

Tebaldi C, Hayhoe K, Arblaster JM, Meehl GA (2006) Going to the extremes. An intercomparison of model-simulated historical and future changes in extreme events. Clim Change 79:185–211

Van der Linden P, Mitchell JFB (eds) (2009) ENSEMBLES: climate change and its impacts: summary of research and results from the ENSEMBLES project. Met Office Hadley Centre, Exeter. http://ensembles-eu.metoffice.com/docs/Ensembles_final_report_Nov09.pdf

Van Meijgaard E, van Ulft LH, van de Berg WJ, Bosveld FC, van den Hurk BJJM, Lenderink G, Siebesma AP (2008) The KNMI regional atmospheric climate model RACMO, version 2.1. KNMI Tech Rep 302, KNMI, De Bilt. www.knmi.nl/bibliotheek/knmipubTR/TR302.pdf

Van Pelt SC, Beersma JJ, Buishand TA, Van den Hurk BJJM, Kabat P (2012) Future changes in extreme precipitation in the Rhine basin based on global and regional climate model simulations. Hydrol Earth Syst Sci 16:4517–4530

Wilby RL, Wigley TML (1997) Downscaling general circulation model output: a review of methods and limitations. Prog Phys Geogr 21:530–548

Wilby RL, Wigley TML, Conway D, Jones PD, Hewitson BC, Main J, Wilks DS (1998) Statistical downscaling of general circulation model output: a comparison of methods. Water Resour Res 34(11):2995–3008

Wyzga B, Zawiejska J (2010) Large wood storage in channelized and unmanaged sections of the Czarny Dunajec River: implications for the restoration of mountain rivers. Folia Geogr Ser Geogr-Phys 41:5–34. Jagiellonian University, IGiGP

Xu C-Y, Singh VP (2004) Review on regional water resources assessment models under stationary and changing climate. Water Resour Manage 18:591–612

Zhang X, Hegerl GC, Seneviratne SI, Stewart R, Zwiers FW, Alexander LV (2013) WCRP grand challenge: understanding and predicting weather and climate extremes. http://www.wcrp-climate.org/gc-extremes-documents

Hydroclimatic Projections for the Upper Vistula Basin

Mikolaj Piniewski, Mateusz Szczesniak, Abdelkader Mezghani and Zbigniew W. Kundzewicz

Abstract In this study, we apply a previously calibrated SWAT model of the Vistula and Odra basins in order to assess hydrological impacts of climate change in the Upper Vistula Basin. Raw projections from an ensemble of nine EUR-11 CORDEX climate model runs (precipitation and minimum and maximum temperatures) assuming an intermediate greenhouse gas emission scenario of 4.5 W/m^2 were adjusted using a quantile-mapping correction approach. We analysed changes between two future horizons 2024–2050 (near future) and 2074–2100 (far future) and a reference period (1974–2000). We found that, for the near future, all climate models agree well about ubiquitous warming on both seasonal and annual scales, while eight models agree about an increase in projected mean annual precipitation and total runoff. For the far future, an increase in temperature, in mean annual precipitation, as well as in the total runoff, is projected using all climate models. Results also highlighted a higher temperature increase in winter than in other seasons and a higher increase in minimum temperature than in maximum temperature. The highest runoff increase is projected in winter, consistently, by all climate models. In addition, we assessed projected changes in high streamflow indicator based on the 90th monthly flow percentile (Q90). Based on the median of climate model simulations, we found that the mean basin-wide increase in monthly Q90 is 6.4 and 15 % for the near future and the far future, respectively. Nevertheless, the range of projected changes in precipitation, runoff and high flows calculated across the whole ensemble remains relatively high and spatial patterns are not fully consistent across different climate models.

M. Piniewski (✉) · M. Szcześniak
Warsaw University of Life Sciences, Warsaw, Poland
e-mail: M.Piniewski@levis.sggw.pl

M. Piniewski · Z.W. Kundzewicz
Potsdam Institute for Climate Impact Research, Potsdam, Germany

A. Mezghani
Norwegian Meteorological Institute, Oslo, Norway

Z.W. Kundzewicz
Institute for Agricultural and Forest Environment, Polish Academy of Sciences, Poznań, Poland

Z.W. Kundzewicz et al. (eds.), *Flood Risk in the Upper Vistula Basin*,
GeoPlanet: Earth and Planetary Sciences, DOI 10.1007/978-3-319-41923-7_16

Keywords Climate change · Temperature · Precipitation · River discharge · Projections · Upper Vistula Basin · Poland

1 Introduction

Anthropogenic global warming has clearly manifested itself as an increase of mean air temperature at various spatial scales, ranging from local to national, regional, continental, hemispheric, and global. The globally averaged combined land and ocean surface temperature has increased by about 0.85 (0.65–1.06) °C, over the period from 1880 to 2012 (IPCC 2013). All 15 individual years of the 21st century (2001–2015), have been among the 16 globally warmest years on record. The year 2015 proved to be the warmest on record, with the global mean temperature exceeding the previous record (established in 2014) by as much as 0.13 °C. Also in Central Europe and Poland, many temperature records were set in 2014 and 2015. Indeed, the heat goes on (Graczyk et al. 2016). The year 2015 has been the first and only year analyzed in thermal classification (since 1966) at the observation station Warsaw-Okecie (cf. www.imgw.pl) marked as "anomally warm", as compared to the reference period 1971–2000.

Since the attribution of the observed warming is well understood and persuading (IPCC 2013), it is possible to produce model-based climatic projections for the future.

The hydroclimatic projections performed in this study used the bias-corrected output from EUR-11 CORDEX experiments, assuming RCP 4.5, (cf. Moss et al. 2010). A quantile mapping method (QMAP) developed by the Norwegian Meteorological Institute was applied as a bias correction procedure for both historical and RCP 4.5 scenario runs (Mezghani and Dobler 2015). Three climate variables (precipitation, minimum and maximum temperature) from nine GCM-run-RCM combinations were available. All data were interpolated on the 5-km grid, regardless of the original RCM resolution. This was the same grid as the one of the CHASE-PL Forcing Data Gridded Daily Precipitation and Temperature Dataset 5 km (CPLFD-GDPT5) (Berezowski et al. 2015a, b), that was used as the forcing data for calibration of the hydrological model SWAT (Soil and Water Assessment Tool) for the Vistula and Odra basins (Piniewski et al. 2016a, Piniewski and Szcześniak 2016).

All bias-corrected data were available for the following three time slices: 1971–2000, 2021–2050, and 2071–2100. Since SWAT requires three years of warm-up period (the period for which different unknown water storages can stabilise and which is not taken into account in model output), all model simulations is available for three 27-year-long time slices: 1974–2000 (hereafter referred to as "historical period"), 2024–2050 ("near future"), and 2074–2100 ("far future"). In this chapter we focus on the projections extracted for the area of the Upper Vistula Basin, i.e. the Vistula upstream of the confluence with the river San. This part of the Vistula River Basin includes nearly the entire Polish part of the Carpathian mountains, with three important right-hand tributaries: Dunajec, Wisłoka and San.

2 Projections of Temperature

Climate model projections agree well about the direction of change of temperature leading to an overall ubiquitous warming across the Upper Vistula Basin and Poland. This refers to both projections for the near future, and for the far future. There is a fairly consistent signal of mean temperature increase for the whole Upper Vistula Basin in all models considered, with a likely range of 0.5–1.7 °C in the near future and 1.2–3.2 °C in the far future. In both future horizons, spatial distribution of change is not pronounced and there are slight differences between the regions when considering individual models albeit, the difference does not exceed 0.5 °C. The increase in minimum temperature is consistently larger (by 0–0.5 °C) than the increase in maximum temperature.

Seasonal projections of temperature change also show ubiquitous warming. Mean temperature change is the highest in winter (DJF) according to the median RCM: 1.3 and 2.9 °C for the near and far future, respectively. For all other seasons the magnitude (for the median RCM) is comparable and is slightly lower than in winter: between 0.8 and 1.1 °C for the near future, and between 1.7 and 1.9 °C for the far future.

3 Projections of Precipitation

Mean annual precipitation in the whole Upper Vistula Basin is projected to increase by 1.0–11.8 % in the near future (with an exception of one climate model for which the mean projected change is negative, −0.1 %) and by 3.8–11.5 % in the far future. This change is not spatially uniform and the spatial patterns differ between individual climate models. Eight out of nine models project an increase in precipitation that accelerates with time. In the far future, only three climate models show a decrease in precipitation for certain sub-basins but the area for which it happens is negligibly small. The magnitude of projected precipitation change for the Upper Vistula Basin is in general lower than for the entire Vistula Basin.

Models do not agree about changes in mean annual precipitation for the Upper Vistula Basin, in the near future. While, in general, most models show increase of mean annual precipitation, with the median change of 3.8 %, there are subareas in the basin where precipitation decrease is projected, in particular in the southern mountainous part. For the far future, all models tend to show increase of mean annual precipitation for the Upper Vistula Basin (median change by 8 %), with some models showing small areas of projected precipitation decrease.

Precipitation change is not seasonally uniform. For different seasons, seasonal distribution of changes is different. For the near future, the median RCM results suggest an increase of winter and spring precipitation by 10.6 and 7.1 %, respectively, while for summer and autumn only by 0.1 and 1.2 %, respectively. In winter and spring, precipitation increase is uniform for individual climate models, whilst in summer and autumn, the direction of change is very uncertain. The magnitude of mean changes in the Upper Vistula Basin is similar to that calculated on the entire Vistula and Odra basins for winter, spring and summer, whilst for autumn it is considerably lower.

For the far future, the median of the RCM simulations suggests an increase for winter and spring precipitation by almost 16 %, while for summer and autumn only by 1.5 and 4.5 %, respectively. Similarly to near future projections, changes in summer and autumn are more uncertain than for winter and spring. Summer seems to be the only season for which the magnitude of changes is significantly lower for the Upper Vistula Basin than for the entire Vistula and Odra basins.

Rising winter temperatures and increasing winter precipitation lead to non-trivial changes in the amount of precipitation falling as snow (assessed based on the SWAT model output). Considering the median of the projections, there is a sharp latitudinal gradient related to elevation change: in southern mountainous parts projected decrease in snowfall is relatively small (approximately 5 and 10 % in the near and far future), while in the northern part it is higher by the factor of four, reaching 20 and 40 %, respectively.

4 Projections of Runoff

Piniewski and Szcześniak (2016) produced hydrological projections driven by nine aforementioned climate models for the Vistula and Odra basins, consisting of 2633 sub-basins, using the SWAT hydrological model, which was calibrated and validated against daily discharge data from 80 "benchmark" catchments (Piniewski et al. 2016a). Figures 1 and 2 show projected changes in mean annual total runoff (i.e. the sum of baseflow, subsurface and surface runoff) for both near and far future, respectively.

Similarly to precipitation, mean annual runoff in the Upper Vistula Basin is projected to increase for eight out of nine climate models in the near future (Fig. 1) and for all nine models in the far future (Fig. 2). Most climate models project an increase in runoff by approximately more than 10 % in the near future, but one model projects a decrease in runoff (by 3.8 %). However, for the far future, the number of models showing an increase in runoff higher than 10 % reaches eight (out of nine in total), four of which project an increase above 20 %. Moreover, projected changes are not spatially uniform and the spatial patterns differ between models. Interestingly, for both future time slices and for all nine models, the Upper Vistula Basin faces a drier (or "less wet") future than the whole Vistula and Odra basins (Figs. 1 and 2). While in the near future there are areas in which runoff decreases (most frequently SE Poland), in the far future this is much less common and is limited mainly to small parts of the mountainous areas in the south.

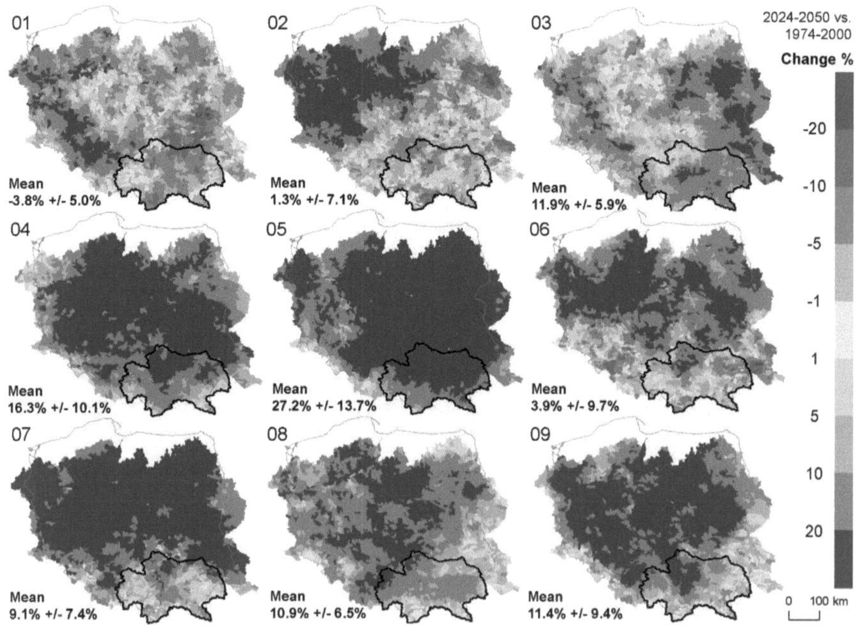

Fig. 1 Mean annual runoff projections for the near future (relative change between 2024 and 2050 and 1974–2000). *Black thick line* delineates the Upper Vistula Basin, whereas *green* labels denote mean and standard deviation of percent changes over sub-basins inside the Upper Vistula Basin (color figure online)

Surprisingly, the highest increase of mean annual runoff for one model in the near future is slightly higher than for the far future (27.2 %, with less uncertainty and 26.2 %, with more uncertainty, respectively).

The seasonal increase in runoff in the Upper Vistula Basin is the largest in winter (17.3 and 37.2 %, respectively) according to the median RCM. In summer and autumn, the projected changes are similar (5–6 % for the near future and 16 % for the far future), whereas spring is the only season in which the change is similar in both periods (increases by 7.7 and 8.9 %, respectively). Comparing these values with the projected increases in seasonal precipitation, one can note that spring is the season for which a response in runoff to increasing precipitation is considerably lower than for other seasons. It is presumably related to the fact that in the historical period the annual runoff peak occurs in spring, while in the future periods (particularly far future) it is shifted towards winter due to decreasing snow melt.

In all seasons apart from winter, runoff projections for the Upper Vistula Basin are considerably different than for the entire Vistula and Odra basins. In particular, in the southern mountainous part in all these three seasons there are substantial areas of runoff decrease, which occurs much less frequently elsewhere.

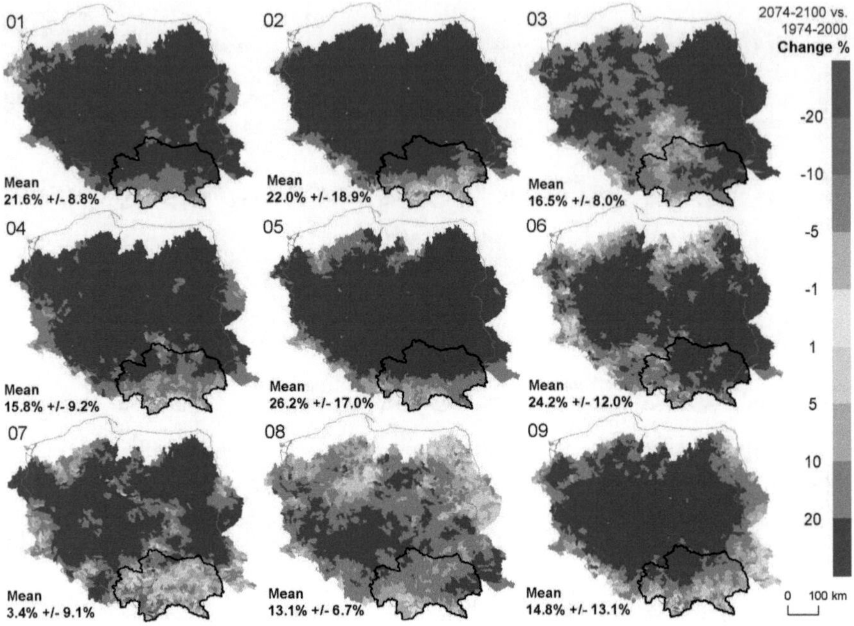

Fig. 2 Mean annual runoff projections for the far future (relative change between 2074–2100 and 1974–2000). *Black thick line* delineates the Upper Vistula basin, whereas *green* labels denote mean and standard deviation of change over sub-basins inside the Upper Vistula Basin (color figure online)

5 Changes in High Flow

Figure 3 shows projected changes in high streamflow indicator, the 90th daily flow percentile (Q90), for the near and far future, respectively, according to the median RCM, for the Upper Vistula Basin. The mean basin-wide increase is 6.4 % and 15 % for the near future and the far future, respectively. Spatial pattern of changes in Q90 is similar for both periods: small increases are prevailing for the right-hand tributaries of the River Vistula, draining high elevation catchments, and higher (for the far future even above 20 %) for the left-hand tributaries characterized by lower elevation. The magnitude of Q90 increase in the Upper Vistula Basin is lower than in the entire Vistula and Odra basins, while the uncertainty related to the direction of change is higher.

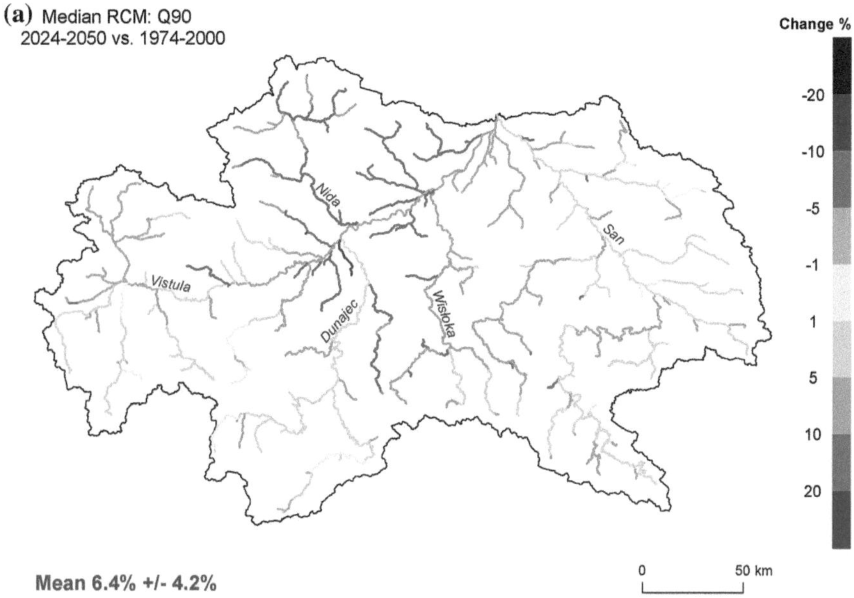

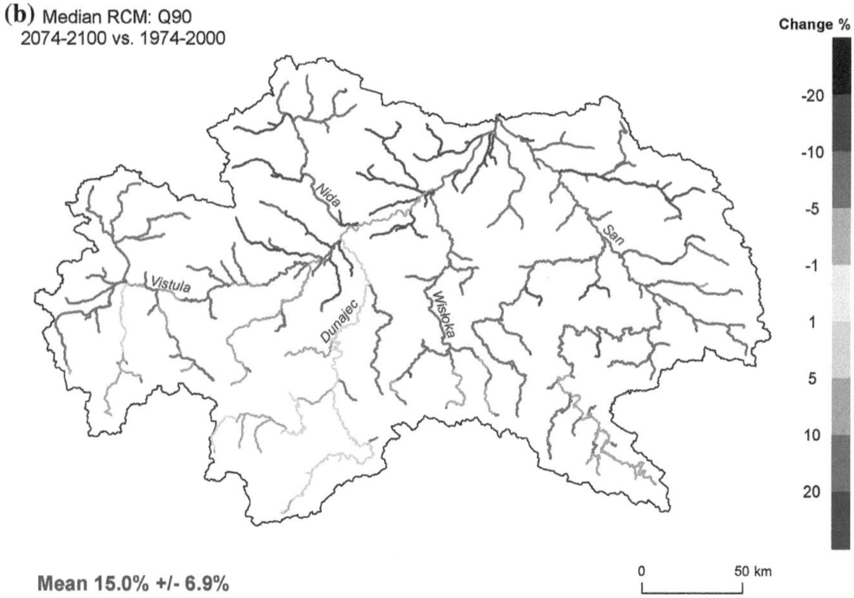

Fig. 3 High (Q90) monthly flow indicator projections for the near (**a**) and far (**b**) future for median RCM

6 Concluding Remarks

Projections of changes in hydroclimatic variables: temperature, precipitation, runoff and high flow, for the Upper Vistula Basin from an ensemble of bias-corrected RCM experiments and the SWAT model show several robust signals:

1. A ubiquitous mean temperature increase in the near and far future, with low spatial variability.
2. A higher temperature increase in winter than in other seasons.
3. A higher increase in minimum temperature than in maximum temperature.
4. An increase in mean precipitation and runoff projected by the majority of climate models in both future periods considered. The relative magnitude of increase (in percent) in runoff is generally higher than the corresponding increase in precipitation, so that an amplification effect can be observed. The change accelerates with time.
5. Smaller magnitude of precipitation and runoff changes in the southern mountainous part of the Basin than in its northern highland part.
6. Seasonally variable precipitation changes according to most of the models: in both future periods the increase is largest in winter and spring.
7. The highest runoff increase is projected in winter, consistently, by all climate models.
8. Changes in precipitation and runoff for the Upper Vistula Basin are different in comparison to the entire Vistula and Odra basins (a smaller magnitude of increase, in general).
9. An increase in high flow projected mainly in the northern highland part of the Basin.

Nevertheless, the range of projected changes based on the climate model ensemble remains relatively high and spatial patterns are not fully consistent across different climate models. For example, although the mean runoff increase is projected consistently by all models in the far future, its range is very wide: from 3.4 to 26.2 %.

Acknowledgments All authors acknowledge support of the CHASE-PL (Climate change impact assessment for selected sectors in Poland) project of the Polish–Norwegian Research Programme in the frame of Project Contract No. Pol-Nor/200799/90/2014. The first author is grateful for support to the Alexander von Humboldt Foundation and to the Ministry of Science and Higher Education of the Republic of Poland. Additionally, ZWK acknowledges support of the FLORIST (Flood risk on the northern foothills of the Tatra Mountains) project, supported by a grant from the Swiss Government through the Swiss Contribution to the enlarged European Union (PSPB No. 153/2010).

References

Abbaspour K (2008) SWAT-CUP2: swat calibration and uncertainty programs—a user manual. page 95. Department of Systems Analysis, Integrated Assessment and Modelling (SIAM), Eawag, Swiss Federal Institute of Aquatic Science and Technology, Dübendorf, Switzerland

Berezowski T, Szcześniak M, Kardel I, Michałowski R, Okruszko T, Mezghani A, Piniewski M (2015a) CPLFD-GDPT5: high-resolution gridded daily precipitation and temperature dataset for two largest Polish river basins. Earth Syst Sci Data 8, 127-139. doi:10.5194/essd-8-127-2016, 2016

Berezowski T, Szcześniak M, Kardel I, Michałowski R, Piniewski M (2015b) CHASE-PL forcing data—gridded daily precipitation temperature dataset (CPLFD-GDPT5). Dataset on 3TU. Datacentrum. doi:10.4121/uuid:e939aec0-bdd1-440f-bd1e-c49_10d0a07

Graczyk D, Pińskwar I, Kundzewicz ZW, Hov Ø, Førland EJ, Szwed M, Choryński A (2016) The heat goes on—changes in indices of hot extremes in Poland. Theor Appl Climatol accepted doi:10.1007/s00704-016-1786-x

Mezghani A, Dobler A (2015) CHASE-PL D2.2 note on bias-adjustment of daily precipitation, minimum, and maximum temperature from regional climate models over Poland using GDPT5 as observations. CHASE-PL Project Report

Moss RH, Edmonds JA, Hibbard KA, Manning MR, Rose SK, van Vuuren DP, Carter TR, Emori S, Kainuma M, Kram T, Meehl GA, Mitchell JF, Nakicenovic N, Riahi K, Smith SJ, Stouffer RJ, Thomson AM, Weyant JP, Wilbanks TJ (2010) The next generation of scenarios for climate change research and assessment. Nature 463(7282):747–756

Piniewski M, Szcześniak M, Kardel I, Berezowski T (2015) CHASE-PL natural hydrology dataset (CPL-NH). Dataset on 3TU.Datacentrum. doi:10.4121/uuid:b8ab4f5f-f692-4c93-a910-2947aea28f42, 2015

Piniewski M, Szcześniak M, Kardel I, Berezowski T, Okruszko T, Srinivasan R, Vikhamar Shuler D, Kundzewicz ZW (2016) Modelling water balance and streamflow at high resolution in the Vistula and Odra basins (in review)

Piniewski M, Szcześniak M (2016) CHASE-PL report on the hydrological modelling of regional climate change scenarios for the Vistula and Odra basins. CHASE-PL Project Report.

Projections of Changes in Flood Hazard in Two Headwater Catchments of the Vistula in the Context of European-Scale Studies

Renata J. Romanowicz, Zbigniew W. Kundzewicz, Hadush K. Meresa, Markus Stoffel, Valentina Krysanova and Joanna Doroszkiewicz

Abstract European-scale flood hazard projections do not give a consistent view of future changes in Central Europe, including Poland. Some studies indicate decreases in the magnitude and frequency of high flows, whilst others show increases. In this chapter, we summarize the current state of knowledge on flood perspectives in Central Europe and Poland under future climatic conditions at the background of large-scale European flood hazard projections and we contrast it with a small-scale study. Projections of changes in flood hazard in two catchments are considered in a multi-scale perspective, and against a background of large-scale: global and European projections, through regional (Central Europe) and national, to local. A discussion on causes of differences in flood-hazard projections and their possible interpretation is included. Among other issues, the uncertainties related to the processes taking part in the computational chain leading to the derivation of projections are listed. Specifically, the possible changes in the 30-year and 100-year return period quantiles of the maximum annual flows in the Dunajec and the Upper Wisla basins, two headwater catchments in the Vistula Basin are presented. The analysis is based on seven driving GCM/RCM projections under the

R.J. Romanowicz · H.K. Meresa · J. Doroszkiewicz
Institute of Geophysics, Polish Academy of Sciences, Warsaw, Poland

Z.W. Kundzewicz (✉)
Institute for Agricultural and Forest Environment, Polish Academy of Sciences, Poznan, Poland
e-mail: kundzewicz@yahoo.com

Z.W. Kundzewicz · V. Krysanova
Potsdam Institute for Climate Impact Research, Potsdam, Germany

M. Stoffel
Climatic Change and Climate Impacts, Institute for Environmental Sciences, University of Geneva, Geneva, Switzerland

M. Stoffel
Department of Earth Sciences, University of Geneva, Geneva, Switzerland

M. Stoffel
Dendrolab.ch, Institute of Geological Sciences, University of Berne, Berne, Switzerland

Z.W. Kundzewicz et al. (eds.), *Flood Risk in the Upper Vistula Basin*,
GeoPlanet: Earth and Planetary Sciences, DOI 10.1007/978-3-319-41923-7_17

RCP4.5 scenario. The results for both catchments are consistent with some of the previous European-scale studies, but do not give a coherent image. At this stage, the only explanation of the differences in the projections of future flood changes in both catchments lies in climatic variability and the uncertainty of the results. The results of this study confirm the view that flood hazard assessment is influenced by multiple climatic and non-climatic factors which introduce uncertainties and whose relative importance is site-specific.

Keywords Flood hazard · Projections · Uncertainty · River Vistula · Poland

1 Introduction

Economic damage caused by river floods has grown considerably in recent decades, at all spatial scales, from local to global. Many river floods with material damage of the order of billions of euros and with multiple fatalities have been recorded in Europe. Since recent floods have reached new heights of stage and discharge records, there is a concern that not only flood damage may have increased, but also flood hazard (Kundzewicz et al. 2016).

Increase of heavy precipitation has been noted in many areas as well but its effect on floods has been difficult to pinpoint. Also studies of change detection in observed high river flows show no convincing and ubiquitous increase of flood hazard, in Europe and world-wide (Kundzewicz 2012; Kundzewicz et al. 2005; Madsen et al. 2014). Nevertheless, some indications of an increasing tendency in the number of large floods in Europe, of considerable magnitude and severity, have been detected recently (Kundzewicz et al. 2013). It is important to stress that reliable determination of flood frequency characteristics requires long-term observations and continuous gauge records. Typically, however, existing time series of records are not long enough, or stations have been re-located over time, which renders analysis and interpretation difficult. In addition, flood trends cannot be easily detected in the observation record because the signal-to-noise ratio is typically low whereas natural variability is high. The failure to detect a ubiquitous rising trend in floods has apparently been a surprise to some experts describing recent flood events as possible harbingers of a rise in flood risk related to climate change. This was exemplified by the sarcastic title chosen by Schiermeier (2003): "*Analysis pours cold water on flood theory*", when referring to failure to detect an increasing trend in a study reported by Mudelsee et al. (2003).

Model-based large-scale projections of changes in river flood frequency indicate increases in the amplitude and frequency of high river flows in most areas of the world (but not necessarily in Europe) in the warming climate (Hirabayashi et al. 2008, 2013; Arnell and Gosling 2016; Dankers et al. 2014). In addition, and in the case of Europe in particular, considerable disagreement has been reported between projections (Lehner et al. 2006; Dankers and Feyen 2009; Rojas et al. 2011, 2012, Alfieri et al. 2015; Roudier et al. 2016).

In this chapter, projections of changes in flood hazard in the two mountainous catchments situated in headwaters of the River Vistula are considered in a multi-scale perspective, and against a background of large-scale, global and European projections, through regional (Central Europe) and national, to local.

2 Review of Large-Scale (Global and European) Flood Hazard Projections

Large-scale, global or continental, projections of changes in flood frequency and intensity have been presented in many publications. Lehner et al. (2006), Dankers and Feyen (2009), Rojas et al. (2011, 2012), Roudier et al. (2016) and Alfieri et al. (2015) report projections for Europe, while Hirabayashi et al. (2008), (2013), Arnell and Gosling (2016), and Dankers et al. (2014) present global projections. It can be noted that large-scale projections may considerably differ between studies (Kundzewicz et al. 2016).

Alfieri et al. (2015) concluded that increases in Q_{100} dominate in most countries in three future periods studied (2006–2035, 2036–2065, 2066–2095), though the frequency of peak flows over threshold is projected to decrease in some areas (NE Europe and southern Spain). The projected increases are significant in all considered 37 countries in time slice of 2080s with values ranging between 18 % (Finland) and 982 % (Iceland). Rojas et al. (2011, 2012) show a dominant increase of frequency of Q_{100} at the end of 21st century for British Isles, France, Italy, Balkan and Carpathians, and decrease in eastern Germany, Poland, southern Sweden, Baltic countries and some rivers in the Iberian Peninsula. Roudier et al. (2016) found a clear North to South gradient in changes of flood hazard, with a moderate to strong increase of flood magnitude south of the 60°N parallel, and a strong decrease in northern Scandinavia and NW Russia. For Poland, these three studies suggest an increase, a decrease and a weak increase in floods, respectively.

In contrast, recent projections of change in flood hazard in Europe reported by Hirabayashi et al. (2013) indicate flood frequency decrease in much of Northern, Central and Southern Europe, and only for a part of western Europe (British Isles, northern France, and part of Benelux), prevailing increases in frequency of Q_{100} are projected. Results of Dankers et al. (2014) show increases in flood frequency (Q_{30}) prevailing in projections for British Isles only. For Poland, both studies project a decrease in floods at the end of this century.

3 Flood Hazard Projections for Central Europe

In line with the continental-scale studies, flood hazard projections do also differ considerably between studies for Central Europe. For instance, results of projections of changes in flood hazard, reported by Rojas et al. (2012) and Alfieri et al. (2015) do not agree on changes in Poland and Eastern part of Germany, notwithstanding the fact that in

Table 1 Information on a direction of projected changes of compared large-scale flood hazard projection studies for Central Europe (based on Kundzewicz et al. 2016, updated)

Study	Coverage of the study	Number of climate scenarios	Number of hydrological models	Direction of change
Roudier et al. (2016)	Europe	11 CORDEX	3: Lisflood, E-HYPE, VIC	↑
Alfieri et al. (2015)	Europe	7 CORDEX	1: Lisflood	↑↓
Dankers et al. (2014)	Global	5 GCMs	9 global HMs	↑W ↓E
Arnell and Gosling (2016)	Global	21 GCMs	1: Mac-PDM.09	↓↑W
Hirabayashi et al. (2013	Global	11 GCMs	11 AOGCMs	↑NW↓SE
Rojas et al. (2012, 2011)	Europe	1 GCM	1: Lisflood	↑↓NE
Dankers and Feyen (2009)	Europe	5 RCMs	1: Lisflood	↑ NW SE↓C
Hirabayashi et al. (2008)	Global	1 GCM	1: MATSIRO LSM	↑↓
Lehner et al. (2006)	Europe	2 GCMs	1: WaterGAP	↓↑

↑ mostly increase
↑ partly increase (in subareas)
↓↑ mostly decrease, in subareas increase
↓↑ mostly decrease, in some subareas increase
↓N ↑S decrease in northern part, increase in southern part

both studies the same hydrological model (LISFLOOD) was used. This is probably due to the different climate scenarios used in these studies: from one GCMs in Rojas et al. (2012) contrasting to seven RCMs in Alfieri et al. (2015).

Uncertainty also shows up in studies at the national scale (Madsen et al. 2014). For instance, disagreements exist between projections of changes in flood hazard over Germany (Kundzewicz et al. 2016) in studies using the same hydrological model, SWIM (Krysanova et al. 2015), but different greenhouse gas scenarios, SRES A1B and RCP8.5, described by Moss et al. (2010), Meinshausen et al. (2011) and Nakicenovic et al. (2000), respectively.

Table 1 summarizes information on projected changes in flood hazard, for the Central European region and illustrates considerable disagreement between particular studies.

4 Interpretation of Differences in Flood-Hazard Projections

An increase of observed record-breaking precipitation events has been detected for most of Europe under global warming (Lehmann et al. 2015). However, considerable problems remain, namely those related to projecting intense precipitation.

Indeed, a warmer atmosphere can retain more water vapour in line with the Clausius-Clapeyron law and model-based projections indicate likely increase of frequency of heavy precipitation in the warming Europe. Seneviratne et al. (2012) illustrated that heavy precipitation will become more frequent in the future, i.e. the median of the projected return period of 20-year, 24-h precipitation will decrease for all three sub-regions of Europe. However, model skill in reproducing extreme storm events and trends, given some change in forcing, is not persuasive. Large uncertainties in the projection of precipitation extremes are associated with uncertainties in models, downscaling techniques and natural variability (Nicholls and Seneviratne 2013). The underestimation of rainfall extremes by the models (often observed when models simulate historical data rather poorly) may be also related to the coarse spatial resolution used in model simulations, suggesting that projections of future changes in rainfall extremes in response to anthropogenic global warming may also be underestimated.

Many sources of discrepancy in flood-hazard projections can be identified, such as:

- uncertainty related to differences in General Circulation Models (GCMs);
- uncertainty related to differences in emission scenarios;
- uncertainty due to Regional Climate Models (RCMs) and downscaling techniques;
- deficiency of climate models in representing intense precipitation;
- problems related to bias correction;
- poor performance of models for extremes;
- uncertainty due to differences in global hydrological models (GHMs) and regional hydrological models (RHMs);
- problems related to extreme value techniques.

It is worth identifying basic assumptions in recent work on flood hazard projections, embracing Eastern Europe, listed in Table 1 (Kundzewicz et al. 2016). The particular studies differ with respect to emission scenarios (whether older, drawing from SRES or newer, based on the concept of RCPs), models (climate models—GCMs, RCMs, global and regional hydrological models—GHMs, RHMs), future horizon of interest, as well as spatial resolution, return period, downscaling technique, bias reduction method, just to name a few.

Differences among the projections may be due to the selection of GCMs, which is generally the largest source of uncertainty in the climate impact studies. Older papers by Hirabayashi et al. (2008) or Rojas et al. (2011, 2012), for instance, considered just one GCM, whilst the newer studies—ensembles of several GCMs/RCMs, up to a study based on 21 GCMs (Arnell and Gosling 2016). Hirabayashi et al. (2013) and Dankers et al. (2014) analyzed results from 11 GCMs and five GCMs, respectively, whereas Alfieri et al. (2015) used seven EURO-CORDEX climate scenarios (combinations of three GCMs downscaled with four RCMs).

Studying changes in river flood frequency requires an estimation of extreme river flows. Even if the notion of 100-year flow (Q_{100}) is used most frequently (as a typical protection level of structural defences), some authors use other return periods, high flow percentiles, or other variables. For instance, Dankers et al. (2014) studied 30-year 5-day peak flow, i.e. a moderately extreme river discharge, whereas Roudier et al. (2016) used both Q_{100} and Q_{10}. Estimates of extreme river flows are often based on extreme value distributions, and are increasingly uncertain at more extreme discharge levels, especially for return periods beyond the length of the data the estimate is based on. This uncertainty is well-known in the field of statistics, and techniques exist to estimate this uncertainty, but in many hydrological studies it is not taken into account. In this respect, an estimate of Q_{10} based on 30 years of simulations can be considered more robust compared to estimation of Q_{100} based on 30 years.

The compared papers deal with different future horizons, mostly 2070–2099 or 2071–2100, but Arnell and Gosling (2016) used 2050s whereas Lehner et al. (2006) has chosen 2020s and 2070s as future horizons. Alfieri et al. (2015) used the future horizons of 2020s, 2050s, and 2080s.

One important reason of differences is related to differences in emission scenarios. Impacts are typically modelled, based on climate-model projections using either of two scenario approaches: Representative Concentration Pathways (RCP), cf. Moss et al. (2010), Meinshausen et al. (2011), or the older IPCC Special Report on Emission Scenarios (SRES), cf. Nakicenovic et al. (2000). Typically, flood hazard projections are based on either SRES scenarios, such as A1B, mostly in older papers, such as Lehner et al. (2006), Hirabayashi et al. (2008), Arnell and Gosling (2016), Rojas et al. (2011) also A2 and B2 in Dankers and Feyen (2009). The Representative Concentration Pathways (RCP), in particular RCP 8.5, were used in newer papers, such as Hirabayashi et al. (2013), Dankers et al. 2014 and Alfieri et al. (2015).

Also control (reference) intervals differed in compared studies, being 1901–2000 for Hirabayashi et al. (2008), 1961–1990 for Rojas et al. (2011, 2012), 1971–2000 for Hirabayashi et al. (2013) and Dankers et al. (2014), and 1976–2005 for Alfieri et al. (2015). The differences in underlying assumptions and methods can explain, to some extent, the differences in projections. Bias correction does not necessarily improve climate scenarios, but adds just another non-linear transformation which may affect projection of extreme precipitation, and hence floods.

The recent study by Roudier et al. (2016) does considerably differ from the other papers reported here, as the authors did not select a specific future time horizon but focused on the definition of the hydrological impacts in a world with a +2 °C change in global mean temperature relative to pre-industrial levels (1881–1910). This is important for the policy world, especially in terms of the outcome of the Paris Climate Summit (COP21, i.e. Conference of Parties of the UN Framework Convention on Climate Change) in December 2015. Describing the impacts of a +2 °C (and possibly—also +1.5 °C) has been explicitly requested in France last year. Roudier et al. (2016) considered three RCPs 2.6, 4.5 and 8.5. The time horizons, for which a +2 °C global warming was expected for different assumptions, were: 2016–2045; 2023–2053; 2027–2056; 2028–2057; 2030–2059; 2042–2071; 2050–2079; 2071–2100.

Usually, in the studies discussed, just one hydrological model was used, except for the work by Dankers et al. (2014), considering nine global hydrological models and Roudier et al. (2016), considering three global hydrological models. Global models can provide consistent and coherent simulations across very large scales and are useful for getting global overviews, but they often have to compromise the model performance at the regional scale and in individual catchments for overall performance. An important reason for uncertainties in projected results is that, in the large-scale studies, usually the global hydrological models are used mostly without any calibration. Some global hydrological models may reproduce the long-term average seasonal dynamics of discharge fairly well, but their ability to reproduce floods seems problematic. Even if some authors show some kind of "validation" for large river basins, they use aggregated results over time and over large spatial areas, where the results may look fairly well. Other authors do not report any validation at all. Should we trust results produced by poorly adjusted tools? Perhaps expecting a consistency is futile.

Some studies rely on land surface models used in GCMs, where issue is about the computation of soil moisture, hillslope runoff, etc. There are some global hydrological models that are not directly coupled with GCMs. But, even those global hydrological models may have computation schemes similar to land surface models of GCMs that are good at representing large-scale water budgets and changes, but less good at representing small-scale peak runoff, i.e. floods in small or upstream rivers. In addition to land surface processes, simulation of rainfall from GCMs and RCMs is better for large spatio-temporal scales of different studies.

Regional Hydrological Models (RHMs) and catchment-specific models, on the other hand, have finer spatial resolution and often incorporate more detailed information on, for example, the topography, soils and water management practices, and it is a standard procedure to calibrate them using observations.

Calibration of hydrological models at catchment-scale is a well-established procedure. Models that are calibrated for a particular gauging station provide more realistic simulations under present and similar climate conditions. However, the assumption that the calibrated model parameters will remain constant into the future may not hold, and is indeed unlikely to be true for parameters that are sensitive to the climate. A hydrological model that is tuned to historic conditions may not always provide plausible projections under different climatic conditions.

Also, good model performance in simulating discharge at the catchment outlet may mask variable performance across the catchment, and indeed in other variables. If a model is calibrated and validated using only a few main criteria (such as Nash and Sutcliffe Efficiency, NSE, and percent bias, PBIAS) and for runoff at the catchment outlet only, then it is, strictly speaking, suitable only for evaluation of the daily, monthly or seasonal dynamics of runoff at the outlet gauge. Representing the results as spatial patterns on maps and also for other variables than runoff assumes the model has some skill which ideally should be demonstrated, for example through validation at intermediate gauges and for other variables. Likewise, if climate change impacts on extremes are investigated, the model performance needs to be evaluated for these extremes using special criteria. However, these rules are not always followed strictly by modellers. Under high-end climate change scenarios for the end of the century and for extreme events using both types of models (global and catchment-scale) is therefore connected with high uncertainty.

5 Flood Hazard Detection and Projections for Poland

Comparing detection of change in individual European countries, Madsen et al. (2014) included Poland, where in general, decreasing trends were detected in both the mean and the variance of annual maximum flow series. The tendency is more pronounced in rivers with a high contribution of winter floods (Strupczewski et al. 2009).

Alfieri et al. (2015) included Poland in their analysis of the projection of change in mean annual exceedance frequency of the 100-year return period peak flow for different European countries and estimation of percentage change between the baseline and the future time slices. In their aggregate view, they would consider the percentage increase between the baseline and the future time slices to read: 127, 86, 94 % for 2020, 2050, and 2080, respectively.

In their inter-comparison of results of national flood projections, Madsen et al. (2014) reported on a Polish study conducted by Osuch et al. (2012) for the Wełna and Orla catchments in western Poland, based on six RCMs from the ENSEMBLES project using the A1B scenario, with quantile mapping applied for the bias correction of climate projections and the lumped, conceptual rainfall–runoff model (HBV). The simulation results by Osuch et al. (2012) showed different directions of change or a lack of statistically significant changes for simulations driven by different RCM/GCMs. That was the only study that applied catchment-scale, lumped hydrological modelling for future flood projections. The other studies mentioned applied pan-European hydrological models Lisflood, E-HYPE and VIC (Rojas et al. 2012; Alfieri et al. 2015; Roudier et al. 2016).

Catchment-scale model projections are important for the adaptation of flood risk management to climate change and may provide different flood indices than the large-scale models. As discussed in Sect. 4, there are many reasons why basin-scale and catchment-scale flood projections may differ. The most important seems to be the uncertainty related to meteorological projections and hydrological modelling at those different scales. The other important issue, related to the scale is spatial and temporal averaging that may have a large impact on estimates of hydrological extremes. The next but not last is the variability of circulation patterns and land-surface-atmosphere feedback depending on local, catchment-scale features of the terrain.

In the next section we illustrate the approach for the derivation of flood hazard projections on a catchment scale for two mountainous catchments situated in the upper course of the Vistula basin with the aim of showing in practice the reasons for differences between regional and local estimates of projected future changes.

6 Flood Hazard Projections for the Two Headwater Catchments of the Vistula

In the foothills of the Tatra Mountains, observational records (Ruiz-Villanueva et al. 2016a, b) do not indicate an increasing flood hazard. Results show, by contrast, evidence of seasonal changes in the magnitude and frequency of high flows in the region. Two headwater catchments of the Vistula, of the rivers Dunajec and Upper Wisla, situated in the same region in the vicinity of the Tatra Mountains are chosen for flood hazard projections at the catchment scale. The catchments belong to the set of semi-natural catchments chosen within the Polish-Norwegian project Climate Change Impact on the Hydrological Extremes (CHIHE) (Romanowicz et al. 2016).

The catchments' locations are shown in Fig. 1. The catchments differ in their flood regime; the Dunajec is rainfall-dominated and the Upper Wisla has a mixed (rainfall and snow-melt) flood regime. Both are mountainous catchments characterized by a large variability of discharge. The geographical characteristics of the catchments are given in Table 2. The averaging of precipitation and streamflow observations was performed over the period 1971–2000.

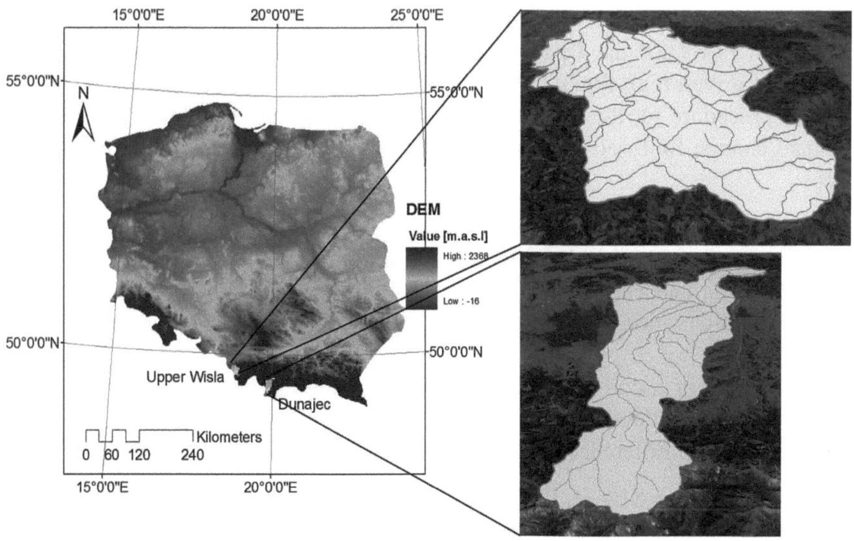

Fig. 1 Location of the Dunajec and the Upper Wisla catchments in Poland

Table 2 Description of the study catchments

Catchment	Gauging station	Area (km^2)	Flood regime	*PM* (mm)	*QM* (m^3/s)	*Qmax* (m^3/s)
Dunajec	Nowy Targ-Kowaniec	681.1	Rainfall	3.1	14.5	383
Upper Wisła	Skoczow	296.5	Mixed	2.6	6.1	238

QM is an average daily value of streamflow; *PM* is an average daily value of precipitation; *Qmax* maximum flow in 1971–2000

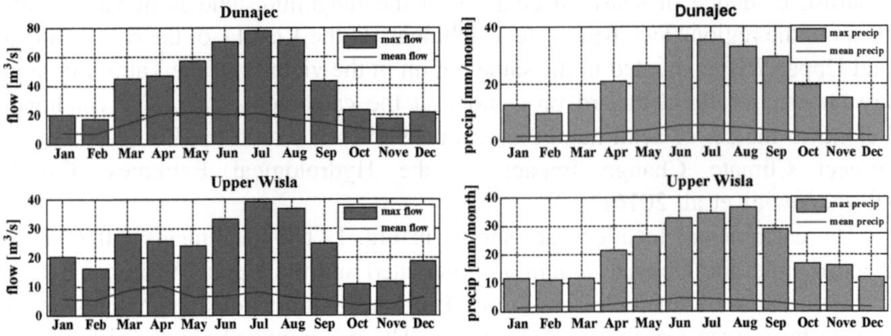

Fig. 2 Observed monthly maximum flows (*left column*) and monthly mean precipitation (*right column*) in the Dunajec (*upper panels*) and the Upper Wisla (*lower panels*) in the period 1971–2000

Figure 2 presents monthly maximum flows and monthly mean precipitation patterns in both catchments. There is a visible, two-modal shape of the mean maximum flows for the Upper Wisla catchment, indicating the presence of the snowmelt driven high flows.

The procedure for the derivation of estimates of changes in flood hazard under climate change followed in this study can be summarized in the following main steps:

1. Preparation of climate projection datasets for assumed climate change scenario for the reference period and the chosen time horizon in the future.
2. Calibration and validation of the hydrological rainfall-runoff model using historical observations of hydro-meteorological variables (temperature, precipitation, flow).
3. Simulation of runoff with the model using climate projections.
4. Assessment of flood risk indices, e.g. in the form of Q_{100} quantiles.
5. Derivation of differences between the future and reference periods.

As explained in Sect. 4, the results of each of these steps contain large uncertainties. In the first step, the uncertainty is related to the choice of climate change scenario, climate model inaccuracy and downscaling techniques. The uncertainty

Table 3 List of GCM/RCM climate models applied

GCM\RCM	RCA4	HIRHAM5	CCLM4-8-17	RACMO22E
EC-EARTH	1	1	1	1
MPI-ESM-LR	1	0	1	0
CNRM-CM5	0	0	1	0

related to the choice of the scenario of future CO_2 emissions is difficult to reduce (Knutti and Sedlácek 2012).

This is why in this study we followed the RCP4.5 (Representative Concentration Pathway) scenario, an "intermediate pathway" in which the global temperature increase in 2100 is estimated to be approximately 2.5 °C relative to the period 1850–1900 (Clarke et al. 2007). The selected combination of climate models consists of three GCMs and four RCMs (Table 3).

The selected available RCMs/GCMs provide seven projections of climatic variables up to 2100, at a resolution of 12.5 km. Analyses of hydro-meteorological conditions were conducted for the whole 1971–2100 period. We followed the study of Alfieri et al. (2015) in the choice of periods for the comparison. Four time intervals are examined: 1976–2005, the so-called reference period, and three future periods: (2006–2035), (2036–2065) and (2066–2095) periods. These periods are called after their median values: "1990s", "2020s", "2050s" and "2080s", respectively.

As a result of simplifications in the description of processes in global climate models, their spatial resolution and downscaling techniques, projections of temperature and precipitation are biased. As discussed in Sect. 4, the bias correction, or de-biasing, of raw projections might lead to undesirable changes in peak flows (Alfieri et al. 2015). On the other hand, some climate model projections do not reproduce correctly the seasonality of meteorological variables (Osuch et al. 2015), and therefore are not physically realistic without bias-correction. In this study we present the climate impact results using raw projections in order to compare with the results of Alfieri et al. (2015).

Projections of an annual maximum precipitation for the period 1971–2100 for the Dunajec and Upper Wisla catchments for seven GCMs/RCMs are presented in Fig. 3. The figure illustrates that there are no visible trends in projected precipitation maxima in either catchment.

Figure 4 presents the annual means and spreads of air temperature in the reference and future periods based on seven climate models for the Dunajec and Upper Wisla catchments. In the case of temperature patterns, there are visible positive trends projected for both catchments, with about 1 °C lower temperatures in the Dunajec than in the Upper Wisla. The mean temperature variability shows large similarity between the catchments and rises by nearly 2 °C by the end of 2100 in comparison with the reference period.

The conceptual rainfall-runoff model HBV (Bergström 1995) was applied to model the catchment response to meteorological forcing. The model was calibrated using daily observations of temperature and precipitation from the period

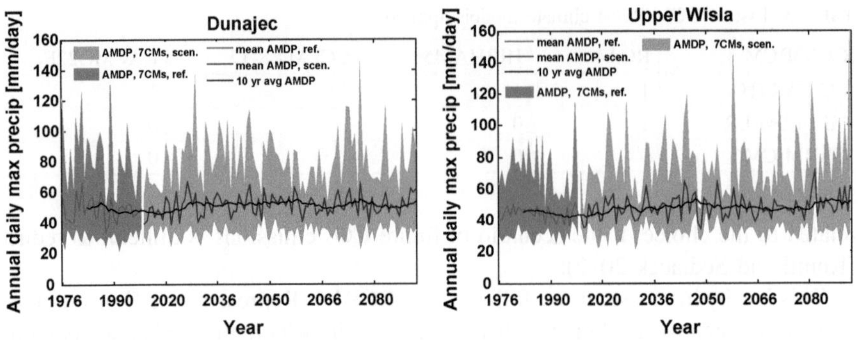

Fig. 3 Annual maximum daily precipitation (AMDP) for the Dunajec (*left panel*) and the Upper Wisla (*right panel*) in 1976–2100 based on seven climate models (CMs) from the GCMs/RCMs ensemble: *shaded areas* present ensemble spreads: *blue* for the reference period 1976–2005; *green* for the period 2006–2100; *lines* show ensemble means of AMDP: *red line* for the reference period, *blue line* for the future period. The *black line* shows a 10-year moving average AMDP (color figure online)

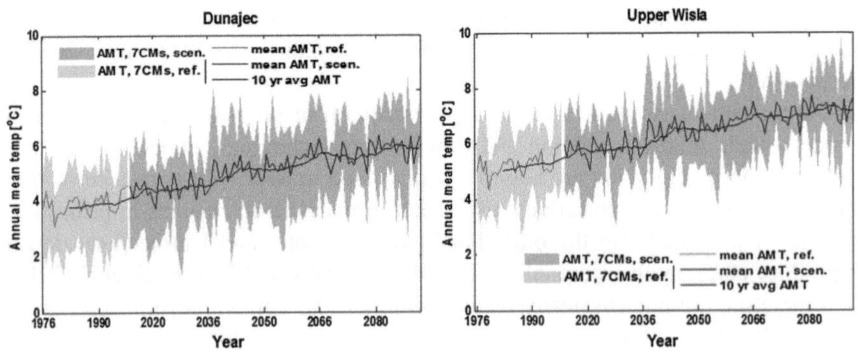

Fig. 4 Projected annual mean air temperature (AMT) for the Dunajec catchment (*left panel*) and the Upper Wisla catchment (*right panel*) in 1976–2100 based on seven climate models (CMs) from the GCMs/RCMs ensemble: *shaded areas* present ensemble spreads: *yellow* for the reference period 1976–2005; *grey* for the period 2006–2100; *lines* show ensemble mean temperature: *red line* for the reference period, *blue line* for the future period. The *black line* shows a 10-year moving average temperature (color figure online)

1971–2000, and validated against data covering the period 2001–2010. Model parameters were optimized using the DEGL (Differential Evolution with Global and Local neighbours) method (Storn and Price 1997). As an objective function the Nash-Sutcliffe coefficient was used (Nash and Sutcliffe 1970). The results of calibration and validation of the HBV model were good, and they were presented in Table 1 of Romanowicz et al. (2016). There is a large uncertainty related to the hydrological model structure and its parameters (Honti et al. 2014) which will not be taken into account in this study, being out of the scope of the present objectives.

Subsequently, the climate projections from seven model combinations were used to run the HBV model, and the series of flow simulations were obtained for each catchment and GCM/RCM model projection for the whole time horizon 1971–2100. The annual maximum values of daily flows were derived and used in flood frequency analysis. The derivation of 100 return period quantiles requires an extrapolation of empirical cumulative distribution function (cdf) based on 30-year periods. We selected GEV distribution to be fitted to each ensemble of the projected annual maximum flows following the discussion presented by Lawrence and Hisdal (2011). The obtained cumulative distribution functions as a function of a return period for Dunajec and Upper Wisla are presented in Fig. 5. It illustrates a large spread of Q_{100} quantiles obtained for different ensemble projections.

In the present study flood indices in the form of Q_{30} and Q_{100} for three future periods, 2020s, 2050s and 2080s are compared with those for the reference period 1990s.

Changes in the Q_{30} and Q_{100} quantiles for three future 30-year periods for the Dunajec and the Upper Wisla relative to the reference period 1990s are presented in Fig. 6. The projected changes in both Q_{30} and Q_{100} quantiles are consistent with each other but the magnitudes of changes in terms of means and confidence bands differ slightly or moderately between the catchments. Following the medians of quantile changes, the Upper Wisla shows slightly decreasing flood conditions in the 2020s, increasing flood conditions by more than 30 % in the 2050s, and again slightly decreasing in the 2080s. However, if the confidence bands are considered, there is an overall tendency to increasing floods in all three periods.

For the River Dunajec, changes in median values are much smaller (less than 10 % for all but Q_{100} in the 2080s) and the directions of change are opposite, with a slight increase, minor decrease and subsequent small increase of mean flood quantiles in the three periods studied. However, if the confidence bands are analysed, there is a tendency to increasing floods in all periods also for this catchment. Some differences in behaviour of both catchments can be caused by different flood regimes, with the Dunajec being rainfall-dominated and the Upper Wisla being of mixed regime, but also it can be due to the local atmospheric circulation patterns. However, if we look at the overall picture, most projections of changes in flood quantiles for both catchments show positive changes, with larger variability for the Upper Wisla than for the Dunajec.

Direct comparison of our findings with the results of Alfieri et al. (2015) is not possible because of too large differences in spatial scales. However, we can conclude that our findings support the tendency of positive changes presented by the large-scale analysis in both cases. The annual sums of precipitation show some variability, but these changes are not very well defined. Therefore at this stage, the explanation of some differences in the projections of future flood changes in both catchments lies in the climatic variability and the uncertainty of results.

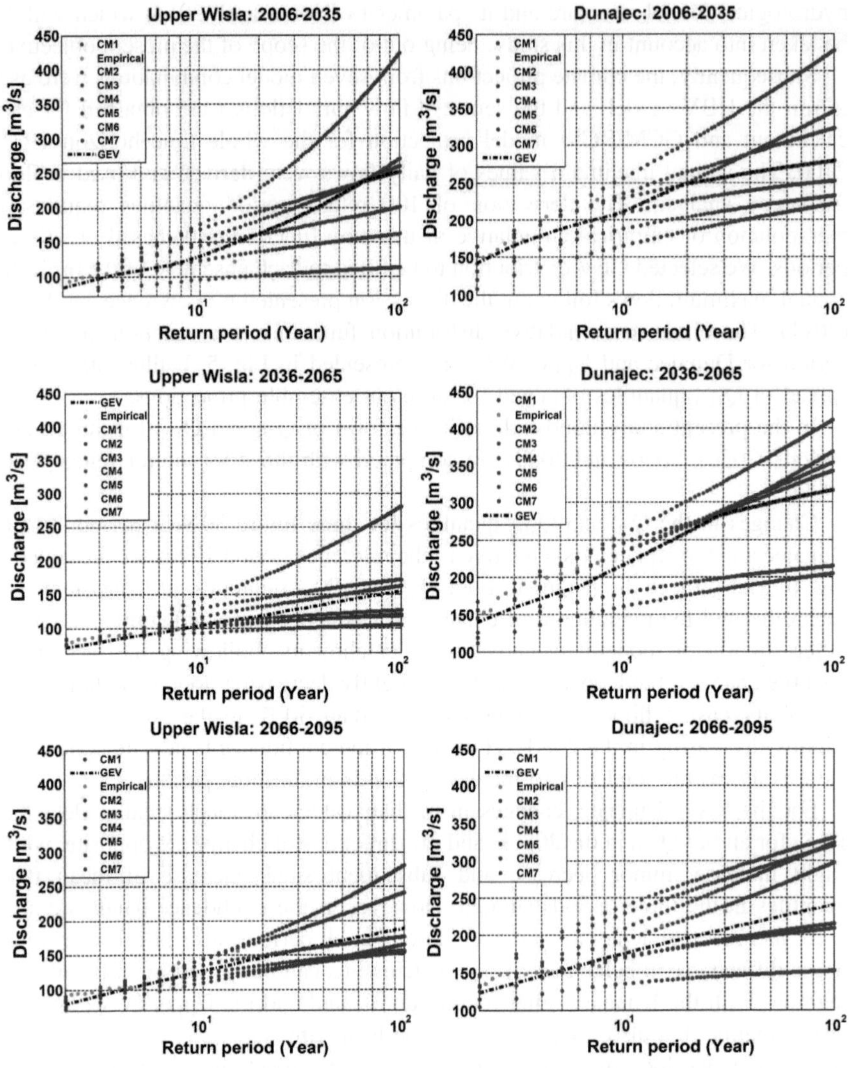

Fig. 5 Cdfs for annual maximum daily flows for the Dunajec (*right panel*) and Upper Wisla (*left panel*) obtained for seven driving GCMs/RCMs projections for three 30-year periods: 2020s, 2050s and 2080s; *blue doted lines* present the GEV fit for each model; median GEVs from the ensembles are shown by *black dashed lines*; *red dots* present ensemble median values of projected annual maximum flows (color figure online)

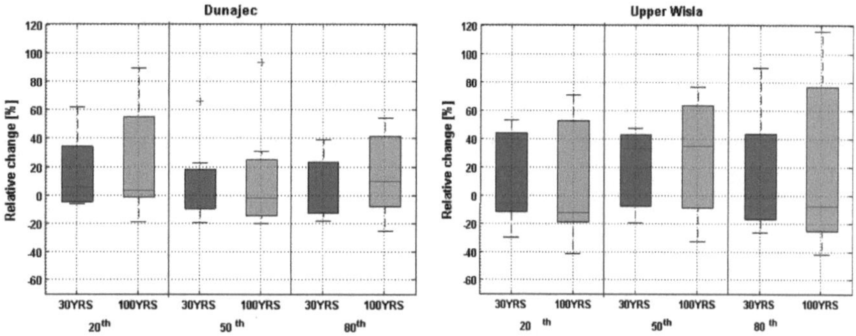

Fig. 6 Estimated changes in Q_{30} and Q_{100} using GEV for 2020s, 2050s and 2080s relative to the reference period 1990s, based on projections from HBV driven by seven GCM/RCMs together with 0.95 confidence bands: Dunajec *left panel* and Upper Wisla *right panel*

7 Impact on Climate Change Adaptation and Flood Risk Reduction

The lack of agreement in projections between studies can be interpreted and understood by scientists, but not readily by stakeholders. Despite the caveats accompanying large-scale flood hazard projection studies (Dankers et al. 2014), stakeholders may look at maps of projections from different sources that strongly diverge in the area of their interest, take results at face value, and become confused. This is how the discrepancy in flood hazard projections is regarded by practitioners in Poland (Kundzewicz et al. 2016).

Stationarity is dead (Milly et al. 2008, 2015), yet, the signal-to-noise ratio in flood hazard projections is low. In other words, the noise dominates and overshadows a weak (if any) signal. Non-stationarity means that a design flood (e.g. a 50-year or a 100-year event) for a particular location, established from historical observations in the reference period, can be dramatically different from a, possibly broad, range of values projected for a future time horizon of importance for adaptation. However, despite the huge uncertainty in flood hazard projections, practitioners and water managers in some European countries and regions already try to incorporate the potential effects of climate change into specific design guidelines, by a precaution-based adjustment, acknowledging increase in intense precipitation in the warming climate (Kundzewicz et al. 2008). An example of a climate change adjustment factor for a design flood is a specified relative increase of a 100-year flood in a specified future time horizon, incorporated in design guidelines. Madsen et al. (2014) compiled information on existing guidelines on climate change adjustment factors on design flood and design rainfall in six European countries (Belgium, Denmark, Germany, Norway, Sweden, UK). Such adjustments were also proposed in The Netherlands (Kundzewicz 2012).

Decision-makers responsible for flood protection and climate adaptation have to be aware of the added uncertainty introduced by enhanced greenhouse forcing

(Kundzewicz et al. 2014). According to the Polish Water Law and the EU Floods Directive (2007), flood hazards and flood risk maps were prepared for three different flood return periods (recurrence intervals): short (≥10 years), medium (≥100 years) and long (≥500 years). The main objectives of national flood adaptation strategy in Poland are to minimize the vulnerability to flood risks associated with changes in climate, and include this issue in the planning phase of future investments. The adaptation policy for Poland in future climate conditions was outlined for the time intervals ending in 2020 and 2070 in two separate papers (IOŚ-PIB 2013a, b), both published by the Ministry of Environment and covering all sectors vulnerable to floods, included in the "White Paper". Vulnerable sectors include water management, urban and rural spatial planning. However, the legislative regulations relating to local spatial planning still have to be established (Doroszkiewicz et al. 2016).

8 Outlook and Concluding Remarks

The vast spread of river flood hazard projections for Central Europe and Poland has to be interpreted with caution, especially by decision makers in charge of climate change adaptation, flood risk reduction, and water resources management. At the present stage, there are some differences in the projections of future flood changes in two analyzed catchments in the Upper Vistula Basin that are likely to be due to the climatic variability and the uncertainty of results. We are far from knowing the future reality, but projections for the future, despite the inherent uncertainty, are important to inform processes of flood risk reduction and adaptation to climate change, sketching the range of possible futures. At the same time, there is no doubt that a better preparedness for existing climate variability is necessary, but this is unlikely to be sufficient for future changes (Field et al. 2012; Kundzewicz et al. 2014) in areas with increasing flood hazard.

In some "problematic" regions, such as Central Europe, rain-floods and snow-floods both influence future flood changes. Researchers may be encouraged to separately investigate rain-caused floods and snow-caused floods in the analysis, and to determine which of these are dominant in particular river basins, for present and future conditions.

Projected climate-driven changes in future flood frequency are complex, depending on the generating mechanism, e.g., increasing flood magnitudes where floods result of increasing heavy rainfall and decreasing magnitudes where floods are generated by less abundant spring snowmelt (Kundzewicz et al. 2010).

For the time being, there is no conclusive and general proof as to how climate change affects flood behaviour. There is a scarcity of studies detecting, in a persuasive way, an influence of anthropogenic climate change on rain-generated peak streamflow trends. Natural variability is strong. Detection and attribution research for river flooding is not easy to carry out. The conventional attribution framework struggles with the low signal-to-noise ratio and uncertain nature of the forced changes (Trenberth et al. 2015).

Flood hazard is influenced by multiple climatic and non-climatic factors, whose relative importance is site-specific. Climatic factors include predominantly changes in intense precipitation and snowmelt, but changes in any component of the hydrological cycle, e.g. precipitation and evaporation also play a role. Non-climatic factors include changes (mostly anthropogenic) in rivers themselves, e.g. modification of river channels, such as construction of water structures—dikes and dams, channel shortening, removal of flood plains, and changes in catchments, such as urbanization, deforestation, drainage of wetlands, and other factors (Hall et al. 2014).

As exemplified by the derived flood indices for two Polish catchments, Upper Wisla and Dunajec, flood hazard projections for future climate are strongly uncertain. However, that uncertainty should not be used as an excuse for a lack of adaptation strategy. In contrast, it should encourage the governments to put more effort in preparing adaptation strategies that can face that uncertainty.

Acknowledgments This work was partially supported by the project CHIHE (Climate Change Impact on Hydrological Extremes), carried out in the Institute of Geophysics Polish Academy of Sciences, funded by Norway Grants (contract No. Pol-Nor/196243/80/2013). The hydro-meteorological data were provided by the Institute of Meteorology and Water Management (IMGW), Poland. Two co-authors (ZWK and MS) acknowledge the support of the FLORIST project (Flood risk on the northern foothills of the Tatra Mountains), via a grant from the Swiss government through the Swiss Contribution to the enlarged European Union (PSPB No. 153/2010).

References

Alfieri L et al (2015) Global warming increases the frequency of river floods in Europe. Hydrol. Earth Syst Sci 19:2247–2260, www.hydrol-earth-syst-sci.net/19/2247/2015/

Arnell NW, Gosling SN (2016) The impacts of climate change on river flood risk at the global scale. Climatic Change 1–15. doi:10.1007/s10584-014-1084-5

Bergström S (1995) The HBV model. In: Singh VP (ed) Computer models of watershed hydrology. Water Resources Publications, Highlands Ranch, CO, pp 443–476

Clarke L, Edmonds J, Jacoby H, Pitcher H, Reilly J, Richel R (2007) CCSP scenarios of greenhouse house gas emissions and atmospheric concentrations and review of integrated scenario development and application. U.S. Government Printing Office, Washington, DC

Dankers R, Feyen L (2009) Flood hazard in Europe in an ensemble of regional climate scenarios. J Geoph Res Atmos 114, D16

Dankers R et al (2014) First look at changes in flood hazard in the inter-sectoral impact model Intercomparison project ensemble. Proc Natl Acad Sci 111:3257–3261

Doroszkiewicz J, Hisdal H, Romanowicz RJ (2016) Guidelines how to quantify and handle uncertainty when adapting to changes in floods under a future climate, report 2. CHIHE, Institute of Geophysics, PAS, Warsaw, Poland

Field CB et al (eds) (2012) Managing the risks of extreme events and disasters to advance climate change adaptation. A special report of working groups I and II of the Intergovernmental Panel on Climate Change (IPCC). Cambridge University Press, Cambridge

Fischer EM et al (2013) Robust spatially aggregated projections of climate extremes. Nat Clim Change 3:1033–1038

Hall J et al (2014) Understanding flood regime changes in Europe: a state-of-the-art assessment. Hydrol Earth Syst Sci 18:2735–2772. www.hydrol-earth-syst-sci.net/18/2735/2014/

Hirabayashi Y et al (2008) Global projections of changing risks of floods and droughts in a changing climate. Hydrol Sci J 53(4):754–772

Hirabayashi Y et al (2013) Global flood risk under climate change. Nat Clim Change 3:816–821. doi:10.1038/nclimate1911

Honti M, Scheidegger A, Stamn C (2014) The importance of hydrological uncertainty assessment methods in climate change impact studies. HESS 18:3301–3317

IOŚ-PIB (2013a) Strategic adaptation plan for areas vulnerable to climate change until 2020 [SPA Strategiczny plan adaptacji dla sektorów i obszarów wrażliwych na zmiany klimatu do roku 2020]. Warszawa: Ministry of Environment

IOŚ-PIB (2013b) Development and implementation of the strategic plan for adaptation for sectors and areas vulnerable to climate change. stage iii—adaptation of sensitive sectors and areas of polish climate change to 2070—project. [Opracowanie i wdrozenie strategicznego planu adaptacji dla sektorow i obszarow wrazliwych na zmiany klimatu. etap iii—Adaptacja wrazliwych sektorow i obszarow Polski do zmian klimatu do roku 2070 -projekt]. Warszawa, Ministry of Environment

Knutti and Sedlacek (2012) Robustness and uncertainties in the new CMIP5 climate model projections. Nat Clim Change Lett. doi:10.1038/NCLIMATE1716

Krysanova V et al (2015) Modelling climate and land-use change impacts with SWIM: lessons learnt from multiple applications. Hydrol Sci J 60(4):606–635

Kundzewicz ZW (ed) (2012) Changes in flood risk in Europe, Special Publication No. 10, IAHS Press, Wallingford, Oxfordshire, UK. 516 +xvi pp

Kundzewicz ZW et al (2005) Trend detection in river flow time series: 1. Annual maximum flow. Hydrol Sci J 50(5):797–810

Kundzewicz ZW et al (2008) The implications of projected climate change for freshwater resources and their management. Hydrol Sci J 53(1):3–10

Kundzewicz ZW et al (2010) Assessing river flood risk and adaptation in Europe—review of projections for the future. Mitig Adapt Strat Glob Change 15(7):641–656

Kundzewicz ZW et al (2013) Large floods in Europe, 1985–2009. Hydrol Sci J 58(1):1–7

Kundzewicz ZW et al (2014) Flood risk and climate change: global and regional perspectives. Hydrol Sci J 59(1):1–28

Kundzewicz ZW, Krysanova V, Dankers R, Hirabayashi Y, Kanae S, Hattermann FF, Huang S, Milly PCD, Stoffel M, Driessen PPJ, Matczak P, Quevauviller P, Schellnhuber H-J (2016) Differences in flood hazard projections in Europe—their causes and consequences for decision making. Hydrol Sci J (in review)

Lawrence D, Hisdal H (2011) Hydrological projections for floods in Norway under a future climate. NVE Report 5-2011, Oslo, Norway

Lehmann J, Coumou D, Frieler K (2015) Increased record-breaking precipitation events under global warming. Clim Change. doi:10.1007/s10584-015-1434-y

Lehner B et al (2006) Estimating the impact of global change on flood and drought risks in Europe: a continental, integrated assessment. Clim Change 75:273–299

Madsen H et al (2014) Review of trend analysis and climate change projections of extreme precipitation and floods in Europe. J Hydrol 519:3634–3650

Meinshausen, M et al (2011) The RCP greenhouse gas concentrations and their extension from 1765 to 2300. Clim Change (Special Issue). http://edoc.gfz-potsdam.de/pik/get/5095/0/0ce498a63b150282a29b729de9615698/5095.pdf

Milly PCD et al (2008) Stationarity is dead: whither water management? Science 319:573–574

Milly PCD et al (2015) Commentary on critiques of "stationarity is dead: whither water management?". Water Resour Res 51(9):7785–7789

Moss RH et al (2010) The next generation of scenarios for climate change research and assessment. Nature 463(7282):747–756

Mudelsee M et al (2003) No upward trends in the occurrence of extreme floods in central Europe. Nature 425:166–169

Nakicenovic N et al (2000) IPCC special report on emissions scenarios. Cambridge University Press, Cambridge

Nash J, Sutcliffe J (1970) River flow forecasting through conceptual models. Part I—a discussion of principles. J Hydrol 10(3):282–290. doi:10.1016/0022-1694(70)90255-6

Nicholls N, Seneviratne SI (2013) Comparing IPCC assessments: how do the AR4 and SREX assessments of changes in extremes differ? Clim Change. doi:10.1007/S10854-013-018-0

Osuch M, Kindler J, Romanowicz RJ, Berbeka K, Banrowska A (2012) KLIMADA Strategia adaptacji Polski do zmian klimatu w zakresie sektora "Zasoby i gospodarka wodna", KLIMADA project, Institute of Environmental Protection. National Research Institute, Warsaw (in Polish)

Osuch M, Romanowicz RJ, Lawrence D, Wong WK (2015) Assessment of the influence of bias correction on meteorological drought projections for Poland. Hydrol Earth Syst Sci Discuss 12:10331–10377. doi:10.5194/hessd-12-10331-2015

Rojas R et al (2011) Improving pan-European hydrological simulation of extreme events through statistical bias correction of RCM-driven climate simulations. Hydrol Earth Syst Sci 15:2599–2620. doi:10.5194/hess-15-2599-2011

Rojas R et al (2012) Assessment of future flood hazard in Europe using a large ensemble of bias corrected regional climate simulations. J Geophys Res Atmos 117:D17109. doi:10.1029/2012JD017461

Romanowicz RJ et al (2016) Climate change impact on hydrological extremes (CHIHE): preliminary results from the Polish-Norwegian project, Acta Geophysica, accepted for publication

Roudier PH, Andersson JCM, Donnelly CH, Feyen L, Greuell W, Ludwig F (2016) Projections of future floods and hydrological droughts in Europe under a +2 °C global warming. Clim Change. doi:10.1007/s10584-015-1570-4

Ruiz-Villanueva V et al (2016) Decadal variability of floods in the northern foreland of the Tatra Mountains. Reg Environ Change 16(3):603–615. doi:10.1007/s10113-014-0694-9

Ruiz-Villanueva V et al (2016b) Variability of flood frequency and magnitude during the late 20th and early 21st centuries in the northern foreland of the Tatra Mountains. This volume

Seneviratne SI et al (2012) Changes in climate extremes and their impacts on the natural physical environment. In: Field CB et al (eds) Managing the risks of extreme events and disasters to advance climate change adaptation. Special report of working groups I and II of the intergovernmental panel on climate change (IPCC). Cambridge University Press, Cambridge

Schiermeier Q (2003) Analysis pours cold water on flood theory. Nature 425:111

Storn R, Price K (1997) Differential evolution—a simple and efficient heuristic for global optimization over continuous spaces. J Global Optim 11(4):341–359. doi:10.1023/A:1008202821328

Strupczewski WG, Kochanek K, Feluch W, Bogdanowicz E, Singh VP (2009) On seasonal approach to nonstationary flood frequency analysis. Phys Chem Earth 34:670–678

Trenberth KE, Fasullo JT, Shepherd ThG (2015) Attribution of climate extreme events. Nature Clim Change 5(8):725–730

Flood Risk Management in the Upper Vistula Basin in Perspective: Traditional versus Alternative Measures

Bartłomiej Wyżga, Artur Radecki-Pawlik and Joanna Zawiejska

Abstract Flood-protection works carried out in the Upper Vistula basin since the late nineteenth century have been based on channel regulation and river embankment leading to fast evacuation of floodwater and a significant reduction in floodwater retention on the valley floors. Such a policy of flood-control management stemmed not only from the unfamiliarity with other methods and a generally technocratic approach to nature but also from the need to protect all arable land adjacent to rivers. This last motivation justified the approach in the early part of the period when farming provided for the existence of most of the society, but it progressively declined in significance with increasing urbanization of the region and economic development of the country. Flood-risk management based on the conventional methods has resulted in a severe degradation of the rivers' ecological quality and increased peak discharges of flood waves recorded in the downstream parts of regulated and embanked river reaches. It is thus a priority to decelerate flood runoff and increase floodwater retention in less developed parts of the valleys in order to reduce flood hazard in spatially concentrated, urbanized areas along rivers. This paper presents alternative measures that either aim at reducing flood hazard at various stages of flood-wave passage through the region or serve to diminish flood risk by preventing development of river-adjacent areas, re-construction of bridges and cessation of the detrimental in-channel gravel mining.

B. Wyżga (✉)
Institute of Nature Conservation, Polish Academy of Sciences, al. Mickiewicza 33, 31-120 Kraków, Poland
e-mail: wyzga@iop.krakow.pl

A. Radecki-Pawlik
Department of Hydraulics Engineering and Geotechnique, University of Agriculture, al. Mickiewicza 24/28, 30-059 Kraków, Poland
e-mail: rmradeck@cyf-kr.edu.pl

J. Zawiejska
Institute of Geography, Pedagogical University of Cracow, ul. Podchorążych 2, 30-084 Kraków, Poland
e-mail: zawiejska.joanna@gmail.com

Z.W. Kundzewicz et al. (eds.), *Flood Risk in the Upper Vistula Basin*,
GeoPlanet: Earth and Planetary Sciences, DOI 10.1007/978-3-319-41923-7_18

Keywords Flood hazard · Flood risk · Flood protection · Flood-control measures · Floodwater retention

1 Introduction

Floods pose a significant threat to human life and property (Merritts 2011). Channelization, bank protection and construction of dam reservoirs were among the measures introduced to limit that threat (Bravard and Petts 1996). A reduction of flood hazard was imperative in mountain and piedmont regions where high precipitation values and considerable slope and channel gradients result in rapid concentration of flood waves and high energy of flood flows, and where settlements are frequently located on valley floors. In the Italian Alps, technical interventions were already introduced in Roman times (Comiti 2012). Large-scale channelization of Carpathian tributaries to the Vistula was initiated in the late nineteenth century in their reaches in the Sandomierz Basin (the San—Szumański 1986; the Dunajec—Zawiejska and Wyżga 2010). In the twentieth century channelization works encompassed almost the whole length of these rivers and a large proportion of the courses of their tributaries.

Flood embankments were constructed in the second half of the nineteenth century along the upper course of the Vistula (Łajczak 1995) and in the 1880s–1890s in the lowest courses of its Carpathian tributaries (Zawiejska and Wyżga 2010). First dam reservoirs on Polish Carpathian rivers were constructed after the catastrophic flood of 1934 (Porąbka Reservoir on the Soła and Rożnów Reservoir on the Dunajec), and several others have been built since the World War II. Until now these interventions (channel regulation, river embankment and dam reservoirs) have been practically the only measures adopted in the Upper Vistula basin to mitigate flood hazard. However, long-term experience in this and other regions has shown that these measures bring a range of adverse effects and they provide only localised protection from flooding, whereas flood risk at the regional scale is not reduced but only transferred downstream (Wyżga and Radecki-Pawlik 2011). In the last decades several methods were indicated that may be more effective in reducing flood hazard and risk at a regional scale than the conventionally used technical measures (European Commission 2003; Wyżga and Radecki-Pawlik 2011).

This paper presents limitations and adverse effects of the "traditional" methods of flood protection and indicates a range of alternative measures which may be more effective in reducing flood hazard and risk at a regional scale.

2 Conventional Methods of Flood Protection

2.1 Channelization Schemes

Natural, vertically stable rivers have channel capacity close to 1.5-year flood flow (Williams 1978). In turn, sizing of regulated channels depends on the character of land use in the adjacent area, with designed flow capacity of the channels ranging from a 2-year flow in forested areas to a 10-year flow in urbanized areas (Hydroprojekt 1987). Channel straightening and narrowing, often to a width significantly smaller than that of the natural channels, resulted in a considerable increase in river transport capacity, while simplification of the natural, morphologically diverse channels reduced their resistance to flow (Wyżga 2001). Consequently, river channelization was followed by channel incision (Fig. 1) (Wyżga 2001; Zawiejska and Wyżga 2010) that further increased flow capacity of the channels. Although the absolute (expressed in metres) amount of the twentieth-century incision of Polish Carpathian rivers was highest in their lower and

Fig. 1 Example of detrimental effects of inappropriate river management practices on channel stability and valley-floor infrastructure. **a** Downstream view of the Czarny Dunajec from the bridge at Chochołów in 2002. The river flows in two braids and its channel is vertically stable. **b** The same reach in 2005. *Arrow* indicates the position of the former left braid that has been blocked and filled up with earth. **c** Deep incision induced by flow concentration in the single channel downstream of the bridge is visible in 2015. **d** Backward erosion induced by deep incision of the single channel threatens the stability of the bridge

middle courses, the largest increase in channel capacity took place in the headwater reaches where the initial channel capacity was relatively small (Wyżga et al. 2016). By increasing channel capacity for flood flows and accelerating their downstream transfer, channel regulation leads to a substantial reduction in floodplain storage of floodwater (Wyżga 1996). In consequence, river channelization does not reduce flood hazard but only shifts it to downstream river reaches and at the same time, it results in increased peak discharges of flood waves in the reaches located downstream of a channelized reach (Wyżga 1997).

River channelization also resulted in a considerable decrease in habitat diversity and, consequently, a reduction in the abundance and diversity of river biota, including fish and benthic invertebrates (Wyżga et al. 2009, 2011). Rapid channel incision was a cause of considerable economic losses reflecting, among others, the need to reconstruct bridges or bank water intakes (Wyżga 2001). To prevent a loss of vertical channel stability resulting from channel regulation, hydraulic drop structures were constructed in many channelized reaches. This limited an uncontrolled increase in flow capacity of the regulated channels but disrupted river continuity for biota, particularly fish migration.

2.2 Flood Embankments

River embankments provide only localised protection from flooding, while flood hazard is shifted downstream where peak discharges increase as a result of reduced floodwater storage within the constricted floodplain. A reduction in the floodplain width leads to an increase in water stages attained at given discharges within the embanked river section (e.g. Punzet 1985). During large floods such high water stages are in these sections associated with large water depths that enhance failure of the embankments. These are usually destroyed not because of overtopping by floodwater but when their toe is washed away as a result of the increased pore pressure from water seepage. Within the Sandomierz Basin, embankments of the Vistula were breached during the floods of 1934, 1960, 1970, 1997 and 2010, resulting in the inundation of managed areas behind the embankments. In this section of the Vistula, protection from flooding provided by the embankments during minor floods appears problematic during large, catastrophic floods.

As the construction of the embankments along the Upper Vistula and its tributaries in the Oświęcim and the Sandomierz basins encouraged intense development of the adjacent areas, their removal is impossible. However, the above observations question the reasons for raising the existing embankments or constructing new ones in the mountain and foothill reaches of Carpathian rivers. Raising the embankments would increase water pressure during flood events and thus likely increase the probability of the embankments being washed away. New embankments in the mountain and foothill valley sections would prevent flooding of less developed, mostly agricultural areas. With the inevitable increase in flood hazard to downstream river reaches, resulting from the construction of embankments, introduction of these measures to protect agricultural land is not justifiable. In the valleys with gravelly alluvium, typified by high hydraulic conductivity (and thus fast water filtration through sediment), water tables on both sides of the embankments quickly become established at the same level. Under such conditions, not only are the embankments ineffective in providing protection from flooding, but they also create a false sense of security, thus encouraging development of the land adjacent to the embankments.

2.3 *Construction of New Dam Reservoirs*

With the ability to capture a proportion of the volume of flood waves, dam reservoirs may significantly reduce flood hazard in downstream river reaches. However, this important role of dam reservoirs in flood hazard reduction is limited as a result of the disruption of bed material transfer to the river reaches downstream from the reservoir. This leads to bed degradation (Kondolf 1997) which, in turn, increases channel capacity and reduces floodwater storage on the valley floor. If the channel downstream from a reservoir becomes incised on a considerable distance, the loss of floodplain storage of floodwaters may largely obliterate the flood-control effect of the dam reservoir. This problem could be resolved by coarse sediment augmentation, in this case transferring bedload material deposited at the river's inlet to the reservoir to the river sections below. This solution is now commonly applied in Germany, France or the USA (Kondolf 1997; Rollet et al. 2008).

An important factor to consider before reservoir construction is its useful lifetime, reflecting the rate of silting of the reservoir conditioned by the amount of suspended sediment transported by the river and the rate of water exchange in the reservoir (Łajczak 1994). Longer useful lifetime is typical of deep reservoirs located in the upper reaches of Carpathian rivers, draining largely forested catchments (e.g. the planned Kąty-Myscowa reservoir on the Wisłoka River). In contrast, shallow reservoirs on small watercourses in foothill agricultural catchments with soils developed on loess and loesslike deposits would have very short useful lifetime and their construction is pointless. Recently, construction of a few reservoirs in such locations has been planned; it seems reasonable to abandon such schemes as the reservoirs, built at high cost, will be silted shortly and are most likely to function only as breeding grounds for mosquitoes.

2.4 *Conventional Flood Control Methods on the Carpathian Rivers—Examples*

Even though it is now evident that the conventional measures of flood control have adversely impacted the environment and shifted flood risk downstream without its

Fig. 2 Rip-rap constructed in 2013 on the concave bank of the Czarny Dunajec at Chochołów protects from erosion a floodplain overgrown with alder and willow shrubs. Buildings that might potentially require protection from erosion are distant from the river bank by more than 100 m, equal to approximately two channel widths at the site

reduction at the regional scale, the already introduced schemes cannot be easily abandoned. However, it is imperative to draw conclusions from the analysis of the past interventions to limit their flood-control use only to the most developed or urbanized valley reaches (Bojarski and Wyżga 2009). Until now the approach to flood protection in the Upper Vistula basin has remained unchanged, with a number of hydrotechnical works done over the last years which were either unnecessary or harmful to the environment and which increased, rather than reduced, flood hazard at the regional and sometimes even local scale (Żelaziński and Wawręty 2005). Below we present examples of such interventions.

Transformation of the multi-thread channel of the Czarny Dunajec at Chochołów into a single-thread one (in 2005) has had no positive economic impact as the cut off and filled side channel was adjacent to fallow land and extensively used meadows. Flow concentration in one of the previous channels induced rapid channel incision and backward erosion that threatens the stability of a nearby bridge (Fig. 1). Reinforcement of the river bank with rip-rap, carried out at Chochołów in 2013, provides protection from erosion to fallow land of no particular economic value (Fig. 2). An incentive to perform these works could have only been the need to spend public funds as there are no structures on the floodplain that require protection from potential bank erosion. At Stróże, a wall constricting the floodplain of the Biała River was erected (2012) in close vicinity of the channel but at a large distance from buildings that may require flood protection. Such a reduction of floodplain width, eliminating floodplain storage of floodwater in this river reach, will inevitably lead to increased peak discharges of flood waves in the downstream reach. At the same time, construction of the wall caused a complete destruction of riparian vegetation and affected riverscape aesthetics (Fig. 3).

Fig. 3 A concrete wall constructed in 2012 along the Biała River at Stróża under the pretext of preventing the valley floor from flooding. The wall construction resulted in complete eradication of riparian vegetation and separated from the river a wide band of the floodplain that did not require protection from flooding

3 Alternative Methods of Flood Hazard and Risk Reduction

Below we indicate measures which may be significantly more effective than the conventional methods in reducing flood hazard and risk at regional scale—the entire basin of the Upper Vistula River.

3.1 Flood Hazard and Flood Risk Maps as a Basis for Improved Management of Valley Floors

EU Floods Directive (European Commission 2007) indicated the adjustment of valley-floor management to the level of flood hazard and risk as a key measure to control potential flood damage (Santato et al. 2013). Developed valley floors undoubtedly need to be protected from flooding, usually through their separation with flood embankments from artificially constricted floodplains. However, in recent decades the Upper Vistula basin has seen progressive encroachment of development and infrastructure onto riverside land. Such areas are next postulated to be included in flood protection schemes (channel regulation, flood embankment construction). This trend tends to neglect that flood protection of all such areas is from the beginning doomed to fail and the floodwater that was prevented from inundating one valley section will most likely, and uncontrollably (levee breeching), inundate another, downstream valley section resulting in larger flood losses.

It is thus essential to stop the trend and replace it with proactive management of riverside land that is adjusted to the actual level of flood risk (Bobiński and Żelaziński 1996).

A premise for the 2015 Flood Risk Management Plans, the recently prepared maps of flood risk and flood hazard present an opportunity for a change in the management of riparian areas. However, using these maps to improve spatial planning in the valley floors has also certain limitations. First, as a result of high costs of their preparation, in Poland the maps were only made for the valleys of major rivers. Such argumentation entirely neglects the fact that preventing development on the land with high flood risk, based on such maps, is the cheapest method of limiting flood losses. Second, in several places local authorities defy indications derived from these materials, arguing that the maps only present the current level of flood hazard and risk that will be diminished in the upcoming years as a result of engineering works carried out in the catchments. However, it should be emphasized that flood hazard and risk assessments are dynamic in character as in the future both may diminish, thanks to the engineering works improving floodwater retention (e.g. construction of dam reservoirs), or increase as a result of channel regulation or incision and construction of flood embankments in the upstream sections of river valleys or because of climate change. This is why decisions on the management of valley floors must be based on the existing maps presenting current assessment of flood hazard and risk. Third, methodology behind the construction of these maps is based on the assumption that flood damage only results from inundation of valley floors or high flow velocities in these areas. This assumption is not adjusted to the character of floods in mountain regions where a considerable proportion of flood damage is associated with the erosion of riverbanks and abrupt lateral shift of the channels during flood events (cf. Hajdukiewicz et al. 2016).

3.2 Spontaneous or Enhanced Re-Establishment of Small Channel Capacities and Floodplain Retention in Headwater River Reaches

Headwater, first- to third-order (sometimes also fourth-order) river reaches have rarely been of particular interest to river managers as these valley sections are usually undeveloped and do not require construction of hydraulic structures or channel regulation. However, it is here where floods originate and, because these headwater reaches total tens of thousands kilometres of river network in the Upper Vistula River basin, the situation in these reaches will shape the flood conditions in the downstream, developed valley reaches. Over the last few decades a significant improvement of vegetation cover on hillslopes (increase in forest cover, a shift from arable land to grassland) occurred in the Upper Vistula River basin, particularly its Carpathian part. In consequence, the delivery of sediment to the headwater channel

Fig. 4 Deep channels, incised to bedrock, are widespread in headwater reaches of streams in the Polish Carpathians as exemplified by Kryściów Stream in the Low Beskid Mountains (*photo credit* Roman Żurek)

reaches considerably decreased. At the same time, headwater streams continued to be cleaned of fallen trees to maximize the amounts of timber harvested from the forests and use channels as routes for its transport. With the reduced supply of sediment to headwater stream channels and a lack of large wood as a major component of hydraulic roughness, many such channels deepened considerably and incised to bedrock (Fig. 4). As the dimensions of headwater channels are generally small, even low amount of bed degradation, in the range of 0.2–0.4 m, resulted in a manifold increase in their capacities and the resultant loss of floodplain storage of floodwaters (Wyżga et al. 2016).

It is thus necessary to allow accumulation of bed material in these channels to diminish their capacities and re-establish the conditions for floodwater storage in their riparian areas. Accumulation of alluvium can be enhanced by allowing spontaneous formation of wood dams from fallen trees (Fig. 5) or artificial formation of such dams in headwater channel reaches (Bojarski et al. 2005; Wyżga 2007). This requires abandonment of the harvest of riparian trees and of the removal of trees fallen to the channels, that would only slightly decrease the amount of timber harvested from the mountain forests. At the same time, this loss can be compensated with better conditions for tree growth, resulting from slowed drainage of groundwater in near-channel slope sections, particularly beneficial for spruce with its shallow root system. This change in management of riparian forest along headwater stream channels can easily be introduced in the state-owned forests (managed by the State Forests agency). Notably, the limited mobility of large wood in headwater channels means the wood will not pose a flood hazard to downstream, urbanized valley reaches.

Fig. 5 Wood dams, either naturally formed by fallen trees or artificially constructed in headwater channels, facilitate bed material storage, reduce channel capacity and promote maintenance or re-establishment of floodplain retention of floodwaters. Kamienica Stream in the Gorce Mountains

3.3 Storage of Stormwater Runoff from Small Paved Catchments

While in mountain catchments the factors responsible for the formation of surface runoff and rapid concentration of flood waves are steep slopes and channel gradients, in urbanized areas they result mostly from the impervious nature of a large proportion of the catchment surface. During storms, paved surfaces (rooftops, roads, parking lots) prevent infiltration and all rainfall can be transformed into runoff, resulting in the formation of flash floods and inundation of vast urban areas (Radecki-Pawlik 2003). To prevent rapid concentration of flood waves that form in such areas, it is necessary to provide retention for storm runoff from small paved catchments. This can be achieved in dry or wet stormwater management ponds. Dry ponds store runoff only during storms and later slowly drain. Wet ponds do not have outlets and store stormwater that later evaporates or infiltrates into the ground (Fig. 6). Apart from stormwater storage, wet ponds can be used for settling pollutants washed from impervious surfaces, particularly road pavements (Wałęga et al. 2013). Over the last years, a number of wet ponds have been constructed along motorways in southern Poland. Storage ponds for storm runoff from small paved catchments were also constructed in some towns in the Upper Vistula River basin (Fig. 6).

Fig. 6 Wet stormwater pond constructed to collect surface runoff from paved surfaces in a housing estate in Kraków

3.4 Measures Increasing Floodplain Retention of Floodwater

Floodplain retention consists in temporary retention of floodwaters inundating a valley floor or slowing down their flow in comparison with that of floodwaters in the channel. The retention thus increases with the increasing proportion of total flow that is conveyed in the floodplain zone of a cross-section and with increasing difference between flow velocities in the floodplain and channel zones (Wyżga 1999). In the twentieth century, incision was the dominant tendency of Polish Carpathian rivers, increasing flow capacity of their channels and substantially reducing the potential for floodwater retention on their floodplains (Wyżga 2008; Wyżga et al. 2016). It is thus necessary to reduce flow capacity of the channels and re-establish the potential for floodplain retention of floodwaters. This can be achieved by either elevating channel beds or lowering the floodplain surface. The first method, inevitably resulting in increased water stages at particular discharges, may only be applied in relatively undeveloped valley sections.

In more developed valley sections, a reduction in channel capacities and the resultant increase in floodplain retention of floodwaters can be obtained by removing a proportion of valley-floor sediments and the formation of a new floodplain at lower elevation (Fig. 7). In many river reaches, such a new floodplain can be constructed on the land owned by the state and managed by the Water Authority (Regional Water Management Board). Where the Authority has no rights to such land or where it is too narrow, the purchase of undeveloped riverside land from local owners will be necessary. The cost of floodplain lowering may be significantly decreased if the works are carried out with the use of gravel mined for aggregate from the re-constructed floodplain. Such floodplain reconstruction may

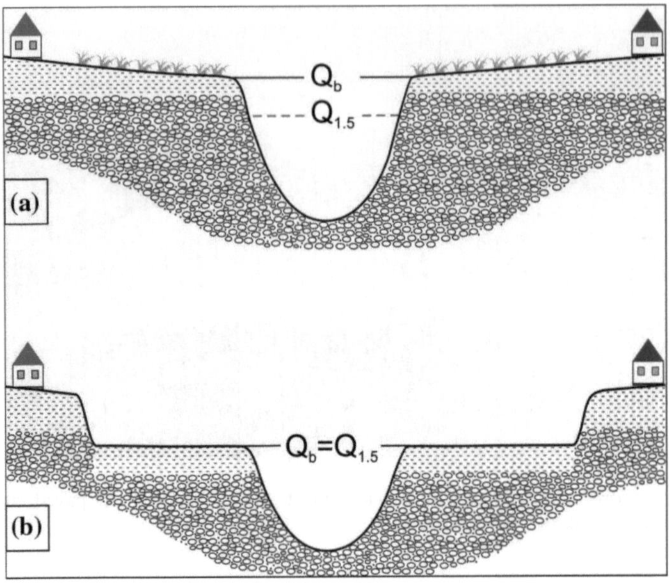

Fig. 7 **a** Incised channel has bankfull capacity (Q_b) larger than that of 1.5-year flow. **b** Re-establishing floodplain retention and channel capacity typical of natural, vertically stable rivers achieved as a result of artificial floodplain excavation at a lower position. In the process, a proportion of the gravelly sediments is removed, while soil fines are replaced over the lowered floodplain

be beneficial not only for flood control but also for the environment as it will enhance re-establishment of riparian habitats, previously lost as a result of channel incision that excessively increased their elevation above the water table.

Establishment of erodible corridors in undeveloped riparian areas can be an efficient way to restore floodplain retention along incised Polish Carpathian rivers (Bojarski et al. 2005). Within the corridors, lowering of the surface of riparian land and the formation of new, low-lying floodplains will be carried out by the laterally migrating rivers. The costs of the functioning of erodible corridors will mainly consist in the purchase of riverside areas from private owners. River restoration projects aimed at the establishment of an erodible river corridor have recently been implemented on the Biała and the Raba rivers (Wyżga and Zawiejska 2012).

3.5 *Measures Increasing Channel Storage of Floodwater*

In the valley of the Upper Vistula and in the lower courses of its tributaries, an important factor of flood hazard is insufficient width of the inter-embankment zone that results in the formation of high water stages during floods, likely to overflow the embankments. This hazard can be reduced by implementing measures increasing channel storage of floodwaters. Importantly, channel incision that occurred on Polish Carpathian rivers during the twentieth century (Wyżga 2008; Wyżga et al. 2016) did not enlarge channel storage as it can only be increased by a concurrent increase of channel capacity and deceleration of downvalley water flow.

Since the 1990s construction of secondary, side channels along the main regulated channel has been introduced in some EU countries (e.g. Schropp 1995; Hornich and Baumann 2008). At flood conditions these side channels carry a proportion of the total flow and their presence allows a significant decrease in water stages attained at particular discharges. Irrespective of their flood-control function, side channels enrich the array of river habitats, in particular re-establishing slow-velocity habitats previously eliminated as a result of channel regulation. Construction of secondary channels need not be expensive if it follows appropriately designed exploitation of gravel within the valley floor or incorporates disused gravel pits.

3.6 Dry Reservoirs

In the upper river reaches, dry reservoirs can play a significant role in the attenuation of flood waves. Dry dams have spillways of relatively small conveyance, which allows undisturbed passage of water at low to medium flows and causes the reservoir to fill at higher flows. Dry reservoirs have some undeniable advantages:

- unlike typical dam reservoirs, their full storage capacity is used to hold floodwater and thus dry reservoirs are very effective in capping large flood waves;
- the bottom of such reservoirs can be used for agriculture as it is inundated only once in a few years during a larger flood;
- they do not disrupt bedload transfer and longitudinal connectivity of the watercourse for biota.

Dry reservoirs were constructed on the streams and rivers draining the Sudetes (Lower Silesia region) in the first half of the twentieth century, especially after the catastrophic flood of 1897 (Lenar-Matyas et al. 2009). Currently, a large dry reservoir is under construction on the Oder River in Racibórz. So far, no such structures have been constructed in the Upper Vistula River drainage basin.

3.7 Polders

In the lower river courses where valleys are wide, floodwater should be effectively stored in polders (detention basins): undeveloped areas within valley floors that can be inundated during larger floods. Separation of a polder from the river by a flood embankment allows its inundation during a flood peak and thus efficient attenuation of the flood wave (Huang et al. 2007). Polders with a single connection to the river are filled with water during the rising limb of a flood wave or at flood peak; the water gradually drains back into the river during flood recession. Elongated polders typically have two connections with the river: that located in the upper section of the valley enables inflow of floodwater which subsequently returns to the river through the connection in the lower section of the valley.

Polders can be constructed in undeveloped portions of valley floors, cut off river bends and large gravel pits. To date, the practice of flood control in the Upper Vistula River drainage basin has not involved construction of polders. The 2015 Flood Risk Management Plan prepared for this area includes construction of polders in the Upper Vistula valley. A study by a pro-environmental organization indicated a potential for construction of nine polders upstream of Kraków with a total volume of 59.3 million m^3, comparable to the flood-control capacity of a large dam reservoir on the Skawa River at Świnna Poręba (Cieżak et al. 2014).

3.8 Compensating the Loss of Floodplain Water Storage Resulting from Channel Regulation or the Construction of Flood Embankments

Formation of a channel with flow capacity larger than that of natural, vertically stable rivers, i.e. 1.5-year flow (Williams 1978) (Fig. 8a, b), and construction of flood embankments—necessary in urbanized valley sections—accelerate evacuation of floodwaters from such a valley section and reduce its storage on the valley floor resulting in increased flood peaks downstream (Wyżga 1997). Therefore, the loss of floodplain water storage resulting from flood protection of developed riparian areas should be compensated for with increased floodwater storage in either upstream or downstream, undeveloped valley sections (Fig. 8c) (Bojarski and Wyżga 2009).

It seems advisable to implement the approach of balancing the loss of floodplain storage of floodwater by the measures re-establishing such storage; each action reducing floodwater storage as a result of channel regulation or the construction of flood embankments must be accompanied by other activities increasing the lost storage potential in other valley sections, including purchase of riparian land. Only then the decrease in floodplain water storage will not result in increased flood hazard at a regional level; at the same time, decisions about channel regulation or valley-floor embanking will have to incorporate not only the costs of such works but also of re-establishment of floodwater storage in other parts of the valley.

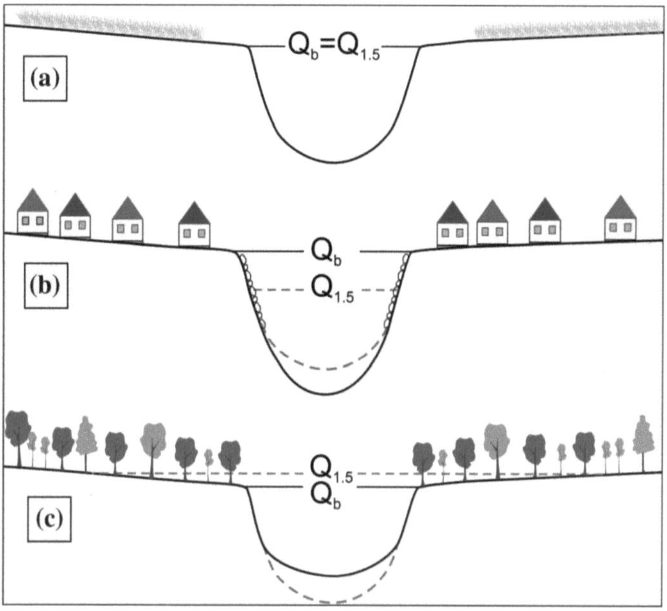

Fig. 8 Flow capacity (Q_b) typifying channels of natural, vertically stable rivers (**a**) and regulated channels formed in highly developed valley sections (**b**), and postulated flow capacity of river channels in undeveloped valley sections (**c**). In the two last situations, the stage attained at a discharge of 1.5-year recurrence interval and the location of the bed of vertically stable channel are indicated by dashed line. Modified after Bojarski and Wyżga (2009)

3.9 Solution to River–Bridge Conflict: Bridge Reconstruction

A large proportion of flood losses result from the destruction of bridges with insufficient conveyance and excessively concentrated flow energy (Jeleński 2004), caused by undermining of their piers, or from damage to bridge construction and elevation of local water level by wood clogging under bridges (Fig. 9a), especially with insufficient span between bridge piers (Wyżga 2007; Hajdukiewicz et al. 2016). Preventing or limiting such losses does not require interventions frequently undertaken in or along the river, such as channel regulation or clearing of riparian forest, but rather bridge reconstruction to accommodate more flow, including an increase in the span between piers or placing the piers outside the channel (Fig. 9b) (Wyżga 2007). Numerical modelling may help to predict the probability of clogging particular bridges with wood and thus to identify bridges that are most vulnerable to clogging and should be rebuilt (Ruiz-Villanueva et al. 2016). Unfortunately, destroyed bridges are still being rebuilt without modifications to significantly improve their functioning under flood conditions. This was the case with the bridge over the Białka River destroyed by the flood in July 2008 and subsequently rebuilt without an increase of its span.

(a)

(b)

Fig. 9 **a** Bridge on the Czarny Dunajec River at Chochołów before 2005, with its piers based on the channel bed. During the flood of 2001 wood jams were deposited on the upstream side of the piers, reducing cross-sectional area of the flow. **b** The bridge at Chochołów rebuilt in 2005. The arch supporting the new bridge is based outside the river channel, which excludes a possibility of channel obstruction by deposited woody debris. After Wyżga and Radecki-Pawlik (2011)

3.10 Introduction and Effective Enforcement of the Ban on in-channel Sediment Mining

In-channel sediment mining leads to sediment deficit in the channels and the destruction of bed armour that prevents entrainment of the underlying bed material (Rinaldi et al. 2005; Wyżga et al. 2010). As a consequence, it results in degradation of channel beds and undermining bridge piers and channel regulation structures (Korpak et al. 2009). The condition of river channels is also essential in reducing flood damages. Undisturbed, long-term development of river channel leads to the stabilization of its vertical position and flow capacity. Such a channel becomes less vulnerable to erosion caused by flood flows. Stability of the banks and bed of mountain streams and rivers is one of the key factors determining flood losses. In-channel gravel mining, especially if uncontrolled, destroys that stability leading to an increase in channel dimensions and the resultant reduction in floodplain storage of floodwaters as well as the change from alluvial channel boundary conditions to bedrock ones (Rinaldi et al. 2005; Wyżga et al. 2010). Ecological consequences of such activities include elimination of river biota caused by the destruction of habitats for benthic invertebrates and disappearance of spawning grounds of lithophilic fish. To reconstruct damaged road infrastructure after floods, large volumes of sediment are mined from Carpathian rivers under the pretext of increasing channel capacities for floodwaters and preventing erosion of concave banks positioned against channel bars. In the light of its negative consequences, sediment exploitation from river channels should be forbidden and the ban should be effectively enforced. Improving the conditions for floodwater transfer and protecting riverbanks from erosion should instead be obtained by bulldozing channel sediment towards the concave banks and concurrent straightening of the thalweg. In case some of the sediment needs to be removed from the channel, the material should be put into the river in another reach to prevent sediment deficit and channel incision.

4 Conclusions

Since the late nineteenth century flood-control measures introduced in the Upper Vistula basin were based on the notion of fast evacuation of floodwater that was associated with a significant reduction in floodwater retention on the valley floors. Such a policy of flood-control management stemmed not only from the unfamiliarity with other methods and a generally technocratic approach to nature (Wyżga 2007) but also from the need to protect all arable land adjacent to rivers as at the time farming provided for the existence of most of the society (Wyżga 2008). At present the negative effects of the conventional methods of flood control are evident and agriculture contributes only a fraction of gross national product. It is thus now a priority to decelerate flood runoff and increase floodwater retention in less developed parts of the valleys in order to reduce flood risk to spatially concentrated, urbanized areas. This paper has presented a catalogue of environment-friendly measures that comply with the best practices of flood protection recommended by the European Commission (2003) and are aimed at such a change in the flood-control policy in the Upper Vistula basin.

References

Bobiński E, Żelaziński J (1996) Czy można przerwać błędne koło ochrony przeciwpowodziowej? Gosp Wodna 4:99–107

Bojarski A, Wyżga B (2009) Przeciwdziałanie zagrożeniom stanu środowiska cieków karpackich – środki zaradcze na tle planowania przestrzennego (Counteracting the threats to the environmental status of Carpathian watercourses—remedial measures in the frame of spatial planning). Czas Techn 10:63–68 (in Polish, with English summary)

Bojarski A, Jeleński J, Jelonek M, Litewka T, Wyżga B, Zalewski J (2005) Zasady dobrej praktyki w utrzymaniu rzek i potoków górskich (Good-practice manual of sustainable maintenance of mountain streams and rivers in southern Poland). Ministerstwo Środowiska, Warszawa (in Polish, with English summary)

Bravard JP, Petts GE (1996) Human impacts on fluvial hydrosystems. In: Petts GE, Amoros C (eds) Fluvial hydrosystems. Chapman and Hall, London, pp 242–262

Ciężak K, Król W, Wawręty R, Żelaziński J (2014) Identyfikacja podstawowych działań rewitalizacyjnych w dolinie górnej Wisły wraz oceną wpływu wybranych rozwiązań na maksymalne przepływy i rzędne zwierciadła wody na przykładzie powodzi z maja 2010 roku. Fundacja Dzika Polska, Warszawa

Comiti F (2012) How natural are alpine mountain rivers? Evidence from the Italian Alps. Earth Surf Proc Land 37:693–707

European Commission (2003) Best practices on flood prevention, protection and mitigation. Brussels

European Commission (2007) Directive 2007/60/EC of the European Parliament and of the Council of 23 October 2007 on the assessment and management of flood risks. O J Eur Union L288/27

Hajdukiewicz H, Wyżga B, Mikuś P, Zawiejska J, Radecki-Pawlik A (2016) Impact of a major flood on mountain river habitats, channel morphology and valley infrastructure. Geomorphology 272C. doi:10.1016/j.geomorph.2015.09.003

Hornich R, Baumann N (2008) River restoration of the River Mur along the border between Austria and Slovenia. In: Gumiero B, Rinaldi M, Fokkens B (eds) 4th ECRR international conference on river restoration. ECRR, Venice, pp 487–496

Huang S, Rauberg J, Apel H, Lindenschmidt KE (2007) The effectiveness of polder systems on peak discharge capping of floods along the middle reaches of the Elbe River in Germany. Hydrol Earth Syst Sci Discuss 4:211–241

Hydroprojekt (1987) Regulacja rzek górskich, wytyczne projektowania. Warszawa

Jeleński J (2004) Droga i jej odwodnienie jako trwałe uszkodzenie doliny i łożyska cieku (Road and its drainage as the permanent damage of the valley and riverbeds of watercourses). Zesz Nauk-Techn SITK RP, Oddz. Kraków 112:115–135

Kondolf GM (1997) Hungry water: effects of dams and gravel mining on river channels. Environ Manage 21:533–551

Korpak J, Krzemień K, Radecki-Pawlik A (2009) Wpływ budowli regulacyjnych i poboru rumowiska na koryta rzek i potoków górskich – wybrane przykłady z rzek karpackich (Impact of channelization structures and in-stream gravel mining on the channels of mountain streams and rivers—examples from the Polish Carpathians). Gosp Wod 7:274–281 (In Polish, with English summary)

Łajczak A (1994) The rates of silting and the useful lifetime of dam reservoirs in the Polish Carpathians. Zeitschr Geomorphol 38:129–150
Łajczak A (1995) The impact of river regulation, 1850–1990, on the channel and floodplain of the Upper Vistula River, southern Poland. In: Hickin EJ (ed) River geomorphology. Wiley, Chichester, pp 209–233
Lenar-Matyas A, Poulard C, Ratomski J, Royet P (2009) Konstrukcja i działanie suchych zbiorników przeciwpowodziowych o różnej charakterystyce i lokalizacji (The construction and action of dry dams with different characteristics and location). Infrastr Ekol Ter Wiej 9:115–129 (In Polish, with English summary)
Merritts D (2011) The effects of variable river flow on human communities. In: Wohl EE (ed) Inland flood hazards: human, riparian, and aquatic communities. Cambridge University Press, New York, pp 271–290
Punzet J (1985) Wezbrania Wisły w obrębie Krakowa – dawniej i dziś (Floods on the Vistula River in the Cracow region—in the past and today). Gosp Wod 8:190–193
Radecki-Pawlik A (2003) Wzrost spływu powierzchniowego z terenu zlewni jako jedna z przyczyn powstawania wezbrań i powodzi (Runoff increase from paved surfaces as one of the reasons of floods). Gosp Wodna 4:154–159 (In Polish, with English summary)
Rinaldi M, Wyżga B, Surian N (2005) Sediment mining in alluvial channels: physical effects and management perspectives. River Res Appl 21:805–828
Rollet AJ, Piégay H, Bornette G, Persat H (2008) Sediment dynamics, channel morphology and ecological restoration downstream a dam: the case of the Ain River. In: Gumiero B, Rinaldi M, Fokkens B (eds) 4th ECRR international conference on river restoration. ECRR, Venice, pp 497–504
Ruiz-Villanueva V, Wyżga B, Mikuś P, Hajdukiewicz M, Stoffel M (2016) Potential wood-related hazards at bridges: the Długopole bridge on the Czarny Dunajec River, Polish Carpathians. In: Koboltschnig G (ed) 13th Congress Interpraevent 2016. Lucerne, Switzerland, pp 610–618
Santato S, Bender S, Schaller M (2013) The European Floods Directive and opportunities offered by land use planning. CSC Report 12. Climate Service Center, Germany
Schropp MHI (1995) Principles of designing secondary channels along the River Rhine for the benefit of ecological restoration. Water Sci Technol 31(8):379–382
Szumański A (1986) Postglacjalna ewolucja i mechanizm transformacji dna doliny Dolnego Sanu (Late Glacial and Holocene evolution and mechanism of transformation of the floor of the lower San valley). Geologia 12(1):5–92 (in Polish, with English summary)
Wałęga A, Radecki-Pawlik A, Kaczor G (2013) Naturalne sposoby zagospodarowania wód opadowych. University of Agriculture in Kraków, Kraków
Williams GP (1978) Bank-full discharge of rivers. Water Resour Res 14:1141–1154
Wyżga B (1996) Changes in the magnitude and transformation of flood waves subsequent to the channelization of the Raba River, Polish Carpathians. Earth Surf Proc Land 21:749–763
Wyżga B (1997) Methods for studying the response of flood flows to channel change. J Hydrol 198:271–288
Wyżga B (1999) Estimating mean flow velocity in channel and floodplain areas and its use for explaining the pattern of overbank deposition and floodplain retention. Geomorphology 28:281–297
Wyżga B (2001) A geomorphologist's criticism of the engineering approach to channelization of gravel-bed rivers: case study of the Raba River, Polish Carpathians. Environ Manage 28:341–358
Wyżga B (2007) Gruby rumosz drzewny: depozycja w rzece górskiej, postrzeganie i wykorzystanie do rewitalizacji cieków górskich (Large woody debris: storage in a mountain river, perception and use in the restoration of mountain watercourses). Instytut Ochrony Przyrody PAN, Kraków (in Polish, with English summary)
Wyżga B (2008) A review on channel incision in the Polish Carpathian rivers during the 20th century. In: Habersack H, Piégay H, Rinaldi M (eds) Gravel-Bed Rivers VI: From process understanding to river restoration. Dev Earth Surf Proc 11, Elsevier, Amsterdam, 525–555

Wyżga B, Radecki-Pawlik A (2011) Jak zmniejszyć zagrożenie i ryzyko powodziowe w dorzeczu górnej Wisły? (How to reduce the flood hazard and risk in the upper Vistula basin?). Gosp Wodna 10:414–421

Wyżga B, Zawiejska J (2012) Hydromorphological quality as a key element of the ecological status of Polish Carpathian rivers. Georeview 21:56–67

Wyżga B, Amirowicz A, Radecki-Pawlik A, Zawiejska J (2009) Hydromorphological conditions, potential fish habitats and the fish community in a mountain river subjected to variable human impacts, the Czarny Dunajec, Polish Carpathians. River Res Appl 25:517–536

Wyżga B, Hajdukiewicz H, Radecki-Pawlik A, Zawiejska J (2010) Eksploatacja osadów z koryt rzek górskich – skutki środowiskowe i procedury oceny (Exploitation of sediments from mountains river beds—environmental impact and evaluation procedures). Gosp Wodna 6:243–249

Wyżga B, Oglęcki P, Radecki-Pawlik A, Zawiejska J (2011) Diversity of macroinvertebrate communities as a reflection of habitat heterogeneity in a mountain river subjected to variable human impacts. In: Simon A, Bennett SJ, Castro JM (eds) Stream restoration in dynamic fluvial systems: scientific approaches, analyses and tools. American Geophysical Union, Washington, pp 189–207

Wyżga B, Zawiejska J, Radecki-Pawlik A (2016) Impact of channel incision on the hydraulics of flood flows: examples from Polish Carpathian rivers. Geomorphology 272C. doi:10.1016/j.geomorph.2015.05.017

Zawiejska J, Wyżga B (2010) Twentieth-century channel change on the Dunajec River, southern Poland: patterns, causes and controls. Geomorphology 117:234–246

Żelaziński J, Wawręty R (2005) Ocena wybranych robot hydrotechnicznych finansowanych z pożyczki Europejskiego Banku Inwestycyjnego. Towarzystwo na Rzecz Ziemi, Polska Zielona Sieć, Oświęcim-Kraków

Stability and Change of Flood Risk Governance in Poland

Piotr Matczak, Jakub Lewandowski, Adam Chorynski, Malgorzata Szwed and Zbigniew W. Kundzewicz

Abstract In Poland floods are the predominant natural disasters entailing large losses. In the paper it is presented how flood risk management is organized in Poland, and how the approaches to flood risk in the country evolve. The focus is on organisation of flood risk management. Public policies and activities of public administration are analysed. Based on data collected within the EU 7th Framework Project STAR-FLOOD, the description of the Polish Flood Risk Management Strategies and Governance arrangements is provided. Next, the dynamics of the Polish flood risk management strategies and governance arrangements and evaluation of arrangements are presented. The reliance on structural defence is indicated, as a stable component of the floor risk management. Also main factors triggering changes and four main turning points opening windows of opportunity for Polish flood management system are identified: two of them are related to socio-political development and two to flood event. The Polish flood risk management system has experienced changes (i.e. establishment of a new crisis management system), while preserving its core functional characteristics at the same time (reliance on structural defence). Sectorial management with little coordination at the level of ministries renders the flood risk management in Poland being more like bargain and competition than collaborative and consensus-driven type of governance. The Polish Flood Risk Governance Arrangement is characterised as having a strong capacity to buffer shock events, as shown by the major 1997 and 2010 floods. An overall weakness of the Polish Arrangement is the lack of coordination and strategic management. Problems with effective spatial planning and a highly fragmented actors structure both within and between sub-arrangements are the main challenges for the Polish flood risk management.

P. Matczak (✉) · J. Lewandowski · A. Chorynski · M. Szwed · Z.W. Kundzewicz
Institute for Agricultural and Forest Environment, Polish Academy of Sciences, Poznań, Poland
e-mail: Matczak@amu.edu.pl

P. Matczak
Institute of Sociology, Adam Mickiewicz University, Poznań, Poland

Z.W. Kundzewicz
Potsdam Institute for Climate Impact Research, Potsdam, Germany

Z.W. Kundzewicz et al. (eds.), *Flood Risk in the Upper Vistula Basin*,
GeoPlanet: Earth and Planetary Sciences, DOI 10.1007/978-3-319-41923-7_19

Keywords Flood risk management · Flood risk governance · Flood defence · Crisis management · Resilience · Poland

1 Introduction

Floods are a natural disaster that causes major losses in many parts of the world. In Poland, floods caused by heavy precipitation, are the predominant natural disasters. They entail large losses and are likely to remain a serious hazard also in future (Dobrowolski et al. 2004). As mentioned in the National Crisis Management Plan, more than 1000 municipalities (out of 2500) are exposed to flood risk. Large river floods result in significant strains for the state budget. The most destructive floods in the last 25 years: in 1997 (mostly on the River Odra and its tributaries) and in 2010 (on both the large rivers in Poland—the Vistula and the Odra) brought material damage in the order of magnitude of €3 billion each (Biedron 2012).

This paper illustrates how flood risk management is organized in Poland, and how the approaches in the country evolve. Thus, only a part of a wide area of flood risk research is covered. Coping with flood risk, understanding its drivers and undertaking proper measures involves complex and costly efforts (Schanze 2006). In last decades, treating flood risk as an external force that causes losses with certain probability has been replaced by a more comprehensive view. The simplistic equation of risk as a multiplication of probability and adverse consequences is extended to a more complex scheme, where flood risk depends on three main drivers: hazard, exposure and vulnerability. In this concept, used also in climate change research in a wider application to various types of natural disasters (Cardona et al. 2012), hazard refers to physical events that may have adverse effects (which, for flood risk, are related to flood depth, and submerged area), exposure is a factor related to the components (mainly population and wealth) of a certain area where hazard may occur, and vulnerability is the propensity of exposed elements to suffer adverse effects (Klijn et al. 2015). Within this approach, managing flood risk and minimizing losses caused by floods refer to all three drivers, although hazard represents mainly the "natural" component (yet human activity can increase the hazard). Influencing (reducing) exposure and vulnerability is the main overall task for flood risk management efforts.

In this chapter, the focus of interest is on organisation of efforts. Thus public policies and activities of public administration are analysed, as they aim at providing security (safety), at an acceptable level, for endangered population. In this perspective, advances of natural sciences (like flood modelling etc.) are inputs to be incorporated into decision making undertaken by responsible public bodies (e.g. crisis management administration). Although the progress in understanding the nature of hydrological processes can be noted, flood risk management has its own dynamics, related to administrative conduct, political decisions, economic situation, media coverage, shifts in discourses etc. (Vetere Arellano et al. 2007).

In this respect, flood risk governance refers to governing and decisions making about public issues including engagement of private and non-governmental sector and stakeholders (McDaniels et al. 1999). It can be treated as distinct from political life and political decisions and comprises public policy making and implementation of policies in the domain of flood risk management (Sarewitz et al. 2003). Flood risk governance engages public administration and utilizes public resources. Its forms, measures and scopes change, being country specific, depending on national traditions, types of risk in a country, changes of flood risk, and public perception of flood risk.

Flood risk is likely to increase due to at least three main factors: (a) urbanisation; (b) climate change; (c) economic growth (Bouwer et al. 2010). These factors have impact (positive and negative) on exposure, vulnerability and hazard. Urbanisation and urban sprawl processes increase the amount of impermeable surface, exceeding capacities of sewage systems in cities (Nirupama and Simonovic 2007; Kowalewski et al. 2013). This elevates the flood risk. Moreover, flood hazard is likely to increase due to the rise of extreme precipitation entailed by climate change. Although the impact of climate change on alteration of the scale and frequency of large river floods is uncertain, climate model simulations indicate a general increase of frequency and amplitude of heavy precipitation. In Poland, increase of heavy precipitation in the future, warmer, climate is likely and it would particularly contribute to the growing problem of urban floods and flash floods (Pinskwar 2010). Flash floods in urban areas already get increasingly frequent and this tendency is likely to be strengthened. Also the economic growth contributes to housing development and welfare accumulation which lead to growing damage (exposure) potential.

Thus, due to increasing flood risk, Polish flood risk governance needs to be adopted. In this chapter, the dynamics of the Polish flood risk governance is presented—to what extend it changes and how much it remains stable?

The chapter draws from deliverables of the EU 7th Framework Project STAR-FLOOD (see: www.starflood.eu for an outline of the project), that focused on policy, administrative and institutional aspects of flood risk management in six European countries. It assesses flood risk governance arrangements from a combined public administration and legal perspective, with the aim to identify factors contributing to make European regions more resilient to flood risk. The project investigates how current flood risk governance arrangements can be strengthened or redesigned to enhance societal resilience to flooding. To this end, it is assessed to what extent governance arrangements support or constrain the diversification of Flood Risk Management Strategies as well as the scope to which such a diversification of strategies enhances societal resilience to flooding.

Several sources of data were utilized in the Polish study reported in this paper. Legal acts and documents were collected and analysed as well as other documents (policies, reports and scientific literature). Fifty four in-depth interviews were conducted with experts and practitioners on both the national and local level—in two locations (the city of Wrocław and the city of Słubice), severely endangered by floods in 1997 and 2010 and one location (Powiat Poznanski) that was affected by the 2010 flood.

In the following Sect. 2, the conceptual framework is introduced, comprising Flood Risk Governance Arrangements (FRGAs) and Flood Risk Management Strategies (FRMSs). In Sect. 3, the description of flood risk management strategies and governance arrangements in Poland is provided. Next, the dynamics of the Polish flood risk management strategies and governance arrangements and evaluation of arrangements are presented, respectively, in Sects. 4 and 5 respectively. Finally, in Sect. 6, conclusions are offered.

2 Notions of Flood Risk Governance Arrangements and Flood Risk Management Strategies

Coping with floods involves significant efforts and expenses. Such activities are organized within administrative, legal and economic systems. These institutional frameworks are called here Flood Risk Governance Arrangements (FRGAs). Following policy sciences they can be defined as institutional constellations consisting of four main components: (i) actors engaged in flood risk management (public and non-public); (ii) discourses (scientific, media and other discussions, disputes, conceptualisations); (iii) rules (legal and other regulations); and (iv) resources (financial and others) used for activities related to flood risk management. FRGAs are shaped by an interplay between actors and actor coalitions involved in all policy domains relevant for flood risk management (including water management, agriculture, spatial planning, disaster management etc.); their dominant discourses; formal and informal rules regulating actors operations; and the power and resources the actors involved possess (Hegger et al. 2014). FRGAs can be identified at different governance levels, from local, to regional, national and international. A national jurisdiction is a basic unit, where an FRGA is present. However, actors are engaged at other levels as well. Moreover, influence of supranational rules (such as the Floods Directive for the EU countries) has growing impact on the national FRGAs.

On the frontline of flood risk management, several strategies can be identified. Drawing from the concept of the emergency management cycle, five categories of Flood Risk Management Strategies (FRMSs) can be distinguished: (i) prevention (spatial planning and other methods to shift people from flood-prone areas); (ii) defence (comprising engineered structures); (iii) mitigation (comprising water retention measures); (iv) preparation (disaster planning and response); and (v) recovery (measures that help to rebuilt after a disaster).

Traditionally, the defence has been the dominant strategy in most of the countries. Still it plays a very significant role. However, despite large investments in structural protection, losses are growing. It undermines the trust in effectiveness of structural measures belonging to the defence strategy. As the result, other strategies are getting attention and ideas of changing flood management strategies, sometimes radically, are called for (Werritty 2006). The changes of the role and strengths of strategies are dependant of the shape and operations of flood risk governance arrangements, within which portfolios of strategies are embedded. Thus, any shifts are complex processes.

2.1 Dynamics of FRMAs and FRGAs

In all countries certain dynamics of both FRGAs and FRMSs can be observed. In fact, shifts of strategies and arrangements are desirable in order to adjust flood risk management to changing circumstances. Yet, the changes are not automatic and they can be constrained. Factors influencing stability and/or change of FRMSs and FRGAs in a country can be divided into five main types: (i) physical circumstances (geographic conditions and types of floods); (ii) institutional and legal factors; (iii) social and economic characteristics (iv) characteristics of agency (actors, their relations, power and actions) and (v) influence of shock events. These five factors may be found within flood-relevant policy domains but they can also be external to them.

Eventually, functioning and changes of FRGAs can be assessed in terms of desired outcomes, taking into account resilience, efficiency and legitimacy of a particular settings of Flood Risk Governance Arrangements and entailed sets of strategies. The set-up of the investigation—from analysis via explanation to evaluation—is presented in Fig. 1.

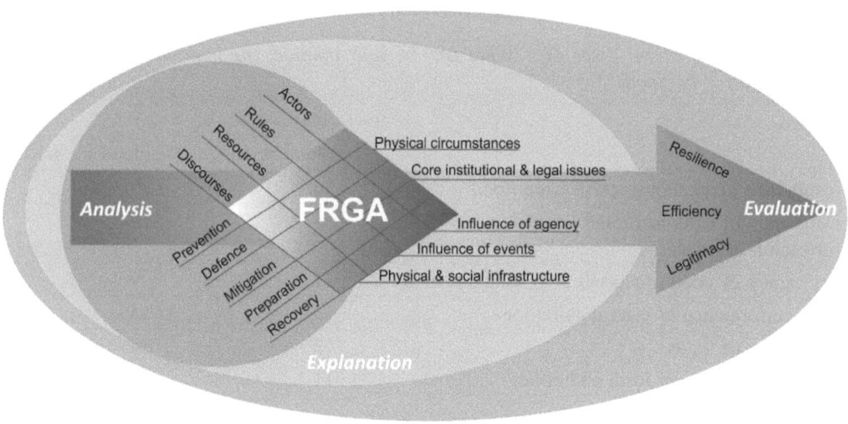

Fig. 1 Conceptual framework of flood risk governance analysis

3 Flood Risk Management Strategies and Governance Arrangements in Poland

3.1 Coping with Flood Risk

Traditionally, structural protection has been developed in Poland for many decades as the most important flood risk protection measure. It revealed to be insufficient, though since major flood losses have been related to dike failures. The large flood event in 1997 showed weaknesses of the existing flood protection system (Chmielewski 2008). Eventually, the flood triggered legislative and organisational reforms in Poland that finalised in 2001 with the Water Act and the Act on Crisis Management in 2007, establishing the crisis management system. Also, several region-wide flood infrastructure programmes (for the Odra River Basin and for the Upper Vistula River Basin) were set up after 1997.

According to the Water Act of 2001, Poland was supposed to prepare a national flood management programme (Kolipinski 2015). Simultaneously, with accession to the EU in 2004, much exertion has been devoted to comply with the obligations stated in both the Water Framework Directive (WFD) and Floods Directive (FD), but with only a partial success (Kitowski 2010). While the National Programme for Municipal Wastewater Treatment has been successfully completed, the flood management programme is not yet finished. Efforts to implement the Floods Directive partially compensate the lack of a national flood strategy. It has been criticised by public opinion (Table 1).

Table 1 Opinion of the Poles on: *"What are, in your opinion, the main reasons for massive damages caused by floods?"*

Reason	(%)	
	Flood in 1997	Flood in 2010
Insufficient preparedness for a large flood	36	39
Deficiency of equipment and its inadequacy for such a disaster	35	14
Deficiency of good organisation in such situations	31	9
Poor performance of central authorities	17	15
Deficiency of funds	12	14
Poor performance of local authorities	11	13

Source CBOS (2010)

3.2 Flood Risk Governance Arrangement in Poland

The current flood risk policy in Poland is a consequence of the existing domestic legal acts and programmes, and the requirements of the European Union legislation. Flood Risk Governance Arrangement is framed in Poland within three main institutional and legal sub-arrangements: (a) structural defence; (b) spatial planning; and (c) crisis management. These sub-FRGAs are characterised by consistent groups of actors, rules, resources and discourses. Some other activities associated with risk management, e.g. collecting rainwater where it falls (i.e. mitigation) or after-flood recovery have been implemented rather ad hoc and cannot be considered as full-fledged sub-arrangements. As a result, the structural defence, with embankments, dikes, dams and reservoirs as typical measures is dominant. Both, the national level of flood risk management as well as the regional/local one (case study of Poznań County, Commune of Słubice and the City of Wrocław) confirms clear dominance of the hydro-technical defence approach. The significance of other sub-arrangements, such as crisis management and spatial planning, has been increasing in Poland. Figure 2 illustrates the Polish FRGA including flood risk management strategies realised within sub-arrangements.

Next subsystem of Polish flood risk management is the one formed by crisis management actors. The development of this sector might be recognised as a consequence of the large flood of 1997 which brought enormous losses and proved shortcomings of crisis management in the country. The role of State Fire Brigades became leading here since the beginning of systemic transition (1989/90), before which the defence approach dominated this arrangement. Within this arrangement, the monitoring of waters is a duty of the Polish Hydro-meteorological Service: Institute of Meteorology and Water Management (IMGW). The municipalities and

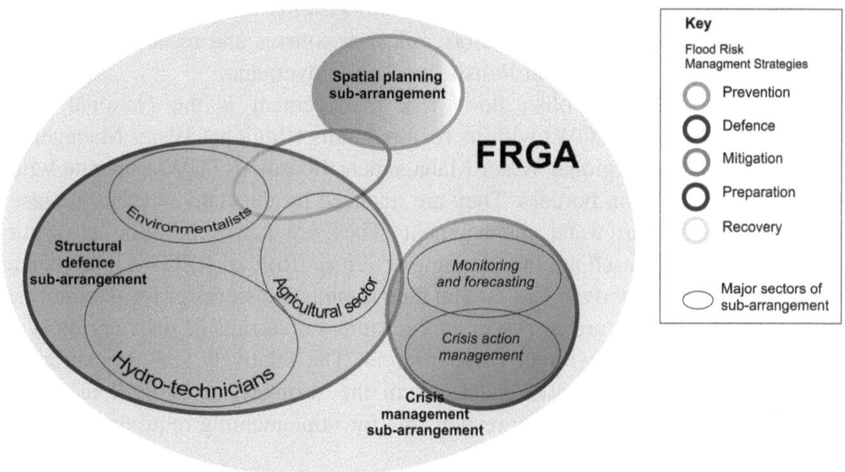

Fig. 2 The Polish flood risk governance arrangement

the State Fire Brigades are mostly responsible for the crisis management planning and actions. The spatial planning sub-arrangement is generally flawed due to an implementation gap. Prioritising economic development makes local governments leaning to invest in flood prone areas. It remains to be seen whether this would considerably change when Flood Hazard and Flood Damage Maps are incorporated into local spatial development plans (if they become incorporated). The amendment to the Water Law, passed by the Polish Parliament in 2015, reduced the possible impact of flood hazard maps on spatial planning decisions re: conditions of development and siting of investments. The new law states that local authorities "can" (rather. than "have to", as before) consider the flood hazard. This means that local authorities are not forced to reduce flood risk by controlling spatial arrangement.

The strategy of flood mitigation is not broadly implemented. Apart from some pilot projects in urban areas, small water retention programmes are undertaken. Nevertheless, impact of small water retention measures on decreasing flood risk can be seen as limited. The strategy of flood recovery is dominated by the state. The flood of 1997 triggered plans to implement a universal flood insurance system. Even though at this time the attempts did not succeed and the insurance system was seen as decreasing the effectiveness of Polish flood risk governance system, nowadays it is developing. This is mostly due to the growth of mortgage market, where insurances are compulsory, and these include flood insurance as a part of a premium package. Nonetheless, the strategy of flood mitigation, as well as the strategy of recovery are developing distantly from the mainstream of the flood risk management system in Poland and cannot be seen as consolidated sub-arrangements.

The Polish Flood Risk Governance Arrangement (FRGA) is fragmented, with rather underdeveloped mechanisms linking sub-arrangements. It is a result of cross-responsibilities (parts of the flood risk management are assigned to different ministries) and a contingent evolution.

Table 2 provides introduction to major sub-systems of Polish flood risk management system, including main actors, rules, discourses and resource for each of the three approaches present in Polish flood risk governance.

The major actor in Polish flood risk management is the National Water Management Board (KZGW) with its regional units (Regional Water Management Boards—RZGWs). Regional Water Management Boards (RZGWs) operate within the range of river basin borders. They are financed by the state and they represent the State Treasury in water management. They are active in flood prevention strategy which is focused around planning activities. Role of RZGWs is to manage larger rivers (with flow over 2 m^3/s) and then issuing decisions for local authorities. The heads of RZGWs are in power of decisions on storing or discharging water from storage reservoirs without compensation. The relatively stable position and funds sufficient for their tasks makes them the strongest actor in Polish FRM. The KZGW is the national body responsible for implementing requirements of the EU Floods Directive.

Another crucial actor in Polish flood risk management is the Regional Authorities on Drainage, Irrigation and Infrastructure (WZMiUW) supervised by

Table 2 Summary of formal sub-FRGAs within the flood policy domain in Poland

Sub-FRGA	Description	Key actors	Key rules/legislation	Key discourses	Resources
Structural defence	Structural measures consisting mainly of dikes and reservoirs, with the aim to protect people against flood risk	• KZGW; • RZGWs; • WZMiUWs; • Regional environment protection departments; • Municipalities	• Water Act	Need to protect people and property	• The state budget; • The EU funds
Crisis management	Preparing crisis management capacities, including preparation, warning, exercises, and operational capacities	• State fire brigades; • Municipalities; • NGOs; • RZGWs; • IMGW	• Act on crisis management	Need to provide security to citizens and to manage crisis	• The state budget
Spatial planning	Regulative capacity, mainly on the municipal level to manage land use	• Municipalities; • RZGWs; • KZGW; • Private sector (developers)	• Act on planning and spatial development	Planning as a tool for local development	• Municipalities' budgets

each Provincial Government. They are undertaking activities within provincial administration borders, contrary to RZGWs duty to work on drainage basins. The WZMiUWs are responsible for development and maintenance of hydro-technical devices (dikes etc.) on small rivers (with flow below 2 m^3/s). They also take part in crisis activities during flood events. Due to their focus on small streams, they are close to the agricultural sector. The WZMiUWs are financed through the ministries but they also receive EU structural grants for investments. Such type of funding contributes to tensions in the sense of weak financial security of maintenance works unlike realisation of new tasks for which purpose substantial funds are available. Apart from resources for maintenance, WZMiUWs also deal with human resources problems. A generation gap in educating of hydro-engineers (replaced at universities with environmental engineering curricula) and a number of retiring professionals brings difficulties in providing necessary and competent personnel.

3.3 Flood Risk Management Strategies in Poland

The defence strategy based on technical and infrastructural approach to flood risk management has been dominating in Poland. Other strategies are present as well, supplementing to a certain extent the dominating defence approach. Table 3 presents the list of flood measures undertaken in Poland.

The domination of structural defence strategy appears in terms of resources. This approach is also strongly centralised.

Table 3 Measures employed in Polish flood risk management strategies

Prevention	Defence	Mitigation	Preparation	Recovery
• Spatial plans; • Expropriation policy	• Dikes; • Retention basins outside area to be protected; • Widening, deepening, dredging of the river bed; • Weirs (used to drain faster or for water retention outside area to be protected); • Water course maintenance	• Separate sewers; • Sustainable drainage systems (SUDS), including green roofs—(weakly implemented); urban green; • Rain water reservoirs (weakly implemented)	• 24/7 monitoring and intervention; • Flood forecasting and warning systems; • Intervention and evacuation plans; • Pool of equipment (sand bags, pumps etc.); • Crisis communication	• Insurance systems; • Repair works; • Solidarity fund

4 Stability and Change of Flood Risk Governance Arrangements in Poland

Previous sections identified Polish flood management system as a result of three separated flood risk governance sub-arrangements which exist within one but scattered FRGA. In this section, dynamics of the overarching FRGA in Poland is assessed. The aim of this assessment is to address the question on what kind of events and under what circumstances changes can be triggered in the system and why? More specifically, for the purposes of this analysis we ask two sub-questions: (i) whether substantial turning points can be indicated in the period from 1990 to 2015, and (ii) what factors that either facilitate and/or constrain changes within flood management system can are observed. Time interval taken into account consisted of 25 years. It starts with 1989/1990 with the fall of the communist system and ends up with 2015.

A general observation shows that the Polish FRGAs can be characterised as rather stable. By 'rather' we mean that several changes that can be observed within the analysed period did not cause fundamental changes into the system. It is partially due to the fragmentation of FRGAs and path-dependency of these institutions.

A significant dependence on hydro-technical infrastructure with lack of consistent alternative results in stability within Polish FRGA. Dominance of this approach towards flood management is strengthened by discourses and expertise. Moreover, fragmented organisation with little integration result in unclear or rather maladjusted financial resources distribution as well as overlapping competences. Perversely speaking, there is more competition for resources than coordinated efforts to decrease flood risk. The dominance of measures based on hydro-technical infrastructure is not a zero-sum game, however, as the growing role of crisis management and planning is not at the expenses of flood protection measures taken. Nevertheless, within the field of crisis management major institutions covering: building and sustaining defence infrastructure; modelling; warning systems operate independently from each other. One could argue that in the arrangement where short-term, investment-driven orientation thrives focus mainly on budget maximisation rather than on strategic planning and recurrent floods is exploited as justification for investments either at national, regional and local level.

Four main turning points can be indicated, as milestones of the Polish Flood Risk management. These four milestones can be considered as windows of opportunity for Polish flood management system. Two of them are related to socio-political development and two to flood event. These are summarized in Table 4.

In-depth investigation of these milestones revealed that each of them hardly ever happens with only one driving force or one change agent, but is very often a combination of competition and bargaining. In particular:

- The fall of the communist system in 1989/90 and related political, economic and legal changes had a profound impact on the FRGA context and it was a

Table 4 Milestones identified in Polish FRGA between 1990 and 2015

Year	Milestone	Characteristic	Measure(s) taken in the aftermath
1989/90	Communism collapse	Systemic transformation of the country	Administrative, economic, legal, political changes
1997	Millennium flood of 1997	Regional and nationwide flood defence infrastructure projects initiated; Development of the crisis management	Water Act of 2001: Act on crisis management of 2007
2004	EU accession	Legislative, economic changes	Transposition of EU laws; EU funds became accessible; Implementation of the floods directive (i.e. development of flood hazard and flood damage maps as well as flood risk management plans)
2010	Flood on the Odra (Oder)/ the Vistula	Test for the previously improved flood protection infrastructure and crisis management established	Act on support to entrepreneurs affected by flood events of 2010

fundamental factor implying change. The rebirth of local government and the general reconstruction of the administration took place after 1989/90. Subsequently, water administration based on the river basin level was established (Kowalczak et al. 2013). Although the change involved redistribution of property rights, tasks and responsibilities among the administration agencies and the emerging role of the private sector, the overall arrangement, based on defence and structural protection remained initially untouched, and was driven by inertia. Although, substantial resources were channelled to the area of improving water quality (construction of water and wastewater treatment plants and of sewage systems), flood management has got less attention and was short of money. One could even argue that it was marginalised. The change of the administrative system played a significant role in flood risk management. The new administration offered a framework for flood risk governance. The crisis management system has been built in accordance with the administration structure of local governments and several flood management responsibilities were allocated to provincial governments.

- Large floods had a double role both involving change and stability. The Millennium flood of 1997 triggered a major reform of the crisis management system. The whole crisis management system was established (covering all hazards, not only floods). Besides creating the administrative structure for crisis management at all levels of administration, the State Fire Brigades significantly increased their operation capacities and their equipment was modernised. Pilot projects of local communities' involvement in preparation were launched. The hydrological modelling and the warning system were advanced. Also, there was an (unsuccessful) attempt to create universal flood insurance system. Nevertheless, main investments were transferred to structural defence and storage reservoirs. In this respect, the main approach remained stable. In spite of that, the flood of 2010 had no changing effects, as the crisis situation was basically managed.

- The accession of Poland to the EU in 2004 involved transposition of EU legislation (particularly, the FD and the WFD). It strengthened the position of some actors (such as NGOs), emphasising environmental concern and the new ideas in flood management (such as the re-naturalisation of rivers). The EU legislation and access to the European funds involved a new rhetoric of flood management and undermined the vested interests in the water sector. Implementation of the Floods Directive brought new activities into Polish FRGA.

Both, the Water Framework Directive (WFD) and the Floods Directive (FD) have had a significant impact on the FRM in Poland. Both directives are sources of novel ideas. The environmental critics of the dominant hydro-technical approach often refer to the FD as a source of support for environmental concerns. Since the 2000s, the dominant hydro-technical approach towards flood risk management in Poland has become criticised more and more vividly. It is contested from two angles: as ineffective in terms of flood management and as harmful for the environment. The first criticism emphasises that the current strategy fails, and losses from floods increase, despite the high expenditure on structural defences. Moreover, it is argued that the defence strategy, based on structural protection, leads to an increase rather than decrease of flood risk. This criticism is expressed by some hydrologists, crisis management experts, and NGOs. The second criticism is expressed mainly by environmental NGOs, and focuses on the damage to the environment caused by the flood protection activities, while not securing flood risk decrease. Re-naturalisation of rivers and similar ideas are proposed instead. Fragmented water and flood management legislation is weak and subsequent amendments to Water Act have undermined its coherence.

Similarly, other actors use elements of the FD and other EU documents to legitimise their current activities and even justify existence of relevant institutions. Some improvement in terms of the prevention strategy would not have occurred without the FD as well. It particularly refers to the obligation to prepare the Flood Hazard and Flood Damage Maps together with Flood Risk Management Plans. These maps, stemming from the FD enhance cooperation between the three sub-arrangements.

In terms of governance, some shifts towards more decentralised governance can be observed. Table 5 depicts different modes of governance (Driessen et al. 2012) of Polish flood risk management sub-systems.

Looking for the reason for stability of the Polish flood risk management, it can be argued that flood management and water legislation are seen as weak, mostly due to numerous amendments of the Water Act, what decreases its execution and effectiveness. This reinforces using the same methods by flood managers and the same beaten paths of proceeding which lead to preservation of the flood risk

Table 5 Modes of governance adopted by sub-FRGAs in Poland (country level)

Sub-FRGA	Mode of governance	Shifts in modes of governance
Structural defence	Centralised	Symptoms of a shift towards decentralised governance can be observed: more active role of stakeholders and NGOs. Nevertheless, centralisation is still dominant
Crisis management	Decentralised	Most action and resources can be found at municipal/county/provincial levels. Certain flexibility of rules close to informal ones can be observed, but at the same time voluntary fire brigades representing the non-governmental sector, are becoming more dependent on municipalities
Spatial planning	Decentralised	Main role is played by municipalities, but private sector is getting importance as a stakeholder influencing the procedures

management system in Poland with strong dominance of hydro-technical paradigm. It also has its reflection in the resource distribution: the dominating sub-system of structural defence is strongly financed from the state and through the EU funds, while, for example, spatial planning is funded from local budgets.

Changes in the flood risk management in Poland have been introduced very slowly and lacked strategic thinking. This was caused by specific features of the Polish system. There was a need for radical transformation of the economic, political, and administrative system after the fall of communism in 1989/90. Preparing a system coping with flood risk fitting to the radically changed legal, political and economic reality of the country appeared to be a complex task and of secondary importance (compared with the other urgent social and economic issues). That's why shock events like the occurrence of large floods or the accession to the EU were factors instrumental to trigger reforms.

5 Evaluation of Arrangements

As mentioned earlier, changes in Flood Risk Governance and entailed strategies can be evaluated taking into account how do they contribute to decreasing flood risk. Three main criteria were taken into consideration, when evaluating flood risk governance arrangements in Poland, and identifying its advantages and disadvantages: (a) resilience; (b) efficiency; (c) legitimacy. Operationally, the concept of resilience comprises three components: capacity to resist; capacity to absorb and recover; and capacity to adapt. Advantages and disadvantages of the Polish Flood Risk Governance Arrangement are presented in Table 6.

The Polish FRGA can be characterised as having a strong capacity to buffer shock events, as shown by the major 1997 and 2010 floods. Both stresses did not shorten financing in the defence-oriented flood risk management. Furthermore,

Table 6 Evaluation of flood risk governance arrangements in Poland

Desired outcome of flood risk governance	Sub-criteria	(+) or (−)	Feature of governance
Resilience	Capacity to resist	+	• Diversification of FRMSs; • A catchment-based approach to water (including flood risk) management; • Improvements of flood warning system to increase lead time; • New infrastructural investments
		−	• Homogeneity of FRMSs (domination of one, traditional, strategy, i.e. defence)
	Capacity to absorb and recover	+	• Prolonged and complicated procedures for maintenance of watercourse
		−	• Inconsistent or overlapping responsibilities (fragmentation) and strong path-dependences
	Capacity to adapt	+	• Broadening of measures
		−	• Low cooperation between the different actors involved in FRM; • FRGAs are more effective in terms of goal attainment rather than problem-solving; • Investments in flood defence infrastructure ('low degree of innovation')
Efficiency (financial and non-financial)		+	• Implementation of CBA at the level of particular investments
		−	• Preference for quick money spending (time pressure) and orientation on the immediate rather than long-term effects
Legitimacy		+	• Public consultation procedures; • Access to information (including information about environment)
		−	• The law has been changed many times in the last 25 years; • Legislative gaps or ambiguities; • Belief in the regulative and performative rule of law; • Zoning implementation gap with weak enforcement of ban on development in flood-prone areas

Advantages are marked as (+) and disadvantages are marked as (−)

recovery and EU funds were reinvested in hydro-technical infrastructure. The floods shocks resulted in certain progress in terms of institutional learning (new approaches and new measures were adopted, particularly with reference to good practices in the European countries), but much inertia and reliance on the well-established approaches could also be observed. In other words, adaptive capacity is feeble mainly due to actors' path dependence. Lack of innovation and little bridging mechanisms are key factors that constrain long-term oriented investments and more assembled actions. As a consequence, effects achieved in one sub-arrangement of flood risk governance have been found to be diminished by actions of another. In general, the resilience of the Polish FRGA has not changed dramatically in the analysed period of 25 years, but development was observed in terms of ability to buffer and recover. Summing up, the Polish FRGA is more effective in terms of goal attainment within different sub-FRGAs rather than problem-solving.

Implementation of cost-benefit analysis (CBA) for particular projects (with its time and resource obligations) is a dominant mode of management, hence easier to fulfil than solving the issue meant to be dealt with. Inadequacy and incoherency of data (absence of indicators) in the Polish FRGA constrains possibilities to conduct comprehensive and independent efficiency evaluation. At the same time, rules (e.g. competences of actors, acts, regulations) make it difficult to track the way in which and to what effect money is being spent.

High discretion in the domain of spatial planning undermines improvements in legitimacy of Polish governance. New construction sites on flood plain areas increase potential financial flood losses, thus delegitimising and undermining principles that are meant to be obeyed.

6 Concluding Remarks

The Polish flood risk management system has experienced changes (i.e. establishment of a new crisis management system), while preserving its core functional characteristics at the same time (reliance on structural defence). Sectorial management with little coordination at the level of ministries renders the flood risk management in Poland being more like bargain and competition than collaborative and consensus-driven type of governance. Operationally speaking, a clear diversification of measures can be acknowledged. Some developments, such as effective inclusion of spatial planning, even though widely supported by both experts and public opinion have not yet been utilized. An overall weakness of the Polish FRGA is the lack of coordination and strategic management. In Poland, this context is characterised by problems with effective spatial planning and a highly fragmented actor's structure both within and between sub-arrangements.

Based on recognition of strengths and weaknesses, a several critical issues can be indicated:

- Water and flood risk management needs to be develop around a system based on economic incentives. Such a development would entail integration of the domains to economise the efforts. Incorporating more market mechanism could help to overcome stalemate dominated by ministries and agencies managing their individual domains.

- Improving efficiency is highly needed. The use of the EU funds enforced implementation of clear accountancy procedures, development of tools that easier enhance efficiency, such as strategy-related cost-benefit analysis. Incorporating pricing in the FRGA and market mechanisms could reframe the relation between the public administration and the private sector and individual users. Hence, a strategy for incentivising private sector involvement would be a logical next step. Particularly, the insurance industry would be incentivised to promote flood resistance measures at the individual property scale and to cooperate with the public administration. Risk-reflective pricing of insurance premiums and therefore risk reduction at the property level would be incentivised to reduce costs.
- Enhancing public participation would not only mean involvement of citizens as clients. This seems difficult (and may turn out to be a dramatic experience) but eventually it could improve the legitimacy of decision-making. Yet, raising awareness of decision makers, politicians, managers, private sector, and citizens at large, and fostering public participation and community involvement on all levels, are needed.
- Since collaborative and regional programmes and watercourse maintenance activities suffer lack of coordination and inefficiency due to division of responsibility, a clarification of property rights in terms of water and flood protection infrastructure are much needed. Simultaneously, more financial transparency should be brought into Polish FRM. Currently it is fragmented.
- It seems that much more expertise is needed for designing an integrated flood risk governance system. Particularly, links between spatial planning and FRM are very important and concrete instruments that are needed for new and existing buildings. The spatial planning system, in order to be a consistent and relevant means for FRM, needs to be more effectively coordinated at the regional level, while this coordination is lacking at present.

Acknowledgments The work described in this publication was supported by the European Union's Seventh Framework Programme through the grant to the budget of the STAR-FLOOD (STrengthening And Redesigning European FLOOD risk practices: Towards appropriate and resilient flood risk governance arrangements) Integrated Project financed by the European Commission under grant agreement no. 308364 within FP7.

References

Biedron I (2012) Powódź 2010—analiza strat i szkód powodziowych w Polsce [The flood 1997: analysis of losses and damages in Poland]. Gospodarka Wodna 4/2012

Bouwer LM, Bubeck P, Aerts JCJH (2010) Changes in future flood risk due to climate and development in a Dutch polder area. Glob Environ Change 20:463–471

Cardona OD, Van Aalst MK, Birkmann J, Fordham M, McGregor G, Perez R, Pulwarty RS, Schipper ELF, Sinh BT (2012) Determinants of risk: exposure and vulnerability. In: Field CB, Barros V, Stocker TF, Qin D, Dokken DJ, Ebi KL, Mastrandrea MD, Mach KJ, Plattner G-K,

Allen SK, Tignor M, Midgley PM (eds) Managing the risks of extreme events and disasters to advance climate change adaptation. A special report of working groups I and II of the Intergovernmental Panel on Climate Change (IPCC). Cambridge University Press, Cambridge, New York, pp 65–108

CBOS (2010) Raport na temat społecznych konsekwencji powodzi [Report on social consequences of flood]. Report no BS/93/2010, Public Opinion Research Center, Warsaw

Chmielewski P (2008) Crisis management in the 1997 flood. In: Bynander F, Chmielewski P, Simons G (eds) The politics of crisis management in transitional Poland 1989/90–1999. Swedish National Defense College and CRISMART, Stockholm

Dobrowolski A, Czarnecka H, Ostrowski J, Zaniewska M (2004) Floods in Poland from 1946 to 2001 origin, extent and frequency. In: Proceedings of the international conference "Risks caused by the geodynamic phenomena in Europe" Polish Geological Institute, Wysowa, 20–21 May 2004, pp 79–56. Wyd. Państwowy Instytut Geologiczny, Warszawa

Driessen PPJ, Dieperink C, Van Laerhoven F, Runhaar HAC, Vermeulen WJV (2012) Towards a conceptual framework for the study of shifts in modes of environmental governance—experiences from The Netherlands. Environ Policy Gov 22:143–160

Hegger DLT, Driessen PPJ, Dieperink C, Wiering M, Raadgever T, Van Rijswick HFMW (2014) Assessing stability and dynamics in flood risk governance, an empirically illustrated research approach. Water Resour Manage 28(12):4127–4142. doi:10.1007/s11269-014-0732-x

Kitowski K (2010) Dyrektywa powodziowa a prewencyjne planowanie przestrzenne [Flood directive and preventive spatial planning]. Przegląd Komunalny, 7, Poznań

Klijn F, Kreibich H, De Moel H, Penning-Rowsell E (2015) Adaptive flood risk management planning based on a comprehensive flood risk conceptualization. Mitig Adapt Strat Glob Change 20(6):845–864

Kolipinski B (2015) Opinia o projekcie ustawy- Prawo wodne (w zakresie problematyki ochrony przeciwpowodziowej i jej powiązań z planowaniem i zagospodarowaniem przestrzennym) [Opinion on draft version of the Water Act (concerning flood prevention and its links with spatial planning and spatial management]. Unia Metropolii Polskich, Warszawa

Kowalczak P, Matczak P, Slavikova L (2013) Transition in water management in the Czech Republic and Poland. Int J Water Gov 1:307–322. doi:10.7564/13-IJWG8

Kowalewski A, Mordasewicz J, Osiatynski J (2013) Raport o ekonomicznych stratach i społecznych kosztach niekontrolowanej urbanizacji w Polsce [A report on economic and social losses caused by uncontrolled urbanisation in Poland]. Fundacja Rozwoju Demokracji Lokalnej, Warszawa

McDaniels TL, Gregory RS, Fields D (1999) Democratizing risk management: successful public involvement in local water management decisions. Risk Anal 19(3):497–510

Nirupama N, Simonovic SP (2007) Increase of flood risk due to urbanisation: a Canadian example. Nat Hazards 40:25–41. doi:10.1007/s11069-006-0003-0

Pinskwar I (2010) Projekcje zmian w ekstremach opadowych w Polsce [Projection of changes in extreme precipitation in Poland]. Monografia KGW PAN, Z. 34

Sarewitz D, Pielke R, Keykhah M (2003) Vulnerability and risk: some thoughts from a political and policy perspective. Risk Anal 23(4):805–810

Schanze J (2006) Flood risk management: a basic framework. In: Schanze J et al (eds) Flood risk management: hazard vulnerability and mitigation measures. Springer, Berlin, pp 1–20

Vetere Arellano AL, De Roo A, Nordvig J-P (2007) Reflections on the changes of the EU policy-making with view to flood risk management. In: Begum S, Stive MJF, Hall JW (eds) Flood risk management in Europe: innovation in policy and practice. Springer, Berlin, pp 433–467

Werritty A (2006) Sustainable flood management: oxymoron or new paradigm? Area 38(1):16–23

Flood Risk Reduction—Opportunities for Learning

Roman Konieczny, Zbigniew W. Kundzewicz and Piotr Matczak

Abstract This chapter examines opportunities for learning in the area of flood risk reduction in Poland, that was devastated by several recent floods, in particular in 1997 and 2010. We can learn from a champion, such as Gilbert F. White, a great visionary who revolutionized flood risk management. We can learn from own past failures of ourselves or from failures of others. However, sometimes, no constructive lessons from own failures are drawn—we learn nothing and we fail again. Finally, we can learn from examples of good solutions and practices, both domestic (such as flood risk reduction initiatives at the grass root level, as well initiatives in the realm of internet and social media) and international (such as US insurance and relocation programs and the Pitt Report, prepared after the 2007 flood in the UK).

Keywords Flood risk · Flood risk management · Flood risk reduction · Adaptation · Learning · Gilbert White

1 Introduction

Growing flood losses and difficulties to deal with floods despite vast research efforts and very high cumulative expenditures on flood protection and management have led to a conclusion that the problem is not in the lack of sufficient knowledge but in "our lack of understanding of the governance and cultural systems and how they are structured and managed and interact with ecological systems, and how we produce science and knowledge for policy" (Pahl-Wostl et al. 2008). In this respect mere

R. Konieczny (✉)
Institute of Meteorology and Water Management—State Research Institute, Cracow, Poland
e-mail: rkoniec@gmail.com

Z.W. Kundzewicz · P. Matczak
Institute for Agricultural and Forest Environment, Polish Academy of Sciences, Poznan, Poland

P. Matczak
Institute of Sociology, Adam Mickiewicz University, Poznan, Poland

Z.W. Kundzewicz et al. (eds.), *Flood Risk in the Upper Vistula Basin*,
GeoPlanet: Earth and Planetary Sciences, DOI 10.1007/978-3-319-41923-7_20

producing more knowledge and collecting more information does not necessarily help us to adapt to flood risk. Changes in routine performance is required in order to adjust measures to challenges (Berkhout et al. 2006). The system of flood risk management is a complex one and a change means a change of a large socio-technical system (Bos et al. 2013). It appears to be difficult and may have several forms. In principle, three types of social learning are differentiated: single-loop learning, involving correction of routines in order to avoid errors; double-loop learning, which involves correcting errors by reconsidering values and policies); and triple-loop learning, which is the most thorough and involves re-designing governance norms and protocols (Armitage et al. 2008).

Learning in practice appears much more complex than in theory (Bos et al. 2013) and it is not clear how a learning process emerges. There are many factors influencing a process, either enhancing or hampering it. Schusler et al. (2003) show that common purpose and collaborative relationship need to be developed as a prerequisite for learning, while several other conditions shape the success of social learning (open communication, diverse participation, unrestrained thinking, constructive conflict, democratic structure, multiple sources of knowledge, extended engagement, and facilitation). Capacity and credibility of engaged actors, mismatches between governance layers and decentralized mode of governance, and lack of coordination can hinder learning (Medema et al. 2015; Johannessen and Hahn 2013). At the same time a small group of professionals can initiate a change (Johannessen and Hahn 2013).

This chapter examines opportunities for learning in the area of flood risk reduction in Poland, that was devastated by several recent floods, in particular in 1997 and 2010, so that improvement of flood risk management is indeed necessary and also requested by the EU Floods Directive. In this chapter, we focus on learning initiation. The role of a breakthrough idea and a leadership are presented, epitomised by Gilbert White. Further, learning potential of failures, such as the 1997 flood in Poland, and a few examples of good solutions and practices from Poland, the USA and Great Britain are presented.

2 Learning from a Champion—Tribute to Gilbert F. White, A Great Visionary

Gilbert F. White (1911–2006) was an American geographer who revolutionized the area of flood protection. The thread of White's professional career, spanning about seven decades, was linked with academia and policy. At the Geography Department of the University of Chicago, he received academic degrees and became Professor and Department Chair. Early in his professional career, he served in the Executive Office of the White House under the administration of President Franklin D. Roosevelt. The New Deal policy, marking the era of optimism, was an innovative way to overcome the crisis of the Great Depression. White reviewed proposed natural resources legislation and presented summaries to the President. It was in the time of large structural flood protection projects, the centralized national planning commission, the Tennessee Valley Authority, and the era of dams, such as the Hoover Dam, whose construction gave much needed employment to 10 thousand people. White continued to inform public policies over many decades, consulting administration of a dozen of US presidents.

2.1 *Promoting the New Approach*

Noting that dikes can provide a false sense of absolute security, White advocated other adjustment measures, in particular a wise use of floodplains (farmlands, wetlands, woodlands, ecosystems) and reduction of unwise reliance on structural means. Since his landmark work—the 1942 dissertation *Human Adjustment to Floods*—White challenged the dominating "levees-only" solution and the belief that natural hazards are best addressed by structural ("hard") defenses. He advocated modifying human behaviour and undertaking adaptation to, or accommodation of, flood hazards. He was indeed a precursor of the concept that has been re-discovered recently as a nature-based solution: living with floods, giving room back to the rivers, or moving out of harm's way (retreating from unsafe areas).

White showed another way to manage water resources and mitigate natural hazards. It is folly to occupy flood plains. Even if dikes are in place, they offer limited safety only—losses soar when extreme flooding overcomes structural barriers. White questioned the attitude of controlling and "bending" nature. Instead of treating natural extreme events as foes, he perceived them as a part of nature, with which one should live in harmony.

His saying "floods are acts of God, but flood losses are largely acts of man" illustrates, on the one hand, the humility in trying to stop the floods in a hopeless struggle with the forces of nature, while, on the other hand, suggests the need for a greater focus on the mistakes humans have made in the areas adjacent to rivers. One cannot win with God. One has to focus on what one can influence.

White's research, studies, publications and activities have changed strategy for reducing the impact of floods in many countries, worldwide. Before someone decides to build a home, office, factory, school, or hospital in a flood-prone area, one should carefully examine the flood risk. If a decision on construction in a flood plain is made, the structure should be rendered immune to the effects of flooding.

In 1939, on a national US conference on the management of natural resources, White (aged 28) came up with a postulate that the federal government should not approve construction of storage reservoirs in California until the state adopts laws limiting development in areas at risk of flooding. Such words sounded like a heresy at that time—only three years earlier, the US Congress passed the so-called Flood Control Act and billions of dollars were being spent on construction of levees and reservoirs. The United States Army Corps of Engineers (USACE) was a monopolist and water engineers were pets of the media and of the American public, almost like modern celebrities.

Gradually, more and more people echoed White's views. Indeed, despite the huge expenditure on technological protection measures in the US (within 30 years since the passing of the Flood Control Act in 1936, some $12b were spent in the US on structural defenses), flood losses continued to rise. The strategy to mitigate the impact of floods in the US adopted in 1936 proved to be ineffective. People simply built more and more on floodplains, tempted by the naïve feeling of being protected.

There was a long-lasting and wide-spread belief that structural flood defenses can reduce flood losses almost completely. In addition, once launched institutional machine, spinning billions of dollars, employing tens of thousands of people, could well defend itself.

White worked on rectification of past errors. For example, noting an irresponsible placement of buildings in the floodplains and parks on high grounds in Tulsa, Oklahoma, he was instrumental in undertaking a corrective action, clearing over 900 buildings from floodplains and converting the areas next to the river into picnic places (Hinshaw 2006).

White's visionary and groundbreaking work had a great impact on thinking about reducing the impact of floods in America, and world-wide. He suggested the following:

- One should delay storm water runoff into rivers and streams in the upper part of a catchment (by taking care of plant cover, ponds and wetlands etc.).
- Areas prone to flooding must be used in such a way that, in case of flooding, losses are as small as possible (limiting housing and industrial development, etc.).
- Buildings constructed in flood prone areas should be resistant to floods.
- Levees, reservoirs and relief canals should be built where they are economically justified.
- During floods, houses should be protected with the help of temporary shields and efficient system of evacuation of people and livestock should be prepared.
- The insurance system should be formed to enable insurance of people and their belongings from flooding and to create a system of aid to flood victims.

These suggestions resulted from White's observations: in some areas, flood effects are significantly lower than in others, despite the similarity of conditions. In these areas the local communities, business owners and institutions used a variety of measures to mitigate the effects of flooding: houses were prepared for flooding, companies had evacuation plans etc. This "discovery" was a result of White's working methods—checking everything in the field, although fieldwork in the flood protection industry was not prevalent.

2.2 *Impact of the New Ideas*

The United States Army Corps of Engineers stated that no scholar of the 20th century had a more significant impact on the Corps than White. Today, after

70 years, the paradigm stating that only technical measures can effectively reduce the impact of floods has not only been an American specificity. One can see that the visions of White influence strategies in many other countries, including the European Union and its Floods Directive.

However, penetration of the White's approach cannot be taken for granted. In Poland, a few years ago, during a discussion devoted to preparations for future floods organized in the city of Sandomierz, a statement made by a known environmental activist, that the destructive role of beavers during the floods in 2010 was exaggerated, raised a controversy. A representative of the Provincial Board of Irrigation and Drainage protested and accused environmentalists of hindering the construction of reservoirs, opposing the cutting of trees and shooting beavers, hence accusing them for flood losses.

In Poland, the spirit of White influences only a minor portion of flood risk management plans, say 3.5 % of the total (value based on results of analysis of 107 flood plans made after 1997). The structural defenses dominate. Yet, most of the land flooded during the recent floods was protected by dikes: during the floods in July 1997 and in 2010, some 670 km and over 1300 km of dikes, respectively, were damaged. This just illustrates failures of structural defenses during heavy floods. How to help the people inhabiting flood-risk areas? Heightening and strengthening the dikes appears inefficient. Capacity to resist can be interpreted with regard to the load, i.e. a load-resistance perspective can be taken. A system withstands load, but not without limits. Perfect, absolute, resistance is not possible. According to a statistical design concept, defenses should withstand a design flood, e.g. 100-year flood, but may fail if the actual flood is much higher. Therefore, a more disaster conscious society needs to be built with better preparedness and safe-fail (safe in failure; this is the core of the concept of resilience) rather than, unrealistic, fail-safe (safe from failure) design principle, cf. Kundzewicz and Takeuchi (1999). However, the discourse on this matter is disappointingly weak. The administration of water and most water engineers hardly cooperate with environmentalists, economists, spatial planners and even with local governments.

White contributed heavily to the practice of floodplain management and flood insurance, and to the research in the area of the interdisciplinary science of natural hazards and preparedness, both nationally and internationally. His ideas precipitated to the mainstream knowledge and practice.[1]

[1]A biography of Gilbert F. White by Robert Hinshaw (2006) is a useful reading for anyone interested in different aspects of floods (Kundzewicz 2007), including those related to the EU Floods Directive. More information about G.F. White can be found at http://www.colorado.edu/hazards/gfw/.

3 Learning from Failures

Learning from one's own failures is less advantageous that learning from failures of someone else. However, it is important to take lessons from a failure, such as the flood management action during the event of July 1997 that unveiled myriads of deficiencies of the flood preparedness system in Poland.

The flood event of July 1997 in drainage basins of two large Polish rivers, the Vistula and the Odra, that came after a long time interval without floods, made Poles aware of how dangerous and destructive a flood can be. It also demonstrated where the weaker and stronger points of the flood defense were and exhibited the most important needs. The flood unveiled the elements in the existing flood protection system where improvements were badly needed. Indeed, every link in the chain of operational flood management (observation–forecast–response–relief) was found to be in need of strengthening.

The structural flood defenses proved to be dramatically inadequate for such a rare, gigantic flood (baptized, somewhat loosely, the Millennium Flood), like the July 1997 event. Some of the reservoirs in the affected basins contributed to mitigating the flooding downstream, but, in general, the existing reservoir capacity was far too low in the context of such a great flood.

During the 1997 flood, organization was also a weak component of the system, especially at the beginning of the flood, even if the factors of the extraordinary scale of the event and the effect of shock are taken into account. Legislation was inadequate; for example, there was a lack of consistent regulations concerning financing of flood action. Financing was based more on ad hockery, than on a systematic approach, according to a well established, consistent, and complete legislation. In fact, the old laws (valid in the communist system) did not work already and the new laws did not work yet. Division of responsibilities and competence was ambiguous. As a result, regional and local authorities were uncertain as to their share in decision making (with important operational and financial implications).

The flood revealed immediate shortcomings of the flood risk management system. The information gap was clearly felt, especially in the first phase of the flood. This resulted from the lack of an automatic observation system, destruction of staff gauges by the flood, communication breakdown, and the evacuation of observers. It also became clear that the interface of flood action and the media was highly non-satisfactory. To considerable extent, crisis services were not adequately prepared to collaboration with the media.

On a more general level, the 1997 flood made apparent that a well-established national policy of limiting the effects of floods, i.e. flood risk reduction policy was clearly lacking in Poland. As a result, determination of the main directions of change and indication of actors responsible for implementation of these changes was missing. Also evaluation criteria allowing determination of strengths and weaknesses of the existing system were poorly devised. The policy did not integrate engineering activities with spatial planning and the ability of local communities to address flood risk reduction. The flood risk management policy was a product of water administration bureaucracy, instead of a product of exchange of ideas, diagnoses and strategic thinking of specialists in various sectors: spatial planners, crisis managers, civil and water engineers, environmentalists, climate change experts, the academia, as well as representatives of local communities.

It is worthy of noting that, after the destructive 2010 flood, the Governor of the Masovian Province (Voivodship), initiated a Flood Safety Program in the Middle Vistula Water Region. This program advantageously differed from earlier planning documents related to the Upper Vistula and the Odra. Unlike the earlier programs, the Middle Vistula program was not focused exclusively on structural defenses (even if it still overestimated their role), but included spatial planning, crisis management, education, architectural and urban conditions as well as technological standards for the floodplains. It integrated activities of local authorities, ranging from provincial offices (and their departments), to counties and municipalities. It generated discussions with environmentalists. A unique, and high-level, human potential was created, including formation of valuable relationships, joint arrangements, and negotiated positions. The program involved an interdisciplinary team of experts. It was a real breakthrough. In brief, the program made it clear that it is necessary, both, to keep water away from people and wealth (with the help of structural defenses) and to keep people and wealth away from water (by control of to human encroaching into flood-prone areas). Unfortunately, this well-auguring program was suspended.

The interface between the crisis services and the media can be regarded as another example of successful learning. It became the subject of several activities (discussions and meetings, including a seminar attended by the President of Poland). Polish Hydrometeorological Service (IMGW-PIB) published a guidebook for crisis management staff in municipalities as well as for IMGW-PIB personnel on how to collaborate with the media and a guidebook for journalists devoted to natural disasters. It can be argued that the flood 1997 initiated a tangible progress in this respect.

3.1 Deficiencies in Learning from Failures

In Poland, the need to modernize and repair the existing levees remains the principal topic of discussion on flood preparedness system. This is reasonable. However, advocating for construction of new dikes is less reasonable given that coping with the maintenance of 8500 km of existing dikes is not an easy task. However, imagination of decision makers is ignited by a vision of large investments and large opportunities. After the flood in 2010, the debate held in Polish Parliament called for a long-term investment plan for the security of hydraulic flood defenses, with adequate funding. Also reservoirs were considered important and the European Cohesion Fund was identified as a possible funding source. This can be interpreted as a single-loop learning, i.e. undertaking corrective actions without challenging the paradigm about the dominance of structural defenses.

Decisions makers need the professional support, with a strong research community that analyze losses and their causes, devise methods and evaluate their effectiveness, and examine reasons for different human stances towards disasters, ways to influence human behavior, methods of communication and education, etc. Such support is lacking in Poland, hence the water administration does not implement novel solutions, does not test them and does not monitor their effectiveness. Awareness of alternative strategies and methods to counteract floods, different from investing in hydraulic structures is marginal.

Lack of learning and particularly lack of engagement of local actors led to limited knowledge about reasons why particular houses and even villages were flooded. As a result, several elements of prepared flood risk reduction plans appear questionable. There are places (e.g. in Ciężkowice on Biała Tarnowska), where a flood dike was designed along the main river, although most of the losses in this area occurred along the tributaries. Apparently, nobody has actually analyzed where losses emerged. There are also places (e.g. Gorzanów on the River Nysa Kłodzka in the Bystrzyca Kłodzka commune, Kłodzko Valley), where the dikes were strengthened except for a segment in the center of the locality, of the length of several tens of meters, just because this dike segment was not present in the registry and decision-makers may not have paid a fact-finding visit to the area of concern. Such cases show the weakness of a top down approach, where local details are likely to be disregarded, undermining the effectiveness of the whole effort. Lack of monitoring of what citizens whose properties are often inundated do in order to reduce flood risk undermines recognition of right and wrong practices.

Limited concern and information about the experience on the local level results in limited capabilities of experts in informing the process of decision making. Poor performance of the administration enforces citizens to help by indicating absurdities and supporting reasonable solutions.

Lack of a policy agreed upon by various stakeholders obscures the division of responsibilities, e.g. indication who is responsible for building standards in flood risk areas, or who is to prepare a guidebook on flood-proofing the buildings, It hinders the development of flood risk management plans and virtually does not render it possible to evaluate the initiatives.

Poor learning is one of the reasons for loose and unstable water and flood risk management policy. Many iterations of discussions about changes in the structure of water management in Poland proposed by the Ministry of Environment exposed the systemic weakness. There have been 36 Water Law amendments so far, since 2001. The actual document has been criticized, virtually, by all interested parties, though for various reasons. For environmentalists, it does not ensure good environmental status of water in the future (as requested by the EU Water Framework Directive). According to the energy lobby, implementation of this proposal would destroy hydropower, while according to the navigation lobby it would exterminate navigation.

The current law directs resources stemming from water charges (received by provincial environmental funds), entirely to structural defenses, leaving aside other important activities, such as: land-use planning (zoning), protection of buildings, buy-out of properties, warnings, etc.

4 Learning from Good Examples

4.1 *Good Examples from Poland*

4.1.1 Flood Risk Reduction at the Household Level

The single-loop learning involves correcting erratic situations. There are several lessons of this type among accommodation owners who suffered losses during the 1997 and the 2010 floods in Poland. They looked for methods to prepare their houses for flooding in the future. Although the state is responsible for protecting citizens against disasters, addressing expectations towards the water administration would not lead to find adequate support there. The flood reduction planning, at the national or regional levels, is based on the premises that most flood losses are caused by flooding of major rivers, so that building dikes and reservoirs can protect people against such events. As a result, flood-affected population, especially where floods are frequent, come to the conclusion that they need to do and, indeed, can do something on their own.

A characteristic example of such activity stems from a place near to a tributary to the river Biała Tarnowska in the southern Poland (flood related studies of that river are reported in Czech et al. 2016 and Hajdukiewicz et al. 2016). In 2010, the river flooded a garage, destroying a car service workshop with expensive equipment, a house and a farm. Following information from the Provincial Board of Irrigation and Drainage that it is not planned to invest in flood preparedness and the repair of the affected river channel, the owners of the garage decided to work on their own. They strengthened the river bank with stone rip-rap fastened by concrete (Fig. 1), heightened the bank creating a dike, prepared dewatering of the terrain and mobile protection of the entrance to the building and to the workshop (Fig. 2), where valuable diagnostic equipment is located. The cost of the entire project is estimated to be of the order of 40 thousand Euros (price levels of 2015). In the case mentioned above, the investors applied to the authorities for permission and respected the resultant restrictions (e.g. as to the ban on usage of some machinery). It may well be that in some other cases the investors did not seek official permissions. They simply improved their flood protection.

Similar actions were undertaken by many local residents exposed to frequent floods. The research carried out in two towns in southern Poland shows that about 60 % of the residence owners (from the pool of 100 households flooded every few years that were surveyed) have taken steps to prevent future losses (Tyszka and Konieczny 2016). Most often it was improvement of rainwater drainage away from the building, or the transfer of central heating boilers and electrical installations to higher elevation (about 50 %). There have also been instances of investing in protecting the banks of nearby streams (23 %) or construction of embankments or walls (16 %), see Fig. 3. Less often, basements or underground garages were eliminated (6 %), or additional upper floor was constructed (6 %).

However, in many areas where floods are rare or where levees are in place, flood preparedness activity of building owners is less pronounced. A survey (over 800 respondents) conducted in the areas at flood risk shows that if insurance is not

Fig. 1 Reinforced stream bank. *Source* Małgorzata Siudak

Fig. 2 Mobile protection on the garage door. *Source* Roman Konieczny

Fig. 3 Examples of a stone wall and an earth dike, protecting houses. *Source* Małgorzata Siudak, Roman Konieczny

Fig. 4 Examples of houses built on artificial earthwork mounds. *Source* Roman Konieczny

counted, on average, only 5–7 % of house owners in flood risk areas take any action in this respect. It is also interesting that the owners respond not only to river flood risk, but also to risks which, according to Polish law, do not fall within the definition of floods, because they are caused by the accumulation of rainwater in undrained areas, surface runoff etc. Quite a popular form of protection against such floods is building on artificial earthwork mound (Fig. 4).

Also local governments undertake activities aimed at preparing for floods. One of relevant examples of increasing popularity is a monitoring-warning system, i.e. a system for monitoring rainfall and water level and warning residents in the likely affected area. The first system of this type in Poland was established in the Klodzko Valley, following a broad criticism of the national flood warning system, which proved to be inefficient during the floods of July 1997. The current shape of the system (Fig. 5) is the result of identification of local needs and discussions with the Institute of Meteorology and Water Management within the international research project[2] devoted to the role of information in reducing the consequences of floods. The system, which has been operating since 2002, currently consists of 29 sensors of water level on the Nysa Klodzka and tributaries, 18 sensors measuring precipitation and system of notification of residents by telephone. This latter system allows conveying information at the rate of up to 700 people per hour while controlling who has been notified and who has not. The measurement data are made available on an openly accessible website on the Internet[3] and updated every 15 min.

A number of local monitoring and flood warning systems have been built in Poland. Some of them consist of a few measuring points, while others, such as the Zywiec county system has a size similar to the system in the Klodzko Valley. The total number of automatic sensors installed by local governments has now exceeded half of the sensors in the State Hydrometeorological Service network.

---

[2]OSIRIS, financed within the EU Fifth Framework Programme.

[3]http://lsop.powiat.klodzko.pl/index.php.

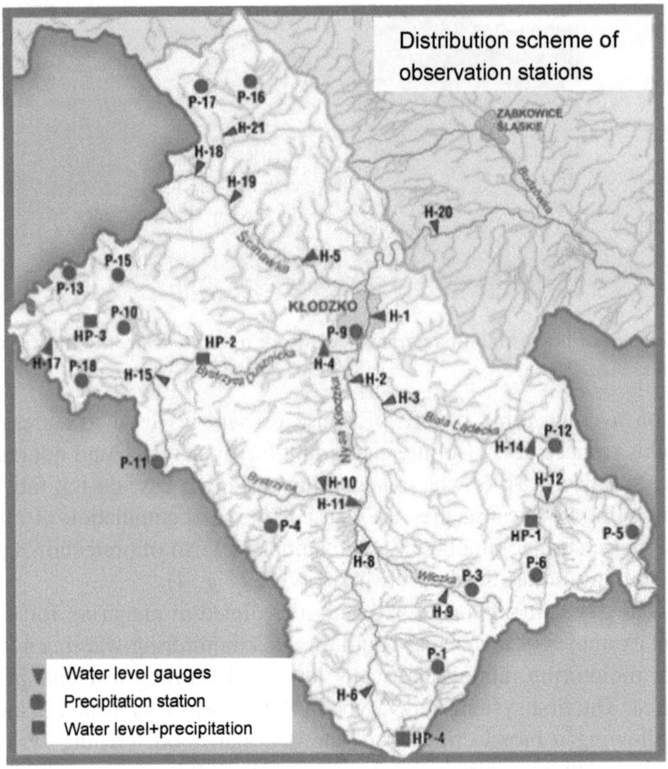

Fig. 5 Distribution of rain and water monitoring system in Klodzko Valley, financed by the Klodzko county (Polish: *powiat*) and municipalities (Polish: *gminy*). *Source* http://www.lsop.powiat.klodzko.pl/index.php

These systems are most frequently based on the ideas put forward by local emergency services and the companies that offer system construction. They are not always the best solutions, and the effective operation is not always guaranteed. Typically, no cost-benefit analysis is undertaken, because resources for system operation and maintenance are scarce or even lacking, so that these systems are often poorly maintained and operated. Identification of the information needs of different vulnerable groups (residents, the elderly and the lonely, handicapped people, entrepreneurs etc.) is not carried out. In result, these groups are not warned sufficiently early and information is sometimes confusing. Also psychological factors affecting people's reactions and the knowledge necessary to respond are not taken into account, so that, in most cases, people do not always respond properly to the warnings. Since these errors are mainly due to lack of knowledge and experience of local governments, this poses new and interesting challenges for institutions dealing with reduction of flood consequences in Poland. In this context, the role that the Institute of Meteorology and Water Management—State Research Institute could play in cooperation with local governments could potentially be very

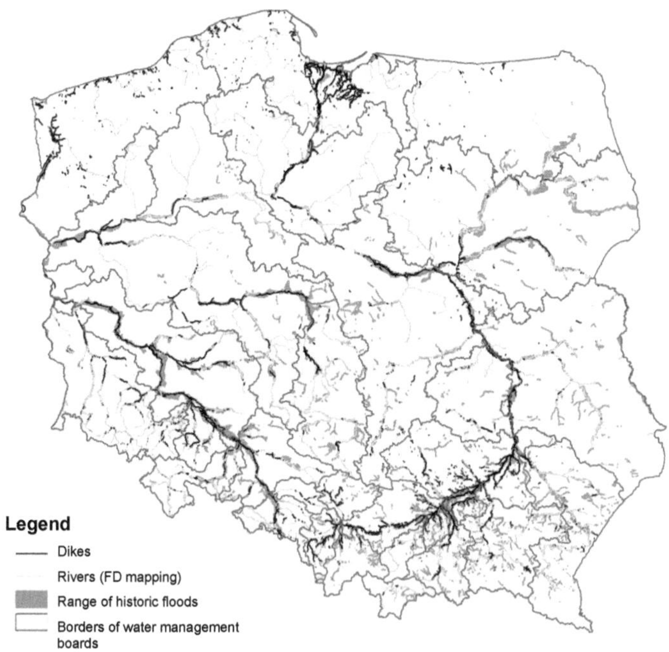

Fig. 6 Areas protected by dikes that were flooded by one of the recent large floods (in 1997, 2001, 2010) (color figure online)

important. This is, after all, the only institution that has experience with the construction and operation of meteorological and hydrological observation systems. Unfortunately, despite successful cooperation with local authorities established after the flood in 1997, the cooperation has recently weakened.

Another problem faced by local authorities is to decide upon what can and what should not be built in areas protected by dikes. This issue raises considerable controversy, because these areas are protected, hence, to some extent, safe. However, taking into account all the great floods that occurred in Poland in the last 20 years, it turns out that the most damage occurred in areas protected by dikes. This is illustrated (Fig. 6) by a map of Poland with two layers of information developed in the preliminary assessment of flood risk: a layer of levees (black line) and a layer of flooded areas in one of the three major floods that occurred in 1997, 2001 and 2010 (marked blue). It is clear that large areas protected by dikes were flooded.

This confirms the well-known wisdom that the dikes are effective for small and medium floods, while in case of large floods they give us only time to escape. The water administration ignored this problem so that there is no central regulation nor recommendations for the areas located behind the dikes. However, some local authorities introduce their own regulations in the framework of local spatial development plans. These are, most commonly, conditions that must be met to obtain a building permission, embracing: a ban on the construction of basements and underground garages or a ban of construction of basements with windows, obligation to use waterproof materials, and to place the residential level at least 1 m above the ground. These principles are also consistent with local traditions. Indeed, old houses, built in these areas decades ago, were built in this way. Sometimes, additionally, a ban on construction of objects that can lead to secondary pollution is also introduced.

However, the described activity is not common. In fact, local governments are trying to pursue mainly those actions that result from the national law or established customs. Paradoxically, in contrast to private owners who undertake several activities to protect their own houses, by surrounding it with a flood wall, a dike, or using mobile protections on windows and doors, for public buildings such actions are very rare, practically non-existent. This lack of impact of grass-root activity on the local authorities would demonstrate a certain weakness of knowledge transfer in Poland. However, the described examples lead to moderately optimistic conclusions. Actions of local authorities can serve the development and strengthen the sense of responsibility regarding preparation for floods and enhancing cooperation. A considerable consequence of this activity of local governments is also its impact on the development of the service sector in this area. There is an increasing number of private companies, that offer construction of systems of monitoring of precipitation and water level, as well as systems of mass notification about the threat.

4.1.2 Floods, Live, on Internet

Spontaneous human activities play an increasingly important role during floods, especially if they support the flow of information at the time when traditional methods fail (as did cellular network during the 1997 flood).

During the disastrous flood in July 1997, the web portal "Voluntary Internet Krapkowice" (in Polish: *Wolontariat Internetowy Krapkowice*) founded by a computer company owner was an important source of flood-related information. Using servers in Warsaw, as well as phones and informants in the field, a group of young people, first in Krapkowice, then—after Krapkowice was flooded—in Opole (also subsequently flooded), prepared professional, well-organized, and reliable information service, instantly reacting to the dynamic development of the situation. The portal had a large audience, including professional media, reaching an astonishingly high number of 300,000 visits (in the time when the number of Internet users was five times lower than now). It showed that civil society initiatives can play a powerful role.

During the flood in May 2010, a blog from Wroclaw ("Wroclaw by choice", in Polish—*Wroclaw z wyboru*), run by Mr. Pawel Andrzejczuk, became the main source of flood-related information for the inhabitants of Wroclaw. One morning, Mr. Andrzejczuk noticed anxiety in a local grocery store. People knew in general terms that a flood was coming (destructive 1997 flood was still in living memory of many inhabitants of Wroclaw), but no specific information was available. Mr. Andrzejczuk returned home and announced on his blog: "I start to write, live, on the flood threat in Wrocław, minute by minute. If you have information, photos or videos from endangered areas, please forward them to me." The first reaction came almost immediately, then in five days the number of e-mails soared to over three thousand (therein 500 sent by Mr. Andrzejczuk), including some five thousand photographs. In the hottest time, mails came every few minutes, day and night. Mr. Andrzejczuk had about 300 permanent informants who reported what was happening in different places, and what were the needs—where manpower was acutely needed and where sand bags were missing. Other groups of volunteers, reacting on information from the blog, moved from place to place, depending on where their help was needed. The news from the blog spread immediately. On the first day (May 20) the blog recorded 27,000 visits, next day 85,000, and the next 140,000. In total, over five days there were 450,000 visits and 2.5 million page views.

The blog of Mr. Andrzejczuk was not the sole source of information for the citizens of Wroclaw during the 2010 flood. Many people used social media. According to the portal Interaktywnie.com, there were 1800 entries with the qualifier "Flood" and 1200 entries with the qualifier "Wroclaw" on Blip, a service modelled on Twitter. Almost a thousand people subscribed to the "Flood Wroclaw 2010" group on Facebook.

4.2 Good Examples Around the World

4.2.1 US Insurance and Relocation Programs

In the USA in 1968, during the presidency of Lyndon B. Johnson, the state offered homeowners and businesses affordable insurance against floods. Since commercial insurance either withdrew from flood-prone areas or requested very expensive premiums, the US Congress approved the National Flood Insurance Program, which is still the basis of the strategy for limiting the effects of flooding in the USA. It was a huge success of the ideas originating from Gilbert F. White (cf. Sect. 2) and a growing number of his followers.

Why was this program unusual? Rather than offering insurance like commercial insurers do, it was a program prompting vulnerable homeowners and entire communities to prepare for the next flood. The program was based on the following principle: a resident can buy insurance if the local authorities subscribe to the National Flood Insurance Program. This means a declaration that:

- areas threatened with a 100-year flood will be charted;
- the government will introduce regulations restricting constructions in those areas.

The program has led to limiting the expansion of construction in flood-risk areas. The beginning was not easy, because the state administration officers commenced implementing it in their way, contrary to the recommendations of the advisory committee (with White, cf. Sect. 2). This resulted in poor response rate—within four years since the establishment of the program only a few of the 20 thousand endangered municipalities joined the program and only four people bought the insurance policy. However, after a few adjustments more than 18 thousand municipalities currently participate in the program. Over all these years, the rules of the program have been modified and supplemented. Currently, the more action is taken by the local authorities, the lower the insurance premium for residents and the greater chance for receiving a government grant for such activities as: construction of warning systems, preparation of plans for flood mitigation, improved rainwater drainage, preparing facilities to flooding, etc.

Another good example, also in the spirit of Gilbert White, is the relocation programme in the USA, after the Midwest flood of 1993 (Kundzewicz 1999). The US Interagency Floodplain Management Review Committee (IFMRC 1994; Galloway 1999) issued a recommendation that federal, state and local governments and those who live or have interest in the flood plain should have responsibility for development and fiscal support of flood plain management activities. The Committee (IFMRC 1994) recommended also that the administration should fund acquisition of properties at risk in the flood plain. The number of families relocated from the vulnerable flood plain locations in the Mississippi basin reached 20,000.

4.2.2 Pitt Report After the 2007 Flood in the UK

Flooding in the summer of 2007 was indeed one of the greatest inundations that the UK has experienced in recent decades. Some 44,600 homes were inundated and the area under water was approximately 42 thousand hectares. There were 13 fatalities related to the flood and some 9000 people were rescued from immediate danger.

The UK State Secretary responsible for the environment, food and rural affairs requested Sir Michael Pitt, leading the planning commission on infrastructure, to prepare a report summarizing the flood of 2007, and to develop suggestions as to necessary changes that could lead to damage reduction during the next floods. The report was a result of extensive research, many discussions with experts, more than 1000 written conclusions, summaries or proposals and a great number of direct meetings with various communities affected by the flood (organized during visits to areas affected by flooding). Another important feature was the collaboration with professional communities involved in reduction of adverse consequences of floods, in various aspects. A preliminary version of the report prepared already in the flood year, 2007, was sent to multiple communities for perusal and consultation. On the basis of comments, the final report was prepared and published in 2008.

According to the statements of Sir Michael Pitt, it is the time to put an end to the dangerous strategy of playing with risk (assuming luck) and to move on to better organize and prepare for floods. According to this report, the English policy to reduce the effects of flooding should be subject to radical and fundamental change. One of the main changes suggested by the report is significantly greater role of local communities in mitigating the effects of floods. The report contains almost one hundred guidelines (recommendations). After consultation with many communities, it is believed that the recommendations are possible to implement, and that their meaning and the anticipated benefits are understandable. Recommendations contained in the report relate to the following issues:

- Flood forecasting, including flood range;
- Improvement of planning and reduction of flood risk and consequences;
- Protective measures and rescue operations during floods;
- Critical infrastructure—particularly water and electricity supply;
- Counselling for people re: securing their families and homes;
- A return to normality (recovery).

From the organizational side, recommendations are mainly related to:

- Strengthening the role of the Environment Agency, as coordinator of all activities;
- Stronger inter-agency cooperation—for example a recommendation for better cooperation between the Met Office and the Environment Agency;
- Dedication and strong support to local communities with respect to limiting the effects of floods.

The UK Parliament and the Government have responded positively to recommendations of the Pitt Report. Annual plan of implementation of these recommendations and a long-term plan have been adopted. Also the law has changed—a relevant strategy was prepared.

5 Concluding Remarks

As shown in this chapter, there are ample opportunities for learning, when building or strengthening the system of flood risk management in Poland. We can learn from past failures of ourselves or of others, paying attention that indeed constructive lessons are drawn. We can also learn from examples of good solutions and practices in the realm of flood risk management at different spatial and organizational levels. In addition to single-loop learning, involving incremental correction of routines in order to avoid errors; without chaining the ruling paradigms, it is recommendable to consider double-loop learning (correcting errors by reconsidering values and policies); and triple-loop learning (re-designing governance norms and protocols).

John Naisbitt, an established authority on trends, who searched for emerging trends and sold his forecasts in quarterly *John Naisbitt's Trend Letter* with a bold advertising statement "Companies who ignore or resist these trends will fall", coined a maxim: "Trends are bottom-up, fads top-down". Certainly, this statement also holds for the area of flood protection in Poland. According to Naisbitt, trends arise where the problems arise, that is "at the bottom" (grass roots level) among citizens in small communities rather than at the level of government documents and strategies. Most of the concepts created in the area of flood protection in Poland by the government are non realizable fads, focusing predominantly on structural defenses—the realm of hydraulic engineering, and replicating errors that had been made in other countries a long time ago. This vision focuses on large investments on structural flood defenses, like flood dikes and water storage reservoirs. These latter cost billions (and raise considerable social and environmental objections). No government is likely to find the necessary funds for construction of large reservoirs in the coming years. Telling people that dikes offer a satisfactory protection against flooding demonstrates either a lack of competence or a lack of honesty. Learning from examples of good solutions and practices at the grass-root level, from people who protect their properties, is advisable.

If floods do not come for long, humans tend to forget the potential danger. Late Canadian hydrologist Vit Klemes (personal communication to ZWK) coined the concept of a hydro-illogical cycle. Everyone knows the term "hydrological cycle", that is cycling of water in the nature. The meaning of hydro-illogical cycle is as follows: when a large flood comes, it mobilizes funds that were indeed so hard to find before the deluge. A number of needed flood protection and preparedness activities are then initiated, yet some time after flood, memory fades. Other needs become more urgent, hence the expenditures on flood risk reduction are cut and comprehensive programmes are reduced or even suspended and abandoned. When a big flood strikes again, it acts as a reminder and triggers anew the mechanism as above. This observation illustrates imminent, and omnipresent, difficulties to the process of learning from floods.

The last years in Poland mark a period of transition. The traditional paradigm of flood protection is losing its validity and the new paradigm has been slowly emerging (cf. Wyzga and Radecki-Pawlik 2011). But what is gradually getting outdated is still strong due to the support of institutions and in some sense a social acceptance. What has been born is still weak, built on the basis of external ideas (the EU Floods Directive), and implemented by some owners of properties at risk of frequent flooding, some local governments and NGOs. However, these experiences do not reach the wider public. The administration responsible for the implementation of the Directive hardly explains the rationale for changes, residents cannot generalize their experience and communicate new ideas. NGOs could, but they are not trusted.

All parties feel a major change, hence the current norms are treated less restrictively. As a result, there are new activities among residents, public administration, NGOs. By regulating the streams near their property and by building their own dikes and flood walls, residents evoke protests from the institutions responsible for such actions, because these activities are technically illegal. At the same time, the public administration makes plans for all areas of the country (programs of flood protection in the basin of the Upper Vistula and of the Middle Vistula, as well as the Odra 2006 Programme), which lack regulatory environment. As the preparation of the plans is not based on standard planning, in order to implement them special decrees of the Cabinet are required. At the same time, the Council of Ministers is unable to enforce the water administration to execute flood protection plans, which are required by the Water Law.

It all gives the impression of chaos—actions undertaken by different entities are based on different values and non-harmonized standards. On the other hand, the experience gained from the implementation of these actions, both positive and negative, can provoke discussing relevant values relevant for the parties, determining common objectives and develop recommendations how to achieve these objectives. An inspirational case for the integration of ideas and knowledge is the experience of the Michael Pitt's committee, examining the English floods in 2007 as well as the Gerald Galloway's Committee examining the flood in 1993 in the US. There are more examples of how the experiences of individuals can contribute to a common strategy. Another interesting examples is the development of water policy in the Netherlands by the interdisciplinary committee composed of independent professionals appointed in 1999 by the Ministry of Transport, Public Works and Water Management and by the Association of Water Boards of the Netherlands. Within a year the committee presented Water Management Policy for the 21st Century, which constituted a foundation for water policy of the country.

In conclusion, it can be stated that in order to implement good practices, existence of good examples and availability of applicable patterns as well as existence of leaders who are able to convince the legislative bodies about reasonable solutions is required. However, good patterns (or practices) can only seldom be applied in a simple way. They must coincide with grassroots experience and culture, and must take into account the interests as well. Scientific literature recognizes nasty and intractable problems (so-called wicked problems) and clumsy and imperfect solutions (Verweij and Thompson 2006). It seems that neither intellectually correct solution without adequate leadership may be effective, nor fair leadership without intellectual backing is likely to result in a fine solution.

Acknowledgments The co-authors wish to acknowledge research projects that have led to the development of the material presented in this publication, namely the project "Social-economic conditioning of social vulnerability to floods, with particular consideration of the role of communication" funded by the National Science Center (NCN), based on decision No. DEC-2012/05/D/HS4/01328 (Jarosław Dzielak, Wojciech Biernacki, Roman Konieczny), the FLORIST Project (FLOod RISk on the northern foothills of the Tatra Mountains), PSPB No. 153/2010, supported by a grant from Switzerland through the Swiss Contribution to the enlarged European Union (Zbigniew W. Kundzewicz) and the STAR-FLOOD (STrengthening And Redesigning European FLOOD risk practices: Towards appropriate and resilient flood risk governance arrangements) project financed by the European Commission under grant agreement no. 308364 within FP7 (Piotr Matczak and Zbigniew W. Kundzewicz). The paper also draws from the material presented at the blog "MokraRobota" founded and maintained by Roman Konieczny.

References

Armitage D, Marschke M, Plummer R (2008) Adaptive co-management and the paradox of learning. Glob Environ Change 18:86–98

Berkhout F, Hertin J, Gann DM (2006) Learning to adapt: organisation adaptation to climate change impacts. Clim Change 78:135–156

Bos JJ, Brown RR, Farrelly MA (2013) A design framework for creating social learning situations. Glob Environ Change 23(2):398–412

Czech W, Radecki-Pawlik A, Wyżga B, Hajdukiewicz H (2016) Modelling the flooding capacity of a Polish Carpathian river: a comparison of constrained and free channel conditions. Geomorphology 272C, doi:10.1016/j.geomorph.2015.09.025

Galloway GE (1999) Towards sustainable management of river basins: challenges for the 21st century. In: Proceedings of RIBAMOD meeting, Wallingford, UK, 25–27 Feb 1998

Hajdukiewicz H, Wyżga B, Mikuś P, Zawiejska J, Radecki-Pawlik A (2016) Impact of a large flood on mountain river habitats, channel morphology, and valley infrastructure. Geomorphology 272C. doi:10.1016/j.geomorph.2015.09.003

Hinshaw RE (2006) Living with nature's extremes—the life of Gilbert Fowler White. Johnson Books, Boulder 340 pp

IFMRC (Interagency Floodplain Management Review Committee) (1994) Sharing the challenge: floodplain management into the 21st century. A blueprint for change. IFMRC, Washington, DC

Johannessen A, Hahn T (2013) Social learning towards a more adaptive paradigm? Reducing flood risk in Kristianstad municipality, Sweden. Glob Environ Change 23(1):372–381

Konieczny R. Blog http://mokrarobota.com/

Kundzewicz ZW (1999) Flood protection—sustainability issues. Hydrol Sci J 44(4):559–571

Kundzewicz ZW (2007) Book review: "Living with nature's extremes—the life of Gilbert Fowler White" by Robert E. Hinshaw (2006) by Johnson Books, Boulder, USA 340 pp. Hydrol Sci J 52(4):840–841

Kundzewicz ZW, Takeuchi K (1999) Flood protection and management: quo vadimus? Hydrol Sci J 44(3):417–432

Medema W, Adamowski J, Orr Ch, Wals A, Milot N (2015) Towards sustainable water governance: examining water governance issues in Québec through the lens of multi-loop social learning. Can Water Resour J/Revue canadienne des ressources hydriques. doi:10.1080/07011784.2015.1088403

Pahl-Wostl C, Mostert E, Tàbara D (2008) The growing importance of social learning in water resources management and sustainability science. Ecol Soc 13(1):24. [Online]. http://www.ecologyandsociety.org/vol13/iss1/art24/

Schusler TM, Decker DJ, Pfeffer MJ (2003) Social learning for collaborative natural resource management. Soc Nat Resour 15:309–326

Tyszka T, Konieczny R (2016) What determines willingness to take preventive actions in areas experiencing severe flooding? In: Large risks with low probabilities: perception and willingness to undertake prevention (in preparation)

Verweij M, Thompson M (eds) (2006) Clumsy solutions for a complex world: governance, politics and plural perception. Palgrave/Macmillan, Basingstoke

Wyżga B, Radecki-Pawlik A (2011) Jak zmniejszyć zagrożenie i ryzyko powodziowe w dorzeczu górnej Wisły? Gospodarka Wodna 10:414–421

Erratum to: Flood Generation Mechanisms and Changes in Principal Drivers

Bartłomiej Wyżga, Zbigniew W. Kundzewicz, Virginia Ruiz-Villanueva and Joanna Zawiejska

Erratum to: Chapter "Flood Generation Mechanisms and Changes in Principal Drivers" in: Z.W. Kundzewicz et al. (eds.), *Flood Risk in the Upper Vistula Basin*, GeoPlanet: Earth and Planetary Sciences, DOI 10.1007/978-3-319-41923-7_4

In the originally published version of this chapter, there were errors in pages 58 and 59 which affected some unquoted lines of the text and proper citation. The erratum chapter and the book have been updated with the changes.

The updated original online version for this chapter can be found at 10.1007/978-3-319-41923-7_4

B. Wyżga (✉)
Institute of Nature Conservation, Polish Academy of Sciences,
al. Mickiewicza 33, 31-120 Kraków, Poland
e-mail: wyzga@iop.krakow.pl

B. Wyżga
Faculty of Earth Sciences, University of Silesia, ul. Będzińska 60,
41-200 Sosnowiec, Poland

Z.W. Kundzewicz
Institute for Agricultural and Forest Environment, Polish Academy
of Sciences, ul. Bukowska 19, 60-809 Poznań, Poland
e-mail: kundzewicz@yahoo.com

V. Ruiz-Villanueva
Dendrolab.ch, Institute of Geological Sciences, University of Bern,
Baltzerstrasse 1+3, 3012 Bern, Switzerland
e-mail: virginia.ruiz@dendrolab.ch

J. Zawiejska
Institute of Geography, Pedagogical University of Cracow,
ul. Podchorążych 2, 30-084 Kraków, Poland
e-mail: zawiejska.joanna@gmail.com

Z.W. Kundzewicz et al. (eds.), *Flood Risk in the Upper Vistula Basin*,
GeoPlanet: Earth and Planetary Sciences, DOI 10.1007/978-3-319-41923-7_21

Erratum to: Flood Generation Mechanisms and Changes in Principal Drivers

Bartłomiej Wyżga, Zbigniew W. Kundzewicz, Virginia Ruiz-Villanueva and Joanna Zawiejska

Erratum to:
Chapter "Flood Generation Mechanisms and Changes in Principal Drivers" in: Z.W. Kundzewicz et al. (eds.), *Flood Risk in the Upper Vistula Basin*,
GeoPlanet: Earth and Planetary Sciences,
DOI 10.1007/978-3-319-41923-7_4

In the originally published version of this chapter [illegible]
[illegible]
[illegible]

[illegible]

B. Wyżga (✉)
Institute of Nature Conservation, Polish Academy of Sciences
al. Mickiewicza 33, 31-120 Kraków, Poland
[illegible]

[illegible]
Potsdam Institute for Climate Impact Research, [illegible]
[illegible] Potsdam, Germany

[illegible]
Institute for Agricultural and Forest Environment, Polish Academy of Sciences
ul. Bukowska 19, 60-809 Poznań, Poland
e-mail: kundzewicz@yahoo.com

V. Ruiz-Villanueva
Dendrolab.ch, Institute of Geological Sciences, University of Bern,
Baltzerstrasse 1+3, 3012 Bern, Switzerland
e-mail: virginia.ruiz@dendrolab.ch

J. Zawiejska
Institute of Geography, Pedagogical University of Cracow
ul. Podchorążych 2, 30-084 Kraków, Poland
e-mail: zawiejska.joanna@gmail.com

Z.W. Kundzewicz et al. (eds.), *Flood Risk in the Upper Vistula Basin*,
GeoPlanet: Earth and Planetary Sciences, DOI 10.1007/978-3-319-41923-7_21

MIX
Papier aus verantwortungsvollen Quellen
Paper from responsible sources
FSC® C105338

If you have any concerns about our products,
you can contact us on
ProductSafety@springernature.com

In case Publisher is established outside the EU,
the EU authorized representative is:
Springer Nature Customer Service Center GmbH
Europaplatz 3, 69115 Heidelberg, Germany

Printed by Libri Plureos GmbH
in Hamburg, Germany